Next Generation Blockchain for Next Generation Society with Futuristic Technologies

Editor

Amit Kumar Tyagi

National Forensic Sciences University
Gandhinagar, Gujarat, India

CRC Press
Taylor & Francis Group
Boca Raton London New York

CRC Press is an imprint of the
Taylor & Francis Group, an **informa** business
A SCIENCE PUBLISHERS BOOK

First edition published 2026
by CRC Press
2385 NW Executive Center Drive, Suite 320, Boca Raton FL 33431

and by CRC Press
4 Park Square, Milton Park, Abingdon, Oxon, OX14 4RN

CRC Press is an imprint of Taylor & Francis Group, LLC

Library of Congress Cataloging-in-Publication Data (applied for)

ISBN: 978-1-041-02607-5 (hbk)
ISBN: 978-1-041-02743-0 (pbk)
ISBN: 978-1-003-62082-2 (ebk)

DOI: 10.1201/9781003620822

Typeset in Times New Roman
by Prime Publishing Services

Preface

The growth of the twenty-first century has seen rapid technological evolution that continuously reshapes every aspect of human life. Among these transformative changes, blockchain technology (BT) stands as one of the most revolutionary innovations which provided transparency, security, and decentralization. The next generation of blockchain systems promises not only to enhance digital trust but also to redefine societal structures by integrating seamlessly with emerging technologies such as artificial intelligence (AI), quantum computing, the Internet of Things (IoTs), and extended reality platforms. Together, these advancements form the foundation of a globally connected, intelligent, and inclusive society.

This book discusses how blockchain can serve as the backbone of tomorrow's digital civilization. It provided the information like how governance, economy, healthcare, education, and environmental systems can be redesigned through distributed architectures that prioritize autonomy, efficiency, and accountability. Hence, by integrating futuristic technologies, blockchain transcends its current boundaries—evolving from a financial tool into an ecosystem of trust that empowers citizens, safeguards data integrity, and fosters collaborative innovation across borders.

As society steps into an era driven by decentralization and digital ethics, the need to understand next-generation blockchain frameworks becomes imperative. This work invites readers, innovators, policymakers, and futurists to explore how blockchain's synergy with advanced technologies can create a resilient, equitable, and intelligent world. The ideas presented here aim to spark dialogue and inspire strategic initiatives toward building a sustainable digital society that reflects the aspirations of a smart and interconnected humanity.

Lastly, we hope that this book will be loved by research community and any corrections are heartly welcome for improving the quality.

Contents

Chapter 1

An Introduction to Blockchain Technology

Methods and Applications

Iram Abrar,[1,*] *Javaid A. Sheikh*[1,*] and *Tanuj Surve*[2]

1. Introduction

Technological advancement is the soul of the new digital ecosystem and has brought forth unprecedented innovation, connectivity, and efficiency. It has changed the fundamentals regarding how communication takes place, transactions are executed, and any decision-making processes occur across industries. Edge-cutting technologies and ways of business and individual interaction are streamlined and redefine how organizations and individuals interconnect in a hyper-connected world. The growing demand for integrity, transparency, and assurance guaranteeing systems that are free from central authorities has been persistent, as the global economy increasingly relies on digital infrastructures. This represents an emerging need for highly secure, flexible, and responsive structures of a sort that can support dynamic shifts in this increasingly changing digital terrain. Blockchains have assured the critical aspects of data security and transparency as the unavoidable building blocks towards that kind of transformational change. Because blockchain uses decentralized structures, removing any single point of failure, they also reduce some of the risks inherent in a centralized system like data breach and unauthorized entry (Theodorakopoulos et al. 2024).

Technology usage indeed manifests in its adaptability to most applications. Blockchain has managed to revolutionize the very traditional systems – from streaming cryptocurrencies, for instance, bitcoin and ethereum, to streamline supply chain management by tracking goods from origin to destination (Kumar and Selvaprabhu 2023). In healthcare, blockchain secures patient records to ensure that they are private and authentic; while in finance, it provides transactions which are secure and transparent (Mohammed et al. 2024, Wu et al. 2024). Secondly, blockchain-enabled smart contracts automate agreements, and reduce the use of intermediaries; hence, there is a considerable reduction in errors by man. Blockchain plays a critical role in facilitating digital identity management, which is transformational and brings opportunities for enhanced security in managing and sharing personal data (Wang et al. 2024). More recently, the

[1] Department of Electronics and Instrumentation Technology, University of Kashmir, Hazratbal, J&K-190006, India.
[2] Department of Management, University of California, Berkeley, USA.
Email: tanujsurve@gmail.com
* Corresponding author: iram.abrar12@gmail.com, sheikhjavaid@uok.edu.in

DOI: 10.1201/9781003620822-1

global push toward sustainability has also been seen in blockchain's application to track carbon credits, monitor energy consumption, and promote ethical sourcing in supply chains, indicating a broader societal contribution (Woo et al. 2021, Boumaiza and Maher 2024). In general, blockchain technology shapes the future of the digital ecosystem, allowing safe, efficient, and transparent execution of processes across several industries. It addresses the very foundational issues regarding trust, security, and scalability and has become an enabler of the next wave of technological change towards innovation in the digital world (Abrar and Sheikh 2024).

A broad overview of architecture, functionalities, and applications of consensus mechanisms in blockchain technology has been presented in the present work. The work further goes into the real-world deployment of these mechanisms across various domains. Additionally, the study addresses both the strengths and limitations of blockchain systems to provide a holistic view of the transformative potential it holds and sets the foundation for possible future developments in this ever-evolving field.

2. Overview of Blockchain Technology

In the aftermath of the economic crisis in 2008, an era of rampant discrepancies across financial industry became apparent. It was time to have more transparent, secure, and reliable systems in place, hence the birth of 'FinTech' a term coined from 'Finance' and 'Technology'. FinTech operates on blockchain, artificial intelligence (AI), and sophisticated data investigation to bridge technical gaps (Schinckus 2020). The initial application of blockchain technology was Bitcoin, a virtual currency that revolutionized digital transactions. While blockchain underpins Bitcoin's cryptographic framework, its use has since expanded to other industries (Watters 2023). Blockchain's versatility has led to a wide array of applications, including e-voting, digital identity management, cross-border payments, crowdfunding, secure gaming transactions, land registration systems, and automated insurance processes such as fraud detection (Casino et al. 2019). So, it is a digital ledger that records orders or data blocks in chronological order. It then grows continuously with new transactions. With every transaction, an asymmetric cryptographic digital signature is generated for the purpose of validation and authentication. The generation of a pair of keys: a public and a private key are involved. The former is known to everyone on the network and is used in decryption, while the latter remains confidential to the owner, and thus enables secure encryption and signing transactions (Gad et al. 2022).

2.1 Blockchain Technology Architecture

Blockchain technology is a combination of various advanced technologies like cryptography (Bhutta et al. 2021, Jonathan Katz and Lindell 2007), mathematical principles, consensus protocols, and smart contracts (Szabo 1997). It runs on a peer-to-peer system, connecting all nodes, thereby providing services to the network. At its core, blockchain functions as a distributed ledger, where blocks are systematically stored and encrypted using cryptographic techniques. In the block body, transactional data is managed using the structure of the merkle tree. Each operation is logged as a leaf node (Merkle 1988, Peng et al. 2021). Such leaf nodes produce hash values, which are then propagated to construct the root node. All such nodes in the system can then easily query transactional data. Moreover, hashes are sensitive to changes in the data across the network, i.e., tampering with transactional data would results in its modification (Gupta and Sadoghi 2019, Abrar and Sheikh 2023). This makes the merkle tree structure a critical component in maintaining information security within the blockchain. The various stages of adding a transaction in blockchain has been depicted in Figure 1.1.

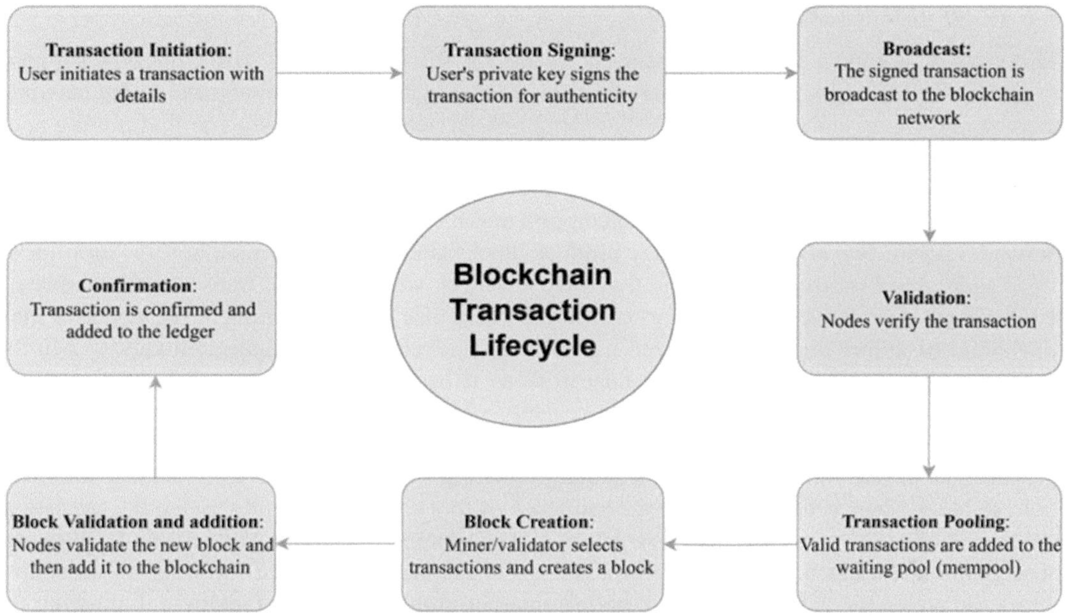

Figure 1.1. Stages of blockchain transaction lifecycle.

2.2 Blockchain Features

Decentralization: Traditional centralized systems relied on third-party agencies, like banks, for the validation and authentication of transactions. This often led to operational bottlenecks and overburdened the central servers. Blockchain eliminates intermediaries by the use of consensus algorithms, making it a decentralized system (Zheng et al. 2017).

Anonymity: Because blockchain is a decentralized system, it allows users to execute transactions with randomly generated addresses, making it impossible for a centralized authority to monitor or record these addresses.

Auditability (Transparency): The distributed ledger in blockchain records transactions, along with their timestamp, so anybody having the network can track and audit those records. All Bitcoin transactions, for instance, are traceable; thus blockchain data is not only transparent but also auditable (Bhutta et al. 2021).

Immutability (Persistency): Once a transaction is recorded in blockchain, it cannot be misrepresented. Every block has a distinct hash (fingerprint) derived from the transaction details, hash of preceding block, and a timestamp. Any attempt to modify the block's data changes the hash, signaling an alteration thereby preserving trustworthiness of the system.

3. Consensus Mechanism

Consensus mechanism plays an important role in securing distributed systems like blockchain (Xiao et al. 2019). The nodes in the blockchain act as host/server and work together to achieve consensus over transactions. In a public blockchain network, where any entity can anonymously become a node, the risk of transaction tampering and fork creation exists. To counter these risks, there are various consensus algorithms which ensure the agreement of entities on a unified version of truth. Based on application requirements, the consensus algorithm is selected. This section explores several consensus protocols and a concise summary of the same has been presented in Table 1.1.

3.1 Proof of Work (PoW)

This protocol is widely used in cryptocurrencies such as Bitcoin (Andreas 2015). It requires miners to perform considerable computational work for validation and to add new blocks to the network, thereby preventing malicious activities like denial-of-service (DoS) attacks. The first mining node to successfully solve challenging mathematical puzzles gains the privilege of adding a new block to the network. PoW ensures fairness by choosing nodes based on their computational power. However, this comes at a cost of large energy consumption and hardware requirements. Sometimes, due to network latency, two mining nodes may produce blocks simultaneously, creating short-term forks. PoW addresses this issue by ensuring that only the block with the initial transaction is validated, while others are removed from the system. In the event that forks do persist, PoW increases their likelihood of failure on a decentralized network. The adversaries are highly unlikely to win the subsequent blocks to perpetuate their false fork as it becomes computationally and financially impossible to reverse the validated blocks.

Even with this strength, PoW has several downsides. Its performance and security are inversely proportional to one another, implying it provides low transaction throughput. It has vulnerabilities to huge block sizes, hence increasing the incidence of fork cases and compromising the security of a network. Finally, its energy intensive nature makes running operations very costly. PoW is also susceptible to certain forms of attacks including eclipse attack and selfish mining. In an eclipse attack, an adversary gains control of a node's internet protocol address, thereby not allowing the target to access legitimate transactions (Valdeolmillos et al. 2020). In selfish mining, a malicious user hides the blocks until he/she has a longer chain than the leader, breaking the integrity of the blockchain (Dedeoglu et al. 2020). These drawbacks motivate the search for alternative consensus methods that balance security, energy efficiency, and performance for a variety of blockchain applications.

3.2 Proof of Stake (PoS)

Proof of Stake is less energy-intensive than PoW. Here the next block validator is determined through a pseudorandom selection process. The factors that influence this selection include staking age, randomization, and the wealth of a node. In PoS, blocks are "forged" instead of mined, and the entities responsible for forging are elected based on their stake, which depends on coins the node owns for the particular block (Jesus 2018). This election process, however, has the potential to result in centralization if one node dominates the stakes. In order to prevent this, techniques such as "randomized block selection" and "coin age selection" are used. Randomized block selection combines the lowest hash value with the highest stake to determine who will be the next forger, while coin age selection depends on stake size and how long it has been held. Coin age is computed as the product of staked coins and the number of days for which they have been staked. After forging, a node's coin age reverts to zero, thus ensuring that no dominance by nodes having large stakes can occur (Irresberger 2018).

When the node is elected, the transactions are verified. If it is legitimate, the node signs and receives the transaction fees as a bonus before committing the block on the network. A node that wants to withdraw its stake and rewards should wait for a certain period in order to ascertain that no fake blocks were entered. Fraudulent behavior is discouraged, as detected fraud leads to penalties, including stake loss and disqualification from future forging opportunities. Despite its advantages, PoS has limitations, such as the costless simulation problem, which occurs since it doesn't require a lot of work and enables attackers to build alternative blockchains. Also, "nothing-at-stake" occurs when participants face no risk of loss and support multiple blockchains, causing forks. In addition to this, PoS is susceptible to subsequent corruption attacks where entities are bribed to forge alternate chains and stake-grinding attacks (Andreina et al. 2020).

3.3 Delegated Proof of Stake (DPoS)

DPoS is a version of PoS proposed by Daniel Larimer that enhances the scalability and efficiency by reducing network validators (Elrom 2019, Zhang and Jong-Hyouk 2020). Currently, it is used in platforms such as BitShares (Yao and Zhu 2017) and Crypti (Garriga et al. 2021). It has high throughput as transactions happen more rapidly and within shorter intervals, thus resulting in the faster generation of blocks. In effect, a recent illustration example of a witness system can achieve thousands of transaction transactions per second within seconds for its confirmation time in the case of BitShares. Witnesses make blocks, validate transactions, and share rewards with their nominators. DPoS employs a separate block creation mechanism, where specific block producers, called witnesses, are entrusted with the duty of producing new blocks and checking transactions. Becoming a witness is possible only through voting for a candidate block producer, and the value of each vote depends on the number of tokens held by that user. Moreover, users can allot their stakes to others so that they can vote on their behalf. Furthermore, every DPoS system limits witnesses who are accountable for proposing blocks. Once witnesses are designated, they generate blocks according to their selection order, after authenticating all transactions. After the witnesses have verified and signed all transactions, a reward is given, usually to the users who nominated them. But in the case of a failed block from the side of a witness, the transaction is never confirmed and no block is produced.

In DPoS, the choice is also available to voters to vote for some other candidates that they think would be more efficient at doing the job, which leads to continuous elections. The voters also elect "delegates" to manage the blockchain. For optimal smooth working of the network, and informed choices, the delegators have to be aware and select efficient witnesses. Even though DPoS certainly offers a much faster and scalable method of handling transactions, it suffers from several potential drawbacks. A high number of witnesses may result in centralization because the users whose stakes are relatively small might not feel motivated to vote because their ballots do not make much difference. Weighted voting provides further risks to network decentralization.

3.4 Proof of Elapsed Time (PoET)

PoET is similar to PoW, however, its consumption of power is much less, making energy consumption and resource usage less intensive. It was proposed by Intel in 2016, for the problem related to "random leader election", since it used a random timer at each node. The node whose timer is shortest creates the next block and takes the reward attached to it (Mahmood and Wahab 2018). The protocol relies on Intel's SGX for safe execution. SGX ensures nodes run trusted codes by generating attestation that is rooted in the CPU. There are two phases of PoET. The first step is verification and network joining, during which the nodes execute the trusted PoET code on an SGX, which generates a key pair and produces attestation, and then successfully joins the network after verification. The second phase is mining lottery, where the nodes receive a signed timer object. Upon expiry, it waits for the timer. The node with the shortest timer object generates the block, submits it to the network, and commits it. Despite its energy efficiency, PoET has its shortcomings. It uses hardware systems like SGX that contradict the concept of trustless operation in blockchain. In addition, PBFT is prone to Sybil attacks wherein malicious users create artificial identities for network exploitation (Douceur 2002).

3.5 Practical Byzantine Fault Tolerance (PBFT)

Castro et al. introduced the PBFT protocol in 1999 (Castro et al. 1999), which facilitates a consensus among the nodes via communication, assuming that all legitimate nodes maintain identical ledgers. The consensus mechanism in PBFT is more energy-efficient than others since it cannot depend on any energy-intensive calculations, thereby reducing the environmental burden on miners. Unlike PoW, PBFT gets collective decisions through agreement due to this passing of messages. It thus encourages participation from all active nodes and makes sure that rewards are distributed equally and minimizes disparities in reward allocation for miners.

For a PBFT protocol to work right, malicious nodes should have a smaller number than that of total nodes in number (x/3 where x represents total nodes). As the network is large, security will be improved, as the possibility of getting one-third nodes as malicious is low. The process of generating blocks in PBFT is divided into four phases. Firstly, the client sends a transaction request to the head node. Secondly, the organizer node groups these requests into a block and broadcasts it to the backup nodes. Thirdly, all backup nodes verify the transactions, forming a valid transaction block, calculating the block hash, and transmitting it to subsequent nodes. Lastly, a node waits for consistent responses from the (r + 1) nodes or two-thirds of the total nodes, in which (r) is the number of faulty nodes. After verifying the consistent responses, the block is added to the node ledger.

However, PBFT is more transaction throughput intensive compared to PoW. Furthermore, the PBFT algorithm is susceptible to Sybil attacks in public networks where a single entity can easily create multiple false identities without the proper certification; this may even compromise a huge part of the network (Vukolić and Vukoli 2015).

3.6 *Proof of Authority (PoA)*

PoA is a type of consensus algorithm in which updation of distributed ledger and authentication of transactions are assigned to a small set of trusted validators (Li et al. 2020). Here, validators are chosen based on reputation rather than staking coins. Therefore, this method is very scalable and appropriate for private or consortium blockchains. PoA ensures transparency as validators must register publicly to validate their identities. Moreover, it is a form of consensus aimed to provide better resistance than PoW against a 51% of the attacks. PoA will force the attacker to seize control over 51% of the nodes that represent the entire network, and with PoW, it forces an attacker to have 51% of computational power. It is harder for a node in a PoA network to be granted access into the authorized nodes since these already have been authenticated, and upon their unavailability, the

Table 1.1. Summary of blockchain consensus protocols.

Consensus Mechanism	Key Features	Advantages	Challenges	Best Use Case
PoW	Mining-based, computationally intensive, fairness via power.	Secure, prevents tampering, DoS attack resistant.	High energy use, low throughput, selfish mining, eclipse attacks.	Cryptocurrencies (e.g., Bitcoin).
PoS	Validators chosen by stake, energy-efficient, no mining needed.	Energy-efficient, reduces hardware needs.	Centralization risk, nothing-at-stake issue.	Energy-saving blockchain networks.
DPoS	Few validators, voted witnesses, high throughput.	Scalable, fast transactions, shared rewards.	Centralization risk, voter disengagement.	High-performance blockchains.
PoET	Random leader election via Intel SGX, low energy usage.	Energy-efficient, fair leader election.	Hardware reliance, Sybil attacks.	Private or consortium blockchains.
PBFT	Message exchange for consensus, low energy use.	Energy-efficient, high throughput, fair rewards.	Scalability issues, Sybil attacks.	Private networks.
PoA	Validators based on reputation, suitable for private chains.	High scalability, resistant to 51% attacks.	Less decentralized, censorship risk.	Private or enterprise blockchains.

network can remove these nodes from an approval process. Although an attacker may try to force a target node to stop, for example, with multiple transactions or blocks, that is hard. Two ways that PoA mechanism gives protection against DoS attacks is by blocking creation authority to only verified nodes; such nodes have capabilities that can survive such attacks, and secondly, whenever a node is inaccessible for a certain duration, it is considered to be removed from the group of nodes responsible for approval and thus eliminates the risk of a DoS attack. However, PoA sacrifices decentralization for efficiency that makes it preferable for large organizations where logistical needs become a concern. Even though PoA provides a set of advantages, there exist reputational manipulations by way of censorship and blacklisting, which compromise the integrity within the network (Liu et al. 2019).

Here, Table 1.1 provides a summary of blockchain consensus protocols in detail.

4. Applications

Blockchain technology is one of the transformative innovations that brings secure, transparent, and decentralized data management capability to sectors. It is applied broadly in domains of finance, healthcare, and supply chain with the aim of increasing efficiency, trust, and automation. This section has discussed some of the key applications that impact industries in modern times, depicted by Figure 1.2. Also, a concise summary of various applications has been presented in Table 1.2.

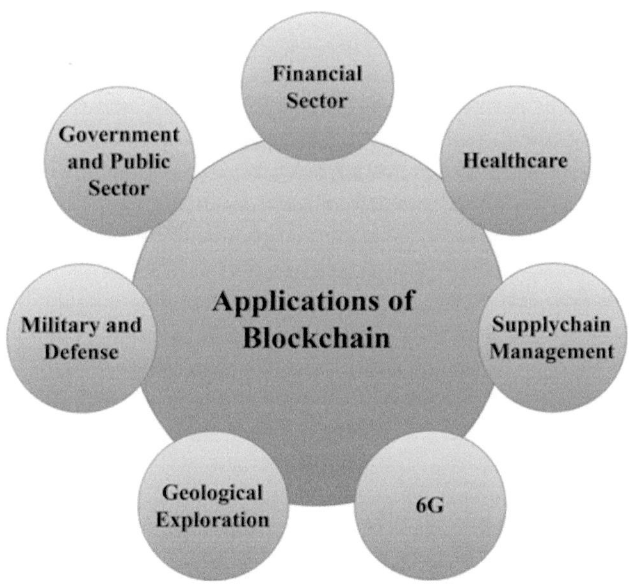

Figure 1.2. Applications of Blockchain Technology.

4.1 Finance Sector

Blockchain technology has emerged as a disruptive force that solves the problems long existing in the finance sector, including over-documentation, excessive transaction fees, inefficiencies of the regulatory bodies, and vulnerabilities in security. Since blockchain removes intermediaries and central authorities, it provides direct transactions between investors and companies without involving middlemen, and thereby reduces the cost of operation (Grima et al. 2021). It aims to provide a robust fraud protection system that can help in providing transparency for trustworthy solutions in streamlining financial procedures. It is, thus, considered a best practice solution in making finance services prompt and efficient in terms of trade finance, speedier settlements, and accuracy of data, besides relieving other significant bottlenecks from the normal bank systems. In the domain of digital asset management, blockchain can provide a very secure and efficient framework for organizing, storing, and transferring digital assets. Features such as two-step verification and tamper-proof ledgers ensure safety and accountability of cryptocurrency assets, reducing risks, and operational costs. Moreover, the ability of blockchain to encrypt transactional records enhances the protection of user data while maintaining transparency (Kumar Singh et al. 2023). Financial institutions also use smart contracts to automatically settle claims and verify identities without the possibility of error or fraud. Blockchain technology is changing global finance by allowing better connectivity and financial freedom; low-cost international payments and fast settlements are a new reality (Acharjee et al. 2023). Further, blockchain has the capability of digitizing financial assets and interfacing with other innovative technologies like distributed ledgers, it opens a new avenue of innovation. Although integrating blockchain applications with legacy systems remains a challenge, cost reduction, efficiency, and security potential achieved in blockchain represent the core of the future financial ecosystem. With financial institutions adopting blockchain-based solutions, operational efficiencies, transparencies, and securities can be experienced in the financial services industry (Geetha et al. 2024).

4.2 Healthcare

By resolving important issues and increasing productivity, blockchain has the ability to revolutionize the healthcare sector. Through decentralized consensus mechanisms and cryptographic algorithms, blockchain will be able to store and share sensitive health data in a secure manner while ensuring privacy and trust (Jayaraman et al. 2019). This technology can enable easy data sharing between healthcare networks allowing for real-time access to patient records without redundant testing. The blockchain-based rights and permissions can help patients take better control of their healthcare data by securely sharing them with authorized parties. Moreover, blockchain technology also helps to trace the supply chain of drugs and make it more transparent. This can greatly reduce the chance of counterfeit medications (Abbas et al. 2020). It can also make clinical trial outcomes more reliable and expedite healthcare insurance claims and reduce frauds. Research into blockchain based health sector revealed that it provides connectivity between different systems, security of data, and patient-centric approaches. According to (Attaran 2022) blockchain enables diverse healthcare systems to connect in a way that improves service delivery. In addition, blockchain enhances the pharma supply chain by improving traceability in tracking drugs and medical devices (Dhingra et al. 2024). When blockchain is introduced in healthcare, this industry has the potential to achieve operational efficiency, transparency, and trust with regard to delivering more transformative solutions in modern health issues.

4.3 Supply Chain Management (SCM)

Indeed, blockchain is revolutionizing the supply chain by adding more transparency, accountability, and efficiency. Complex logistics operations of supply chains can be transformed owing to the decentralized and immutable nature of blockchain, providing a reliable method of recording and authenticating every process of the supply chain (Han and Fang 2024). Blockchain, therefore, provides a shared ledger that is tamper-proof for the easy cooperation of suppliers, manufacturers, distributors, and retailers. It would mean that every transaction or transfer is recorded and verifiable in a transparent manner, thereby, reducing the dispute and dependence on the intermediaries. Traceability is a crucial SCM activity, especially for perishable goods, high-value products, or products with ethical and legal concerns. The blockchain technology enables the tracing of any product from the source to the consumer in real time, thereby complying with the legal aspects and quality and authentication. For instance, in the food industry, it can enable products to be tracked from the farm to the consumer. This increases the standards of safety. Consequently, if some issues like contamination occur, then there is a rapid response to the situation. Secondly, blockchain ensures accountability because each participant has records of every activity performed. Thus, it is easy to find out who to blame in case of non-compliance or failure (Rakibul Hasan Chowdhury 2024). Though its immutability discourages tampering and fraud since records are permanent. However, its implementation in supply chain management is faced with challenges in incorporating blockchain with pre-existing IT systems, scalability issues, and achieving universal standards that will enable interoperability.

4.4 6G Networks

Since 6G networks are targeted at providing massive device interconnectivity and dynamic data exchanges, blockchain's decentralized and immutable characteristics can provide an effective solution in the management of cryptographic keys that ensure against eavesdropping and man-in-the-middle (MITM) attacks. Blockchain can hold public keys safe within tamper-proof blocks for the efficient management of keys. Key management is the core process in authentication and data encryption within 6G environments (Hewa et al. 2020). In addition, smart contracts provide decentralized mechanisms for access control and authentication; therefore, less reliance on central systems exists to ensure sensitive data protection against unauthorized access (Abdel et al. 2022).

From the security and privacy perspective, blockchain helps to increase trust and accountability due to its transparent nature and non-repudiation properties. In order to guard against threats such as hijacking and eavesdropping, blockchain is useful in reliability and integrity within 6G networks. Cross-domain interoperability and secure, efficient identity verification can also be assisted by decentralized identity management systems based on blockchain (Garzon et al. 2022). Using blockchain, technologies can share resources efficiently since transactions are automatically facilitated through smart contracts, while there are fewer or no middlemen (Le et al. 2023). Thus, with all these capabilities, the integration of blockchain can play an integral role in developing secure, efficient, and reliable 6G next-generation networks.

4.5 Geological Exploration

The potential of blockchain in geological exploration is tremendous because this technology may make sharing of data more open, secure, efficient, and trustworthy. With a decentralized ledger system, it can record geological data through an immutable chain of blocks, incorporating core sample analyses, mineralogy reports, as well as results in exploration. It grants better authenticity and integrity to the data all along its lifetime, so that the record may be traceable, tamper-resistant, and straightforward. As an example, blockchain can track the origin of minerals from their extraction to processing, and even to delivery with the guarantee of ethical standards, while combining environmental and social requirements. Smart contracts make agreements in geologic exploration very smooth, for example, joint ventures or mineral rights with the payment of royalties. For instance, when a mining company receives an allotment of the deposit, it automatically receives royalties paid in accordance with terms that the owners of the lands set through smart contracts. This also ensures interaction between geologists, mining companies, and other regulatory agencies, which is securely conducted with less possibility of unapproved access to data and any form of alteration. Mineral asset tokenization can provide fractional ownership and frictionless trade, affording equal chances to everyone interested in investing in the minerals sector. Decentralized storage of geology maps, surveys, and other types of geospatial data ensures genuine and tamper-proof information related to the exploration. Cryptocurrencies and blockchain-based fundraising options like ICOs and STOs act as alternative fundraising mechanisms for exploration projects (de Andrés et al. 2022, Parmentola et al. 2022). Blockchain technology therefore ensures transparency, accountability, and sustainable practices to be aligned with a revolutionary method of data management and cooperation in geological exploration.

4.6 Military and Defense

For years, cyber-attacks and exploitation of cybersecurity systems have revealed weaknesses not only in military but also in civilian networks, making them extremely vulnerable to military systems (Peres et al. 2020). Such challenges demand long-term and cost-effective defense policies. Blockchain is still largely untested in the military domain, yet it has the potential to solve security flaws by removing the single-point-of-failure paradigm. Military applications of blockchain include intrusion detection, unmanned aerial vehicle (UAV) management, and encrypted communications. While these are in nascent stages of development, some of the initial practical implementations for military operations include defense logistics and security improvements. UAVs, or drones, are commonly used for traffic monitoring, package delivery, and real-time object detection within the domain of smart transportation. These applications need large data collection and transfer from drones to distant data centers and demand high computation power, reliable connections, and rapid transmission speed. However, the increased number of drone deployments poses challenges of security vulnerabilities, congestion in air traffic, and probable crashes. Blockchain technology could solve these issues by providing safe communication, privacy, and UAV authentication.

Current solutions for drone communication face issues like unreliability in transmission, delays, and bandwidth limitations. These problems can be overcome by blockchain's decentralized framework that provides secure and efficient communication, as well as robust authentication services (Singh and Venkatesan 2021). For instance, the integration of blockchain with fog and cloud computing systems may potentially assist in the enhancement of operations by UAVs as a result of assured secured availability of data and better modes of authentication (Chen et al. 2021). In vehicular ad hoc networks (VANETs), blockchain allows anonymous authentication and fast re-authentication of the vehicle, so safe information on the road could be shared among vehicles in real-time (Maria et al. 2021). In addition, blockchain is utilized for the safe storage and transmission of application data in VANETs without relying on a central authority. In addition, blockchain is utilized for the safe storage and application of data transmission in VANETs without relying on a central authority (Kudva et al. 2021). Military and defense systems may apply blockchain technology in terms of providing cybersecurity, reliable communication, and robust data management. Hence, it holds great promise as a solution to the new evolving challenges of modern defense operations.

4.7 Government and Public Sectors

Blockchain can revolutionize the operations of the government and the public sector with better transparency, security, and effectiveness, especially in voting systems and public document management. Owing to its decentralization and immutability properties, government problems of trust and integrity in public administration can be dealt with easily. Additionally, it can transform voting processes in elections with higher security and accessibility; traditional systems of voting may be exposed to limitations like voter fraud, ballot tampering, and lack of transparency, which result in the loss of public confidence. Blockchain-based voting systems can record votes as immutable transactions, so once a vote has been cast, it cannot be modified or deleted, and thus, such immutability ensures integrity within the electoral process, whereas methods of cryptography maintain anonymity of the voters and confidentiality of ballots. Apart from physical voting, blockchain allows for distance and electronic voting. Voters may use personal computers, laptops, or even cell phones to make their choices. This accessibility will reduce logistical barriers and costs. Moreover, increased participation in distant areas or when pandemics prevail will be possible. Besides voting, blockchain enhances record management of records in public domain such as property titles, licenses, educational qualifications, and health records (Kumari et al. 2024, S et al. 2024).

Blockchain uses a safe, distributed ledger that helps authenticate information with no possibility of unauthorized access or alteration. This means efficiency and accuracy in public services, reduced administrative burdens, and reduced risks for data loss or corruption. Furthermore, blockchain improves inter-agency cooperation through the use of the same, secure kind of platform from which records are potentially accessed in real-time, and thus further streamlines government operations. However, there are still some challenges when implementing blockchain within the government. Firstly, technical know-how and huge investments in infrastructure are needed. Secondly, updates in legal frameworks to cover issues on privacy, security, and ethics are required. Thirdly, cultural adaptation on the part of citizens and public officers is significant in succeeding in the adoption of blockchain technology at full maturity. Though blockchain technology has been proven transformative on many grounds, yet it requires a certain level of openness and shared control that will make the adoption challenging for most organizations (Wamba et al. 2024). Table 1.2 discusses about Comparative analysis of blockchain applications across different Sectors.

Table 1.2. Comparative analysis of blockchain applications across different Sectors.

Application	Key Benefits	Challenges	Examples
Finance	Fraud protection, faster settlements.	Legacy system integration.	Smart contracts.
Healthcare	Secure data sharing, drug traceability.	Interoperability issues.	Patient data control.
Supply Chain	Transparent tracking, reduced disputes.	IT system integration.	Food traceability.
6G Networks	Secure key management, authentication.	Complexity, resources.	Decentralized identity.
Geological	Ethical tracking, transparent data.	Tokenization hurdles.	Mineral origin tracking.
Military	Cybersecurity, UAV authentication.	Scalability challenges.	Blockchain in UAVs.
Government	Transparent voting, secure records.	High costs, legal updates.	Blockchain voting systems.

5. Limitations

Though blockchain technology holds transformative powers yet it has significant weaknesses in its structure, damaging not only its scalability but also its security, efficiency, and usability. These limitations need to be addressed for its full-scale implementation and effective execution. Some of these limitations have been discussed in this section. Table 1.3 presents a brief summary of these limitations.

5.1 Scalability and Performance

Perhaps the most evident weakness of blockchain is its scaling problem. Blockchains, particularly those based on the PoW consensus mechanism, have problems with low transaction throughput and very high latency. This creates complexity, slows down the processing time, and causes delays during peak transaction times, as every transaction on the blockchain network needs to be recorded. Another significant challenge is ensuring the scale of blockchain networks scale with increased transaction volumes without degradation in performance (Guo et al. 2020). Several methods have been proposed in order to solve scalability, which includes sharding and layer-two protocols (for example, Lightning Network for Bitcoin), splitting off a blockchain, or facilitating off-chain transactions, which then ease congestion and increase speed.

5.2 Data Privacy and Security

Though blockchain guarantees data security due to immutability and transparency, it is inherently open, and so privacy can be an issue. However, sensitive information in the blockchain could be available for all the entities involved, creating a problem, especially in areas requiring confidentiality, for example, the health and financial sectors. Balancing the open nature of blockchain with the need for data privacy requires intricate solutions (Karras et al. 2023). Zero-knowledge proofs and homomorphic encryption come as some potential solutions. The former enables verification of data without revealing what that data is whereas the latter allows computations on encrypted data without data decryption to ensure privacy (Wibowo and Sandikapura 2019).

5.3 Interoperability

Interoperability is the biggest problem for the blockchain system. Each blockchain network has a unique consensus mechanism, data structures, and protocols that don't facilitate a smooth exchange of data or assets across systems. Since these are not standardized, interoperability exists only between the peers, thereby affecting the broader usage of the blockchain systems (Ahsan et al. 2021). Other such initiatives include interoperability-focused frameworks like Polkadot and Cosmos, which facilitate blockchains speaking to each other, allowing them to exchange data very seamlessly. Middleware also serves the purpose of interface in the existing systems.

5.4 Regulatory and Compliance Issues

The decentralized and immutable nature of blockchain may conflict with existing regulatory frameworks. For instance, one finds it problematic to follow the data protection rules of the General Data Protection Regulation (GDPR) and California Consumer Privacy Act (CCPA) because the immutability aspect of blockchain cannot agree with data deletion or modification in case a customer requests it (Theodorakopoulos et al. 2024). Secondly, well-defined laws for blockchain applications create potential liabilities from a legal standpoint. Hybrid blockchain models that combine the features of public and private blockchains are being developed to overcome these problems. Regulatory sandboxes are also adopted for testing blockchain solutions in sandboxed environments, with the aim that developers can achieve compliance before massive-scale deployment.

5.5 Cost and Resource Intensity

Setting up blockchain systems requires high-cost infrastructure, power consumption, and highly skilled manpower. Proof-of-work-based blockchains require a lot of computational resource inputs, i.e., they consume significant amounts of power and incur much operational cost, which may discourage budget-constrained organizations (Xu et al. 2019). Costs can be reduced by switching to energy-efficient consensus mechanisms such as PoS or DPoS. Cloud-based blockchain services are scalable and cost-effective, thus requiring less upfront investment.

5.6 Governance and Consensus Challenges

Effective governance and consensus mechanisms are essential for maintaining the stability and inclusivity of blockchain networks. Conflicting interests among stakeholders may lead to conflict and delay in decision-making. In addition, the implementation of consensus algorithms, including PoS and DPoS, increases the complexity in achieving fairness, transparency, and security (Fahlevi et al. 2023). New governance models like decentralized autonomous organizations use smart contracts to facilitate automated decision-making and stakeholder participation. Research on consensus mechanisms like BFT continues to address these governance challenges.

5.7 Greedy Miners and Mining

Blockchain networks are susceptible to attacks from greedy and conspiring miners. A well-known example is the 51% attack, where nodes controlling 51% or more of the network's hashing power can reverse blockchain's state and transactions. Recent research (Eyal and Sirer 2018), however, indicates that nodes with less than 51% computing power can also compromise the network. Selfish miners for example, mine private branches and withhold broadcasting their blocks until certain conditions are met. Once exposed, these private branches, being longer than the publicly mined chain, are accepted by the network, resulting in wastage of resources spent by honest miners on the public chain. Since the selfish miners face less competition while mining their private chain, they gain a larger share of the revenue. This encourages rational miners to participate in the selfish pool for economic rewards, increasing the pool's strength to above 51%. Some models proposed are truth state (Saad et al. 2019) and detective mining (Lee and Kim 2023), which try to minimize the security weaknesses created through selfish mining attacks.

5.8 Nothing-at-Stake

Some of the weak consensus protocols used are PoS, proof of history that consume less energy and are vulnerable to nothing-at-stake attacks. This vulnerability arises when validators do not risk losing anything for supporting multiple competing branches simultaneously. In such scenarios, attackers can exploit this by working on multiple branches, creating network instability. Vitalik Buterin proposed the slasher algorithm to address this issue (Buterin 2014). This algorithm deters nothing-at-stake attacks by requiring validators to lock in a deposit for a specific timeframe. If validators try to sign multiple chains, their deposit is lost, which deters such attempts.

5.9 Cryptanalytic Attacks

Cryptanalytic attacks are difficult to carry out, but are a major threat to blockchain technology. They target the cryptographic algorithms that underpin blockchain systems. For example, private keys generated using the elliptic-curve digital-signature algorithm (ECDSA) are widely used in blockchain networks like Hyperledger Fabric, and are vulnerable to key-based attacks (Hartwig 2016). These attacks often occur due to poor randomness in generating private keys or through a leakage of private keys. Highly randomized generation processes for private keys are, therefore, recommended (Bos et al. 2014). Quantum computing increases these risks further. Algorithms for quantum computing can break into the conventional system used in blockchains, threatening its security. For instance, a study on quantum attacks against blockchain is suggested in (Yin et al. 2018), along with an anti-quantum transaction system to be developed to overcome these threats. Though blockchain has the potential to become a groundbreaking phenomenon in the future, it currently lacks certain essential elements. For example, issues with scalability, privacy, interoperability, regulatory compliance, costs, and governance signify the need for more innovative solutions and continuous research. Addressing these limitations will unlock the full potential of blockchain systems, making them truly powerful, secure, efficient, and transformative.

Table 1.3 Provides the detailed information about Blockchain constraints and the potential resolutions.

Table 1.3. Blockchain constraints and potential resolutions.

Limitation	**Details**	**Proposed Solutions**	**Challenges**
Scalability	Low throughput, high latency.	Sharding, layer-2 protocols.	Complex implementation.
Privacy and Security	Privacy issues in sensitive sectors.	Zero-knowledge proofs, encryption.	Balancing privacy and transparency.
Interoperability	Limited cross-chain communication.	Polkadot, Cosmos, middleware.	Lack of standardization.
Regulatory Issues	Conflicts with GDPR, CCPA.	Hybrid blockchains, sandboxes.	Legal and decentralization conflicts.
Cost & Resources	High costs and power use in PoW.	PoS, DPoS, cloud-based solutions.	Adoption barriers for small players.
Governance	Stakeholder conflicts.	DAOs, better algorithms (BFT).	Resistance to decentralization.
Greedy Miners	51% attacks, selfish mining.	Truth state, detective mining.	Network coordination.
Nothing-at-Stake	Validators support multiple chains.	Slasher algorithm.	Penalizing malicious validators.
Cryptanalytic Attacks	Key vulnerabilities, quantum threats.	Randomized keys, quantum-resistant.	Transition to new cryptography.

6. Future Directions

Blockchain technology is always evolving, and its potential to change industries across the digital landscape is still growing. These promises of future growth need to address the inherent strengths and weaknesses with the emergence of new opportunities, focusing on developing scalability, its integration with next-generation technologies, and ethical considerations in a regulation that is anticipated to shape this future development track. Scalability is a must for blockchain's widespread acceptance. Techniques such as sharding, which split blockchain networks into smaller, manageable segments, and layer-two solutions like lightning networks, improve transaction throughput and reduce latency. Sidechains are also being developed to enhance interoperability, making it possible for blockchains to communicate seamlessly. These innovations are also important for ensuring the efficiency of blockchain when implemented in these industries globally that are more high-speed, highly secure transactions-industries, with high speed or secure transactions industries. Achieving this will require groundbreaking applications driven by the integration of advanced technologies into blockchain systems including AI, the Internet of Things and quantum computing. AI is believed to optimize operations related to blockchains and upgrade the functionality of smart contracts, and similarly, blockchains can provide secure Internet of Things ecosystems with data integrity and authenticity. The emergence of quantum computing shall challenge blockchain systems with quantum-resistant cryptographic techniques against threats to preserve the security strength of blockchains over the long run.

Blockchain, in the financial sector is expected to bring about a lot of transformative change through decentralized finance and programmable currencies. The developments will end reliance on intermediaries, saving costs and enhancing operational efficiency. It is also expected to help improve SCM, offering unmatched transparency, traceability, and accountability in monitoring commodities and reducing environmental impacts, as well as, ensuring ethical sourcing. Healthcare applications of blockchain point towards the safety management of patient data, which is sensitive and allows real-time access across systems in healthcare seamlessly. This not only enhances patient outcomes but also facilitates global collaboration in medical research. Moreover, since blockchain is an immutable technology, it supports the traceability of drugs, eliminating the risks associated with counterfeit medicines and building confidence in pharmaceutical supply chains. The blockchain-big data integration can offer immense potential for environmental monitoring and smart city management. Blockchain's transparent and tamper-proof ledger can ensure the authenticity of real-time environmental data, while big data analytics can uncover patterns and predict changes. In smart cities, blockchain can enhance resource distribution, transportation networks, and energy management systems, leading to sustainable urban environments.

As distributed ledger technology gains more widespread acceptance, ethical and regulatory issues must be addressed. Fully balanced transparency and privacy data protection must be adopted, such as GDPR for global data protection, and to ensure ethical use of blockchain systems. More importantly, universal standards will have to be established and collaboration fostered among stakeholders to ensure that responsible blockchain deployment occurs. Lastly, blockchain is capable of bringing about noteworthy socio-economic change by promoting financial inclusion and governance accountability. For instance, blockchain-based secure digital identities for underserved populations aid distribution platforms can improve access to basic services. Challenges such as high energy consumption, integration with existing systems, and a lack of skilled professionals must be addressed to unlock blockchain's full potential.

7. Conclusion and Future Scope

Blockchain has emerged as a transformative force in the modern digital ecosystem by addressing the critical challenges of security, transparency, and trust across different domains. Its decentralized and immutable nature ensures an end to single point failures in handling such data, assuring reliable and tamper-proof data management. It has several industrial applications in domains such as finance, healthcare, SCM, and defense, thereby creating solutions for efficiencies in operation, while the integrity of the data is maintained. Despite challenges such as scalability, high resource consumption, and regulatory difficulties, ongoing developments in blockchain technology continue to yield new applications and solutions.

The world is blessed with an open gateway toward an efficient, more secure, bright future through this ever-increasing interconnectivity between networks. As such, developing collaboration on the existing barriers could make it possible to redefine the process of blockchain technology even on the very traditional systems of transparency and robustness in its digital space. Innovation and transformation have endless possibilities for blockchain. It can be integrated with other technologies like AI, IoT, and big data, which can unleash new applications that go from predictive analytics to the efficient management of resources. With maturing technology, it can form the cornerstone of a secure, transparent, and inclusive digital economy and transform industries and societies across the globe.

Acknowledgements

This work has been supported by the Ministry of Electronics and Information Technology; Government of India under its Visvesvaraya Ph. D Scheme vide implementation no: PhD-02/2022/59.

References

Abbas, Khizar, Muhammad Afaq, Talha Ahmed Khan and Wang-Cheol Song. (2020). A blockchain and machine learning-based drug supply chain management and recommendation system for smart pharmaceutical industry. Electronics 9(5): 852. doi:10.3390/electronics9050852.

Abdel Hakeem, Shimaa A., Hanan H. Hussein and HyungWon Kim. (2022). Security requirements and challenges of 6G technologies and applications. Sensors 22(5): 1969. doi:10.3390/s22051969.

Abrar, Iram and Javaid A. Sheikh. (2023). Intelligent blockchain technologies for disaster management: challenges, research gap and solutions. International Journal of Blockchains and Cryptocurrencies 4(4): 311–25.

Abrar, Iram and Javaid A. Sheikh. (2024). Current trends of blockchain technology: architecture, applications, challenges, and opportunities. Discover Internet of Things 4(1): 1–17.

Acharjee, Purnendu Bikash, Manish Kumar, Gopal Krishna, Kamalakar Raminenei, Read Khalid Ibrahim and Malik Bader Alazzam. (2023). Securing international law against cyber attacks through blockchain integration. pp. 2676–81. In 2023 3rd International Conference on Advance Computing and Innovative Technologies in Engineering (ICACITE). IEEE. doi:10.1109/ICACITE57410.2023.10182646.

Ahsan, Ali Shabbir, Ahsan. (2021). Blockchain and big data: exploring convergence for privacy, security and accountability. Sage Science Review of Educational Technology 4(2): 53–68.

Andreina, Sebastien, Jens Matthias Bohli, Ghassan Karame, Wenting Li and Giorgia Azzurra Marson. (2020). Pots: A secure proof of tee-stake for permissionless blockchains. IEEE Transactions on Services Computing 15(4): 2173–87.

Andrés, Pablo de, David Arroyo, Ricardo Correia and Alvaro Rezola. (2022). Challenges of the market for initial coin offerings. International Review of Financial Analysis 79(January): 101966. doi:10.1016/j.irfa.2021.101966.

Attaran, Mohsen. (2022). Blockchain technology in healthcare: challenges and opportunities. International Journal of Healthcare Management 15(1): 70–83. doi:10.1080/20479700.2020.1843887.

Bhutta, Muhammad Nasir Mumtaz, Amir A. Khwaja, Adnan Nadeem, Hafiz Farooq Ahmad, Muhammad Khurram Khan, Moataz A Hanif et al. (2021). A survey on blockchain technology: evolution, architecture and security. Ieeexplore.Ieee. Org 9: 61048–73. https://ieeexplore.ieee.org/abstract/document/9402747/.

Bos, Joppe W., J. Alex Halderman, Nadia Heninger, Jonathan Moore, Michael Naehrig and Eric Wustrow. (2014). Elliptic curve cryptography in practice. pp. 157–175. In International Conference on Financial Cryptography and Data Security. Berlin, Heidelberg: Springer Berlin Heidelberg.

Boumaiza, Ameni and Kenza Maher. (2024). Harnessing blockchain and IoT for carbon credit exchange to achieve pollution reduction goals. Energies 17(19): 4811. doi:10.3390/en17194811.

Buterin, Vitalik. (2014). Slasher: A Punitive Proof-of-Stake Algorithm.

Casino, Fran, Thomas K. Dasaklis and Constantinos Patsakis. (2019). A systematic literature review of blockchain-based applications: current status, classification and open issues. Telematics and Informatics 36(March). Pergamon: 55–81. doi:10.1016/J.TELE.2018.11.006.

Castro, Miguel and Barbara Liskov. (1999). Practical byzantine fault tolerance. pp. 1–14. In Proceedings of the Third Symposium on Operating Systems Design and Implementation.

Chen, Chin-Ling, Mao-Lun Chiang, Yong-Yuan Deng and Wei Weng. (2021). A traceable firearm management system based on blockchain and IoT technology. Symmetry 13(3). https://www.mdpi.com/1025472.

Dedeoglu, V., Jurdak, R., Dorri, A., Lunardi, R. C., Michelin, R. A., Zorzo, A. F. et al. (2020). Blockchain Technologies for IoT. Studies in Big Data 60. Springer Science and Business Media Deutschland GmbH: 55–89. doi:10.1007/978-981-13-8775-3_3/COVER.

Dhingra, Sarthak, Rakesh Raut, Koteswararao Naik and Kamalakanta Muduli. (2024). Blockchain technology applications in healthcare supply chains—A review. IEEE Access 12(1): 11230–57. doi:10.1109/ACCESS.2023.3348813.

Douceur, John R. (2002). The Sybil Attack. Lecture Notes in Computer Science (Including Subseries Lecture Notes in Artificial Intelligence and Lecture Notes in Bioinformatics) 2429. Springer Verlag: 251–60. doi:10.1007/3-540-45748-8_24.

Elrom, Elad. (2019). The Blockchain Developer. The Blockchain Developer. doi:10.1007/978-1-4842-4847-8.

Eyal, Ittay and Emin Gün Sirer. (2018). Majority is not enough. Communications of the ACM 61(7): 95–102. doi:10.1145/3212998.

Fahlevi, Mochammad, Moeljadi, Siti Aisjah and Atim Djazuli. (2023). Corporate governance in the digital age: a comprehensive review of blockchain, AI, and big data impacts, opportunities, and challenges. In E3S Web of Conferences, edited by R. Isnanto, Hadiyanto, and B. Warsito, 448: 02056. doi:10.1051/e3sconf/202344802056.

Gad, Ahmed G., Diana T. Mosa, Laith Abualigah and Amr A. Abohany. (2022). Emerging trends in blockchain technology and applications: a review and outlook. Journal of King Saud University - Computer and Information Sciences 34(9). Elsevier: 6719–42. doi:10.1016/J.JKSUCI.2022.03.007.

Garriga, Martin, Stefano Dalla Palma, Maxmiliano Arias, Alan De Renzis, Remo Pareschi and Damian Andrew Tamburri. (2021). Blockchain and cryptocurrencies: a classification and comparison of architecture drivers. Concurrency and Computation: Practice and Experience 33(8). John Wiley & Sons, Ltd. doi:10.1002/CPE.5992.

Garzon, Sandro Rodriguez, Hakan Yildiz and Axel Kupper. (2022). Decentralized identifiers and self-sovereign identity in 6G. IEEE Network 36(4): 142–48. doi:10.1109/MNET.009.2100736.

Geetha, B. T., Kafila Kafila, S. Tulasi Ram, Dr. Amar Prabhakar Narkhede, Ahmad Y. A. Bani Ahmad and Mohit Tiwari. (2024). Creating resilient digital asset management frameworks in financial operations using blockchain technology. pp. 1–7. In 2024 International Conference on Trends in Quantum Computing and Emerging Business Technologies. IEEE. doi:10.1109/TQCEBT59414.2024.10545263.

Grima, Simon, Murat Kizilkaya, Kiran Sood and Mehmet ErdemDelice. (2021). The perceived effectiveness of blockchain for digital operational risk resilience in the european union insurance market sector. Journal of Risk and Financial Management 14(8): 363. doi:10.3390/jrfm14080363.

Guo, Leiyong, Hui Xie and Yu Li. (2020). Data encryption based blockchain and privacy preserving mechanisms towards big data. Journal of Visual Communication and Image Representation 70(July): 102741. doi:10.1016/j.jvcir.2019.102741.

Gupta, Suyash and Mohammad Sadoghi. (2019). Blockchain transaction processing. Encyclopedia of Big Data Technologies. Springer International Publishing, 366–76. doi:10.1007/978-3-319-77525-8_333.

Han, Yanhu and Xiao Fang. (2024). Systematic review of adopting blockchain in supply chain management: bibliometric analysis and theme discussion. International Journal of Production Research 62(3): 991–1016. doi:10.1080/00207543.2023.2236241.

Hartwig, Mayer. (2016). ECDSA security in Bitcoin and Ethereum: A research survey. Blog.Coinfabrik, 1–10.

Hewa, Tharaka, Gurkan Gur, Anshuman Kalla, Mika Ylianttila, An Bracken and Madhusanka Liyanage. (2020). The role of blockchain in 6G: Challenges, opportunities and research directions. pp. 1–5. In 2020 2nd 6G Wireless Summit (6G SUMMIT). IEEE. doi:10.1109/6GSUMMIT49458.2020.9083784.

Irresberger, Felix. (2018). Coin Concentration of Proof-of-Stake Blockchains. Leeds University Business School Working Paper. https://papers.ssrn.com/sol3/papers.cfm?abstract_id=3293694.

Jayaraman, Raja, Khaled Salah and Nelson King. (2019). Improving opportunities in healthcare supply chain processes via the internet of things and blockchain technology. International Journal of Healthcare Information Systems and Informatics 14(2): 49–65. doi:10.4018/IJHISI.2019040104.

Jesus, Emanuel Ferreira. (2018). A survey of how to use blockchain to secure internet of things and the stalker attack. Security and Communication Networks. https://www.hindawi.com/journals/scn/2018/9675050/.

Jonathan Katz and Yehuda Lindell. (2007). Introduction to Modern Cryptography - Jonathan Katz, Yehuda Lindell - Google Books.

Karras, Aristeidis, Anastasios Giannaros, Leonidas Theodorakopoulos, George A. Krimpas, Gerasimos Kalogeratos, Christos Karras et al. (2023). FLIBD: A federated learning-based IoT big data management approach for privacy-preserving over Apache Spark with FATE. Electronics 12(22): 4633. doi:10.3390/electronics12224633.

Kudva, Sowmya, Shahriar Badsha, Shamik Sengupta, Ibrahim Khalil and Albert Zomaya. (2021). Towards secure and practical consensus for blockchain based VANET. Information Sciences 545: 170–87. doi:10.1016/j.ins.2020.07.060.

Kumar C. Vinoth and Poongundran Selvaprabhu. (2023). An examination of distributed and decentralized systems for trustworthy control of supply chains. IEEE Access 11: 137025–52. doi:10.1109/ACCESS.2023.3338739.

Kumar Singh, Rajesh, Ruchi Mishra, Shivam Gupta and Archana A. Mukherjee. (2023). Blockchain applications for secured and resilient supply chains: a systematic literature review and future research agenda. Computers & Industrial Engineering 175(January): 108854. doi:10.1016/j.cie.2022.108854.

Kumari, Divya, Nitish Veni, Pranith Kumar M, Manoj H. and Hrishikesh Purohit. (2024). Votereum: Blockchain based secure voting system. pp. 580–84. In 2024 4th International Conference on Pervasive Computing and Social Networking (ICPCSN). IEEE. doi:10.1109/ICPCSN62568.2024.00097.

Le, Yuwei, Xintong Ling, Jiaheng Wang, Ruiwei Guo, Yongming Huang, Cheng-Xiang Wang et al. (2023). Resource sharing and trading of blockchain radio access networks: architecture and prototype design. IEEE Internet of Things Journal 10(14): 12025–43. doi:10.1109/JIOT.2021.3135414.

Lee, Suhyeon and Seungjoo Kim. (2023). Rethinking selfish mining under pooled mining. ICT Express 9(3): 356–61. doi:10.1016/j.icte.2022.03.003.

Li, Xiaoqi, Peng Jiang, Ting Chen, Xiapu Luo and Qiaoyan Wen. (2020). A survey on the security of blockchain systems. Future Generation Computer Systems 107(June). North-Holland: 841–53. doi:10.1016/J.FUTURE.2017.08.020.

Liu, Xuefeng, Gansen Zhao, Xinming Wang, Yixing Lin, Ziheng Zhou, Hua Tang et al. (2019). MDP-based quantitative analysis framework for proof of authority. pp. 227–36. In Proceedings - 2019 International Conference on Cyber-Enabled Distributed Computing and Knowledge Discovery, CyberC 2019. Institute of Electrical and Electronics Engineers Inc. doi:10.1109/CyberC.2019.00046.

M. Andreas. (2015). Mastering Bitcoin: Unlocking Digital Cryptocurrencies - Andreas M. Antonopoulos - Google Books.

Mahmood, Waqas and Abdul Wahab. (2018). Survey of Consensus Protocols. ArXiv Preprint ArXiv:1810.03357. Elsevier BV. doi:10.2139/ssrn.3556482.

Maria, Azees, Vijayakumar Pandi, Jeatha Deborah Lazarus, Marimuthu Karuppiah and Mary Subaja Christo. (2021). BBAAS: Blockchain-based anonymous authentication scheme for providing secure communication in VANETs. Security and Communication Networks 2021. Hindawi Limited: 1–11. doi:10.1155/2021/6679882.

Merkle, Ralph C. (1988). A digital signature based on a conventional encryption function. Lecture Notes in Computer Science (Including Subseries Lecture Notes in Artificial Intelligence and Lecture Notes in Bioinformatics) 293 LNCS. Springer Verlag: 369–78. doi:10.1007/3-540-48184-2_32/COVER.

Mohammed, Mazin Abed, Abdullah Lakhan, Dilovan Asaad Zebari, Mohd Khanapi Abd Ghani, Haydar Abdulameer Marhoon, Karrar Hameed Abdulkareem et al. (2024). Securing healthcare data in industrial cyber-physical systems using combining deep learning and blockchain technology. Engineering Applications of Artificial Intelligence 129(March): 107612. doi:10.1016/j.engappai.2023.107612.

Parmentola, Adele, Antonella Petrillo, Ilaria Tutore and Fabio De Felice. (2022). Is blockchain able to enhance environmental sustainability? A systematic review and research agenda from the perspective of sustainable development goals (SDGs). Business Strategy and the Environment 31(1): 194–217. doi:10.1002/bse.2882.

Peng, Li, Wei Feng, Zheng Yan, Yafeng Li, Xiaokang Zhou and Shohei Shimizu. (2021). Privacy preservation in permissionless blockchain: a survey. Digital Communications and Networks 7(3). Elsevier: 295–307. doi:10.1016/J.DCAN.2020.05.008.

Peres, Ricardo Silva, Xiaodong Jia, Jay Lee, Keyi Sun and Armando Walter Colombo. (2020). Industrial artificial intelligence in Industry 4.0-systematic review, challenges and outlook. IEEE Access 8: 220121–220139. https://ieeexplore.ieee.org/abstract/document/9285283/.

Rakibul Hasan Chowdhury. (2024). Automating supply chain management with blockchain technology. World Journal of Advanced Research and Reviews 22(3): 1568–74. doi:10.30574/wjarr.2024.22.3.1895.

Saad, Muhammad, Laurent Njilla, Charles Kamhoua and Aziz Mohaisen. (2019). Countering selfish mining in blockchains. pp. 360–64. In 2019 International Conference on Computing, Networking and Communications (ICNC). IEEE. doi:10.1109/ICCNC.2019.8685577.

Schinckus, Christophe. (2020). The good, the bad and the ugly: an overview of the sustainability of blockchain technology. Energy Research & Social Science 69(November): 101614. doi:10.1016/j.erss.2020.101614.

Singh, Jaya and Venkatesan, S. (2021). Blockchain mechanism with Byzantine fault tolerance consensus for internet of drones services. Transactions on Emerging Telecommunications Technologies 32(4): e4235. John Wiley & Sons, Ltd. doi:10.1002/ETT.4235.

Sumathy, S., Karthick Murugan, S., Latha, J., Mageshwari, G., Allen Joseph Gilbert Andrews and Hema Kumar, M. (2024). Blockchain based voting system for secure and transparent electoral processes. pp. 395–400. In 2024 10th International Conference on Advanced Computing and Communication Systems (ICACCS). IEEE. doi:10.1109/ICACCS60874.2024.10717101.

Szabo, Nick. (1997). Formalizing and securing relationships on public networks. First Monday 2(9). doi:10.5210/fm.v2i9.548.

Theodorakopoulos, Leonidas, Alexandra Theodoropoulou and Constantinos Halkiopoulos. (2024). Enhancing decentralized decision-making with big data and blockchain technology: a comprehensive review. Applied Sciences 14(16): 7007. doi:10.3390/app14167007.

Valdeolmillos, Diego, Yeray Mezquita, Alfonso González-Briones, Javier Prieto and Juan Manuel Corchado. (2020). Blockchain technology: a review of the current challenges of cryptocurrency. Advances in Intelligent Systems and Computing. Springer Verlag, 153–60. doi:10.1007/978-3-030-23813-1_19.

Vukolić, Marko and Marko Vukoli. (2015). The Quest for Scalable Blockchain Fabric: Proof-of-Work vs. BFT Replication The Quest for Scalable Blockchain Fabric: Proof-of-Work vs. BFT Replication. Inter-National Workshop on Open Problems in Network The Quest for Scalable Blockchain Fabric: Proof-of-Work vs. BFT Replication. 10. doi:10.1007/978-3-319-39028-4_9ï.

Wamba, Samuel Fosso, Serge-Lopez Wamba-Taguimdje, Qihui Lu and Maciel M. Queiroz. (2024). How emerging technologies can solve critical issues in organizational operations: an analysis of blockchain-driven projects in the public sector. Government Information Quarterly 41(1): 101912. doi:10.1016/j.giq.2024.101912.

Wang, Feng, Yongjie Gai and Haitao Zhang. (2024). Blockchain user digital identity big data and information security process protection based on network trust. Journal of King Saud University - Computer and Information Sciences 36(4): 102031. doi:10.1016/j.jksuci.2024.102031.

Watters, Casey. (2023). Digital gold or digital security? Unravelling the legal fabric of decentralised digital assets. Commodities 2(4): 355–66. doi:10.3390/commodities2040020.

Wibowo, Satriyo and Tesar Sandikapura. (2019). Improving data security, interoperability, and veracity using blockchain for one data governance, case study of local tax big data. pp. 1–6. In 2019 International Conference on ICT for Smart Society (ICISS). IEEE. doi:10.1109/ICISS48059.2019.8969805.

Woo, Junghoon, Ridah Fatima, Charles J. Kibert, Richard E. Newman, Yifeng Tian and Ravi S. Srinivasan. (2021). Applying blockchain technology for building energy performance measurement, reporting, and verification (MRV) and the carbon credit market: a review of the literature. Building and Environment 205(November): 108199. doi:10.1016/j.buildenv.2021.108199.

Wu, Hanjie, Qian Yao, Zhenguang Liu, Butian Huang, Yuan Zhuang, Huayun Tang et al. (2024). Blockchain for finance: A survey. IET Blockchain 4(2): 101–23. doi:10.1049/blc2.12067.

Xiao, Yang, Ning Zhang, Jin Li, Wenjing Lou and Thomas Hou, Y. (2019). Distributed Consensus Protocols and Algorithms. Books.Google.Com. Vol. 25.

Xu, Chenhan, Kun Wang, Peng Li, Song Guo, Jiangtao Luo, Baoliu Ye et al. (2019). Making big data open in edges: a resource-efficient blockchain-based approach. IEEE Transactions on Parallel and Distributed Systems 30(4): 870–82. doi:10.1109/TPDS.2018.2871449.

Yao, X. and Zhu, T. (2017). Blockchain Is to Create a New Ecology of Cross-Border Payment. Financ. Expo.

Yin, Wei, Qiaoyan Wen, Wenmin Li, Hua Zhang and Zhengping Jin. (2018). An anti-quantum transaction authentication approach in blockchain. IEEE Access 6: 5393–5401. doi:10.1109/ACCESS.2017.2788411.

Zhang, Shijie and Lee Jong-Hyouk. (2020). Analysis of the main consensus protocols of blockchain. ICT Express 6(2): 93–97. https://www.sciencedirect.com/science/article/pii/S240595951930164X.

Zheng, Zibin, Shaoan Xie, Hongning Dai, Xiangping Chen and Huaimin Wang. (2017). An overview of blockchain technology: architecture, consensus, and future trends. pp. 557–64. In Proceedings - 2017 IEEE 6th International Congress on Big Data, BigData Congress 2017. Institute of Electrical and Electronics Engineers Inc. doi:10.1109/BigDataCongress.2017.85.

Chapter 2
Blockchain Technology for the Modern Era

*Sangeeta Rani** and *Vikram Singh*

1. Introduction

Blockchain is a shared, immutable, distributed ledger across a computer network's nodes. It is a peer-to-peer network that stores transactions and monitors assets in a business network. Due to its decentralization and immutable nature, it was used in cryptocurrency systems but now its use is not limited to cryptocurrency. With the help of the immutability feature, a block can't be changed, so trusted third parties like auditors or other intermediaries are not needed, thus reducing the effort or chances of making mistakes. This technology makes use of an advanced database facility that provides a transparent information-sharing model within any business network. With this technology, users can confirm transactions without the need for a central clearing authority (Imteaj et al. 2021).

1.1 Structure of Blockchain

Data is stored in the blockchain in the form of a chain of connected blocks. Information is chronologically consistent in this system. No one can delete or modify the information present in the blocks without consensus. Therefore, this technology is prone to building an unalterable or immutable database for monitoring orders, payments, accounts, etc. Built-in mechanisms are furnished that turn aside unauthorized transactions that create inconsistencies in the ledger or database. Blocks are basic constructs of blockchain, and each block contains a set of transactions, metadata, and mining information. Each block in a blockchain is linearly connected to the next and is secured cryptographically. The first one is called the genesis block. All blocks that are confirmed and validated are derived from the genesis block. Each block is, in turn, connected to the next one and creates a chain. To insert a new block into the chain consensus algorithm is used. Proof of work (PoW) and proof of stake (PoS) come under the list of consensus algorithms. Each block is split up into two parts, that is, the header and the body (Yan 2022, Zhou et al. 2023).

The header: This section consists of the hash of the previous block hash address, timestamp, nonce, and merkle root, and the body consists of the list of all transactions. Figure 2.1 shows the structure of the Block Header.

Department of Computer Science & Engineering, Chaudhary Devi Lal University, India.
* Corresponding author: sangeetarani@cdlu.ac.in

DOI: 10.1201/9781003620822-2

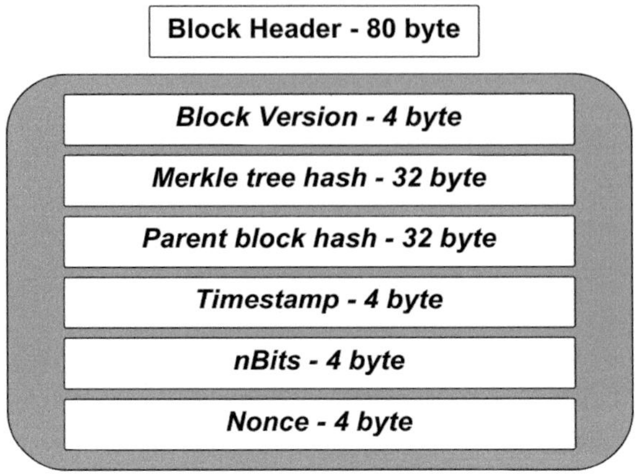

Figure 2.1. Block Header (source: Moosavi et al. 2024).

***Previous block hash address*:** Each hash address is produced using the cryptographic hash algorithm SHA 256. This helps to identify each block in the blockchain structure easily. It helps to locate any type of modifications in the blocks. Every block in the chain provides a reference to a previous block with the help of hash functions. It helps to create an unbreakable link between the blocks in the chain and makes it impossible to tamper with the block without making changes in subsequent blocks (Moosavi et al. 2024).

***Timestamp*:** A timestamp is used to record the occurrence of a particular event so that everyone working on the project can see the time of the event. It is done by storing the digitally recorded moment when the block has been mined and used to validate the transactions. It gives the date and time of day for a specific event.

***Nonce*:** Nonce is a number that is used by the mining process and is known as a number used only once. Miners alter the nonce and create different permutations to generate the correct block hash in the sequence. To find out the right combination of nonce and other values requires computational power.

***Merkle root*:** This tree gives a structured method to encode the data so that verification can be easy and also for security purposes. It consists of the hashed transaction hashes within the transaction. Hashes are utilized for transaction verification. In this each hash is further hashed.

***Body*:** This part comprises a set of transactions. The maximum number of transaction depends upon the block size, which may vary in blockchains, depending on network congestion and communication overhead.

1.2 Key Principles of Blockchain Technology

Blockchain is an innovative technology that is based on the concept of immutability, decentralization, transparency, security, and automation in the digital era. It is best used for organizations that rely on trust, transparency, and security without depending upon a central authority (Moosavi et al. 2024).

***Immutability*:** It is an immutable public digital ledger; transactions cannot be altered after recording a transaction. Immutability is achieved through cryptographic hashing and the chaining of blocks. It gives an assurance of an unforgettable and tamper-resistant record of transactions; implants trust among participants; and provides a transparent and auditable history of events. All transactions in blockchain are recorded consistently and are time- and date-stamped.

***Decentralization*:** Decentralization is a facet of blockchain technology. A single entity is not responsible for controlling the data, rather it distributes the ledger across a network of nodes. This feature of blockchain eliminates the need for intermediaries and central authorities, offering participants to run secure and transparent transactions. No intermediary fee is required.

***Transparency*:** The prominent feature of blockchain is transparency. Every user on the network can access a similar copy of the distributed ledger, confirming that transactions are visible and verifiable. Data is stored identically at more than one location because it works as a distributed ledger. The ledger is updated automatically, so the transactions are done instantly and transparently. In the absence of blockchain, each organization has to maintain a separate record. Due to this feature, participants with permission perceive the entire history of a transaction which is the concept of providing full transparency. It eliminates most of the chances of fraud.

***Security*:** Cryptography is a keystone of blockchain development, securing the integrity and privacy of data. Blockchain is secure due to the encryption feature. The record is encrypted end-to-end, and can't be altered therefore the blockchain helps to turn aside fraud and unauthorized access. Data is recorded on a network of computers, rather than a central server, which makes information difficult for hackers to view data. The authenticity of a transaction is verified and confirmed using cryptography. A pair of keys are used in cryptography, known as public and private keys and each participant has this pair of cryptographic keys. A public key is used for identification and a private key is for secure access. This cryptographic infrastructure makes blockchain a foundational technology with applications spanning from finance and supply chain to healthcare and beyond and ensures that transactions are verifiable, tamper-resistant, and confidential. The consensus mechanism is another important element that describes how nodes agree on the state of the blockchain. There are two popular mechanisms available PoW and PoS. In Proof of Work, nodes strive to solve complex mathematical problems and demonstrate computational effort as proof of validity. In Proof of Stake, participants with a significant stake in the network validate transactions by considering ownership rather than computational power. Both mechanisms prevent malicious activities. The combination of security features makes blockchain a potent solution for shielding sensitive data and transactions (Yan 2022).

1.3 The Significance of Blockchains

Traditional database systems provide solutions for businesses and organizations and work as a backbone for data storage and management. In the modern era of digital transformation, these systems face several difficulties, where data integrity, scalability, and security are to be achieved.

***Traditional database technologies*:** These technologies face several challenges when recording finance-related transactions. In the case of property selling, the property is transferred to the buyer by swapping money for the property. Both buyer and seller can separately record the payment transactions, but neither party can be fully trusted. It would be very easy for the seller to claim that they did not get the money which they did, and for the buyer to claim that they sent the money when in fact they had not done so. To handle these issues, an intermediary must be there to validate transactions. A centralized system generates a single point of susceptibility and also integrates the transactions. If the central database was corrupted, both parties would face significant losses. Centralization is the major challenge in traditional databases. In traditional databases, a centralized authority is used to manage the database, which makes it vulnerable to cyberattacks, data breaches, and single points of failure. The entire system can fail if the central server does not work properly, which will cause operational as well as financial losses. External auditing or third-party intermediaries are required in the case of databases to ensure data integrity and to prevent fraud, which adds extra cost and complexity.

Blockchain: A blockchain deals with these concerns by providing a distributed, tamper-proof system to store transactions. The blockchain resolves the difficulty by creating a shared ledger for the buyer and the seller in the case of a property sale scenario. All transactions are automatically updated in the shared ledger and must be approved by both parties. Any tempering in past transactions will distort the full ledger. These aspects of blockchain technology steer its use in various sectors: for example, in digital currency like Bitcoin. Blockchain is a segregated and distributed ledger in which data is present across several nodes on the network. There is no centralised system so it eliminates the chances of a one-point failure, and makes the system more adaptable to attacks and disruptions. In addition, the use of cryptographic techniques ensures the concept of data security and immutability which means once a transaction is recorded, it cannot be modified. This concept gives surety to data integrity without the need for external audits or intermediaries.

2. Techniques in Blockchain

This section discusses the underlying methods and techniques before delving into the types of blockchain networks, the opportunities and challenges within them, and their applications in various domains of public life.

2.1 Cryptography

Blockchain technology ensures security, immutability, and trust by using cryptography techniques. Hashing, digital signatures, and the Merkle tree are used for cryptography.

Hashing: The hashing technique is used in blockchain that is based on a hash function. It is an operation used in cryptography that converts any input data, irrespective of size, into a fixed-length output, which is called a hash address. The hash function assures the validity and integrity of data within the system. Different ways of calculating hash addresses are available, and that creates a unique hash for each block. Based on the previous hash address, the address of the next block is calculated, which makes a series of blocks that are very difficult to modify. The reason for that is that if there is a change to any block, the new hash address will change, and it will break the order and be visible to every node in the network. This makes the use of hash functions in blockchain technology important. It is a secure method of protecting the data against unauthorised access.

Digital signatures: Digital signatures make the system trustworthy and safe. They are used to provide veracity and integrity of transactions. The digital signature is used when things are processed online without the need for paperwork. In this method, the digital signature is used instead of a classic signature. Moreover, digital signatures are like digital fingerprints and they contain timestamps. The motive is to authenticate and maintain probity. Authentication means to verify that the information is accessed only by authorized users and integrity ensures whether the authorized user has worked with the transaction and the data retains integrity, that is, data has not been amended by someone halfway and it is identical to the authentic data sent by the sender. This is achieved by utilizing a public key infrastructure to create a unique digital fingerprint. Every transaction is digitally signed with the sender's private key. Then by using the sender's public key digital signature, it is verified by the receiver. The receiver can be assured that no tampering has been done with the transaction and it is intact and authentic, in case of a valid digital signature. They impart a method for resolving disputes, certifying that transactions are not tampered with, and disallowing unauthorized access. Bitcoin, Ethereum, and Ripple frequently utilize digital signatures. The working of the digital signature is represented in Figure 2.2.

Merkle tree: Blockchain technology uses an underlying cryptographic data structure known as Merkle trees to safeguard the validity and integrity of the data. It encodes the transaction data efficiently and securely. It is either quoted as a hash tree, or a binary tree, where every node on the leaf node represents the block of data, and the non-leaf node is the hash of every child node. The

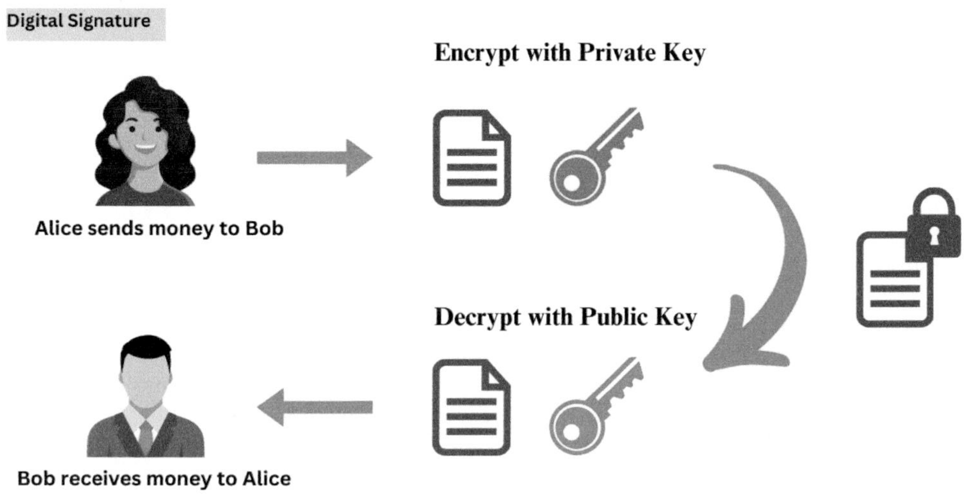

Figure 2.2. Digital Signature (source: (Moosavi et al. 2024)).

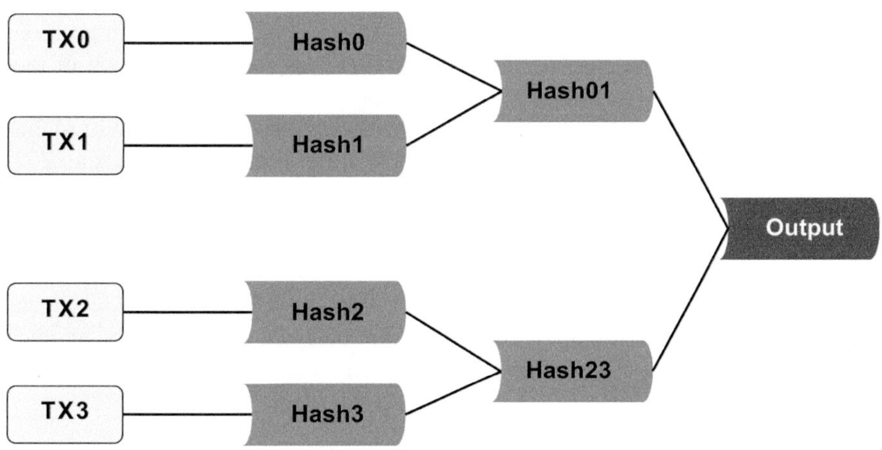

Figure 2.3. Merkle Tree Root Hash (source: Moosavi et al. 2024).

hash of the complete merkle tree is verified against the hash recorded in the blockchain header for validating the correctness of a data block. Figure 2.3 depicts the Merkle tree.

After the matching of hashes, data probity can be confirmed, which indicates that the data has not been tempered with. A vast amount of data can be verified without storing the full dataset. These play a pivotal role in Bitcoin as they impart quick verification and guard the history from unethical users. Merkle trees become a part of modern cryptographic protocols and are used in Ethereum and Litecoin.

2.2 Consensus Mechanism

Blockchain technology utilizes consensus algorithms to confirm that every participant on the network concedes on the present state of the ledger. Multiple nodes work together to insert and authenticate new transactions into the distributed ledger. Cryptography is utilized to fortify the transactions and ensure that only permissioned nodes can get involved in the process of validation. Various consensus methods, including Proof of Work (PoW), Proof of Stake (PoS), Delegated Proof of Stake (PoS), and many more are implemented in networks. These mechanisms are discussed in the following subsections (Huang et al. 2023).

Proof of work: This algorithm was initially proposed by Dwork and Naor in 1993 to contend with spam mail and to gain access to shared resources. This protocol was used as a consensus algorithm in Bitcoin in 2008 to prevent double-spending attacks and to provide secure transactions. This was the first consensus algorithm adopted in the blockchain network. The intention behind this is to encourage miners to utilize their computational power to calculate a complex puzzle from a mathematical issue. The miner live streams the block to the network when the problem is solved and other miners can check the correctness of the solution. The average time consumed is 10 minutes to produce a new block in Bitcoin. This method is used by Bitcoin, Dogecoin, and Litecoin (Cao et al. 2020).

Proof of stake: It was originally thought of by King and Nadal in 2012 and was first implemented by Peercoin cryptocurrency. This method has evolved as a low-cost, low-energy-consuming in comparison to PoW. This selection process is used instead of a competitive process based on the stake of validators. Miners in the PoW system are equivalent to validators in the PoS system. It involves assigning the responsibility of initiating new blocks to a candidate node extent to the number of virtual currency tokens held. When a new block is proposed it is verified by the network and added to the blockchain upon reaching the consensus. The number of coins used for proving its ownership is called the "coin age" metric. This is the product of the currency amount by the period of its possession. The Validator validates the block after the selection by putting a bet on it and gets a bonus relative to the placed. The PoS system is not implemented 100% in peercoin currency generation, rather it utilizes the combination of PoW and PoS for the formation of the currency. Therefore, Peercoin is an amalgamation system that merges the characteristics of both PoW and PoS. Nxt cryptocurrency is based on purely PoS. Ethereum is also drafting to shift to the PoS in its 2.0 version. This updated version is to be known as the Casper protocol (Kalita et al. 2024).

Delegated proof of stake: This decentralized consensus algorithm is produced by Larimer. This method is based on the election process for decentralized voting. On behalf of stakeholders, delegates are appointed to secure the network. These delegates randomly create blocks turn-wise. If a representative is unable to generate a block properly, it mislays its fame. As a result, stakeholders do not vote in favor of that representative and then that delegate is replaced by another delegate. Blockchains like BitShares use versions of DPoS (Lashkari and Musilek 2021).

3. Types of Blockchain Networks

Blockchain networks are classified in subsequent ways. Let's dive deep into these networks with pros, cons, and ideal use cases. The classification of networks is shown in Figure 2.4.

Public blockchain: A public blockchain is a type of blockchain that is accessible to anyone and allows anyone on the network to access and verify it. Cryptocurrency, like Bitcoin, originated in a Public blockchain that uses the concept of disseminated ledger technology. These networks deal with the issues that arise with centralization, less security and transparency. Every user on the network has the freedom to use and sign in to the blockchain platform to become an authorized node. Information is not recorded in one place in a distributed ledger rather it is distributed on a node-to-node network. Due to this distributed nature, the means of verifying the originality of data is required. The way of doing this is known as consensus mechanism, where users in the blockchain agree on the present state of the ledger. The two common consensus methods are: Proof of Work and Proof of Stake. In this network, users can access both current and historical records and perform mining activities that involve compound computations to prove transactions and add to the ledger. No valid information can be tampered with as the source code is open for everyone, and users are at liberty to verify the transactions, search for errors, or propose modifications.

A public network is appropriate for businesses that prioritize transparency and trust, such as social support groups or non-governmental organizations. However, due to its public nature, public network

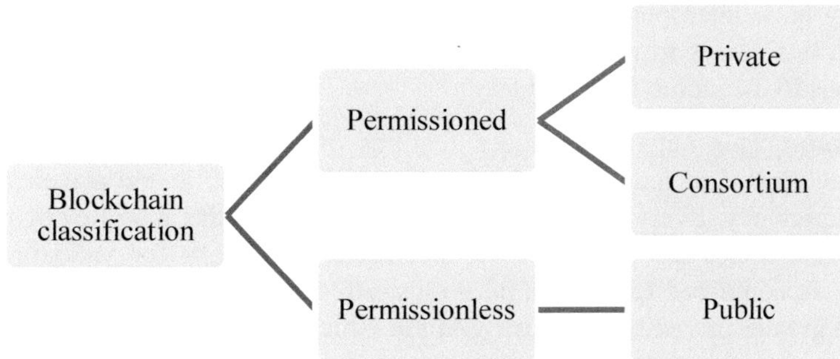

Figure 2.4. Blockchain Classification (source: Moosavi et al. 2024).

may not be the preferred choice for private businesses. Public blockchain networks provide the facility of transparency. Security protocols and methods are used to make public blockchain networks secure. It is perfect for organizations founded on truthfulness, including non-governmental organizations and social support groups. It is not an ideal choice for private businesses due to its openness.

Security is a significant challenge for public blockchains. Some developers address this by implementing a competitive, distributed system for validating, proposing blocks, and rewarding participants, while some rely on it by using a collateralized system. Another challenge is the lack of complete privacy and anonymity. Public blockchains make transaction amounts and associated addresses visible to anyone. Users lose their anonymity in this case when the owner of these addresses is identified.

***Private blockchain*:** A private blockchain is confined to a category of users that have been granted authorization to access the network. It is called a permissioned blockchain network. It is a distributed ledger technology that is restricted to a specific group of participants, within a single organization or consortium. The key features of a private blockchain are privacy and control. Private blockchains are implemented within the bounds of a specific business or consortium, whereas permissionless networks are fully transparent. In private networks authorized users are authenticated to grant access and validate transactions. Restricted and controlled access is available in this network, as the worker has the right to overrule, modify, or delete entries present on the blockchain. Permissioned users can view the transactions on the network. Only authorized participants are allowed to join the network. It is the opposite of a public blockchain, as private blockchains use a specialized form of blockchain technology that can work within a specific organization or consortium. Public blockchains are open to anyone with full transparency, whereas private blockchains operate on a permission basis, allowing only authorized participants to access and validate transactions. The framework of private networks offers greater security, privacy, and control, making it appealing to those businesses seeking to optimize operations while ensuring confidentiality (Swathi et al. 2021).

The benefits of using private blockchains are that they help to reduce fraud by identifying and then confirming membership and giving access privileges to known organizations. Due to the restrictions on the number of users, this type of blockchain has higher throughput, which provides the facility for faster business transactions. A private blockchain is best suited for applications that demand data privacy and control over information sharing. Since the number of users in a blockchain is few, it consumes less amount of energy and material, making it a good option compared to a public blockchain. Private blockchains enhance privacy for businesses, as they regulate access to their data and can ensure that sensitive information remains protected from unauthorized parties.

The challenge of private blockchains is that they may lack transparency because access is restricted to a limited group of participants. It can make it difficult for businesses to showcase transparency and accountability to stakeholders. With fewer participants, the network becomes more vulnerable to hackers, who can potentially take control and manipulate the data. Private blockchains can be cost-effective, and can also be expensive to operate and maintain in some scenarios. Private blockchains may have interoperability challenges with other blockchains including public ones. This can restrict businesses from interacting with other networks and may require complex and costly integrations to overcome these issues.

***Hybrid blockchain*:** This network includes features of both permissioned and permissionless networks. It is delineated to reduce the shortcomings of both public and private blockchains and enhance their supremacy. Everybody is allowed to take part in the activities of public networks. Meanwhile, private nodes are governed by specific organizations for validating and checking transactions. These combined features build a most effective hybrid blockchain and can facilitate faster work with greater privacy and security. At the same time, public nodes provide the facility of decentralization and transparency that makes it impossible for any one group to obtain supervision over the network. Two components are used for making this network a private and public blockchain. After granting access to the user on a hybrid blockchain platform, users are free to participate in the activities. The details of the users remain confidential from other participants on the network. The information of the user is disclosed only to the specific party they are communicating with on the network. Businesses and organizations use the KYC facility to confirm that the process of identification process is completed accurately.

The benefit of using a hybrid blockchain is that it increases transparency, and lowers transaction costs, with expenses potentially as low as $0.01 per transaction. Private blockchains are perfect for handling privacy-related issues, but they lack effective communication channels with external parties. Businesses using hybrid blockchains can securely protect sensitive information while ensuring smooth interaction with shareholders. Hybrid blockchains are resistant to 51% of attacks since hackers cannot gain sufficient access to the network to cause attacks. Hybrid blockchains offer faster operations and are easier to manage. In comparison to private blockchains, these blockchains provide harsh data security protocols and more stretchability. Unlike other networks that adhere to a rigid infrastructure, they offer pliability to customize the ledger according to individual and organizational requirements.

***Consortium blockchain*:** Consortium blockchain emerges as an innovative model that bridges the gap between unauthorized access of public networks and the restriction on private ones. It offers a communal platform where multiple organizations can work together, maintaining control over the network that ensures a higher level of privacy and efficiency. Consortium networks are absolutely fine for organizations that need to communicate securely without sacrificing autonomy or control. It is a semi-decentralized network managed by more than one organization. In this blockchain, several organizations can act as nodes, enabling them to exchange information or perform mining. Each organization, such as a financial institution, may have its private blockchain, and an authorized group of nodes is responsible for managing the consensus process. These blockchains operate as permissioned networks, meaning users must receive authorization and an invitation to join. These blockchains are collaboratively handled and maintained with the help of organizations or entities. Resilience and service continuity are two of the most important elements for managing consortium blockchain networks, in cases when companies and institutions depend on networks for critical operations. The consortium blockchain gives the facility of data transfer but does not handle cryptocurrency creation or management, because they are typically managed by public blockchains. Resilience means the network's ability to remain operational even if some of its nodes fail. This is achieved by incorporating mechanisms from the initial design of the network to ensure that the entire system continues to function without interruption. Imagine it like emergency generators in a hospital: if the power supply fails, the hospital can continue its daily operations smoothly. In the same way, a blockchain network needs backup systems, so that if one or more nodes stop operating, others can seamlessly take over and maintain operations.

The benefit of using consortium blockchains is data privacy, like private blockchains, only a select group of individuals can access the datasets. Due to fewer nodes and a lighter network load, consortium blockchains enhance data privacy. It requires approval from specific nodes, in case of any changes to the network's information, ensuring strong data protection. Controlled user groups can also speed up the process of consensus-building.

Establishing and managing a consortium blockchain can be complex and requires specialized technical expertise. These blockchains are managed by a group of participants, there is a risk of conflict of interest. Participants might prioritize their own goals over the collective interests of the network, which leads to governance challenges and disputes. A consortium blockchain requires a carefully designed governance framework to ensure all participants remain aligned with the network's objectives and shared interests. However, it can be difficult to achieve, especially when participants have differing priorities or conflicting interests.

4. Opportunities and Challenges

Blockchain technology offers many opportunities in the real-life domains of classified documents, high-security vaults, and cryptocurrencies. To tap the real potential of blockchain, certain challenges must be dealt with. This section discusses the opportunities offered by blockchain and the challenges therein (Nguyen et al. 2019).

4.1 Opportunities

***Enhanced security*:** One advantage of blockchain technology in the bank sector is also that it can contribute to data security. Traditional banking systems are vulnerable to cyber-attacks and data breaches, resulting in significant financial loss and reputational harm. The decentralized technique of storage on blockchain makes it close to unimaginable for cybercriminals to change or corrupt information (Al-Farsi et al. 2021).

***Increased transparency and trust*:** Each organization has to maintain a separate database without blockchain, but in the case of blockchain, a distributed ledger is maintained to store identical information in multiple places. All users on the network can view the same information, at an identical point in time, that further used to provide reliability. Blockchain provides the facility of a distributed ledger that gives a transparent and auditable transaction record. Finance-related transactions take advantage of this distributed ledger feature as it allows real-time observation and monitoring of financial transactions. This feature has helped to reduce the resources required and the consequence is enhanced efficacy of compliance procedures (Monrat et al. 2019).

***Increased efficiency and speed*:** Conventional methods use manual processes that are susceptible to human error, time-consuming, and many times require trusted intermediaries. By well-organising these processes, transactions can be finished more efficiently. Data related to transactions is recorded on the blockchain instead of being stored on paper. Due to its distributed ledger facility, there's no requirement to maintain several ledgers; therefore, accommodation and settlement can be accelerated.

***Smart contracts*:** Smart contracts are used to automate transactions, which speed up the process and result in increased efficiency. After the pre-specified conditions are met, the next step in the transaction is automatically activated, which lessens human intervention and dependency on mediators to verify the completion of the agreement terms.

***Immutability*:** Immutability is that once a transaction or record is stored on a chain, it cannot be altered or deleted. All transactions in the blockchain are time and date stamped; any permanent record can be accessed at any time and used to trace details over time; which assures reliable examination of information. Manual filing is error-prone, that is, more easily corrupted or retired.

***Decentralization*:** Blockchain works on the principle of decentralization. It does not rely on a central authority; instead, it is a distributed, decentralized, public ledger across the network. In this decision-making and control is not in the hands of the central authority rather they are distributed equally to remove the chances of bias.

4.2 Challenges

This section discusses the issues that arise in implementing blockchain technology. Figure 2.5 illustrates the challenges that arise in blockchain networks

***Scalability*:** The total Number of users in Coinbase until October 2017 was 11.7 million. The average transactions have changed rapidly as it gained more popularity and a lot of people were getting used to it. It impacted the processing speed of the transactions as more people meant more people accessing the network, which in turn made the whole system tedious. Scaling the technology for a large number of concurrent users is still a technical challenge for the blockchain industry. Many of the complex algorithms are used in processing just a single transaction on the blockchain technology. Blockchain networks also tend to be slow and inefficient because validating transactions requires significant computational resources. This complicates blockchain networks in applications where rapid transaction processing speeds are needed. Traditional blockchains such as Bitcoin and Ethereum use consensus algorithms such as Proof of Work (PoW) and Proof of Stake (PoS), which can be slow and resource-intensive (Al-Farsi et al. 2021).

***Energy consumption*:** Heavy power usage stands as a major challenge with blockchain technology applications. Performing blockchain transaction validation processes requires intensive computer power that leads directly to elevated energy requirements. Blockchain technology introduces environmental concerns because it produces substantial carbon emissions while also affecting the environment during its use. Most blockchain networks consume massive volumes of power. Cryptocurrency Bitcoin relies on Proof of Work for consensus processes but users must resolve complex mathematical problems that need extensive energy-consuming calculations because of their demanding processing requirements. Proof of demand blockchain systems require substantial amounts of energy both to execute transactions and operate the network. Blockchain networks

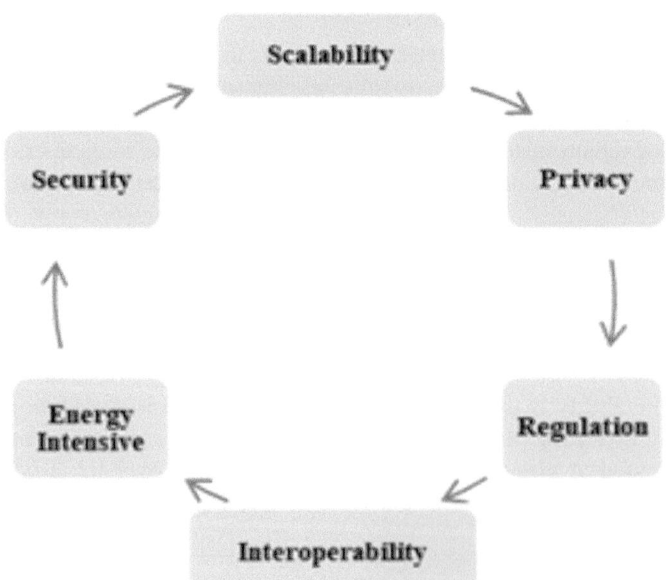

Figure 2.5. Blockchain Issues and Challenges (source: Moosavi et al. 2024).

implementing Proof of Stake demonstrate reduced energy usage in comparison to other consensus mechanisms used for network operation. The lowered energy consumption of the Ethereum 2.0 platform shows a positive movement in blockchain network power management. The future success of blockchain requires continuous community support for advanced, sustainable energy-efficient solutions (Rajasekaran et al. 2022).

***Governance and regulatory issues*:** Through blockchain technologies, immutable and transparent mechanisms, organizations find regulation enforcement and auditing more effective. Blockchain delivers accurate transaction documentation as an organizational tool that helps organizations meet legal and regulatory standards. Verifying blockchain data and transactions supports auditors in decreasing the opportunities for both fraud and mistakes among financial records. The emergence of blockchain technology has met regulatory challenges because existing regulation is constantly reshaping. The process of new technology comprehension has global regulatory bodies and governments across the world that are developing rules. The unclear regulatory landscape generates unpredictability, which signifies a major issue faced by investors and businesses, because they require detailed regulatory projections about upcoming impacts on blockchain operations (Abbas et al. 2023).

Companies demonstrate reluctance to fully invest in blockchain projects because of the lack of regulatory indications. The expanding body of regulations worries them because possible modifications could make their projects illegal or technically complicated. Nurturing an environment requires mandatory clear and continuous supportive regulatory practices. Blockchain technology demonstrates robust functionality, which allows for safe adoption across multiple industries in this supportive environment. The operation of networks under the blockchain system depends on blockchain governance through rules combined with decision systems. The implementation of governance structures regularly encounters various obstacles. Blockchain decentralization creates an environment void of central authority decision powers, but disagreements regarding implementations may result in blockchain fork divisions. Forks within blockchain networks establish split paths with varying protocols that cause navigation confusion while breaking network coherence. The network suffers from two separate blockchains whenever two factions fail to agree on a chain update. Such partitions between blockchain entities can fracture system reliability and destroy institutional trust. Blockchain requires stable operations and consistency, so maintaining effective governance models becomes crucial to prevent or control forks.

5. Applications

Blockchain technology provides creative solutions in numerous sectors, including financial services, supply chain management, healthcare, government sector, intellectual property management, and IoT (Alqahtani and Algarni 2023),

5.1 Financial Services

Without taking the help of traditional banks public blockchains run a range of financial services, such as trading, lending, and borrowing. This technology furnishes a trustful and immutable record of all transactions by recording financial and non-financial transactions. It assures that transactions are precisely recorded and can be authenticated by all parties indulged in the process.

***Cryptocurrency systems*:** Many businesses initiated over the last few years offer distributed exchange of cryptocurrency that results in faster and cheaper transactions. In this distributed exchange, investors are not required to deposit their assets as with the central control system. Blockchain-based exchanges deal in cryptocurrency. *Bitcoin* is an application of blockchain technology. This digital currency has no centralized control. In the beginning, Bitcoins were initially designed to build financial transactions online; nowadays these can be converted into global currencies such as euros and USD and are known as digital assets. Without the need of any intermediaries Bitcoin and Ethereum do node-to-node transfers of digital currencies (Hashemi et al. 2020).

***Cross-border payments*:** Using blockchain technology, money transfer is affordable and fast as compared to traditional money transfer systems. It is veracious in the case of cross-border transactions. Money transfers can take days using conventional methods whereas using takes a small amount of time in the modern U.S. financial system. Ripple is used as a payment protocol that is used to provide secure, rapid, and reasonable cross-border transactions. It ensures the reliability and security of transactions. Ripple is used by financial institutions to reduce the complexity and intricacies of international payments and to enhance reliability.

5.2 Supply Chain Management

This technology can remarkably upgrade supply chain management by laying out real-time tracking of transactions, ensuring transparency, and reducing inefficiencies. It allows stakeholders to trace products and verify authenticity throughout the supply chain. Public blockchains improve transparency and traceability in supply chains, allowing stakeholders to verify the launch and journey of products. Companies like VeChain leverage blockchain technology for this purpose (Khanfar et al. 2021, Kim and Shin 2019).

***Tracking and transparency*:** IBM Food Trust works on the concept of supply to enhance transparency and traceability. The launch and movement of food products are recorded at every step of the chain and ensure data is accurate and sealed. It upgrades the safety of food and builds the trust of the consumer (Rodrigues et al. 2021).

5.3 Healthcare

This technology is prominently used in the healthcare sector for drug authentication, recording medical records, expedited credentialing, and real-time insurance information (Cekerevac and Cekerevac 2022).

***Secure patient records*:** Blockchain is used to store patient records securely, so that they are only accessed by authorized users. This is a blockchain-based system meant for recording electronic medical records to magnify patient privacy, information security, and coordination. MedRec, developed by MIT is an example of this. This gives a safe, decentralized method to record and share patient data that helps doctors and medical professionals to analyze the reports of patients accurately with up-to-date information by recording the information on blockchain. This way of storing the records speeds up the system for fetching records of patients which helps in timely treatment.

***Drug traceability*:** This provides a medicine traceability system that is used to quickly identify and remove fake medicines from the market. Every transaction involving medicines may be documented on the blockchain as they go through the supply chain. It consists of details about the people involved in the financial transaction, as well as, the location and timestamp. The status of the medicines may be automatically updated by the smart contract, which can also follow their journey from producer to distributor to drugstore. Smart contracts work on the data gathered by the sensors that help to provide the information that medicines are maintained properly along the supply chain.

5.4 Government Sector

Blockchain technology has numerous benefits in the government sector by protecting data, systematic processes, and reducing the chances of fraud. Additionally, it helps to build accountability and transparency in the system. A distributed ledger is secured by using cryptographic techniques that make it possible for individuals, businesses, and governments to share resources without the need for third-party users. This method curtails the chances of a single point of failure and works as a shield against sensitive data owned by government entities and citizens.

***Secure voting systems*:** Such measures help to validate that voters participate only once, and keep votes safe from tampering attacks. Keep Personal identities on blockchain systems to help develop voting possibilities based on this technology. Access to voting can become simpler than using smartphone buttons through this system. The expenses related to election administration would dramatically drop if the system is used. Public networks provide secure and trustful voting systems which help decrease election fraud and enhance voter trust in the results of democratic processes. The use of digital voting receives support through numerous programs which aim to advance this method.

***Public service delivery*:** Land Registration comes under the category of public service delivery. The application of blockchain technology for land registration cuts out numerous cost sections and reduces verification periods, while enabling customizable modular products. The system expands access to fractional property investments while maintaining proof of a fractional property's origin during trading.

***Digital identity*:** Identity management is a keystone of governmental services. The storage of digital identity on blockchain serves to manage government benefits, including welfare programs and social security benefits and medicare. Blockchain technology decreases operational costs and fraud occurrences simultaneously. Digital distribution of funds can happen faster because blockchain technology enables payments on its platform. A digital identity system eliminates the requirement for people to depend on one specific profile provider which helps decrease documentation requirements. The number of identity solutions found in public blockchains allow users to control their identities independently from centralized administrative entities: uPort and Civic represent two projects that belong to this category.

***Intellectual property management*:** The innovative powers of blockchain are reflected in its real-life applications. There are some specific circumstances in which technology is being constructively used in IP rights management. It provides a shield against intellectual property by giving safe and transparent information of ownership and rights. This provides a transparent method for tracking the ownership changes or transfers of patents, trademarks, copyrights, and other forms of IP (Luo 2022, Song et al. 2021).

***Protection of copyrights*:** In blockchain, authenticity and ownership are given by using a timestamp that can't be modified. It is used to verify the origin and ownership at a specific moment. *Blockai* is a blockchain company that helps artists and creators to meet their diverse needs. The timestamp is used to document the creation of new works and protect them from copyright infringement. Creators journal their work on the blockchain and make certain that their rights are protected and verifiable (Song et al. 2021).

***Patent filing*:** Inventors and businesses convert their trademarks or patents into digital assets such as non-fungible tokens (NFTs). The unique identification number is assigned to each non-fungible token which makes it easier to track unauthorized usage, secure storage, and enhance asset protection. Non-fungible tokens represent ownership of distinct digital assets, such as art or collectibles on public blockchains. Ethereum is the main platform used for minting, creating, and trading NFTs.

5.5 Internet of Things

Blockchain technology provides the feature of storing passwords and other information on a distributed network instead of a centralized system. Due to its immutability, it offers protection against data tampering. Several corporations have sketched out to expose the blockchain's potential. In addition, the Internet of Things unfolds innumerable opportunities for businesses to execute smart operations. Most devices surrounding us are now furnished with sensors, supplying data to the cloud. So, merger of IoT and blockchain technologies can make the system competent (Krichen et al. 2022).

Device communication: Blockchain integrated with IoT amplifies security and provides transparency to IoT ecosystems. This framework gives a scalable and decentralized domain for applications based on IoT platforms. Data is shared without using any central network. Data is shared using blockchain's distributed network of computer systems which allows communications to be tamper-proof.

Security enhancements: Blockchain technology uses a more comprehensive way to verify events that are recorded in a database, verified by different sources, and then added to a shared ledger and distributed to all servers. Blockchains and IOT are a matching set; together they provide many benefits including the ability for smart devices to work autonomously and have no centralized authority. It takes the idea of a smart home and elevates it to an entirely new level, solving security issues and removing centralized infrastructure with blockchain. Biometric authentication and facial/voice recognition systems are the use cases of this technology.

6. Conclusion and Future Perspectives

This section concludes the chapter by discussing the future outlook of the turf of blockchain technology.

6.1 Conclusion

The innovative blockchain technology now introduces new industry possibilities through its combination of transparency and cryptographic security while decentralizing authority. The technology serves applications across multiple sectors from finance to supply chain management to healthcare to governance. Blockchain technology transforms transactions between parties by removing the necessity of third parties while establishing trust between both parties, which allows simpler process workflows, reduced prices, and enhanced operational efficiency. The groundbreaking features of this technology represent vast opportunities for future implementation but its adoption process faces significant hurdles. The widespread implementation of blockchain remains restricted because important issues including scalability and energy usage and governance and regulatory compliance operate as barriers. Leading consensus methods such as Proof of Work (PoW) create excessive computational requirements that produce operational inefficiencies, as well as, environmental worries. The challenge of governing decentralized systems intensifies because decision power spreads across numerous operational entities. The completely transparent structure of blockchain platforms leads to privacy problems, which create issues in areas that require vigorous data protection regulations.

New evidence shows that insufficient global regulatory standards produce market ambiguity between businesses and developers. The widespread classification and regulatory challenges with cryptocurrencies as a main blockchain solution have caused different international governing bodies to establish varying approaches. The difficulties in implementing blockchain highlight the necessity of joint work between blockchain developers alongside industry stakeholders and governmental officials. The development of *Ethereum 2.0* and Layer 2 solution illustrate substantial progress in solving blockchain scalability and energy consumption problems. The transformative capabilities of blockchain technology demand continuous work to mitigate its known shortcomings. The blockchain ecosystem will advance through collaborative efforts and innovative solutions to become a powerful sustainable system that will bring efficiency alongside trust and transparency developments across industries.

6.2 Future Perspectives

The massive potential of blockchain technology covers numerous industries that can transform their transaction operations and system management practices. The growing blockchain evolution will extend its operational range beyond digital currencies to control supply chains and advance healthcare operations, as well as financial and government activities. Through supply chain management blockchain technology delivers traceable and transparent operations that aid businesses to track products across supply networks, thus decreasing fraud and enhancing system efficiencies. The financial industry will continue advancing decentralized financial systems (DeFi) that establish direct peer-to-peer secure trading while lowering costs and building up economic participation among all groups (Xie et al. 2023).

The convergence of blockchain technology with 5G networks, together with AI and IoT systems, creates fresh opportunities for protected IoT device operations and connected urban development. Government entities along with enterprises are experimenting with blockchain technologies to build voting platforms and systems, while protecting intellectual property and confirming the identities of users. Ongoing research into Proof of Stake (PoS) and Layer 2 scaling methods will make blockchain more scalable and at the same time reduce its energy usage and environmental impact. When industries implement blockchain on a massive scale, they will reshape the way people interact with technology.

References

Abbas, H. H., Kareem, M. I. A. and Gheni, H. M. (2023). A survey on security and policy aspects of blockchain technology. Telkomnika (Telecommunication Computing Electronics and Control) 21(2): 353–362. https://doi.org/10.12928/TELKOMNIKA.v21i2.24738.

Al-Farsi, S., Rathore, M. M. and Bakiras, S. (2021). Security of blockchain-based supply chain management systems: Challenges and opportunities. Applied Sciences (Switzerland) 11(12): Article 5585. https://doi.org/10.3390/app11125585.

Alqahtani, A. M. and Algarni, A. (2023). A survey on blockchain technology concepts, applications, and security. International Journal of Advanced Computer Science and Applications 14(2): 1–10. https://doi.org/10.14569/IJACSA.2023.0140296.

Cao, B., Zhang, Z., Feng, D., Zhang, S., Zhang, L., Peng, M. et al. (2020). Performance analysis and comparison of PoW, PoS, and DAG-based blockchains. Digital Communications and Networks 6(4): 480–485. https://doi.org/10.1016/j.dcan.2019.12.001.

Cekerevac, Z. and Cekerevac, P. (2022). Blockchain and the application of blockchain technology. MEST Journal 10(2). https://doi.org/10.12709/mest.10.10.02.02.

Guo, H. and Yu, X. (2022). A survey on blockchain technology and its security. Blockchain: Research and Applications 3(2): Article 100067. https://doi.org/10.1016/j.bcra.2022.100067.

Hashemi Joo, M., Nishikawa, Y. and Dandapani, K. (2020). Cryptocurrency, a successful application of blockchain technology. Managerial Finance 46(6): 767–779. https://doi.org/10.1108/MF-09-2018-0451.

Huang, R., Yang, X. and Ajay, P. (2023). Consensus mechanism for software-defined blockchain in internet of things. Internet of Things and Cyber-Physical Systems 3: Article 100029. https://doi.org/10.1016/j.iotcps.2022.12.004.

Imteaj, A., Hadi Amini, M. and Pardalos, P. M. (2021). Introduction to blockchain technology. pp. 1–10. In SpringerBriefs in Computer Science. Springer. https://doi.org/10.1007/978-3-030-75025-1_1.

Kalita, K. P., Kharbhih, J. C., Boro, D. and Bhattacharyya, D. K. (2024). An enhanced blockchain consensus mechanism using proof-of-work and proof-of-stake. pp. 455–463. In Lecture Notes in Electrical Engineering, 1061 LNEE. Springer. https://doi.org/10.1007/978-981-99-4362-3_46.

Khanfar, A. A. A., Iranmanesh, M., Ghobakhloo, M., Senali, M. G. and Fathi, M. (2021). Applications of blockchain technology in sustainable manufacturing and supply chain management: A systematic review. Sustainability (Switzerland) 13(14): Article 7870. https://doi.org/10.3390/su13147870.

Kim, J. S. and Shin, N. (2019). The impact of blockchain technology application on supply chain partnership and performance. Sustainability (Switzerland) 11(21): Article 6181. https://doi.org/10.3390/su11216181.

Krichen, M., Ammi, M., Mihoub, A. and Almutiq, M. (2022). Blockchain for modern applications: A survey. Sensors 22(14): Article 5274. https://doi.org/10.3390/s22145274.

Lashkari, B. and Musilek, P. (2021). A comprehensive review of blockchain consensus mechanisms. IEEE Access 9: 52865–52880. https://doi.org/10.1109/ACCESS.2021.3065880.

Luo, L. (2022). Application of blockchain technology in intellectual property protection. Mathematical Problems in Engineering 2022: Article 4641559. https://doi.org/10.1155/2022/4641559.

Monrat, A. A., Schelén, O. and Andersson, K. (2019). A survey of blockchain from the perspectives of applications, challenges, and opportunities. IEEE Access 7: 117134–117151. https://doi.org/10.1109/ACCESS.2019.2936094.

Moosavi, N., Taherdoost, H., Mohamed, N., Madanchian, M., Farhaoui, Y. and Khan, I. U. (2024). Blockchain technology, structure, and applications: A survey. Procedia Computer Science 237: 645–658. https://doi.org/10.1016/j.procs.2024.05.150.

Nguyen, C. T., Hoang, D. T., Nguyen, D. N., Niyato, D., Nguyen, H. T. and Dutkiewicz, E. (2019). Proof-of-stake consensus mechanisms for future blockchain networks: Fundamentals, applications and opportunities. IEEE Access 7: 142705–142717. https://doi.org/10.1109/ACCESS.2019.2925010.

Rajasekaran, A. S., Azees, M. and Al-Turjman, F. (2022). A comprehensive survey on blockchain technology. Sustainable Energy Technologies and Assessments 52: Article 102039. https://doi.org/10.1016/j.seta.2022.102039.

Rodrigues, E., Lourenzani, W. and Satolo, E. (2021). Blockchain in supply chain management: Characteristics and benefits. BAR - Brazilian Administration Review 18(spe): e200065. https://doi.org/10.1590/1807-7692bar2021200065.

Song, H., Zhu, N., Xue, R., He, J., Zhang, K. and Wang, J. (2021). Proof-of-contribution consensus mechanism for blockchain and its application in intellectual property protection. Information Processing and Management 58(3): Article 102507. https://doi.org/10.1016/j.ipm.2021.102507.

Swathi, P. and Venkatesan, M. (2021). Scalability improvement and analysis of permissioned blockchain. ICT Express 7(3): 317–321. https://doi.org/10.1016/j.icte.2021.08.015.

Chapter 3

Blockchain Technology in Smart Cities

Current Status and Future Directions

Varsha Bhatia,[1,*] *Bhavesh Bhatia*[2] and *Vivek Jaglan*[3]

1. Introduction

The rapid urbanization of the global population has accelerated the need for innovative solutions to manage cities more effectively and sustainably. Smart cities, powered by the Internet of Things (IoT) devices, artificial intelligence, and advanced data analytics, have emerged as a promising paradigm for addressing urban challenges. Within this context, blockchain technology presents itself as a crucial enabler for secure and transparent urban operations.

Smart cities generate massive amounts of data from various sources, including sensor networks, public transportation systems, energy grids, and citizen interactions. The decentralized nature of blockchain technology offers a robust framework for managing this data while ensuring security, immutability, and transparency. By implementing blockchain-based solutions, smart cities can create trustworthy systems for various urban services, from public transportation and energy management to citizen identification and voting systems (Tripathi et al. 2023).

Integrating blockchain technology within the smart city infrastructure has several advantages, especially in some specific aspects of urban management. Most importantly, safety and privacy become paramount since urban data is kept safe through cryptographic means, and the risk of attacks is minimized by using blockchains. Recording all transactions as well as every operation in the cities brings about accountability and transparency due to the immutability characteristic of the blockchain, which in turn fosters trust between city administrators and citizens (Shen and Pena-Mora 2018). Blockchain maximizes all kinds of work and makes the service delivery process more straightforward through smart contracts involving almost all areas, from billing for public utilities, to processing permit applications, and managing public transportation systems. By enabling secure voting systems, transparent feedback mechanisms, and decentralized decision-making platforms, blockchain technology empowers citizen engagement in urban governance (Khanna et al. 2021).

[1] Sunstone Eduversity, Pune, Maharashtra, India.
[2] University of Western Ontario London, Ontario, Canada.
[3] City Institute of Higher Learning, Moncton, New Brunswick, Canada.
Emails: Bhaveshb@pm.me, jaglanvivek@gmail.com
* Corresponding author: varsha.bhatia.in@gmail.com

DOI: 10.1201/9781003620822-3

The share of blockchain technology in urban infrastructure becomes increasingly significant. This paper examines the current applications, challenges, and future possibilities of blockchain integration in smart cities, providing insights into how this technology can facilitate in creating more sustainable and citizen-centric urban environments.

2. Blockchain

Blockchain is an information technological innovation, that has introduced transformational living around the areas of electronic record maintenance and transaction processing. This paper describes how to derive different categories of blockchain systems and their fields of application. From the moment it was pioneered through Bitcoin in the year 2008, blockchain has extended beyond applications in cryptocurrency. It has thus become a primary building block in all manners and forms of digital innovation. Some formal definitions of blockchain are given below

Definition 1: "Blockchain is a shared, immutable ledger that facilitates the process of recording transactions and tracking assets in a business network. An asset can be tangible or intangible (What Is Blockchain? 2021)."

Definition 2: "A blockchain is essentially a distributed database of records or public ledger of all transactions or digital events that have been executed and shared among participating parties (Crosby et al. n.d.)."

Definition 3: "Blockchain technology is a decentralized, distributed, and public digital ledger that is used to record transactions across many computers so that any involved record cannot be altered retroactively (Bitcoin 2008)."

Definition 4: "A blockchain is a growing list of records, called blocks, that are linked together using cryptography. Each block contains a cryptographic hash of the previous block, a timestamp, and transaction data (Wang et al. 2018)."

Working of blockchain: The general working of blockchain is established through complex mapping between cryptography and distributed computing on consensus mechanisms. At its core, blockchain operates as a distributed ledger block as its atomic unit having a unique hash of the transaction data

Table 3.1. Blockchain types.

Types of Blockchain	Characteristics	Explanation	References
Public blockchain	Open participation Complete decentralization Consensus-based validation High transparency	Operates as a permissionless, decentralized network accessible to anyone, ensuring transparency through distributed consensus mechanisms.	(Eigelshoven et al. 2020)
Private blockchain	Restricted access Centralized control Higher performance Enhanced privacy	Operates within a controlled environment managed by a single organization, ensuring better privacy and performance.	(Sarmah 2018)
Consortium blockchain	Shared governance Limited access Improved scalability Balanced decentralization	A hybrid approach managed by multiple organizations, balancing decentralization and scalability.	(Dib et al. 2018)
Hybrid blockchain	Flexible architecture Customizable access Scalable performance Adaptable security	Combines features of public and private blockchains, offering tailored solutions for complex use cases.	(Alkhateeb et al. 2022)

and a timestamp. It takes shape once a transaction enters the blockchain process and gets distributed widely to verification nodes. The nodes authenticate the transaction as per preset consensus methods. Valid transactions fill a new block and attach it to the chain—via a unique cryptographic hash link—to the previous block. Each node in the network has and keeps a full copy of this chain, making it redundant in terms of storage and secure. Thus, such a decentralized system requires no intermediaries, but it retains the transparency and immutability of records. Smart contracts could even further automate this procedure through their execution of predefined conditions, automatically thereby, improving efficiency and reducing human intervention.

Based on the characteristics of operations, blockchain can be classified into four types. Table 3.1 depicts the types and characteristics of a blockchain.

3. The Architectural Components of Blockchain for Smart Cities

The architecture of a blockchain for smart cities is designed to ensure secure, transparent, and efficient management of urban infrastructure and services. This section delves into the detailed components of blockchain architecture tailored for smart cities (Fiore and Mongiello 2023). Figure 3.1 depicts blockchain layers architecture tailored for smart cities.

Data Layer

The data layer is the foundational element of the blockchain architecture, responsible for storing and managing the data generated by smart city components. This layer ensures data integrity, security, and immutability. The block is an atomic unit of a blockchain. The blocks are linked to form a chain. The structure of a block typically includes the following:

Block header: This contains version number, timestamp, previous block hash, Merkle root, difficulty target, and nonce.

Block body: Includes list of transactions performed in the block (Mohammed and Hashim 2023).

Merkle trees are used to efficiently verify the integrity of data within a block. Each leaf node in a Merkle tree represents the hash of a transaction, and the root node represents the hash of the entire tree (Wang et al. 2021). This structure permits quick verification of transaction data without the need for downloading the entire blockchain. The authenticity and integrity of the transactions are ensured by digital signature. They are generated using cryptographic algorithms, such as the Elliptic Curve Digital Signature Algorithm (ECDSA), which provides a secure way to verify the identity of the transaction sender (Arya et al. 2021).

Network Layer

This layer ensures that transactions and blocks are propagated efficiently across the network, maintaining synchronization and consensus. Nodes are the devices that are linked in the blockchain network, validating transactions and maintaining the ledger.

Nodes can be categorized as Full Nodes or Light Nodes.

The full node: It maintains a blue print of the blockchain and validates all transactions and blocks.

The light node: It maintains a partial copy of the blockchain and relies on the full nodes for transaction validation.

A decentralized peer-to-peer (P2P) network allows nodes to communicate directly with each other. This network topology ensures resilience and fault tolerance.

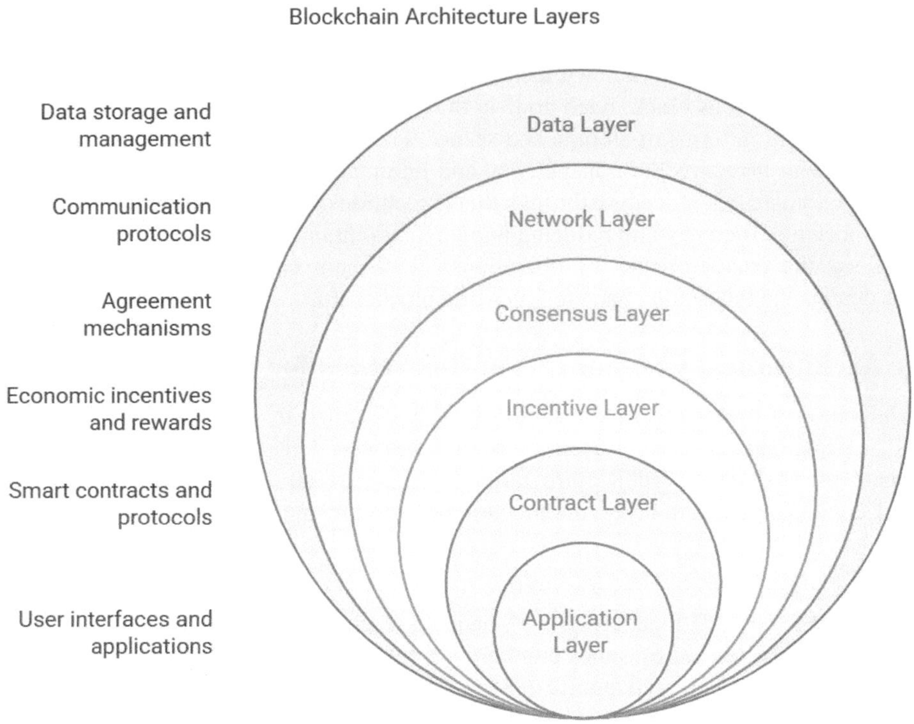

Figure 3.1. Blockchain architecture layer.

Propagation protocols ensure that transactions and blocks are efficiently disseminated across the network. Key protocols include:

Gossip protocol: The nodes share transactions and blocks with their peers, who then share them with their peers, creating a cascading effect.

Flooding protocol: The nodes broadcast transactions and blocks to all their peers simultaneously, ensuring rapid propagation.

Consensus Layer

This layer confirms that all the nodes approve of the state of the blockchain, maintaining a single source of truth. Various consensus algorithms can be employed, each with its own advantages and trade-offs. A few consensus algorithms are briefly explained below (Dong et al. 2023)

Proof of Work (PoW): The nodes engage in computational competitions, solving complex mathematical challenges to validate blockchain transactions. The successful node shares its solution network-wide for peer verification. Upon validation, the winning node earns newly created cryptocurrency as a reward for its computational contribution.

Proof of Stake (PoS): This mechanism selects validation nodes based on their cryptocurrency holdings in the network. These validator nodes earn transaction fees while proposing and adding new blocks. The selection probability directly correlates with the size of their stake in the system.

Delegated Proof of Stake (DPoS): A DPoS is a variation where cryptocurrency holders vote to select transaction validators and block proposers. These elected delegates must maintain network integrity or risk losing their position through community voting mechanisms. The system incentivizes positive network behavior through this democratic approach.

Practical Byzantine Fault Tolerance (PBFT): This is an advanced consensus protocol engineered to manage Byzantine faults - situations where nodes might fail or exhibit malicious behaviour. The system implements multiple communication rounds between nodes to achieve consensus, prioritizing network security and dependability.

Incentive Layer

This layer provides mechanisms to instigate nodes to participate in the network, ensuring its sustainability and security. Incentives can be positive (rewards) or negative (penalties).

Nodes receive rewards for validating transactions and adding new blocks to the chain. Rewards can be in the form of newly minted cryptocurrency or transaction fees. The reward mechanisms include:

Block rewards: The nodes receive a fixed amount of cryptocurrency for adding a new block to the chain.

Transaction fees: The nodes receive a portion of the transaction fees included in the block they validate.

The nodes may face penalties for malicious behavior, ensuring network integrity and security. The penalty mechanisms include:

Slashing: In PoS and DPoS systems, a portion of a node's stake is confiscated if it acts maliciously.

Reputation systems: Nodes that act maliciously may have their reputation reduced, affecting their ability to participate in future consensus rounds

Contract Layer

The contract layer enables the execution of smart contracts, which are self-executing contracts with the terms of the agreement directly written into code. These contracts are written in programming languages, such as Solidity for Ethereum Smart. Contracts automate processes, reduce intermediaries, and enhance efficiency (Chikov et al. 2023). Their key features include (Lawrence 2024):

Deterministic execution: Smart contracts execute in a deterministic manner, ensuring consistent results across all nodes.

Immutability: Once the smart contracts are deployed, they cannot be altered, ensuring their integrity and security.

Turing completeness: Smart contracts can perform complex computations and logic, enabling diverse applications.

Smart contracts rely on components called oracles to connect to the real world. An oracle acts as a trusted messenger between the digital and physical worlds (Beniiche 2024). Input oracles bring real-world information into the blockchain—like feeding weather data or stock prices into smart contracts. On the other hand, output oracles, enable smart contracts to trigger actions in the physical world, such as sending signals to automated systems or updating external databases. This synergy between smart contracts and oracles creates a powerful system that can automate and streamline countless processes in progressively digital world.

Application Layer

The application layer in blockchain technology works as a crucial interface between end-users and systems in delivering innovative solutions for smart cities through various services. In the case of identity management, blockchain leads to self-sovereign (no control of another user). The owner has control over his or her digital identities and access to his or her personal data (Stockburger et al.

2021). This system has decentralized identifiers that are globally resolvable and cryptographically verifiable, along with verifiable credentials ensuring authenticity and integrity.

In the supply chain operation, blockchain safeguards transparency through cunning internal mechanisms like trace tracking and real-time monitoring of goods on transit (Sunny et al. 2020). The use of smart contracts for transaction automation is based on predefined conditions like proof of deliveries. The energy sector derives great benefits from blockchain, by way of peer-to-peer energy trading, where the consumer directly sources excess energy, while demand response management ensures the optimization of energy distribution through fully automated programs.

1. Blockchain-based electronic health records provide better health care services with data security and interoperability. This technology equally speeds up clinical trials as well as application supply chain tracking. In waste management, blockchain offers systems for transparent tracking; and via smart contracts, incentivizes recycling with the aim of promoting an increasingly sustainable urban ecosystem (Bułkowska et al. 2023). It supports the running of, what is known as the circular economy, through an efficient and integrated system, that facilitates tracking and trading of recycled materials, and at the end square-gaiting sustainable urban development.

4. Smart City Systems and Key Challenges

Smart cities signify sophisticated ecosystems of entire networks that integrate with other technological systems and improve life in the cities and manage resources much better. Technology connects and empowers these systems for creating housing systems that operate on the Internet of Things device connectivity, digital technologies, and data analyses to mitigate the emerging urbanism challenges for the betterment of the quality of life that citizens enjoy (Bhatia and Jaglan 2024).

Blockchain facilitates a revolutionary approach to medical data management through integrated IoT networks and mobile technologies (Anderson et al. 2023). The system creates secure, interoperable networks connecting healthcare providers. Blockchain-enabled healthcare systems can reduce administrative costs, while significantly improving patient care coordination (Isravel et al. 2023).

Transportation infrastructure benefits from blockchain through sophisticated IoT-based solutions and sensor networks. Blockchain-enabled transportation systems can reduce traffic congestion through automated traffic management and real-time route optimization. The technology enables secure communication between various transportation nodes, enhancing emergency response capabilities and urban mobility efficiency (Miron et al. 2025).

In manufacturing, blockchain technology revolutionizes production processes through internet-connected machinery and comprehensive IoT integration. Blockchain-enabled manufacturing systems demonstrate a 40% improvement in operational efficiency (Javaid et al. 2021). The implementation of digital twins and advanced analytics, secured through blockchain, enables predictive maintenance and optimal resource utilization while reducing the environmental impact.

Smart energy systems leverage blockchain to create integrated networks managing electricity, heating and cooling infrastructure. Blockchain-enabled energy systems can achieve better resource utilization compared to traditional systems. The technology enables secure peer-to-peer energy trading and optimal load balancing across various urban sectors, supporting the transition to sustainable energy solutions (Wongthongtham et al. 2021).

Agricultural applications of blockchain in smart cities focus on optimizing farming operations through integrated digital technologies. Blockchain-enabled smart farming systems can increase crops, while reducing resource consumption. The technology enables secure data collection from IoT devices, drones, and AI systems, facilitating data-driven decision-making in farm management (Reddy et al. 2024).

These implementations collectively demonstrate blockchain's potential to create more efficient, sustainable, and resilient urban environments. The technology's ability to ensure secure, transparent,

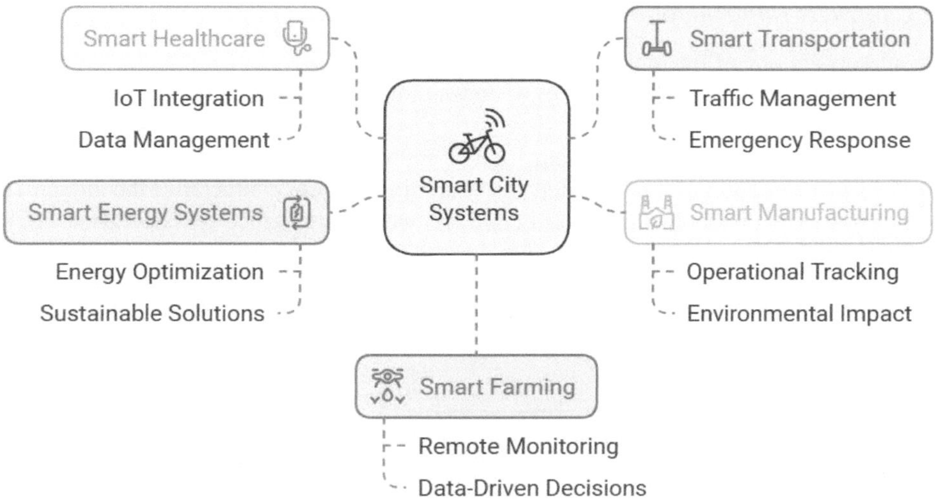

Figure 3.2. Blockchain application areas for smart city systems.

and automated operations across various sectors positions it as a crucial enabler of smart city development.

The following Figure 3.2 shows a few of the common areas where blockchain can be used for smart city systems like smart transportation, smart manufacturing, smart farming, smart health care and smart energy systems.

5. Implementing Blockchain in Smart Cities System

Blockchain technology has revolutionized the way digital infrastructures function with respect to the levels of security, transparency, and efficiency that it provides. A broad, detailed understanding of the various components of an application on blockchain systems and their implementation strategies is given in this paper.

Infrastructure for Data Collection

"Collection of data" is the foundational pillar of blockchain implementation. Modern-day blockchain systems use IoT sensors and edge computing devices to collect real-time data (Anuradha and Rani 2022). Sophisticated means of validation of data integrity before entry into the blockchain continue to characterize these systems. In addition, linking the automated data collection points along with quality assurance protocols has increased the reliability of blockchain systems.

The Implementation of Blockchain Architecture

The topology of networks and configurations of nodes is tedious in the architecture design of blockchain systems. In a study (Alam 2023), the findings demonstrate how the appropriate consensus mechanism and block structures are chosen largely workout their performance results in the system. The architecture of distributed ledger must accommodate transaction validation protocols and chain integrity. The work done in (Sestrem et al. 2020) depicts how an efficiently implemented blockchain architecture can shorten transaction processing time.

Smart Contracts Implementation

Smart contracts are one of the major components of blockchain systems these days: self-executing contracts whose terms are written into lines of code automate a great number of processes and bring down human interaction (Nzuva 2019).

A strong development of challenging logical approaches; creation of automation execution rules; and the incorporation of sophisticated error handling features characterize the implementation. The study showed that understanding smart contracts well could reduce cost operations. Figure 3.3 depicts the steps for the implementation of blockchain in smart cities.

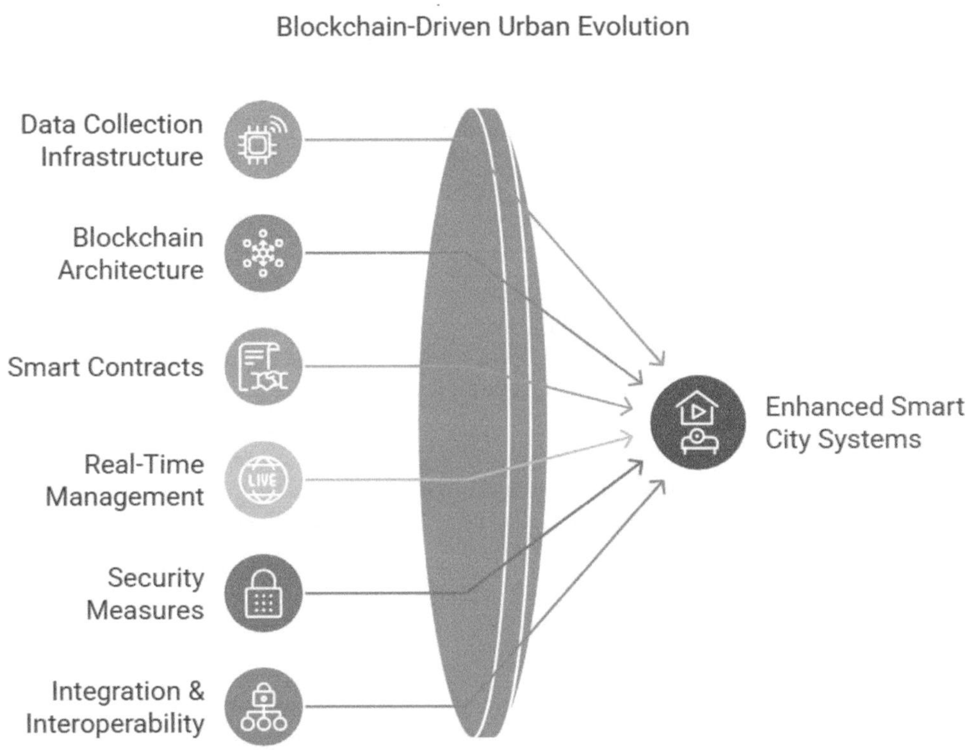

Figure 3.3. Implementation of blockchain in smart city systems.

Real-Time Operation Management

Sophisticated real-time operation management is what effective blockchain systems need. This includes live monitoring systems, performance tracking tools, and automated response mechanisms (Kharche et al. 2024). Resource allocation and load balancing systems ensure optimal performance, while system optimization protocols maintain efficiency. According to studies effective real-time management can improve response time within the system.

Security and Privacy Measures

Security represents the most important concern in blockchain implementation. Modern systems adopt multi-layer encryption protocols and rigid access control systems (Punia et al. 2024). Mechanisms to protect data privacy and advanced authentication ensure system integrity. Implementing effective measures of security will prevent the maximum number of possible breaches.

Integration and Interoperability

All these factors greatly contribute to the success of any blockchain architecture, among which most consider integration and interoperability. More emphasis on standard best practices development of APIs and platform compatibility has emerged. The protocol for data exchange and system integration must ensure smooth channeling in communication between different platforms without affecting efficiency and security.

A careful, planned and staggered approach in all these components, while implementing blockchain technology is recommended. To achieve success, integration of data collection, infrastructure with architectural designs, smart contracts, operations management, security measures, including interoperability protocols, should be underway (Kenaza et al. 2023). As the technology gets enhanced, so too will the implementation strategies.

6. Blockchain to Support Smart Cities' Operations

Smart Healthcare

Smart healthcare leverages mobile devices and IoT technologies to dynamically manage healthcare data, creating an integrated network connecting hospitals, pharmacies, and insurers. This connectivity supports informed decision-making and efficient resource allocation, ensuring better healthcare delivery. By emphasizing preventive care and personalized treatment, smart healthcare systems aim to enhance the quality of life and increase citizen longevity, transforming traditional healthcare into a more accessible, efficient, and patient-centric model (Al-Assaf et al. 2024). Figure 3.4 depicts the benefits of using blockchain in smart healthcare systems.

Blockchain is a rapidly emerging technology with significant potential to transform healthcare. Its ability to enable seamless and efficient data sharing among network members and providers supports the development of cost-effective therapies and advanced treatments for various diseases, driving progress in the sector. While already gaining supremacy in finance and logistics, blockchain offers exciting opportunities for healthcare, enhancing areas like supply chain management, research, and patient-provider relationships. As a critical driver of digital transformation, it holds immense promise for improving healthcare quality and life outcomes (Stoumpos et al. 2023).

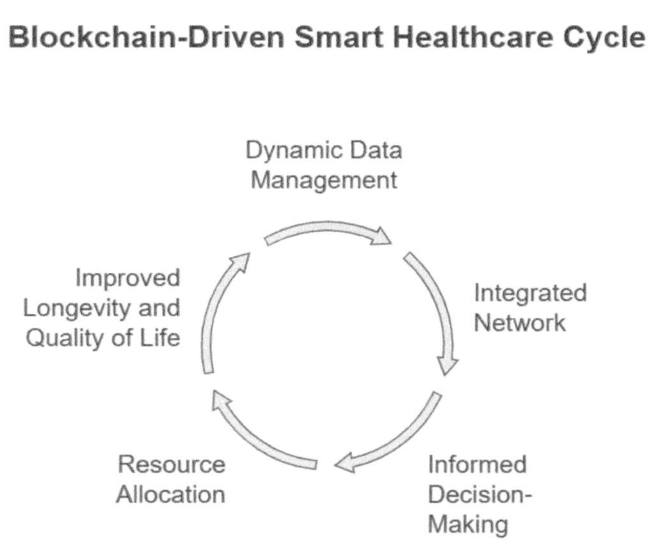

Figure 3.4. Blockchain driven smart healthcare cycle.

Smart Transportation

The incorporation of blockchain technology in intelligent transportation systems (ITS) represents a groundbreaking approach to managing urban mobility and traffic flow. This sophisticated system brings with it a comprehensive data collection infrastructure, that incorporates various IoT sensors strategically placed throughout the transportation network (Bhatia et al. 2022). These sensors continuously gather real-time data about traffic flow, vehicle movements, weather, road status, and parking availability, creating a rich data ecosystem that forms the foundation of an intelligent transportation network (Laña et al. 2021).

At the core of this system lies a carefully designed blockchain architecture that processes and manages this vast amount of data. The system utilizes a permissioned blockchain network, where distributed nodes across traffic management centers work in conjunction with edge computing nodes to ensure real-time processing capabilities. This architecture implements smart contracts that automate crucial traffic management functions, including traffic signal optimization, dynamic lane management, and emergency vehicle routing. The system's ability to execute these contracts automatically and transparently ensures efficient traffic flow and rapid response to changing conditions.

Real-time operations management forms a critical component of the blockchain-based ITS. The system continuously analyses traffic patterns and implements dynamic routing algorithms to optimize traffic flow. When congestion is detected, the system can automatically adjust traffic signal timing, modify lane directions, and update speed limits to maintain smooth traffic movement. In emergency situations, the system proves particularly valuable by automatically clearing routes for emergency vehicles and coordinating rapid response efforts.

Security and privacy considerations are two important aspects for every ITS implementation. The system employs robust end-to-end encryption and privacy-preserving computation techniques to protect sensitive data. Vehicle identification remains anonymous while maintaining the ability to verify compliance and manage traffic violations effectively. Access control mechanisms and secure key management ensure that only authorized entities can access and modify the system's data and operations (Kim 2018).

The integration capabilities of blockchain-based ITS extend to existing transportation infrastructure and related systems. The technology seamlessly connects with emergency services, weather forecasting systems, public transportation networks, and parking management systems (Elassy et al. 2024). This comprehensive integration enables a holistic approach to urban mobility management, where different components of the transportation network work together harmoniously. Figure 3.5 shows the integration of blockchain in smart traffic management systems.

The benefits of this implementation are substantial and far-reaching. Cities experience enhanced traffic management with reduced congestion and delays, while emergency response times improve significantly. Safety increases through faster accident detection and better weather-related traffic management. Environmental benefits include reduced emissions and lower fuel consumption through optimized traffic flow. Economically, cities benefit from reduced operational costs, more efficient resource allocation, and automated revenue collection systems.

With the advent of technologies, these systems will become increasingly sophisticated, further improving the quality of urban transportation management and citizen mobility.

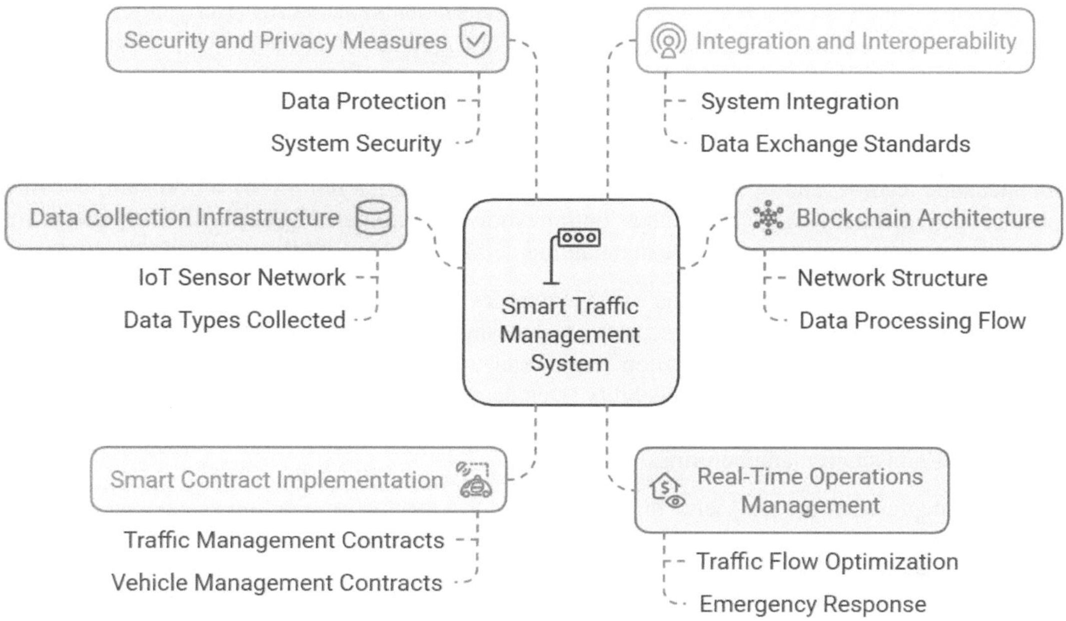

Figure 3.5. Integration of blockchain in smart traffic management systems.

Smart Manufacturing

The convergence of blockchain technology with smart manufacturing represents a transformative approach to industrial production systems. In this integrated framework, blockchain serves as a secure, transparent foundation for managing complex manufacturing operations while ensuring data integrity and traceability throughout the production lifecycle.

Monitoring and control of the production process: The system begins with comprehensive production monitoring through internet-connected machinery. Smart sensors and IoT devices continuously collect real-time data about equipment performance, production rates, quality metrics, and environmental conditions. This data is securely recorded as an immutable record on the blockchain. Hence, blockchain maintains a record of all manufacturing operations. Machine learning algorithms analyze this data stream to deal with issues and challenges in production processes.

Supply chain and inventory management: Blockchain technology transforms supply chain management in manufacturing by creating transparent, traceable records of raw materials and components (Prakash 2024). Smart contracts automate inventory management, triggering automatic reordering when supplies reach predetermined levels. Each component's journey through the production process is tracked with precise timestamps and location data, enabling real-time visibility into the manufacturing pipeline.

Quality assurance and compliance: The integration of blockchain in quality control processes ensures that every product meets specified standards. Digital twins, powered by blockchain-verified data, provide virtual representations of physical products and processes, enabling simulation and optimization before actual production begins (Soori et al. 2023). Quality inspection results, compliance certificates, and regulatory documentation are permanently recorded on the blockchain, creating an auditable trail of product quality and safety compliance.

Resource optimization and environmental impact: Advanced analytics powered by blockchain data help optimize resource utilization, reducing waste and environmental impact. The system monitors energy consumption, material usage, and emissions in real-time, automatically adjusting production parameters to maintain efficiency while meeting environmental standards (Hariyani et al. 2025). Smart contracts enforce sustainability protocols and ensure compliance with environmental regulations.

Integration with robotics and automation: Blockchain technology coordinates robotic systems and automated production lines, ensuring precise synchronization of operations. Robot performance data, maintenance records, and operational parameters are securely stored on the blockchain, enabling predictive maintenance and continuous optimization of automated processes. This integration enhances production efficiency while maintaining detailed records of all automated operations.

Decentralized decision making: The system enables decentralized decision-making across the manufacturing facility through consensus mechanisms and smart contracts (Leng et al. 2023). Different departments and production units can coordinate activities autonomously while maintaining transparency and accountability. This decentralized approach improves response times to production issues and enables more flexible manufacturing processes. Figure 3.6 depicts the role of blockchain in smart manufacturing.

The integrated approach yields high benefits in the form of production efficiency, minimum downtime, good quality control, and a smaller environmental footprint. Blockchain technology will ensure that transparency and traceability are given to manufacturers to ensure compliance with regulatory requirements and to build trust among customers and partners. It decreases operational costs while increasing the lifetime of the equipment, and ensures product quality through automated quality control processes. Apart from maintaining operational efficiency of machine or instruments, detecting defects and abnormalities in critical systems is also mandatory (Bhatia et al. 2024).

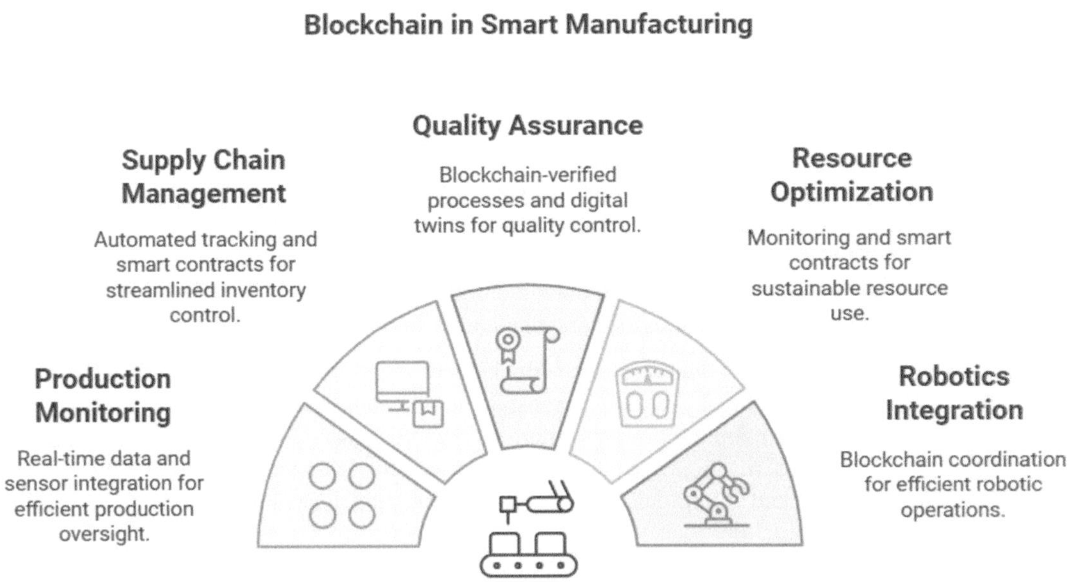

Figure 3.6. Blockchain in smart manufacturing.

Smart Energy Systems

Sector integration and management: A smart energy system implements blockchain by establishing a cohesive network that unites various energy sectors. Through an integrated platform in the form of a unified blockchain, electricity distribution, heating networks, and cooling systems can be managed

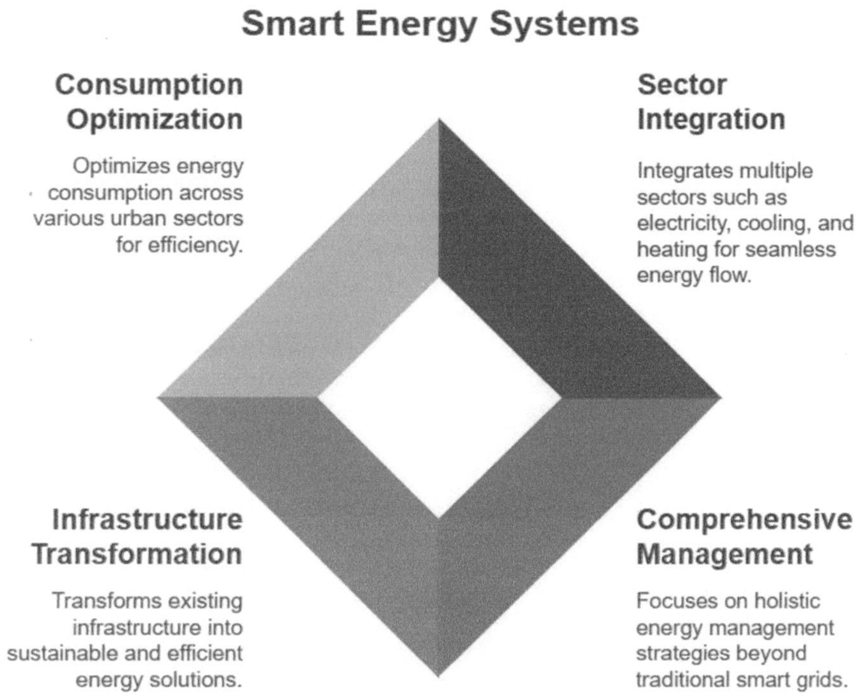

Figure 3.7. Blockchain in smart energy systems.

simultaneously. Smart contracts ensure automatic coordination between various energy flows in multiple sectors to realize optimum resource exploitation (Zhao et al. 2023). It allows monitoring in real-time across various connected sectors regarding demand and supply of energy for the purpose of balancing and effective load distribution.

Advanced energy management: Beyond the smart grid, blockchain-based energy systems allow for sophisticated management. The technology creates a decentralized marketplace for energy resources where consumers and producers can trade directly. Smart meters connected to the blockchain network continuously record energy consumption and generation data, creating transparent and immutable records. The system implements dynamic pricing mechanisms through smart contracts, encouraging energy efficiency and optimal consumption patterns (Cruz et al. 2018). Figure 3.7 shows major areas where blockchain can be used in smart energy systems.

Infrastructure renewal: Blockchain technology supports the conversion of the current infrastructure to a green infrastructure. This is through the integration of renewable sources with grid-based traditional energy sources, leading to a hybrid network. Switching between energy sources is enabled depending on the available sources and the demand through the smart contracts. The infrastructure has a two-way flow; thus, prosumers are enabled to participate in the energy market (Lim et al. 2018).

Smart Farming

The capability of blockchain technology would become parallel with the practice of online digitization-, IoT, drone, and AI technology, that is suitable as applicable to agricultural matters. So, farmers may very well monitor their crops and fields in real-time and discover information needed for proper decision-making. The perception is that farmers would be able to measure their resources differently and predict the possible risks or damage. This would eventually improve the overall management of the farm. This process ends with blockchain, which will secure and give transparency to all agricultural-related records from seed to shelf. Because of this, food safety will be increased through identity-enhanced traceability (Dagher et al. 2018). The end goal of such a system if used in agriculture is to increase crop quality and quantity, while also optimizing labor and helping build a more sustainable, efficient agricultural system (Saberi et al. 2018). Figure 3.8 describes the areas where blockchain can be used in smart farming.

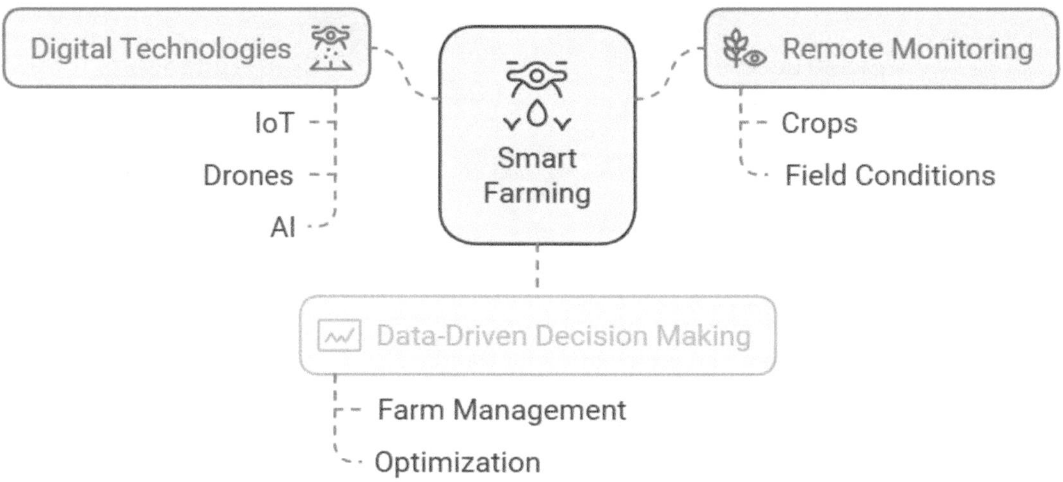

Figure 3.8. Blockchain in smart farming.

7. Challenges for Blockchain Implementation in Smart Cities

Infrastructure Limitations

Old urban infrastructure can act as a hurdle for the successful implementation of blockchain technology. Obsolete systems may not have the bandwidth, processing power, and security features required to support the demands of a decentralized ledger (Zhang et al. 2018). Significant investment is often needed to modernize existing infrastructure, such as upgrading power grids, telecommunications networks, and data centers to meet the increasing needs of blockchain applications (Andoni et al. 2019). Integration challenges arise when trying to integrate blockchain-based solutions with existing urban systems. This can be very technical and may require a lot of effort to ensure compatibility and data flow between different platforms.

Security and Privacy

Citizen data that may be considered sensitive should be well protected. The technology itself in blockchain is secured, but additional security should be in place to prevent hacking into the information and unauthorized access of personal information. Cyberattacks and threats targeting infrastructure like power, transportation, and water supply are considered the biggest risks (Vishwakarma et al. 2024). Protecting the integrity and resilience of such systems ensures safety and functionality. Widespread collection and usage of citizen data raise privacy issues. Data privacy regulations need to be clearly formulated, and strict data protection must be in place to protect rights and prevent personal information misuse (Wang et al. 2019).

Social Inclusion

Accessibility of all citizen groups is very essential for the effective implementation of blockchain-based solutions in smart cities. This includes accessibility for marginalized communities, people with disabilities, and those with minimal access to technology (Esmaeilian et al. 2018).

Bridging the digital divide is essential to maximize the rewards of blockchain technology. Developing user-friendly solutions for non-tech-savvy residents is crucial for widespread adoption. Intuitive and easy-to-use interfaces are necessary to make blockchain-based services accessible and understandable for all citizens, regardless of their technical expertise (Taylor et al. 2020). Hence, providing digital literacy training, increasing access to affordable internet connectivity, and developing user-friendly interfaces for all residents.

Technical Integration

Interoperability between separate smart systems is the biggest challenge. Effective communication and data exchange across different blockchain-based applications, such as those in energy management, transportation, and healthcare, will be vital for the success of the overall smart city ecosystem (Casino et al. 2019a).

To ensure data sharing and interoperability between various platforms, standardization across platforms is important (Belchior et al. 2021). Standardization of data format and protocol between different systems can ensure effective information exchange. Seamless communication among the smart city infrastructure is crucial.

Effective tackling of all these challenges will be key for realizing the full benefits of this transformative technology to upgrade the urban life style of all citizens. The idea is that addressing the above challenges requires innovative approaches and careful planning for the intelligent implementation of such systems in smart cities. Technologies like blockchain and Edge AI are emerging as prospects for such solutions. They ensure extraordinary security, transparency, and efficiency. Figure 3.9 illustrates the summary of challenges for blockchain implementation in smart cities.

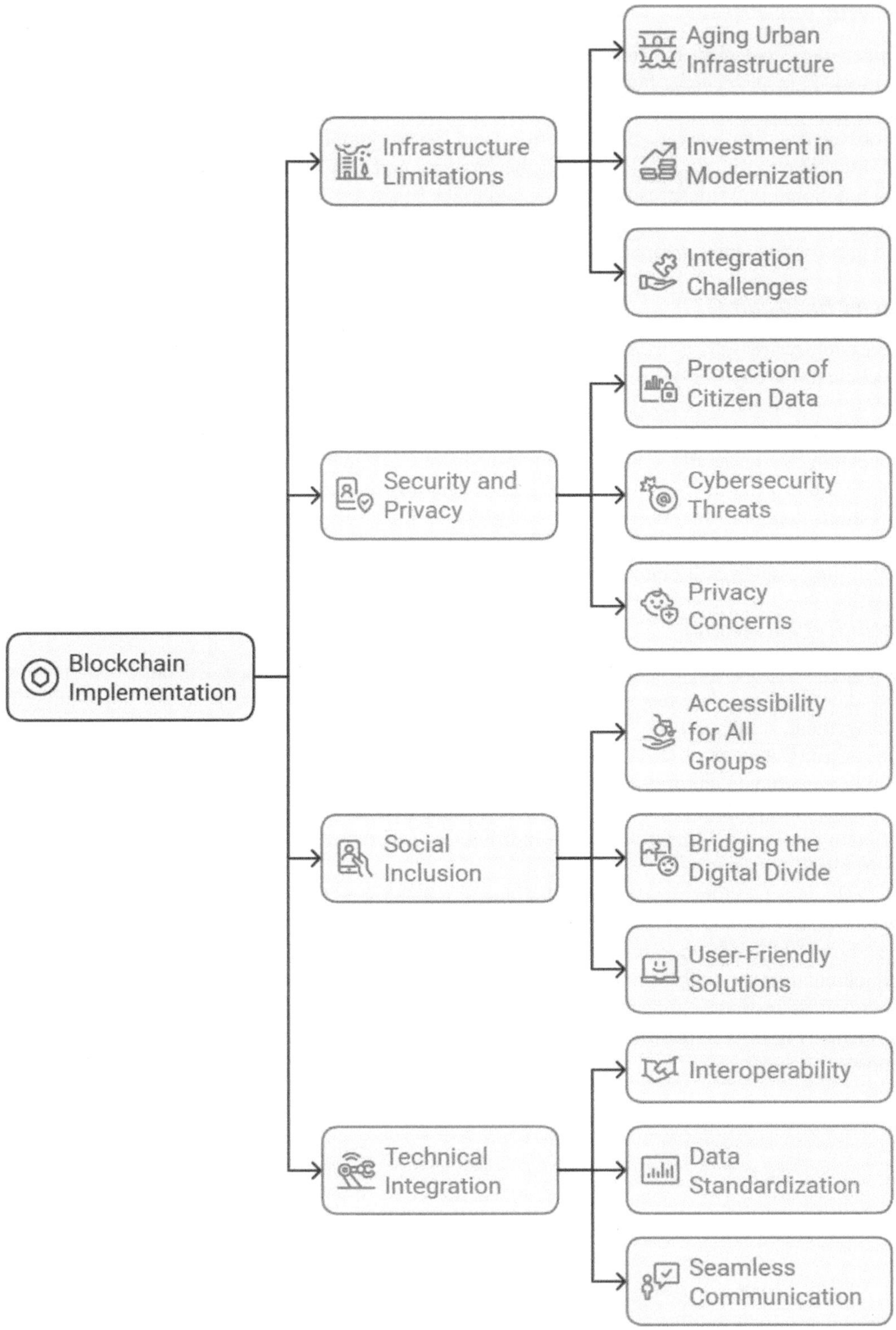

Figure 3.9. Challenges for blockchain implementation in smart cities.

8. Comprehensive Analysis of Blockchain Platforms and Applications in Smart Cities

Blockchain platforms are comprehensive digital infrastructures that provide the foundation for developing and deploying decentralized applications and services. These platforms offer the necessary tools, protocols, and frameworks for implementing blockchain-based solutions in smart cities. The analysis is organized primarily by three major platform categories: Hyperledger, Ethereum, and Cross-platform solutions. Table 3.2 depicts the three platforms their strengths and weakness and their common use cases.

Table 3.2. Common blockchain platforms.

Platform Type	Strengths	Weaknesses	Primary Use Cases
Ethereum	Robust smart contract capability Large developer community High security through PoW/PoS Public transparency Well-established ecosystem	High transaction costs (gas fees) Scalability limitations Higher energy consumption- Limited privacy features- Network congestion	Decentralized applications Energy trading Supply chain tracking Public services Digital identity
Hyperledger Fabric	High performance and scalability Privacy controls No cryptocurrency required- Modular architecture Multiple programming languages	Limited public accessibility Complex implementation Requires technical expertise Smaller developer community Higher setup costs	Healthcare records- Financial services Government services- Identity management- Supply chain
Cross-Platform	Interoperability between chains Flexible implementation- Enhanced scalability Custom solution design- Multi-chain advantages	Complex integration Higher development costs Security challenges Coordination overhead-Standards compatibility	Data exchange Multi-stakeholder systems Hybrid solutions - Enterprise integration Cross-chain services

Table 3.3 presents a structured overview of blockchain implementations across various domains in smart cities, emphasizing the relationship between platforms, applications, and their operational characteristics.

The Hyperledger platform shows its potential in handling sensitive data applications. In identity management, it uses a Permissioned blockchain with Proof of Authority (PoA) consensus, thus ensuring high security and low latency in digital identity verification and access control systems. (Honar Pajooh et al. 2021) In healthcare, too. Hyperledger maintains HIPAA compliance while managing electronic health records and clinical data through its private, permissioned architecture (Mohan and Sujihelen 2023).

Ethereum serves many applications that vary their consensus mechanisms. In the supply chain, it uses Permissioned Proof of Work to track products in real-time and manage logistics (Aslam et al. 2023). In the energy trading sector, Ethereum's Permissionless Proof of Stake helps in efficient P2P trading of energy and management of the energy grid. Waste management systems also happen to adopt this PoW mechanism of Ethereum to track recycling and resource management schemes (Sivaram and Saravanan 2024).

The Cross-platform category addresses interoperability challenges through hybrid architectures. It facilitates data exchange across different blockchain networks while maintaining standard protocols. Smart contracts implementation across multiple chains enables automated governance and policy enforcement, ensuring seamless integration of various smart city services. This all-encompassing framework shows how different blockchain platforms are optimized for specific use cases while maintaining security, efficiency, and scalability in smart city applications.

Table 3.3. Overview of blockchain implementations across various domains.

Platform Type	Application Domain	Implementation Details	Key Features	Performance Metrics
Hyperledger	Identity Management (Cruz et al. 2018), (Lim et al. 2018)	Permissioned, PoA	- Digital Identity Verification - Biometric Data - Access Control	High Security, Low Latency
	Healthcare (Dagher et al. 2018), (Zhang et al. 2018)	Permissioned, PoA	- EHR Management - Patient Data - Clinical Records	HIPAA Compliant, Private
Ethereum	Supply Chain (Saberi et al. 2018)	Permissioned, PoW	- Product Tracking - Inventory Management - Logistics	Real-time Updates
	Energy Trading (Andoni et al. 2019), (Wang N. et al. 2019)	Permissionless, PoS	- P2P Energy Trading - Grid Management - Usage Monitoring	High Throughput
	Waste Management (Esmaeilian et al. 2018), (Taylor et al. 2020)	Permissioned, PoW	- Recycling Tracking - Waste Collection - Resource Management	Environmental Impact
Cross-Platform	Data Exchange (Casino et al. 2019b)	Hybrid Architecture	- Interoperability - Data Sharing - Standard Protocols	System Integration
	Smart Contracts (Belchior et al. 2021)	Multi-chain Support	- Automated Execution - Policy Enforcement - Compliance	Automated Governance

9. Open Research Issues

Major research issues include scalability and efficiency concerns for blockchain for smart cities. Most types of blockchains may take too long in the processing of transactions to be able to support high-throughput applications like real-time traffic management or energy grid operations. In addition, any delay in transaction processing might not allow a timely response or decision to be taken. Interoperability and integration would be great challenges as integrating present urban infrastructures and legacy systems with blockchain would be highly complicated. Setting data standards and protocols would be crucial for enforcing interoperable applications of the smart city ecosystem.

Security and privacy concerns will be paramount as it will involve securing the sensitive citizen data with special reference to the health and also finance sectors. This could be taken care of by coming up with such privacy-enhancing technologies such as zero-knowledge proofs and also homomorphic encryption to protect secretive information from using the advantages of blockchain (Zhou et al. 2024).

Effective governance and regulation promote the responsible adoption of blockchain in smart cities. Clear, legal, and regulatory frameworks will, therefore, ensure that blockchain applications are developed and deployed responsibly, while consumer protection is ensured (Zhuk 2025). Along with that, it would be imperative to explore more creative ways of decentralized governance models for blockchain networks within smart city settings to assure fairness, transparency, and community participation.

Lastly, social and ethical issues must be addressed. Access to such services should be equitable among citizens, irrespective of economic status, or technology literacy, etc. It is also crucial to assess the potential impact of blockchain technology on employment in various sectors of the smart city and developing strategies for workforce retraining and upskilling.

10. Conclusion

These cities appear to develop rapidly and provide persistent hosts of challenges, especially overcrowding and environmental issues. This is why smart cities require effective engagement of technologies in solving problems such as those posed by the conventional situation in which services are poorly delivered and lives are affected negatively. However, increasing numbers of cyberattacks targeting these smart systems have become a serious matter.

Most of the crucial issues would be resolved by the use of blockchain technology such as security, interoperability, and transparency. In this article, it was discussed how blockchain can be useful for its application in various aspects of a smart city including healthcare, transportation, energy management, and farming. In a similar way, the advantages and disadvantages of blockchains in those areas have been critically analyzed and presented in an integrated manner for the benefit of researchers and policymakers. Thus, though still in its infancy, blockchain can very well hold the promise of ushering in a new way in which we may go ahead to build and manage smart cities. However, while working in these fundamental areas, challenges remain that when covered will put us on course toward making better, more secure, more efficient, and more equitable urban environments for all.

References

Alam, S. (2023). The current state of blockchain consensus mechanism: issues and future works. International Journal of Advanced Computer Science and Applications (IJACSA) 14(8): Article 8. https://doi.org/10.14569/IJACSA.2023.0140810.

Al-Assaf, K., Bahroun, Z. and Ahmed, V. (2024). Transforming service quality in healthcare: a comprehensive review of Healthcare 4.0 and its impact on healthcare service quality. Informatics 11(4): 96; https://doi.org/10.3390/informatics11040096.

Alkhateeb, A., Catal, C., Kar, G. and Mishra, A. (2022). Hybrid blockchain platforms for the internet of things (IoT): A systematic literature review. Sensors 22(4): Article 4. https://doi.org/10.3390/s22041304.

Anderson, C., Carvalho, A., Kaul, M. and Merhout, J. W. (2023). Blockchain innovation for consent self-management in health information exchanges. Decision Support Systems 174: 114021. https://doi.org/10.1016/j.dss.2023.114021.

Andoni, M., Robu, V., Flynn, D., Abram, S., Geach, D., Jenkins, D. et al. (2019). Blockchain technology in the energy sector: A systematic review of challenges and opportunities. Renewable and Sustainable Energy Reviews 100: 143–174. https://doi.org/10.1016/j.rser.2018.10.014.

Anuradha, M. P. and Lino Fathima Chinna Rani, K. (2022). Chapter Fourteen—Blockchain technology for IoT edge devices and data security. pp. 379–412. *In*: Raj, P., Saini, K. and Surianarayanan, C. (eds.). Advances in Computers (Vol. 127). Elsevier. https://doi.org/10.1016/bs.adcom.2022.02.011.

Arya, J., Kumar, A., Singh, A., Mishra, T. and Chong, P. H. J. (2021). Blockchain: Basics, Applications, Challenges and Opportunities. https://doi.org/10.13140/RG.2.2.33899.16160.

Aslam, M., Jabbar, S., Abbas, Q., Albathan, M., Hussain, A. and Raza, U. (2023). Leveraging Ethereum platform for development of efficient tractability system in pharmaceutical supply chain. Systems 11(4): Article 4. https://doi.org/10.3390/systems11040202.

Belchior, R., Vasconcelos, A., Guerreiro, S. and Correia, M. (2021). A survey on blockchain interoperability: past, present, and future trends. ACM Comput. Surv. 54(8): 168:1–168:41. https://doi.org/10.1145/3471140.

Beniiche, A. (2024, September 6). A Study of Blockchain Oracles. ResearchGate. https://doi.org/10.48550/arXiv.2004.07140.

Bhatia, V., Arora, R. and Sidharth, S. (2024). Deep Learning for Automated Aircraft Surface Defect Detection. 2024 International Conference on Intelligent Computing and Emerging Communication Technologies (ICEC), 1–10. 2024 International Conference on Intelligent Computing and Emerging Communication Technologies (ICEC). https://doi.org/10.1109/ICEC59683.2024.10837564.

Bhatia, V. and Jaglan, V. (2024). Integrating artificial intelligence and IoT for smart cities. pp. 267–278. In Reshaping Intelligent Business and Industry. John Wiley & Sons, Ltd. https://doi.org/10.1002/9781119905202.ch17.

Bhatia, V., Jaglan, V., Kumawat, S., Siwach, V. and Sehrawat, H. (2022). Intelligent Transportation System Applications: A Traffic Management Perspective (pp. 419–433). https://doi.org/10.1007/978-981-16-2422-3_33.

Bitcoin: A Peer-to-Peer Electronic Cash System | Satoshi Nakamoto Institute. (2008, October 31). https://nakamotoinstitute.org/library/bitcoin/.

Bułkowska, K., Zielińska, M. and Bułkowski, M. (2023). Implementation of blockchain technology in waste management. Energies 16(23): Article 23. https://doi.org/10.3390/en16237742.

Casino, F., Dasaklis, T. K. and Patsakis, C. (2019a). A systematic literature review of blockchain-based applications: Current status, classification and open issues. Telematics and Informatics 36: 55–81. https://doi.org/10.1016/j.tele.2018.11.006.

Casino, F., Dasaklis, T. K. and Patsakis, C. (2019b). A systematic literature review of blockchain-based applications: Current status, classification and open issues. Telematics and Informatics 36: 55–81. https://doi.org/10.1016/j.tele.2018.11.006.

Chikov, I. A., Koliadenko, S. V., .Supryhan, V. A., Tabenska, O. I., .Nitsenko, V. S. and Holinko, O. V. (2023). Smart contracts and business process automation: The Technical Aspect 5: 186–192.

Crosby, M., Pattanayak, P., Verma, S. and Kalyanaraman, V. (n.d.). Blockchain: Beyond Bitcoin.

Cruz, J. P., Kaji, Y. and Yanai, N. (2018). RBAC-SC: Role-based access control using smart contract. IEEE Access 6: 12240–12251. https://doi.org/10.1109/ACCESS.2018.2812844.

Dagher, G. G., Mohler, J., Milojkovic, M. and Marella, P. B. (2018). Ancile: Privacy-preserving framework for access control and interoperability of electronic health records using blockchain technology. Sustainable Cities and Society 39: 283–297. https://doi.org/10.1016/j.scs.2018.02.014.

Dib, O., Brousmiche, K.-L., Durand, A., Thea, E. and Hamida, E. (2018). Consortium Blockchains: Overview, Applications and Challenges.

Dong, S., Abbas, K., Li, M. and Kamruzzaman, J. (2023). Blockchain technology and application: An overview. PeerJ Computer Science 9: e1705. https://doi.org/10.7717/peerj-cs.1705.

Eigelshoven, Felix, Ullrich, André and Bender, Benedict. (2020). Public blockchain – a systematic literature review on the sustainability of consensus algorithms. In Proceedings of the 28th European Conference on Information Systems (ECIS), An Online AIS Conference, June 15–17, 2020. https://aisel.aisnet.org/ecis2020_rp/202.

Elassy, M., Al-Hattab, M., Takruri, M. and Badawi, S. (2024). Intelligent transportation systems for sustainable smart cities. Transportation Engineering 16: 100252. https://doi.org/10.1016/j.treng.2024.100252.

Esmaeilian, B., Wang, B., Lewis, K., Duarte, F., Ratti, C. and Behdad, S. (2018). The future of waste management in smart and sustainable cities: A review and concept paper. Waste Management 81: 177–195. https://doi.org/10.1016/j.wasman.2018.09.047.

Fiore, M. and Mongiello, M. (2023). Blockchain for smart cities improvement: an architecture proposal. pp. 1–5. 2023 7th International Conference on Computer, Software and Modeling (ICCSM). https://doi.org/10.1109/ICCSM60247.2023.00009.

Hariyani, D., Hariyani, P., Mishra, S. and Sharma, M. K. (2025). A literature review on transformative impacts of blockchain technology on manufacturing management and industrial engineering practices. Green Technologies and Sustainability 100169. https://doi.org/10.1016/j.grets.2025.100169.

Honar Pajooh, H., Rashid, M., Alam, F. and Demidenko, S. (2021). Hyperledger fabric blockchain for securing the edge internet of things. Sensors 21(2): Article 2. https://doi.org/10.3390/s21020359.

J, A., Isravel, D. P., Sagayam, K. M., Bhushan, B., Sei, Y. and Eunice, J. (2023). Blockchain for healthcare systems: Architecture, security challenges, trends and future directions. Journal of Network and Computer Applications 215: 103633. https://doi.org/10.1016/j.jnca.2023.103633.

Javaid, M., Haleem, A., Pratap Singh, R., Khan, S. and Suman, R. (2021). Blockchain technology applications for Industry 4.0: A literature-based review. Blockchain: Research and Applications 2(4): 100027. https://doi.org/10.1016/j.bcra.2021.100027.

Kenaza, T., Messai, S., Debicha, I. and Sehaki, M. (2023). A secure and interoperable architecture for blockchain/IPFS assisted electronic health record access control and sharing. Research Square. https://doi.org/10.21203/rs.3.rs-3209163/v1.

Khanna, A., Sah, A., Bolshev, V., Jasinski, M., Vinogradov, A., Leonowicz, Z. et al. (2021). Blockchain: Future of e-Governance in smart cities. Sustainability 13(21): Article 21. https://doi.org/10.3390/su132111840.

Kharche, A., Badholia, S. and Upadhyay, R. K. (2024). Implementation of blockchain technology in integrated IoT networks for constructing scalable ITS systems in India. Blockchain: Research and Applications 5(2): 100188. https://doi.org/10.1016/j.bcra.2024.100188.

Kim, S. (2018). Chapter Two—Blockchain for a trust network among intelligent vehicles. pp. 43–68. *In*: Raj, P. and Deka, G. C. (eds.), Advances in Computers (Vol. 111). Elsevier. https://doi.org/10.1016/bs.adcom.2018.03.010.

Laña, I., Sanchez-Medina, J. J., Vlahogianni, E. I. and Del Ser, J. (2021). From data to actions in intelligent transportation systems: a prescription of functional requirements for model actionability. Sensors (Basel, Switzerland) 21(4): 1121. https://doi.org/10.3390/s21041121.

Lawrence, E. (2024, December 3). Blockchain and smart contracts for secure and transparent transactions. ResearchGate. https://www.researchgate.net/publication/386337137_Blockchain_and_smart_contracts_for_secure_and_transparent_transactions.

Leng, J., Zhong, Y., Lin, Z., Xu, K., Mourtzis, D., Zhou, X. et al. (2023). Towards resilience in Industry 5.0: A decentralized autonomous manufacturing paradigm. Journal of Manufacturing Systems 71: 95–114. https://doi.org/10.1016/j.jmsy.2023.08.023.

Lim, S. Y., Fotsing, P. T., Almasri, A., Musa, O., Kiah, M. L. M., Ang, T. F. et al. (2018). Blockchain technology the identity management and authentication service disruptor: A survey. International Journal on Advanced Science, Engineering and Information Technology 8(4–2): 1735–1745. https://doi.org/10.18517/ijaseit.8.4-2.6838.

Miron, R., Hulea, M., Muresan, V., Clitan, I. and Rusu, A. (2025). Integrating blockchain technology into mobility-as-a-service platforms for smart cities. Smart Cities 8(1): Article 1. https://doi.org/10.3390/smartcities8010009.

Mohammed, M. and Hashim, A. (2023). Blockchain technology, methodology behind it, and its most extensively used encryption techniques. Al-Salam Journal for Engineering and Technology 2: 140–151. https://doi.org/10.55145/ajest.2023.02.02.017.

Mohan, M, S. and Sujihelen, L. (2023). An efficient chain code for access control in hyper ledger fabric healthcare system. E-Prime - Advances in Electrical Engineering, Electronics and Energy 5: 100204. https://doi.org/10.1016/j.prime.2023.100204.

Mulligan, C., Zhu Scott, J., Warren, S. and Rangaswam, J. (2018, April). World Economic Forum. https://www.weforum.org/publications/blockchain-beyond-the-hype/.

Nzuva, S. (2019). Smart contracts implementation, applications, benefits, and limitations. Journal of Information Engineering and Applications 9(5): 63.

Prakash, A. (2024). Blockchain technology for supply chain management: enhancing transparency and efficiency. International Journal for Global Academic & Scientific Research 3(2): Article 2. https://doi.org/10.55938/ijgasr.v3i2.73.

Punia, A., Gulia, P., Gill, N. S., Ibeke, E., Iwendi, C. and Shukla, P. K. (2024). A systematic review on blockchain-based access control systems in cloud environment. Journal of Cloud Computing 13(1): 146. https://doi.org/10.1186/s13677-024-00697-7.

Reddy, B. H. R. D., Veedu, A. K. and BM, B. (2024). Blockchain Technology in Agriculture. pp. 1–7. 2024 8th International Conference on Computational System and Information Technology for Sustainable Solutions (CSITSS). https://doi.org/10.1109/CSITSS64042.2024.10816748.

Saberi, S., Kouhizadeh, M., Sarkis, J. and Shen, L. (2018). Blockchain technology and its relationships to sustainable supply chain management. International Journal of Production Research 57: 1–19. https://doi.org/10.1080/00207543.2018.1533261.

Sarmah, s. (2018). Understanding Blockchain Technology. 8: 23–29. https://doi.org/10.5923/j.computer.20180802.02.

Sestrem Ochôa, I., Augusto Silva, L., de Mello, G., Garcia, N. M., de Paz Santana, J. F. and Quietinho Leithardt, V. R. (2020). A cost analysis of implementing a blockchain architecture in a smart grid scenario using sidechains. Sensors 20(3): Article 3. https://doi.org/10.3390/s20030843.

Shen, C. and Pena-Mora, F. (2018). Blockchain for cities—a systematic literature review. IEEE Access 6: 76787–76819. https://doi.org/10.1109/ACCESS.2018.2880744.

Sivaram, T. and Saravanan, B. (2024). Recent developments and challenges using blockchain techniques for peer-to-peer energy trading: A review. Results in Engineering 24: 103666. https://doi.org/10.1016/j.rineng.2024.103666.

Soori, M., Arezoo, B. and Dastres, R. (2023). Digital twin for smart manufacturing, A review. Sustainable Manufacturing and Service Economics 2: 100017. https://doi.org/10.1016/j.smse.2023.100017.

Stockburger, L., Kokosioulis, G., Mukkamala, A., Mukkamala, R. R. and Avital, M. (2021). Blockchain-enabled decentralized identity management: The case of self-sovereign identity in public transportation. Blockchain: Research and Applications 2(2): 100014. https://doi.org/10.1016/j.bcra.2021.100014.

Stoumpos, A. I., Kitsios, F. and Talias, M. A. (2023). Digital transformation in healthcare: technology acceptance and its applications. International Journal of Environmental Research and Public Health 20(4): Article 4. https://doi.org/10.3390/ijerph20043407.

Sunny, J., Undralla, N. and Madhusudanan Pillai, V. (2020). Supply chain transparency through blockchain-based traceability: An overview with demonstration. Computers & Industrial Engineering 150: 106895. https://doi.org/10.1016/j.cie.2020.106895.

Taylor, P., Steenmans, K. and Steenmans, I. (2020). Blockchain technology for sustainable waste management. Frontiers in Political Science 2. https://doi.org/10.3389/fpos.2020.590923.

Tripathi, G., Ahad, M. A. and Casalino, G. (2023). A comprehensive review of blockchain technology: Underlying principles and historical background with future challenges. Decision Analytics Journal 9: 100344. https://doi.org/10.1016/j.dajour.2023.100344.

Vishwakarma, A. K., Patro, P. K., Acquaye, A., Jayaraman, R. and Salah, K. (2024). Blockchain-based peer-to-peer renewable energy trading and traceability of transmission and distribution losses. Journal of the Operational Research Society 0(0): 1–23. https://doi.org/10.1080/01605682.2024.2441224.

Wang, C., Zeng, J., Huang, Q. and Zuo, Z. (2021). A review of blockchain layered architecture and technology application research. Wuhan University Journal of Natural Sciences 26: 415–428.

Wang, H., Zheng, Z., Xie, S., Dai, H.-N. and Chen, X. (2018). Blockchain challenges and opportunities: A survey. International Journal of Web and Grid Services 14: 352–375. https://doi.org/10.1504/IJWGS.2018.10016848.

Wang, N., Zhou, X., Lu, X., Guan, Z., Wu, L., Du, X. et al. (2019). When energy trading meets blockchain in electrical power system: the state of the art. Applied Sciences 9(8): Article 8. https://doi.org/10.3390/app9081561.

What Is Blockchain? | IBM. (2021, July 8). https://www.ibm.com/topics/blockchain.

Wongthongtham, P., Marrable, D., Abu-Salih, B., Liu, X. and Morrison, G. (2021). Blockchain-enabled peer-to-peer energy trading. Computers & Electrical Engineering 94: 107299. https://doi.org/10.1016/j.compeleceng.2021.107299.

Zhang, P., White, J., Schmidt, D. C., Lenz, G. and Rosenbloom, S. T. (2018). FHIRChain: Applying blockchain to securely and scalably share clinical data. Computational and Structural Biotechnology Journal 16: 267–278. https://doi.org/10.1016/j.csbj.2018.07.004.

Zhao, N., Zhang, H., Yang, X., Yan, J. and You, F. (2023). Emerging information and communication technologies for smart energy systems and renewable transition. Advances in Applied Energy 9: 100125. https://doi.org/10.1016/j.adapen.2023.100125.

Zhou, L., Diro, A., Saini, A., Kaisar, S. and Hiep, P. C. (2024). Leveraging zero knowledge proofs for blockchain-based identity sharing: A survey of advancements, challenges and opportunities. Journal of Information Security and Applications 80: 103678. https://doi.org/10.1016/j.jisa.2023.103678.

Zhuk, A. (2025). Beyond the blockchain hype: Addressing legal and regulatory challenges. SN Social Sciences 5(2): 11. https://doi.org/10.1007/s43545-024-01044-y.

Chapter 4

Spam Detection Using Fuzzy Machine Learning Techniques

Sayanee Ghosh and *Aswani Kumar Cherukuri**

1. Introduction

In today's digital world, communication tools such as SMS, social media, and email are essential for personal and professional interactions. However, the rise in the use of these platforms has led to a surge in unwanted and often harmful messages designed to trick or harm users. Beyond disrupting communication, spam poses serious security threats, including the distribution of malware, fraud, and phishing scams (Shaaban et al. 2022). Creating an effective spam detection system has become increasingly challenging as spammers evolve their tactics to bypass traditional detection methods. The dynamic and complex nature of spam frequently makes it difficult for conventional procedures to keep up, making it difficult to discern between spam and genuine messages (Hameed et al. 2021). As a result, there is a pressing need for advanced methods to tackle this issue. One promising approach is the combination of fuzzy logic and machine learning, which is particularly effective in managing the uncertainty and vagueness often encountered in spam detection tasks (Ayo et al. 2024). Fuzzy logic systems effectively manage the uncertainty involved in spam classification by enabling models to make decisions similar to those of humans. When combined with machine learning techniques, these algorithms can be trained to identify spam patterns more accurately (Nosrati et al. 2023). Despite the potential benefits of fuzzy machine learning in spam detection, challenges such as scalability and real-time performance remain significant hurdles (Salman et al. 2024). Many fuzzy models struggle to analyze large datasets swiftly, especially as the volume of digital communication continues to grow, leading to delays and performance issues. Our research aims to tackle these challenges by developing a fuzzy machine learning-based spam detection model specifically designed for improved scalability and real-time performance. The goal is to create a system that can efficiently manage a significant volume of spam messages across various platforms, all the while ensuring high accuracy and quick detection. This research seeks to boost the effectiveness and reliability of spam detection systems by leveraging the benefits of fuzzy logic and advanced machine learning methods, ultimately improving the security and integrity of digital communication networks.

School of Computer Science Engineering and Information Systems, Vellore Institute of Technology, Vellore, Tamil Nadu.
Email: sayaneeghosh138@gmail.com
* Corresponding author: cherukuri@acm.org

DOI: 10.1201/9781003620822-4

2. Problem Statement

While fuzzy machine learning techniques hold significant promise for spam detection, there are still a number of challenges that must be addressed, including scalability, effectiveness, and real-time performance. One major issue is that current models often struggle to manage the large amounts of data produced in real-world situations. Furthermore, many existing fuzzy-based models are not designed for real-time processing, which is crucial for both preventing harm and ensuring a positive user experience.

3. Literature Survey

In the paper presented by Shaaban et al. (2022), a dynamic deep ensemble model was introduced to identify whether SMS messages are spam or not. To automatically extract features, convolutional and pooling layers were employed and two random forests and two highly randomized trees were utilized for classification. The model surpassed the result of the deep neural networks and conventional classifiers with an accuracy of 98.38%, followed by precision, recall, and F1-score metrics by applying ensemble methods, including boosting and bagging. In the subsequent research efforts, the focus has been on enabling multilingual detection and dealing with messages that contain images, and this approach has been effective in minimizing the risks posed by SMS phishing. Jacob et al. (2022) introduced the MOGA–CNN–DLAS method, which integrates convoluted-pooling-fully connected CNN layers with a multi-objective optimization process to improve Twitter spam recognition. The method successfully classified the tweets as malicious or normal spam with an improvement of about 17%, 13%, 10%, and 19% on accuracy, precision, recall, and F-score respectively, and a reduction of 19%, 16%, and 21% on RMSE, MAD, and MAE when compared to the benchmarked schemes on three datasets namely, Twitter100k, ASU Twitter and SNAP Twitter. Another study focused on social fuzzy systems, specifically, multi-objective optimization. Vahid et al. (2023), using a double layer Bayesian technique suggested a hybrid method for spam classification that supplements naïve-bayesian classification by learning better features. An ensemble of five techniques outperformed the baseline metrics for best overall performance. This language-independent technique works for Chinese, Arabic, Persian, and so on. Salman et al. (2022) presented a novel fuzzy logic-based performance metric for spam email detection aggregating recall, accuracy, and precision into one multi-criteria metric. To evaluate how BERT and LSTM models perform when compared to each other, three datasets namely Enron, Lingspam, and PU were selected as benchmarks. To compare their mathematical expression in fuzzy logic easily, decision rules were created. The results indicated that BERT was superior on Lingspam, while LSTM outperformed Enron and PU. Moreover, many more metrics were newly added. XML and GAN types of methods were tried instead of another method, and the fuzzy metric was applied to many areas as future prospects. In this paper, Zhijie et al. (2020) proposed an approach for Twitter spam detection, based on user traits, content, activity, and relationships. In the I2FELM model, Cholesky factorization based fuzzy weights, combined with a composite kernel function, have been implemented to handle data imbalance. The model performed satisfactorily over both balanced and unbalanced datasets, with hidden nodes being self-organized. Further research has tried to find ways of distinguishing spam, where less labeled data has been available, through semantic and emotion analyses, improved feature selection, and semi-supervised learning. To eliminate any ambiguity, Chatterjee and Namin (2021) proposed the fuzzy Dempster – Shafer Theory (DST) in web spam detection, which improves the selection of evidence through the application of fuzzy rules. Nevertheless, there were limitations in the use of DST on large datasets with conflicting evidence. The idea of adding multilayer neural networks to Deep DST was also proposed in order to solve the problems of computational complexity and improve the quality of the decision. In order to reduce complexity, further studies may be related to the improvement of the neural networks model and the narrowing of the evidence space. Ayo et al. (2024) employed a fuzzy inference approach to create a hybrid correlation-based deep learning model for classifying spam emails. In this study, the authors employed a hybrid feature selection technique that combined rule evaluation, CfsSubsetEval and genetic algorithm in determining the optimal features from the spambase dataset. A deep learning model was built using the specified features and spam severity was classified using the fuzzy logic system. The model gave improved F1 scores of 96.5% and 96.4% for the test sets, 94% accuracy and comparatively lower misclassification in comparison with other methods. Intervening research may focus on fuzzy logic variables development and enhancing the deep learning architecture for enhanced spam classification. Hameed and Ali (2021) presented the first novel model in the field, which employed fuzzy rules and the content of spam SMS messages for spam message classification. To select important rules, the model implemented a binary PSO after relevant features were extracted from the data of spam SMS. The model achieved 98.8% recall, 90.8% precision, 94.6% F-measure, 98.5% accuracy, and 0.9863 AUC. This fuzzy-based technique achieved high rates of detection and showed a good relation between the classified features and messages. Further studies may focus on the use of larger datasets and explore the message spam detection system in other languages. Chrismanto et al. (2023) have analyzed the pairs from SpamID-Pair using emoji attributes, to improve social media spam comment detection utilizing such post-comment pairs. In an experiment it performed fourteen machine learning models. Its performance, when enhanced with post-comment pairs, ranged from 0.7% to 2.11%, and the accuracy due to increased detection was 4%–6% using emoji features. The best overall results had the soft voting ensemble approach, where SVM-RBF, RF, and ET algorithms appeared to be the best techniques. More research is necessary to further investigate the contextual relationships between postings and comments for better accuracy. For comparing multiple machine learning models and anti-spam systems, and to investigate the dynamic nature of SMS spam, Salman et al. (2024) have developed a large SMS dataset. Some deep learning models and anti-spam programs obtained 90% and 80% precision scores in the classification of spam based on the results, and all the models successfully identified authentic communications as ham. It points out the weaknesses of current anti-spam policies and suggests future research directions. This study underlined that SMS spam is still a big problem that needs further effort in developing effective detection systems. Makkar et al. (2023) presented an efficient cognitive framework for web spam detection using a machine learning and fuzzy approach. This ensemble technique increased the overall accuracy and performance of individual classifiers, while reducing the stress on the ranking module of a search engine. For experiment purposes, the dataset used was that of a preprocessing web spam, more CIoT applications should be involved in future research to enhance web spam detection. A hybrid approach to financial risk assessment was put forth by Kaur et al. (2024) that integrated machine learning and fuzzy logic to enhance interpretability and predictive power. The results of experiments show that this model outperforms the usual methods and results in accurate predictions. Since the system is modular, it may be adapted for a number of domains. This method, therefore, significantly advanced the field of financial risk assessment, combining the predictability of machine learning with the explainability of fuzzy logic. Future research will be concentrated on the increase of its scalability and application, focusing on the possibility of interdisciplinary cooperation in solving tough problems. Khushal et al. (2024) divided Polycystic Ovary Syndrome (PCOS) into three classes by applying machine learning and fuzzy logic to a biological dataset. The fuzzy model showed better accuracy and diagnostic range as compared to the binary classification. It proved to be computationally efficient, too. Other diagnostic types and domains may be included in future work with this technique. Kanza et al. (2024) presented an IoT-based system integrating machine learning and fuzzy logic for early Cardiovascular disease (CVD) diagnosis. The machine learning method, particularly the Random Forest model, is quite accurate in predicting CVD. Using heart rate and oxygen levels, it can provide early detection of individual health checkup using fuzzy logic. It also provides easy accessibility with early detection and individualized care through this method. This study may demonstrate the strength that AI can make in CVD management for healthier outcomes, though further validation is needed. Ero et al. (2024) incorporated an OT system into the LPBF machine with a machine learning framework to predict porosity flaws during manufacture. It assigned porosity scores by rightly predicting keyhole and fusion defects by using SOM, fuzzy modeling, and the U-Net model. The strength of the method is its ability to accommodate various materials and conditions and the ability to collect ground truth data automatically, thus reducing human error. The technique utilizes fuzzy rules and adaptive thresholds for detection in enhancing the part qualifying process. A novel decision support system, comprising a Rules Module and a Classification Module, was presented by Carvalho et al. (2023) for the diagnosis of electrical system defects. The presented rules were made easier using fuzzy variables, and it utilized a Decision Tree classification and rule creation. Experimental results on two datasets of electrical faults show that it outperforms other models. Future studies may focus more on such white-box classifiers, rule simplification strategies, and explainable AI methodologies for a better improvement in situational awareness and accuracy of electrical system analysis. This study by Dhafar et al. (2023) discussed several deep learning-based fuzzy logic models and strategies that are focused on improving the performance of deep learning. It has highlighted the use of fuzzy logic in manufacturing and aviation industries and deep learning in fields such as sentiment analysis, healthcare, and self-driving automobiles. To improve the results, the research proposed a hybrid deep neuro-fuzzy system that integrates ANFIS and DNN classifiers. The study concludes that the fusion of deep learning with fuzzy systems has much promise and will be used as a future hybrid model to deal with tasks like text classification with sentiment analysis. For pitch control of wind turbines, Sierra-Garcia and Santos (2021) proposed a hybrid approach that combines deep learning neural networks with a fuzzy logic controller for accurate wind estimation. The method solved wind measurement errors and improved the control performance. The primary findings included the effectiveness of low dropout coefficients, better performance of the Adam solver, and negligible impact of batch size on training. Potential focuses of future research can include predicting vibrations for floating wind turbines and validating the model on real turbines. Kamruzzaman et al. (2022) presented a fog-computing-based remote healthcare application, targeting the reduction of resource expenses and latency for enabling improvement in patient monitoring. There, it focused on using an average of one fog apparatus per institution to manage electromagnetic graphics data. Apart from addressing their concerns regarding resource management scheduling, the study makes further performance analysis concerning lost patients and healthcare benefits from using the system. Sensor nodes, which had acquired biopotential signals at 1000 samples per second, yet which consumed very small amounts of energy, depicted the potentials of fog computing in the healthcare domain, therefore, near-source processing of the data was realized. In order to study beta-thalassemia carrier patients, Ibrahim et al. (2024) developed a late fusion-based machine learning model that is intelligible both to professionals and non-experts. The average overall precision is 96 per cent. Users can input their data to assess the status of thalassemia. Federated learning, the use of LSTM and other algorithms, could further improve it in phases-thalassemia phases like Alpha Max, Alpha Min, Beta Max, and Beta Min. Logic-oriented fuzzy neural networks that were introduced by Alateeq and Pedrycz (2024) with the approach of focusing on a design that uses AND/OR neurons to describe systems while attempting to balance interpretability with accuracy. In addition to listing problems and potential solutions toward making dependability better, it followed the development of formal logic through to the current logic-oriented networks. There is still much work to be done on huge datasets and still more towards interpretable models in achieving accurate predictions. The authors believe that logic-oriented neurocomputing has promise in dealing with the problems of reliability of intelligent models. Kumar et al. (2023) designed human movement identification in video streams, using the proposed approach of dynamic frame extraction along with the detection of anomalous activities. It gained up to 95.04% accuracy in identifying abuse and assaults using a pre-trained model, InceptionV3 CNN, with the capabilities of LSTM for both capturing the local and long-time dependencies and the usage of fuzzy logic framework to enhance fast frame extraction. The complexity of the UCF-crime scene was slightly greater than UCF-50, while performance could still differ across datasets. Future versions would ensure the real time suitability for handheld devices while further enhancing detection in unclear videos. Nilashi et al. (2023) proposed a machine learning approach for predicting the Unified Parkinson's Disease Rating Scale (UPDRS) in order to enhance the diagnosis of Parkinson's disease (PD). It used the DBN for prediction, while integrating the approaches of EM, DL, and ANFIS. The approach has less time complexity with a better accuracy as compared to other approaches. Online learning and optimization algorithms are necessary for further research on PD datasets that are large enough to make more accurate predictions. Rajawat et al. (2023) proposed a CNN and fuzzy logic framework for the automatic identification of depression, referred to as FFL, which further enhances the multimodal representation quality by the fusion of temporal and spatial information. Experiments proved effective results on detecting AVEC 2014 and AVEC 2017. Future improvements could involve changing parts of this framework to modify its settings in order to identify some other diseases utilizing datasets related to this comparison. Overall, the stress tracker can mainly track stress or anxiety status changes in patients as well as elderly patients. Singh et al. (2023) designed the fuzzy logic approach-based system of a medical diagnostic framework in detecting the HBV virus. The multilevel fuzzy inference system scrutinized various risk indicators while discriminating between infected patients and those who are non-infected. Its performance criteria include 94% accuracy, 91.2% sensitivity, 96.29% specificity, and 95.45% precision. Therefore, future developments should develop a system based on the neuro-fuzzy approach capable of combining all features that are considered desirable in the neural networks and fuzzy logic to achieve a better result. For reducing computing time delay, Ravishankar et al. (2023) proposed cyst prediction in an early stage, employing fuzzy CNN-based approach exploiting 320 training images along with 120 testing photos. Result of such an approach will help medical professionals in finding proper prognosis and treatment. Temporal limitations will be incorporated to improve integrity within the current future work. Xiang et al. 2024 presented a deep image feature learning method that outperformed conventional feature extraction techniques, with the use of fuzzy rules in a layered learning structure. DIFL-FR was shown to clearly exhibit transparent and robust nonlinear feature learning on both small and big datasets. Further research will be focused on optimization techniques such as supervised and semi-supervised learning while studying the impact of depth of the models. The use of fuzzy logic and deep learning in biometric systems, as proposed by Al-Sukeinee and Khudeyer (2024), improved fingerprint classification accuracy, as well as, that of images to routinely reach above 99% mostly. The focus for this study was placed upon the prominence of standard data sets on benchmarking, and viable hybrid strategies like coupling CNNs to Mamdani fuzzy rules. Other studies are necessary to solve real-time process issues with real-world considerations and ethical implications. Exploring additional ways of using transformers within novel hybrid models, the extension of hybrid model algorithms to diagnose in healthcare was also identified as part of further work.

4. Proposed Work

The spam detection system uses a hybrid method that integrates a Long Short-Term Memory (LSTM) model for initial spam probability scoring with fuzzy logic to refine classification. It starts by loading the Trec 2007 email dataset, which includes labeled emails with both spam and ham (non-spam) messages. Each email goes through preprocessing to clean and normalize the text. This process includes tokenizing the email messages, removing unwanted characters, converting the text to lowercase, and eliminating stopwords. Since LSTM models need numerical input, the text from the emails is tokenized. Each word is mapped to an integer, and sequences of words are represented as lists of these integers. A vocabulary size is established, retaining only the most frequent words. As LSTMs expect inputs of uniform length, the tokenized sequences are padded to a fixed length. Emails shorter than the maximum length are padded with zeros, while longer emails are truncated. These steps ensure that the LSTM model receives well-structured input data. After preprocessing, the dataset is divided into training and testing subsets, typically in an 80–20 ratio. The training set is used to train the LSTM model, while the test set is kept for evaluating the model's performance after training.

The LSTM model is built to handle the tokenized sequences of emails. LSTMs are selected for their capability to capture long-term dependencies in sequential data, which makes them suitable for identifying patterns in email content that indicate spam. The model is trained using the padded sequences from the training dataset. Throughout the training process, it learns to assign a spam probability score ranging from 0 to 1 for each email. A higher score suggests a greater chance of the email being spam. Once the LSTM is trained on the training set, it produces spam probability scores for the test set. These scores reflect the likelihood of each email being spam based on the patterns it learned from the training data.

Fuzzy logic is used to enhance spam classification. The fuzzy inference system takes into account three inputs, such as the spam probability score from the LSTM model, the length of the message, and the number of hyperlinks in the email. The Mamdani inference method assesses predefined fuzzy rules that define the relationships between these inputs and the resulting spam classification.

The fuzzy output, which represents a fuzzy spam classification score, is then converted into a clear value through a process called defuzzification. The Centroid method is employed for this purpose, determining the center of gravity of the area under the fuzzy membership function. This process results in a single numeric score that indicates the final classification decision. Afterward, this crisp fuzzy score is turned into binary labels (0 for ham and 1 for spam) based on a predetermined threshold. The predictions are then compared against the true labels from the testing dataset to evaluate the model's performance. Various metrics, including accuracy, precision, recall, and F1 score, are calculated to measure the effectiveness of the spam detection system. Figure 4.1 shows the proposed system architecture.

5. Module Description

5.1 Data Collection

The data collection module gathers a dataset that comprises labeled instances of spam and non-spam (ham) emails. The model's training and testing are based on this dataset. In this case, we utilize the Trec 2007 email dataset, which contains the email text (in the origin column) and labels (in the label column), where 1 indicates spam and 0 indicates ham. Once loaded into a DataFrame, the data is reviewed to confirm its contents and structure. Table 4.1 lists the data characteristics.

5.2 Data Preprocessing

Data pre-processing is a crucial step to clean, tokenize, and prepare the data for input into the LSTM model. This module performs several modifications to ensure that the input data is consistent and clear, as emails often include special characters, URLs, punctuation, and other non-text elements. The first task is to extract the main message content of the email, separating it from unnecessary metadata like headers, sender information, and technical details. By concentrating only on the message body, the model can better identify patterns that indicate whether an email is spam or not.

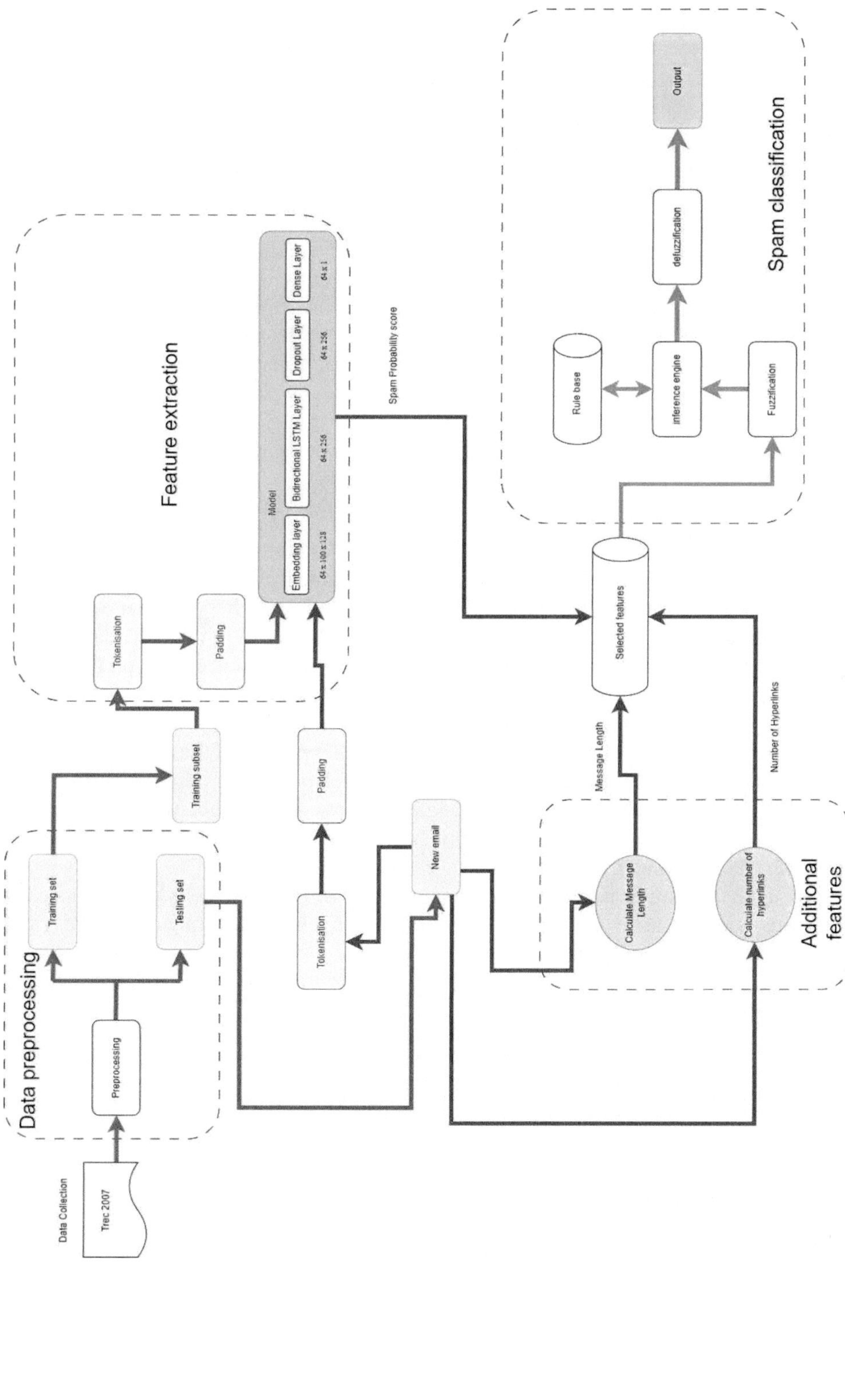

Figure 4.1. Architectural diagram.

Table 4.1. Characteristics of the dataset.

Attribute	Details
Size	75,419 emails
Labels	1 (spam), 0 (ham)
Spam size	50,199
Ham size	25,220
% of Spam	67%
Email Structure	Raw emails with headers and body
Multipart Content	Yes (HTML, plain text, attachments)
Year	2007

The data is further refined by stripping away non-textual elements like HTML tags, punctuation, special characters, and other distractions. By eliminating these interruptions, the model can work with a clear and noise-free version of the text, which facilitates the identification of words, phrases, and linguistic patterns that are often linked with spam or authentic emails.

The final stage creates a neatly organized and polished dataset by placing the processed content into a designated column. This step is crucial as it improves the model's ability to recognize spam indicators and minimizes the possibility of overfitting by removing extraneous data that could distort the training process.

5.3 Feature and Label Preparation

To create a clear structure for classification tasks, features and labels are organized independently, while preparing data for a spam detection model. The labels specify whether a message is spam or ham, while the features usually consist of cleaned email messages, which provide the data that the model will analyze to detect spam. This separation allows the model to concentrate on the content of the messages as input and strive for an accurate categorization as output.

The dataset is typically divided into training and testing portions, usually in an 80–20 split, to ensure a reliable assessment framework. This implies that a smaller portion of the data is reserved for testing, while the majority is used for training, enabling the model to discover important patterns. By testing the model on unseen examples, we gain an insight into how well it might perform in practical applications. This process results in four groups namely, training data for the model, training data labels, testing data, and test result evaluation labels.

This type of dataset splitting is crucial for precisely assessing the model's performance. By clearly defining the training and testing phases, it lowers the danger of overfitting and improves the model's ability to generalize to new data.

5.4 Text Tokenization and Padding

The initial steps in preparing text data for neural network processing involve text tokenization and padding. Because neural networks need numerical inputs, tokenization converts every word in a dataset into a numerical format, which is crucial for compatibility with the model. This approach only numerically represents the most common words in the dataset, leading to a more streamlined and pertinent vocabulary that minimizes noise from rare phrases and makes computation more efficient.

Once the sequences have been tokenized, they are padded to guarantee that every input sequence is of the same length. Padding allows the model to handle each entry in a consistent manner by restricting all sequences to a predetermined length, which avoids discrepancies among inputs. This uniformity minimizes errors that might arise from different input lengths and enables effective batch processing during training.

Tokenization and padding collaborate to systematically convert text into a format that neural networks can utilize to identify meaningful patterns from consistent, structured input data.

5.5 Feature Extraction and LSTM Model Architecture and Training

Data Sequencing

The model starts with a straightforward stack of layers known as a sequential architecture. Every layer in this arrangement connects directly to the one that follows it. Due to its simplicity, this structure is perfect for deep learning models where data moves sequentially through the layers, this is especially relevant for text input processed by LSTM layers.

Embedding Layer

The embedding layer transforms each word into a dense vector that captures its semantic meaning, carrying out the initial feature extraction. In this vectorized representation, semantically related words are positioned closer together, effectively lowering high-dimensional text data into a more compact, lower-dimensional space. The LSTM then receives this meaningful and streamlined input from the embedding layer, which encodes word associations efficiently.

Bidirectional LSTM for Contextual Awareness

The LSTM layer processes the word embedding sequence and identifies patterns or temporal connections within the data. It effectively retains significant information and long-term dependencies over several time steps, learning how words interact, based on their sequence. The LSTM features a memory cell along with input, forget, and output gates that regulate what data is remembered or discarded. This makes it possible for the LSTM to keep and use essential information from earlier in the sequence. Additionally, the bidirectional LSTM captures both forward and backward dependencies, enabling it to comprehend the context of each phrase and identify nuances and trends that point to spam. It examines each word in the sequence, but in contrast to conventional feature extraction methods, it considers the chronological order of the words and their relationships. At the conclusion of the sequence, it produces a summarized version of the complete email. This output, known as a feature vector, encapsulates the main connections and patterns identified within the word sequence. This feature vector is subsequently sent to the dense layer to finalize the last classification operation.

Dense Output Layer for Final Prediction

For binary classification, a Dense layer with a single unit and a sigmoid activation function is used. This setup generates a probability ranging from 0 to 1, where values near 1 denote spam and those near 0 signify ham (non-spam). The complex, high-dimensional feature space of the LSTM is distilled into a single probability score that reflects the possibility of spam through the sigmoid function. This score encapsulates the retrieved features, such as word patterns, sequence relationships, and contextual dependencies, simplifying the binary classification process.

Dataset Preparation and Labeling

The email dataset is split into training and testing subsets before the training process begins. This division allows the model to assess its performance on data it hasn't encountered before while learning from a portion of the data. Every email is categorized as either spam or non-spam, giving the model the necessary ground truth to differentiate between the two categories.

Training Process with Epochs and Iterative Updates

The model processes its training data over several iterations, known as epochs. During each epoch, it modifies its internal parameters based on the differences between its predictions and the actual labels. This optimization process gradually enhances the model's weights, boosting its ability to classify data accurately. With each epoch, the model becomes better at recognizing the key features that distinguish spam from non-spam.

Evaluation and Probability Assignment

The model can evaluate new emails by providing a probability score that reflects the possibility of them being spam after it has been trained. This score enables threshold-based classification, for instance, an email is deemed spam if the likelihood exceeds 0.5. The probability score serves as an advanced metric that can be utilized to adjust spam filtering rules in accordance with the specific needs of different applications.

Fuzzy Logic for Classification Refinement

Fuzzy Variables and Membership Functions

To determine how much a specific value fits into one or more of these categories, the system establishes three fuzzy input variables namely, spam probability, message length, and the number of links. It subsequently assigns particular membership functions to these variables. Below is a definition of these variables:

Spam probability: There are three levels of spam probability namely, low, medium, and high. Each category is assigned a range on a scale from 0 to 1, lower values denote a lower chance of being spam, while higher values imply a greater likelihood. For instance, a likelihood of 0.3 might be classified somewhere between "low" and "medium."

Message length: The three categories of message length are short, medium, and long. They define the word count ranges for each type. Short emails contain no more than fifty words, while medium and long emails gradually increase in length. These fuzzy sets aid in distinguishing between longer messages, which are often found in spam, and shorter ones, which are common in casual communication.

Number of links: There are three categories for the number of links such as few, moderate, and many. A high number of links might imply spam or promotional content, while a low number of links could suggest personal emails.

Spam and ham are the two categories that define the output variable in spam classification, determining whether an email is authentic or unwanted. The trimf (triangular membership function) is utilized to represent all fuzzy sets, as it makes the process of establishing boundaries easy.

Fuzzification

The process of fuzzification helps the system manage the subtleties in data by converting clear, quantitative inputs into degrees of membership for fuzzy sets. This approach enables the system to perceive inputs through overlapping categories with descriptive labels like "low," "medium," and "high," rather than viewing input values in rigid, binary terms. These labels reflect imprecision or uncertainty by indicating a range rather than a specific value.

This method ascertains the degree of membership of each input using triangle membership functions.

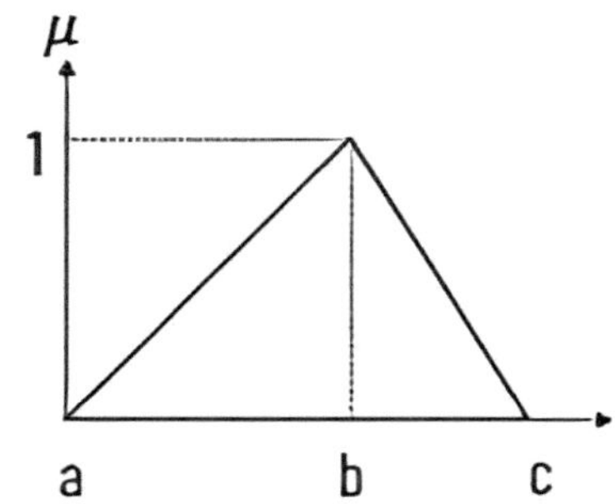

Figure 4.2. Triangular membership function.

The equation for the triangle membership functions, is presented below:

$$\mu A(x;[a,b,c]) = \begin{cases} 0, if \;\; x \leq a \frac{x-a}{b-a}, if \;\; x \in [a,b] \frac{c-x}{c-b}, if \;\; x \in [b,c] \; 0, if \;\; x \geq c \end{cases} \tag{4.1}$$

In the Equation 4.1, the degree of membership of an input x within a fuzzy set A is expressed through the triangle membership function, denoted as μA (x; [a, b, c]). This membership function is very well liked because it is straightforward, utilizing just three points that are a, b, and c to create a triangle on a graph.

The letter a is the left boundary of the triangle in Figure 4.2. The membership degree is zero when the input value x is at or below a. This indicates that the fuzzy set does not encompass values at this boundary. b represents the peak of the triangle. The membership degree reaches 1 when the input value x is precisely at b. The values nearest to b are the most indicative of the fuzzy set, as this denotes the highest membership within the set. c represents the right boundary of the triangle. The membership degree is once more zero when the input value x is at or above c. This indicates that the values in this scenario do not belong to the set.

Knowledge base: The number of fuzzy sets defined for each antecedent variable helps calculate how many fuzzy rules are required for the fuzzy inference system. Table 4.2 displays the fuzzy rule predicates along with classification result. For the membership functions, we have three categories for spam probability input (low, medium, and high), and for message length input, also there are three categories (short, medium, and long). Additionally, for the number of links input, we define three membership functions (few, moderate, many). To calculate the total number of rules, the following formula is used:

Total Rules = (Number of Membership Functions for *spam_prob_input*) × (Number of Membership Functions for *msg_length_input*) × (Number of Membership Functions for *num_links_input*)

After entering in the values: Total Rules = 3×3×3 = 27.

This means the system utilizes 27 if-then rules to account for every possible combination of the fuzzy inputs. The following is the outline:

Low probability of spam: All combinations of message length and link count are considered ham, indicating a low probability of spam.

Medium spam probability: Brief messages with any number of links are classified as spam, suggesting that short communications with links are more prone to being spam. Spam can also be identified in medium and lengthy texts with a medium spam probability based on various link counts.

High spam probability: There is strong evidence that spam is prevalent in this category, as any combination with a high spam probability is likely to be spam. The Mamdani Fuzzy Inference

Table 4.2. Fuzzy rules.

Rule No.	spam_prob	msg_length	num_links	spam_classification
1	low	short	few	ham
2	low	short	moderate	ham
3	low	short	many	ham
4	low	medium	few	ham
5	low	medium	moderate	ham
6	low	medium	many	ham
7	low	long	few	ham
8	low	long	moderate	ham
9	low	long	many	ham
10	medium	short	few	ham
11	medium	short	moderate	spam
12	medium	short	many	spam
13	medium	medium	few	ham
14	medium	medium	moderate	spam
15	medium	medium	many	spam
16	medium	long	few	spam
17	medium	long	moderate	spam
18	medium	long	many	spam
19	high	short	few	spam
20	high	short	moderate	spam
21	high	short	many	spam
22	high	medium	few	spam
23	high	medium	moderate	spam
24	high	medium	many	spam
25	high	long	few	spam
26	high	long	moderate	spam
27	high	long	many	spam

Method evaluates each rule in the system, using logical operators (AND/OR) to determine the least number of memberships in each rule's antecedent. The outcome of each rule is then activated in proportion to the lowest membership degree of the rule.

Defuzzification and output: Lastly, defuzzification is the process used to convert the fuzzy output of an inference system into a clear, distinct value. This step is critical for interpreting fuzzy results quantitatively, such as assigning a score between 0 and 1 that indicates the probability of a message being spam. One widely employed defuzzification method in this spam detection system is the Centroid Method (also known as the Center of Gravity). The centroid of a fuzzy set, C, is calculated by:

$$C = \frac{\int \mu(y) \cdot y dy}{\int \mu(y) dy} \tag{4.2}$$

In Equation 4.2, μ(y) represents the combined membership function after all rule implications have been applied. Consequently, the probability of an email being spam is distilled into a single score. If this score exceeds a set threshold (for instance, 0.5), the email is classified as spam, if not, it is deemed ham.

This setup enhances classification accuracy compared to a straightforward threshold method by enabling more nuanced classification. It effectively manages emails with different spam characteristics by assessing how closely they align with conventional spam or ham patterns.

6. Results and Metrics

The results are examined utilizing a variety of performance metrics following the model's training and evaluation. This section shows the performance of both the hybrid LSTM-fuzzy model and the only LSTM model.

6.1 LSTM Model Performance

The spam probabilities are converted into binary values, where 1 indicates spam and 0 indicates ham, by using a threshold of 0.5. Then performance metrics such as accuracy, precision, recall, and F1 score are calculated and displayed for the LSTM Model Performance to evaluate how well the model distinguishes between spam and ham. Table 4.3 provides the performance metrics results using LSTM.

Table 4.3. Output of the performance metrices using LSTM.

Accuracy	0.9425086184036064
Precision	0.9438136100578727
Recall	0.9349056038357806
F1 Score	0.9543593071432137

6.2 Hybrid Model (LSTM + Fuzzy Logic) Performance

Fuzzy logic enhances the LSTM's classifications in the Hybrid Model Performance section by adding extra parameters such as message length and link count. Metrics that compare the hybrid model to the LSTM-only model indicate improvements in recall, accuracy, and precision. The hybrid model successfully lowers false positives, especially in ambiguous situations, such as emails containing many links but having a moderate spam score. A key benefit of the hybrid approach is its flexibility in managing borderline cases. Table 4.4 provides the output of the hybrid model performance.

In Figure 4.3, the bar chart demonstrates the benefits of the hybrid model by providing a clear comparison of the accuracy, precision, recall, and F1 score between the two model.

Table 4.4. Output of the performance metrices for hybrid model.

Accuracy	0.9929593288892218
Precision	0.9934889986528963
Recall	0.9959486833220796
F1 Score	0.9947173204450939

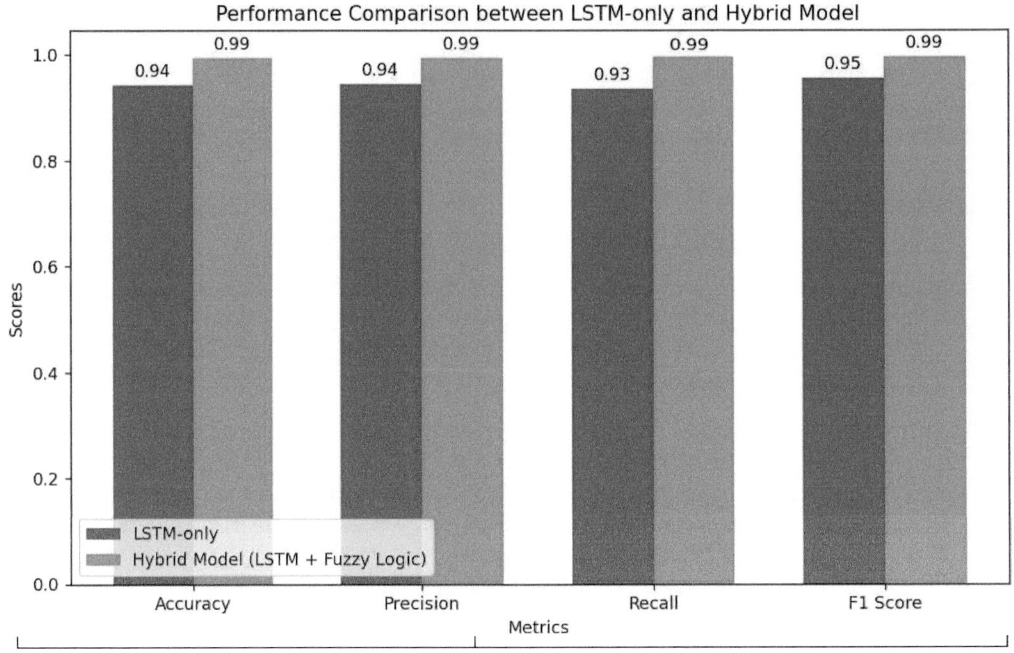

Figure 4.3. Performance comparison between LSTM and hybrid model.

7. Conclusion

This project effectively implemented fuzzy logic alongside Long Short-Term Memory (LSTM) networks to develop a more effective spam detection system. The hybrid model combines the best characteristics of both approaches, that is, fuzzy logic enhances classification by taking into account some additional contextual elements such as message length and the number of links, while the LSTM network identifies complex patterns within the email content.

This integration allows the model to better manage complicated and ambiguous situations that conventional models may struggle to classify accurately. The performance evaluation shows that the hybrid model routinely beats the LSTM-only model in terms of accuracy, precision, recall, and F1 score.

This suggests that the hybrid approach can decrease false negatives and false positives, resulting in a more reliable and effective system. A significant drawback of current methods is that the model can adjust to emerging spam strategies by dynamically altering fuzzy logic rules.

8. Future Work

To enhance the categorization context, future development of this spam detection system could focus on integrating features such as sender reputation and domain analysis. Additionally, employing ensemble learning might boost accuracy by combining insights from multiple models, while establishing a dynamic fuzzy rule adaptation mechanism would allow the model to better respond to changing spam trends.

Optimizing the model for real-time deployment and scaling it to identify spam across various platforms, including social media and SMS, would further improve its practical usability.

Moreover, improving explainability through visual aids or explainable AI techniques would help consumers comprehend classification decisions, promoting transparency and trust. When combined, these enhancements would significantly increase the system's resilience, adaptability, and overall effectiveness.

References

Alateeq, M. and Pedrycz, W. (2024). Logic-oriented fuzzy neural networks: A survey. Expert Systems with Applications 125120. https://doi.org/10.1016/j.eswa.2024.125120.

Al-Sukeinee, R. J. and Khudeyer, R. S. (2024). Review: Deep learning and fuzzy logic applications. Engineering and Technology Journal 09(06). https://doi.org/10.47191/etj/v9i06.09.

Ayo, F. E., Ogundele, L. A., Olakunle, S., Awotunde, J. B. and Kasali, F. A. (2024). A hybrid correlation-based deep learning model for email spam classification using fuzzy inference system. Decision Analytics Journal 10: 100390. https://doi.org/10.1016/j.dajour.2023.100390.

Carvalho, D., Melo, J. F. V., Silva, E. G., Wesley, J., Passos, D. and De Mattos Neto, P. S. G. (2023). Fault detection and classification in electrical systems: a machine learning and fuzzy logic-based system. https://doi.org/10.1109/wcnps60622.2023.10345098.

Chatterjee, M. and Namin, A. S. (2021). A fuzzy Dempster–Shafer classifier for detecting Web spams. Journal of Information Security and Applications 59: 102793. https://doi.org/10.1016/j.jisa.2021.102793.

Chrismanto, A. R., Sari, A. K. and Suyanto, Y. (2023). Enhancing spam comment detection on social media with emoji feature and post-comment pairs approach using ensemble methods of machine learning. IEEE Access 11: 80246–80265. https://doi.org/10.1109/access.2023.3299853.

Ero, O., Taherkhani, K., Hemmati, Y. and Toyserkani, E. (2024). An integrated fuzzy logic and machine learning platform for porosity detection using optical tomography imaging during laser powder bed fusion. International Journal of Extreme Manufacturing 6(6): 065601. https://doi.org/10.1088/2631-7990/ad65cd.

Hameed, S. and Ali, Z. (2021). SMS spam detection based on fuzzy rules and binary particle swarm optimization. International Journal of Intelligent Engineering and Systems 14(2): 314–322. https://doi.org/10.22266/ijies2021.0430.28.

Ibrahim, M., Abbas, S., Fatima, A., Ghazal, T. M., Saleem, M., Alharbi, M. et al. (2024). Fuzzy-based fusion model for β-thalassemia carriers prediction using machine learning technique. Advances in Fuzzy Systems 2024: 1–11. https://doi.org/10.1155/2024/4468842.

Jacob, W. S. and P, J. D. R. (2022). Multi-objective genetic algorithm and CNN-based deep learning architectural scheme for effective spam detection. International Journal of Intelligent Networks 3: 9–15. https://doi.org/10.1016/j.ijin.2022.01.001.

Kanza, R. A., Rasyid, M. U. H. A. and Sukaridhoto, S. (2024). Efficient early detection of patient diagnosis and cardiovascular disease using an IoT system with machine learning and fuzzy logic. International Journal of Computing and Digital Systems 15(1): 183–199. https://doi.org/10.12785/ijcds/160115.

Kaur, T., Rani, K., Thakur, P. and Talwandi, N. S. (2024). Enhanced decision support system for financial risk assessment using hybrid fuzzy logic and machine learning. https://doi.org/10.1109/iccica60014.2024.10584834.

Khan, S. A., Iqbal, K., Mohammad, N., Akbar, R., Ali, S. S. A. and Siddiqui, A. A. (2022). A novel Fuzzy-Logic-Based Multi-Criteria metric for performance evaluation of spam email detection algorithms. Applied Sciences 12(14): 7043. https://doi.org/10.3390/app12147043.

Khushal, R. and Fatima, U. (2024). Fuzzy machine learning logic utilization on hormonal imbalance dataset. Computers in Biology and Medicine 174: 108429. https://doi.org/10.1016/j.compbiomed.2024.108429.

Kumar, M. and Biswas, M. (2023). Abnormal human activity detection by convolutional recurrent neural network using fuzzy logic. Multimedia Tools and Applications 83(22): 61843–61859. https://doi.org/10.1007/s11042-023-15904-x.

Makkar, A., Ghosh, U., Sharma, P. K. and Javed, A. (2023). A fuzzy-based approach to enhance cyber defence security for next-generation IoT. IEEE Internet of Things Journal 10(3): 2079–2086. https://doi.org/10.1109/jiot.2021.3053326.

Nilashi, M., Abumalloh, R. A., Yusuf, S. Y. M., Thi, H. H., Alsulami, M., Abosaq, H. et al. (2023). Early diagnosis of Parkinson's disease: A combined method using deep learning and neuro-fuzzy techniques. Computational Biology and Chemistry 102: 107788. https://doi.org/10.1016/j.compbiolchem.2022.107788.

Nosrati, V., Rahmani, M., Jolfaei, A. and Seifollahi, S. (2023). A weak-region enhanced bayesian classification for spam content-based filtering. ACM Transactions on Asian and Low-Resource Language Information Processing 22(3): 1–18. https://doi.org/10.1145/3510420.

Parida, S. M., Pande, S. D., Challa, N. P. and Panigrahi, B. S. (2024). Enhancing audio accessory selection through multi-criteria decision making using fuzzy logic and machine learning. EAI Endorsed Transactions on Internet of Things 10. https://doi.org/10.4108/eetiot.5364.

Rajawat, A. S., Bedi, P., Goyal, S. B., Bhaladhare, P., Aggarwal, A. and Singhal, R. S. (2023). Fusion fuzzy logic and deep learning for depression detection using facial expressions. Procedia Computer Science 218: 2795–2805. https://doi.org/10.1016/j.procs.2023.01.251.

Ravishankar, T. N., Jadhav, H. M., Kumar, N. S., Ambala, S. and N, M. P. (2023). A deep learning approach for ovarian cysts detection and classification (OCD-FCNN) using fuzzy convolutional neural network. Measurement Sensors 27: 100797. https://doi.org/10.1016/j.measen.2023.100797.

Salman, M., Ikram, M. and Kaafar, M. A. (2024). Investigating evasive techniques in SMS spam filtering: a comparative analysis of machine learning models. IEEE Access 1. https://doi.org/10.1109/access.2024.3364671.

Shaaban, M. A., Hassan, Y. F. and Guirguis, S. K. (2022). Deep convolutional forest: A dynamic deep ensemble approach for spam detection in text. Complex & Intelligent Systems 8(6): 4897–4909. https://doi.org/10.1007/s40747-022-00741-6.

Sierra-Garcia, J. E. and Santos, M. (2021). Deep learning and fuzzy logic to implement a hybrid wind turbine pitch control. Neural Computing and Applications 34(13): 10503–10517. https://doi.org/10.1007/s00521-021-06323-w.

Singh, D., Rakhra, M., Aledaily, A. N., Kariri, E., Viriyasitavat, W., Yadav, K. et al. (2023). Fuzzy logic based medical diagnostic system for hepatitis B using machine learning. Soft Computing. https://doi.org/10.1007/s00500-023-08894-3.

Chapter 5

A Chaotic Map-Based Authentication and Key Agreement Scheme for Cloud Environments

S. Muthurajkumar[1,*] and *Dhanush Gupta V. K.*[2]

1. Introduction

Cloud services, due to their many benefits, such as concurrent resources, automation, scalability, and remote access, have grown in popularity over the past few years. An innovative storage called cloud storage has become evident as a result of the development of the cloud computing idea. If data is stored in the cloud, it can be easily and economically accessed from anywhere in the globe. The online distribution of computing resources, including programmers, storage settings, networks, and processing power, is the basis of the cloud computing idea. An innovative storage called cloud storage has become evident as a result of the development of the cloud computing idea.

While using cloud storage has numerous advantages, such as dependability and effectiveness, it also poses new security and confidentiality concerns to one's personal information. The fast network, data management, and confidentiality options are absent from a public cloud. This system, which makes problems with data privacy and confidentiality worse, makes using cloud services less appealing to the general public. The degree of services, including confidentiality, dependability, and efficiency, is taken into consideration when establishing communication that is secured in a cloud system. Additionally, private clouds enable enterprises to design their own security measures without having to rely on cloud provider security guidelines.

The cloud storage system needs to be protected from insider threat in addition to the security protocols. An authentication technique is needed for using the cloud storage system's data access interface to prevent unauthorized people from enrolling and utilizing the information. For protecting breaches of information in the cloud computing environment, only the password and username are ineffective. Data access therefore essentially needs adaptable, multi-level and secure systems.

The usage of computing resources by organizations and individuals is being revolutionized by the rapidly developing technology known as cloud computing. Rather than being hosted locally

[1] Department of Computer Technology, MIT Campus, Anna University, Chennai, Tamil Nadu.
[2] Thomas Lord Department of Computer Science, Viterbi School of Engineering, University of Southern California, Los Angeles.
Email: danushgupta21@gmail.com
* Corresponding author: muthurajkumarss@gmail.com

DOI: 10.1201/9781003620822-5

on a user's computer or server, the cloud computing paradigm of computing involves delivering resources, such as processing capacity, storage, and applications, to online users as needed.

Few main merits of cloud computing are:

Scalability: Users can quickly scale their processing capabilities up or down with cloud computing as needed. As a result, organizations don't need to make expensive hardware upgrades in order to swiftly and easily change their IT infrastructure to accommodate changing demands.

Cost savings: By removing the need for cost-intensive hardware and software purchases and maintenance, cloud computing can help businesses save money. Instead, consumers pay on a pay-per-use basis for the computer resources they actually use.

Accessibility: With cloud computing, users can utilize their applications and information from any location in the world as long as they have an internet connection. This facilitates remote working for employees and enterprises operating from different places.

Security: Cloud providers often have strong security measures in place to guard against dangers like hacking and data breaches and to safeguard the data and applications of their users. Cloud providers frequently have more advanced security measures than what a single company or user could put in place on their own.

Cloud computing services come in a variety of forms, including IaaS, or infrastructure as a service, which gives consumers on-demand access to computer infrastructure, including servers, storage, and networking. PaaS, or platform as a service, gives users a platform for creating, running, and maintaining their own apps without having to worry about the supporting infrastructure. SaaS or software as a service, gives consumers access to pre-built programmers like email, Customer Relationship Management (CRM), and accounting software.

The act of verifying a user's or system's identity is referred to as authentication. Since it ensures that only those with authorization are able to view confidential information or resources, it is an essential part of security. There are various kinds of authentication techniques. The most popular type of authentication is password-based, which requires users to input a username and password to log into a system or programmer. Users ought to be obligated to change their passwords on a regular basis, and they should be strong and complex. For access, clients must provide two separate kinds of identification resources using the more secure two-factor authentication (2FA). This can be something the user has (like a security token or mobile phone) or something they know (like a password). Biometric identity verification uses physical characteristics consisting of fingerprints, recognition of facial features, or recognition of voice to confirm a user's identity. Although this is a very secure method of authorization, it can be costly to install and cause privacy issues. Digital certificates are used in certificate-based authentication to confirm the identity of individuals or systems. Online banking and e-commerce applications frequently employ this safe method of authentication.

To prevent unauthorized access and safeguard sensitive information, systems and applications must employ strong authentication techniques and frequently review and update their security procedures. The best practices for creating and managing passwords should also be explained to users, and they should be urged to utilize 2FA or other stronger forms of authentication wherever available. To prevent unauthorized access or theft, plain text or data must be encrypted and transformed into ciphertext, a coded or jumbled form. Modern data security must include encryption, which is used to safeguard sensitive information like financial information, personal information, and trade secrets. Using a mathematical method, encryption transforms plaintext into ciphertext. An algorithm is used to scramble the data using a key, which is a string of characters. The key must be used to undo the encryption process and turn the ciphertext back into plain text in order to decrypt the data. A popular cryptographic algorithm for hashing and storing passwords is bcrypt. The algorithm produces a secure hash of the password using a "salt" value, rendering it difficult for intruders to determine the password using a rainbow table attack. Before the password is hashed, a random string of characters is added as a salt. Even if an identical password is used by two people, this random salt value guarantees that their password hashes will differ, rendering it difficult for intruders to determine the password. When a user tries to log in, the salt value can be used to verify the password because it is stored in the database with the password hash.

A popular and safe password hashing method, bcrypt has undergone rigorous testing and has shown to be impervious to attacks. Password hashing is just one part of password security, so organizations should also take other steps to protect the safety of their systems and user data, including robust password guidelines, multiple-factor authorization, and frequent security audits.

Image-based information hiding is a method for concealing data inside of images without changing how they appear. Image steganography tries to make it more difficult for unauthorized users to access the secret information by concealing its existence. The fundamental idea underlying it, is to hide information in the least significant bits of an image's pixel values. This is so that the least important elements of a pixel value can be changed without affecting how the image looks as a whole. The procedure can insert concealed data into the image by altering the least significant bits. The two benefits of this technique are its capacity to conceal a significant quantity of the data as part of an image, and the fact that the presence of a concealed message is not readily apparent.

2. Literature Survey

Sultan et al. (2018) offered a system that showed a safe inter-cloud authorization system that used encryption with ciphertext-policy attributes (CP-ABE). Using web apps that perhaps registered with a different service provider, the proposed approach enabled the ability for data owners to access files stored on service provider-managed cloud storage systems. Users of a web application could access the stored files by using one-time access tokens that the data owner could issue. The protocol did not provide authentic data access that enables strong verification of the legitimate clients prior to doing the different data activities, such as storing, updating, and sharing data. The work is based on the cloud environment (Muthurajkumar et al. 2023).

Tiwari et al. (2019) proposed a scheme with an authorization strategy based on biometrics for secure access to data exchange and storage. The use of a proxy re-encryption key by the cloud server and the production of an authentication token by the data master during decryption restricts user accessibility and enables adaptive sharing of information under the management of a data administrator. It has been noted that it was created utilizing intricate cryptographic processes, which have a high cost for computing and communication and make them unfeasible for use in environments with limited resources.

Roy et al. (2017) proposed a method that offered mobile user authorization, safe exchange of keys, confidentiality of users, and untraceability features in a dispersed portable cloud computing environment. The protocol used fuzzy extractor operations, bitwise XOR operations, and efficient one-way cryptographic hash functions. The cost of energy consumed using their protocol was found out to be very high and not reliable for use in the pacing environment.

Zhou et al. (2019) elucidated an authentication method for cloud servers and IoT-based systems. In order to achieve the highest efficiency, the authentication strategy used crypto-modules like the one-way hash function. The major limitation of the proposed method was the computation cost and the storage cost required to implement the protocol. Even though the energy consumed was comparatively less, the storage and computation costs were relatively high.

Mo et al. (2019) proposed a model that suggested an efficient secure two-factor anonymized authorization for users' mechanism. In addition to offering bidirectional authorization between cloud computing and mobile devices, the suggested system also satisfied established security evaluation standards. The research offered robust and effective key agreement for MCC and ID-based ECC with anonymized two-factor authorization without pairing. The suggested approach had a drawback as the energy consumed was very high compared to the other proposals. Also the computation cost on the user side was very high.

Junejo et al. (2021) proposed a scheme where any entity that needed to be a member of a Fog-IoT system should initially enroll with the organization based on a set of qualities in the proposed secure integrated framework. In an Attribute Based Encryption (ABE) scheme, the attributes were also used to produce confidential keys that are used to encrypt the data. The Server Client (SC) implemented an ABE scheme that is based on Elliptical Curve Cryptography (ECC) in order to apply strong authentication and access control in Fog-IoT systems. The suggested strategy had certain drawbacks because it only employed a single factor authentication technique. Moreover, registration was required for each person, which was an unsuitable manual process.

Sharaf and Shilbayeh (2019) presented a reliable cloud-based Electronic Health Record (EHR) system that ensured the privacy and security of health information stored in the cloud, depending on tiered multi-authority CP-ABE to implement access control regulations. The system allowed healthcare professionals and patients to share EHRs at a high level of integration, interoperability, and sharing. The suggested plan avoided using the conventional encryption method because it was inappropriate for the cloud environment.

Srinivas et al. (2020) had devised a method which used a novel user identification method for secured medical information authentication in the cloud. Both the client and wearable sensor node established a confidential session key that was applied for subsequent firm communications after successful mutual authentication between the two parties. The proposed approach offered session-key security and shielded functional attacks, according to authoritative security analysis based on the Real or Random (ROR) model and formal security verification based on the Automated Validation of Internet Security Protocols and Applications (AVISPA) tool. Although the overall computation cost was lower than any other current schemes' costs, it was still required to provide superior performance and security.

Zhong et al. (2022) presented a mutual authentication and key agreement technique based on the elliptic curve certificate - free cryptography for peer-to-peer cloud, along with an effective transfer of information model between cloud service providers. In order to preserve the privacy of service providers and customers, the method also leveraged server anonymity. To track rogue cloud servers, the technique offered identity traceability. The proposed work blocked multiple users from sharing data across different cloud servers.

Chaudhary et al. (2018) developed a brand-new, lattice-based strong encryption system for smart healthcare that, at various points, utilized both a light-weight key transfer and a method of authentication. A mutual authentication system based on lattices was created to validate requests made to multiple end users and cloud storage. For the purpose of transferring data between numerous users and cloud storage, a data encryption system was developed. For the purpose of issuing permissions to end users, an access rights verification system was created. Most of the system-reported solutions rely heavily on computing to secure smart healthcare data, which may not be appropriate given the limited resources of contemporary smart devices.

Akram et al. (2020) proposed an ECC-based three-factor user authorized key agreement system that was cost-effective to implement and withstood the main security concerns like internal threat, impersonator attack, and credential alteration assaults. Additionally, it used a random oracle model to formally analyze the proposed scheme's security. The suggested scheme had somewhat more communication bits than the related methods.

Amin et al. (2018) presented an architecture suitable for dispersed cloud environments, and on its foundation, a smartcard-based verification protocol had been suggested, allowing the enrolled user to safely access all private data from all the private cloud servers. The system made use of the BAN logic model and AVISPA tool to improve the suggested protocol. Furthermore, casual cryptanalysis verified that the protocol was safe from all security risks. Although the protocol was designed to be lightweight, it may still introduce additional overhead and complexity that can negatively impact the efficiency of the system.

Fang et al. (2020) presented an evolutionary game-based approach for collaborative authorization. Based on the evolutionary game model, they conducted game theoretical analyses. Also they designed an ESS-based algorithm and conducted experiments both on simulated as well as real datasets. The results of the proposed strategy were efficient at an acceptable level of resource consumption. The drawback of the proposed system was that only homogeneous neighboring nodes were discussed. Unquestionably, much more organized explorations are necessary to study heterogeneous neighboring nodes.

Goutham Reddy et al. (2016) proposed a mobile network improved lightweight authorization with key-agreement protocol based on smart cards and elliptic curve encryption. Because it completed a straightforward unidirectional hash function, authentication of the message code, and XOR operation, the suggested protocol was small and appropriate for real-time applications. Concatenation and an exclusive-OR function required relatively little work, whereas elliptic curve point multiplication, symmetrical operations, and asymmetrical operations do.

Ghaffar et al. (2023) presented an architecture where symmetric encryption was used in the proposed protocol to control and define the accessibility of authorized users, as well as, to securely distribute data to users. The symmetric key encryption technique also included the file's contents into the metadata to ensure data protection. To enable safe encrypted data access, the suggested protocol implements a symmetric decryption method. It has been noted that they were created utilizing complicated cryptographic processes, which have a high cost for computing and communication and make them unfeasible for use in environments with limited resources. Furthermore, some protocols prevented them from providing common security features.

Harn et al. (2022) proposed Hilbert Curve Encoding which was used to encrypt the user's message and the locations for processing privately by the dependable authority. The system also suggested hybrid H N G M N encoding, which merged Grey encoding and the Hilbert curve encoding. According to the experiment's findings, in terms of user reaction time, token remaining percentage, and execution time, the encoding approaches were superior than Hierarchical Encoding and comparable to Grey Encoding. The major drawbacks of the work was that it required high memory management and high complexity. The work does not support address grid cells and several spatial methods are used, such as 2D range trees.

Shen et al. (2019) presented Alperin-Sheriff and Peikert's straightforward and tight noise analysis method which was used to create the first identity-based fully privacy-preserving encryption scheme from identity-based encryption and lattice-based cryptography. Additionally, it created a successful multi-profile completely homomorphic encryption technique by elaborating a brand-new ciphertext under one authentication key to an expanded one under a combination key, allowing ciphertexts under many identities to be similarly evaluated. The fact that the suggested schemes were only levelled homomorphic systems like the GSW-IBFHE scheme was a negative.

Amalarethinam et al. (2015) proposed a brand-new Magic Rectangle-based picture encryption technique. According to the procedure, the simple image was divided into single-byte blocks, and the block was then substituted for the value of MR. Additionally, the user choose at random the Magic Rectangle control parameters. After that, the image was encrypted using public key cryptography techniques like RSA, etc. The model improved memory efficiency and added a layer of security to the public key technique. The work's main flaw was that it took more time to construct the magic rectangle and compress the photographs.

Zhang et al. (2022) proposed SPADE, a method of encrypted data deduplication, that did not rely on vulnerable key servers and released clients from the key administration issue. The paper proposed a password-based stacked encoding system and a password-based authorization system and incorporated them into SPADE to allow users to gain access to their information only using the passwords. The protocol was designed to prevent dictionary guessing attacks. Without independent key servers, the suggested system cannot provide the same functionality as SPADE with the same security assurance. The key management issue was incompatible with the suggested password-hardening approach.

Deng et al. (2020) proposed a model by combining two encryption techniques, namely identity-based encryption and identity-based broadcast encryption, an identity-based encryption transformation model is created. The scheme provided the following attractive features such as identity-based data storage, cross-domain encryption transformation and strong security guarantee. The drawback of this proposed approach was it introduced additional complexity to the encryption and decryption process, which can increase the likelihood of errors and make the system more difficult to implement and manage.

Munir et al. (2021) carried out a few cryptographic attacks to extract the key from the auxiliary cryptosystem. Using a chosen - plaintext attack and a single known plaintext ciphertext combination, the key was quickly and efficiently recovered. The diffusion-based encryption algorithm's weakness was shown by the attacks' minimum execution times. The only flaw in the proposed strategy was the diffusion process. The initial system that was put out included conditional shift, hamming distance, and chaos implementation. However, the bitwise XOR and hamming distance operations had the same behavior, ending the effects of one another after diffusion.

Chang et al. (2020) proposed a method of cancellable multi-biometric authentication. Utilizing a secret key produced from another biometric model, a revolutionary bitwise encryption method converted the biometric template into a secure copy. Also, the paper has introduced algorithms for bitwise encryption which were defined over keyed-hash function and block cipher-based encryption. The method reduced recognition accuracy compared to traditional biometric authentication systems, as the biometric data was transformed in a way that made it less distinct and more prone to errors.

Li et al. (2022) presented a method using a public key tree (PKTree) and a PEKS scheme, a PEKS version known as Hierarchical Public Key Encryption with Keyword Search (HPEKS) that enabled a partial-generic design. The approach was based on a more advanced HPEKS method with improved security that combined symmetric and public key encryptions. The security of the work was analyzed using the Random Oracle model. The major drawback was that it was not faster than the previous schemes like BSS-PKE/PEKS scheme and HL-PEKS scheme. The computational cost of the DHPEKS scheme's AES encryption technique was not taken into account in the trial results because it was unpredictable and influenced by the size of the plaintext.

Xiong et al. (2022) proposed a comprehensive analysis of three representative public-key encryption methods that included search capabilities. These were plaintext - checkable encryption, public-key encryption with equality testing, and public-key encoded with keyword search. In order to aid both novice and experienced researchers in understanding the schemes, the paper presented it from a variety of angles. The drawback of the method was that it was computationally intensive, especially when dealing with large amounts of data, which can result in increased latency and reduced system performance.

Zhou et al. (2019) created an accessible through search public-key encryption with cryptographic inverse firewalls that was implemented using the JPBC library. Lacking protected channels, the protocol can withstand a selected keyword assault and an algorithmic replacement attack. The protocol has a large cost and transmission benefit and is immune to malevolent insider attacks in cloud environments using the KGA and ASA. In order to test whether the IBE-based/CL-PKE-based PEKS method can be effective against a randomly selected ciphertext attack, the scheme did not employ the CRF to resist exfiltration attempts.

3. Proposed System of Remote Data Authentication in Cloud Environments

The main goal of the following work is to implement a simple and power-saving mechanism for utilizing a cloud storage environment to access and share data. To provide authenticated data access, the work has the introduction of the password and smart based authentication method. It enables the client to be uniquely identified. A system is developed to enhance the security by providing a multilevel authentication-based data access system. The proposed system can be mainly divided into 3 modules: registration, data access control and cloud data storage.

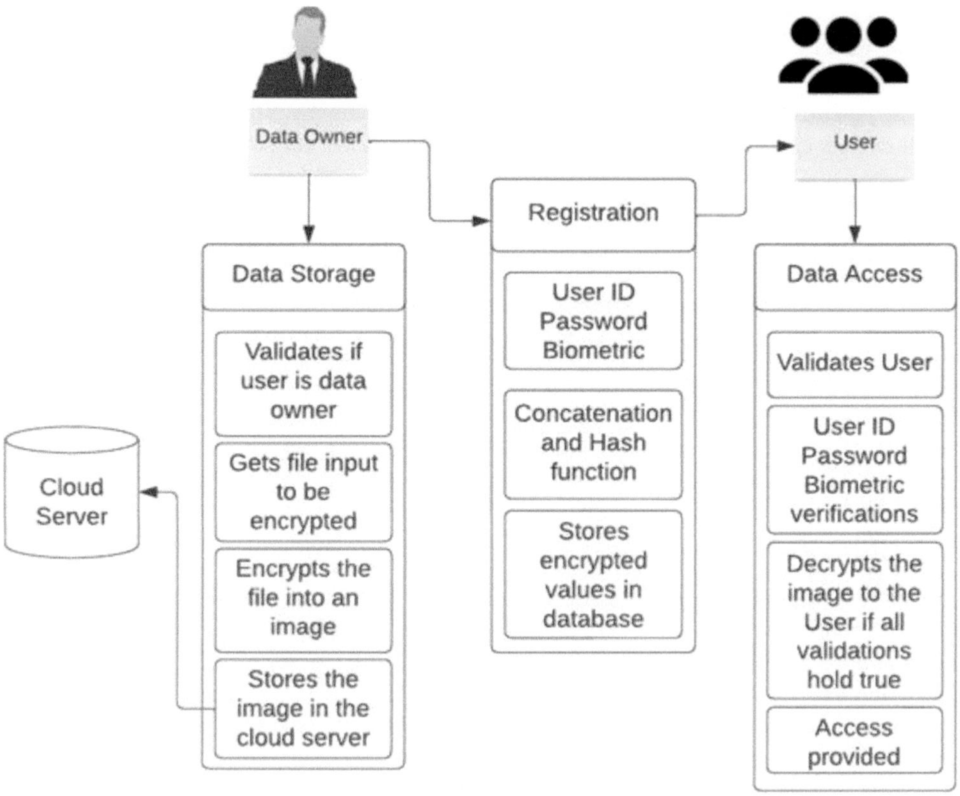

Figure 5.1. Proposed architecture diagram.

Registration: The data owner has lone access to the registration panel. User ID, password and biometric authentication is required for a user to register. Using the credentials, they are encrypted by concatenating and using hash functions. Multiple hashes are performed and each value is stored in a database used for authentication by the cloud server.

Data access control: To access data from the cloud server, primarily login and authentication needs to be done. Using concatenation, XOR operations and hash functions on the credentials, numerous encryptions are performed and communicated to the cloud server. The cloud authenticates the user and if valid decrypts the image stored in the cloud server, else the session is discarded.

Cloud data storage: In this module, verification is done to check if the entered user is the data administrator by entering the credentials. The credentials are concatenated and hashed to be stored and later used for validation by decrypting the encryption used. The file to be encrypted and the image to be stored are communicated to the cloud server where the encryption occurs.

The two entities in the architecture design depicted in Figure 5.1 are the data administrator and the data users. The proposed system has three modules where the data storage model is handled by the data administrator and the user interacts with the data access module. Primarily, the users need to get registered with the cloud server. The users need to get the approval from the data owner to create a login authentication for themselves. A user needs to have a username, password and fingerprint credentials for sign up. These data are encrypted using low-cost schemes and stored into the database for authentication by the server. For the data access module, the flow starts from the verification of the user, that is, if the client is authenticated to utilize the information stored. The authentication takes place with two credentials, username and password, and an additional layer consisting of the fingerprint verification of the user. If the authentication of the user checks all the verification, the user is allowed access to get the decrypted version of the data needed. Moving on to the data storage module, it undergoes the same set of authentication mechanisms for the validation of the data owner. After access is granted, the data owner feeds the input of the file that needs to be stored which gets encrypted and further used in the cloud server for storage.

The major tools that are used in the development of the system are:

Amazon Web Services (AWS): The extensive platform for cloud computing offered by Amazon is called Amazon Web Services (AWS). Organizational tools like computing capacity, storage of databases, as well as content distribution services are available through AWS services.

Google Colab (Python IDE): Collaboratory, or "Colab" for short, is a product of the research arm of Google. Data analysis, instructing, and machine learning are three areas where Colab excels. Using a web browser, anyone may write and run any code written in Python. Colab is an online service for Jupyter notebooks that enables no-cost use of computer features, which includes GPUs, and doesn't require any installation.

4. Algorithm and Implementation

The registration segment is totally under the control of the admin. It is not available to the public. For the registration, the admin has to manually ride the system with the credentials of username, password and fingerprint of the approved user as shown in Figure 5.2. These credentials will be encrypted and stored in the database so that it is secure and cannot be easily hacked by trespassers as shown in Figure 5.3.

```
Welcome, please select an option
Login | Signup:Signup
Enter a username:danush
Create password:danush21042
Confirm Password:danush21042
Enter fingerprint:danush_left.BMP
User created successfully!
```

Figure 5.2. Registration of user.

```
database.txt ×
1 danush, b'$2b$12$HAWSJrlvcO6kNFkp/TLnUuq4vojdh7u5gbi6kBNjTa7PNvBeVOsLa'
```

Figure 5.3. Encrypted user credentials.

Algorithm 5.1: BCRYPT Salt Encryption Algorithm
Input: Plaintext password, Randomly generated salt value **Output:** Encrypted password
1: Create a random salt **2:** Append the salt to the password **3:** Salted password hash **4:** Store salted password in database
Pwd1 = Pwd1.encode('utf-8') Pwd1 = bcrypt.hashpw(Pwd1, bcrypt.gensalt())

The encryption used in this phase is the BCRYPT algorithm. For user login, Bcrypt works nicely. It uses salt to encrypt byte code. The password can be made more secure by adding a certain amount of characters, known as salt, at random. A password string is encoded for encryption into machine-level bytes during the procedure. Raw strings cannot be read by Bcrypt; only bytes can. Typically, byte code is encoded using the 'utf-8' encoding. It is one of the greatest hashing algorithms since it is deliberately slick, which also serves as a security measure. A password string is less vulnerable to brute force assaults when salt is added. The genuine password can't simply be discovered and extracted from the salt by intruders.

The second module is the data access module. This is the phase where the client communicates with the cloud server. It is more like a client-server interaction. To begin with, the user has to verify his identity to check the integrity of the client. The client has to input the respective username and password credentials as shown in Figure 5.4. The system obtains the record using the client's provided username. Since the password is stored in an encrypted manner within the database, it needs to be decrypted and mapped with the input password. Hence, the input password would be converted to byte language and the hashed password would be passed as parameters to the checkpw function of the bcrypto module which validates and returns the Boolean value.

Algorithm 5.2: BCRYPT checkpw( ) Algorithm
Input: Plaintext password to be checked, stored BCRYPT-encrypted password **Output:** Boolean value
1: Extract salt from hashed password **2:** Hash input password with extracted salt **3:** Compare the generated hash with the stored hash
h = h.encode('utf-8') bcrypt.checkpw(pwd.encode(), h)

Moving on to the second level of authentication, biometric verification is introduced. Fingerprint authentication is the prime mode of verification which cannot be bypassed that easily. The flow diagram of the fingerprint authentication is shown in Figure 5.5. In this phase, openCV and kNN match are being used to calculate the score of t match between the stored fingerprint and the input fingerprint obtained from the user. If the score of the match is on the higher slide, access will be granted as shown in Figure 5.6.

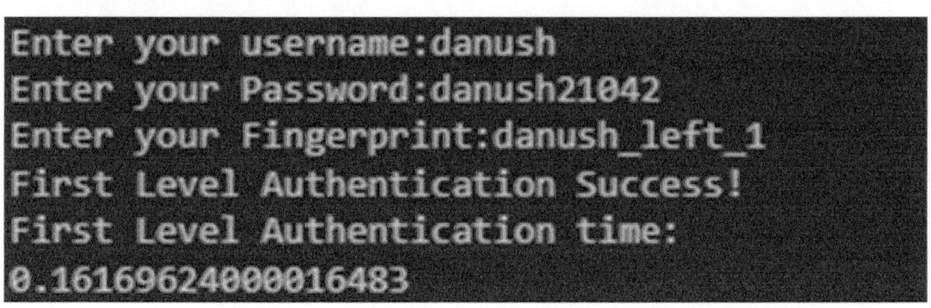

Figure 5.4. First level of authentication.

Once all the user validations hold true, the user is given permission to utilize the cloud server's information. But direct access is not provided further in order to tighten the security feature. The user will receive the data in an encrypted format so that nose pokers will not be able to get access to the files with ease.

The third module is the cloud data storage module. This is the phase where the data owner solely interacts with the cloud server. Only the data owner has privileged access to this module. To begin with, the data owner has to verify his identity to check the integrity of the incoming request.

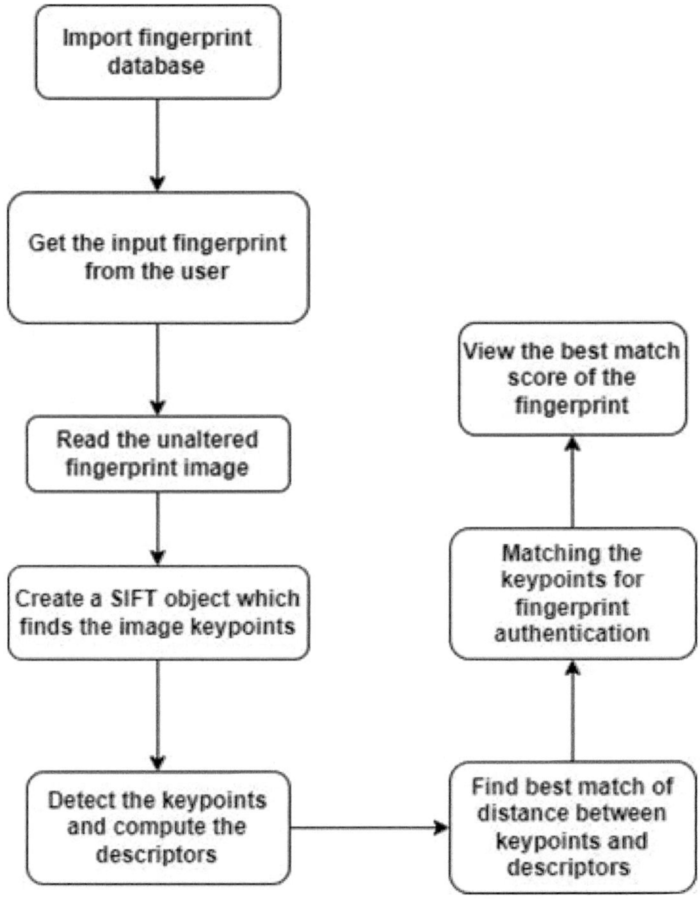

Figure 5.5. Flow diagram of fingerprint authentication.

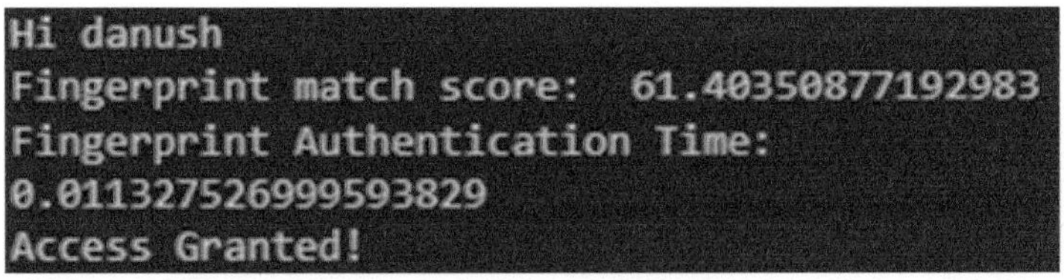

Figure 5.6. Second level of authentication.

The owner needs to enter the respective username and password credentials. An additional level of security of the fingerprint authentication is also done. Both these algorithms use the same encryption and validation algorithms as the user validation shown in section 4.2.

Image-based information hiding entails secret information being concealed in digital images in an approach that is undetectable to the naked eye. Information concealing using images aims to hide the existence of the concealed information in the image, so that it can be transmitted or stored without detection. It is commonly used for various purposes, such as digital watermarking, copyright protection and hiding malware or conducting cyber-attacks. The flowchart for the image-based information hiding is shown in Figure 5.7.

After all the validations hold true, the data owner is granted access to the cloud server to use. The owner can upload the data file which will be stored in a unique encrypted format in the cloud server as shown in Figure 5.8.

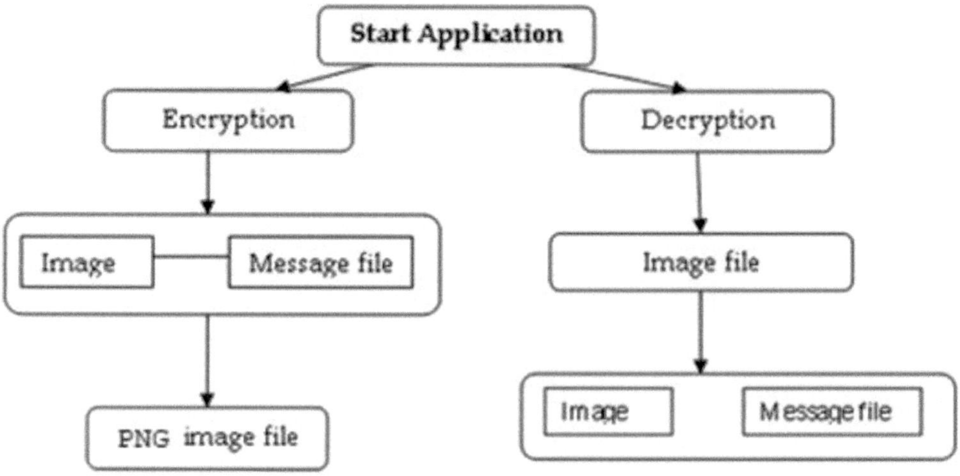

Figure 5.7. Flowchart for image-based information hiding encryption algorithm.

Algorithm 5.3: Image-based Information Hiding Encryption Algorithm

Input: Data file to be encrypted, PNG Image on which data is to be hidden
Output: PNG Image encoded with encrypted data

1: ASCII value of every character present in the information file is used.
2: Value is transformed to a binary of 8 bits.
3: Read three pixels from the image, a total of 9 RGB values, i.e., 3 X 3 values are used.
4: A single character is stored as an 8-bit binary using eight RGB values.
5: Binary data and the associated RGB value are contrasted.

```
    If the binary digit is equal to 1
        RGB value is translated as odd
    Else if the binary digit is equal to 0
        RGB value is translated into even
```

6: Ninth value specifies whether or not further pixels are to be parsed.

```
    If additional information needs to be embedded
        Ninth value becomes even
    Else
        Ninth value becomes odd
```

7: Continue doing this until the image has encoded all of the data.

```
for i when in the range of data length:
p = [v for v in id.__next__()[:3] + id.__next__()[:3] + id.__next__()[:3]]
    for j when in the range of 0 to 8:
    if dl[i][j] is equal to '0' and p[j] is odd:
        p[j] = p[j] – 1
    else if dl[i][j] is equal to '1' and p[j] is even:
      if p[j] is not equal to 0:
        p[j] = p[j] – 1
      else:
        p[j] = p[j] + 1
```

```
Access Granted!
Enter file to be encrypted: excel.csv
Enter image file: sample.png
Encryption time:
0.04626935999840498
```

Figure 5.8. Time taken for image-based information hiding encryption algorithm.

The user, after all verification validations hold true, needs to get the decrypted version of this encrypted image-based information hiding data as shown in Figure 5.9.

Algorithm 5.4: Image-based Information Hiding Decryption Algorithm

Input: PNG Image encoded with encrypted data
Output: Decrypted data file

1: Three pixels are read from the image, a total of 9 RGB values, i.e., 3 X 3 values are used.
2: The secret data is revealed by the initial eight RGB values, and the final value tells us whether we should move further or not.
3: For the first eight numbers
    If value is odd
        Binary bit is set to 1
    Else
        Binary bit is set to 0
4: If final value is even
        Continue parsing three pixels at once
    Else
        Decryption is done

```
while True:
    p = [v for v in id.__next__()[:3] + id.__next__()[:3] + id.__next__()[:3]]
    b = ''
    for i = p[:8]:
        if i is even:
            b = b + '0'
        else if i is odd:
            b = b + '1'
    d = d + c(int(b, 2))
    if p[-1] is odd:
        return d
```

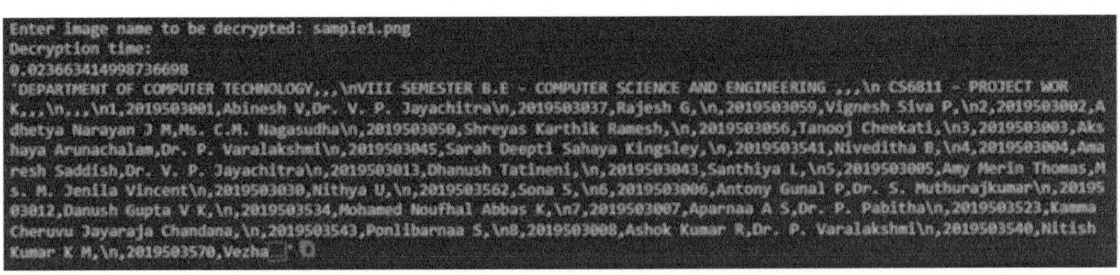

```
Enter image name to be decrypted: sample1.png
Decryption time:
0.023663414998736698
'DEPARTMENT OF COMPUTER TECHNOLOGY,,,\nVIII SEMESTER B.E - COMPUTER SCIENCE AND ENGINEERING ,,,\n CS6811 - PROJECT WOR
K,,,\n,,,\n1,2019503001,Abinesh V,Dr. V. P. Jayachitra\n,2019503037,Rajesh G,\n,2019503059,Vignesh Siva P,\n2,2019503002,A
dhetya Narayan J M,Ms. C.M. Nagasudha\n,2019503050,Shreyas Karthik Ramesh,\n,2019503056,Tanooj Cheekati,\n3,2019503003,Aks
haya Arunachalam,Dr. P. Varalakshmi\n,2019503045,Sarah Deepti Sahaya Kingsley,\n,2019503541,Niveditha B,\n4,2019503004,Ama
resh Saddish,Dr. V. P. Jayachitra\n,2019503013,Dhanush Tatineni,\n,2019503043,Santhiya L,\n5,2019503005,Amy Merin Thomas,M
s. M. Jenila Vincent\n,2019503030,Nithya U,\n,2019503562,Sona S,\n6,2019503006,Antony Gunal P,Dr. S. Muthurajkumar\n,20195
03012,Danush Gupta V K,\n,2019503534,Mohamed Noufhal Abbas K,\n7,2019503007,Aparnaa A S,Dr. P. Pabitha\n,2019503523,Kamma
Cheruvu Jayaraja Chandana,\n,2019503543,Ponlibarnaa S,\n8,2019503008,Ashok Kumar R,Dr. P. Varalakshmi\n,2019503540,Nitish
Kumar K M,\n,2019503570,Vezha
```

Figure 5.9. Image-based information hiding decryption algorithm.

5. Results and Discussion

The implementation of the system has ended in results that have significant implications for data security and management in cloud. Firstly, the system has demonstrated significant time saving compared to existing systems. This is due to the system's efficient performance and streamlined processes. The system has been designed to minimize computing time and storage resources, which has resulted in a faster, more responsive system that can handle large amounts of data.

Secondly, the system has significantly improved security aspects by providing a multilevel authentication scheme. This means that data is protected by multiple layers of security, reducing the ease with which attackers can get sensitive information. The authentication scheme is designed to confirm users' identities before granting access to the system, ensuring that the data is only accessible to authorized users. Finally, the system protects the data by encrypting the stored data in the cloud server. This ensures that even if attackers gain access, they won't be able to view or utilize the information. The encryption process uses advanced cryptographic algorithms, making it virtually impossible for attackers to break the encryption. To ensure that the implemented modules function as intended, various test cases and scenarios have been designed and executed. The test results have been averaged and documented, and the system has been found to perform well under various conditions. This ensures that the system is reliable and can be trusted to handle sensitive data.

Table 5.1 shows the time taken for each level of authentication, that is, username and password validation, including the encryption and decryption time, and fingerprint verification.

The first level of authentication, that is, the username and password verification encompassed with the encryption and decryption of the same from the database takes approximately 0.16169 seconds, as shown in Figure 5.4. Similarly, the time taken for the second level of authentication, that is, fingerprint verification is relatively less as compared with the other systems. This second level of authentication takes approximately 0.01132 seconds as shown in Figure 5.6. In total, the complete time taken for the validation of the user is around 0.17301 seconds which is proven to be less than the other system models shown below in Figure 5.10.

Table 5.1. Time taken for authentication at each level.

Aspect	Time Taken (Seconds)
First level (User Credential Validation)	0.16169
Second level (Fingerprint Verification)	0.01132
Total (Authentication)	0.17301

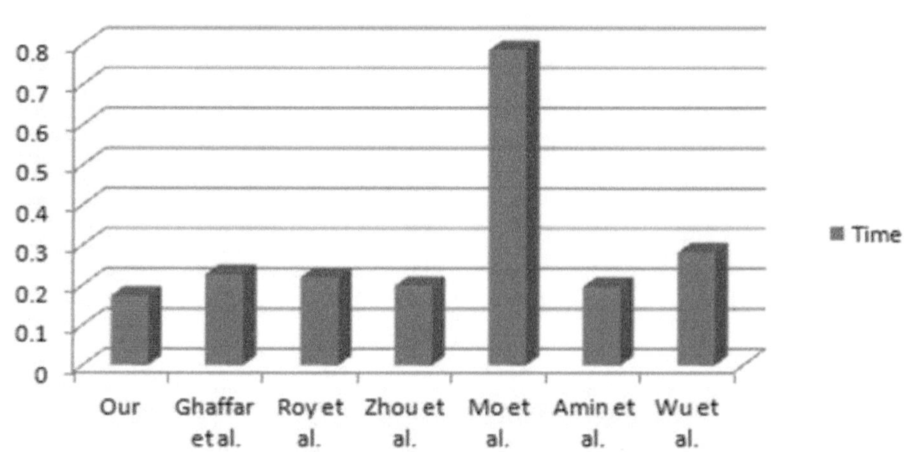

Figure 5.10. Comparison of computation costs.

Table 5.2. The number of user requests denied by symmetric encryption and decryption, as well as the proposed technique.

Experiments	No. of user request	No. of request denied by Symmetric Encryption and Decryption Algorithm	No. of request denied by proposed method
Exp 1	100	4	5
Exp 2	200	7	10
Exp 3	300	11	13
Exp 4	400	13	15
Exp 5	500	16	18

Table 5.2 shows the period of time it takes to encrypt and decrypt data before it can be safely stored on cloud servers. On average, the time taken for encrypting the data file into the image takes approximately 0.04626 seconds as shown in Figure 4.7. Similarly, the period of time it takes for decrypting the image file to retrieve the information encoded takes approximately 0.02366 seconds. In total, the encryption and decryption process take 0.06992 seconds on the whole. Record everything to explain the outcomes of the experiment.

Table 5.2 compares the number of requests refused by symmetric encryption and decryption, as well as the suggested technique, in five experiments with varying numbers of client requests. The client requests had an 18:2 ratio of real to fraudulent user requests. According to Figure 5.11, the suggested model outperforms the symmetric encryption and decryption paradigm in terms of user restriction. The model has a 96% accuracy in detecting fraudulent use of intelligent systems, with efficient key sharing strategies proposed and applied in this research.

Figure 5.11 depicts the security level of the proposed model's allowed number of authorized users. This chart shows that the access authorization of the proposed algorithm is more than the secured cloud map reduction algorithm model. Furthermore, 5% of the users were denied access. As a result, the data security in this job is improved. This is because temporal limitations are used to check the users' behaviors.

Figure 5.12 compares the memory needs for symmetric encryption decryption with the suggested model over time intervals (t1, t2). According to the implementation carried out in this model, there is a difference of 10% memory consumed in the proposed approach due to the compression made

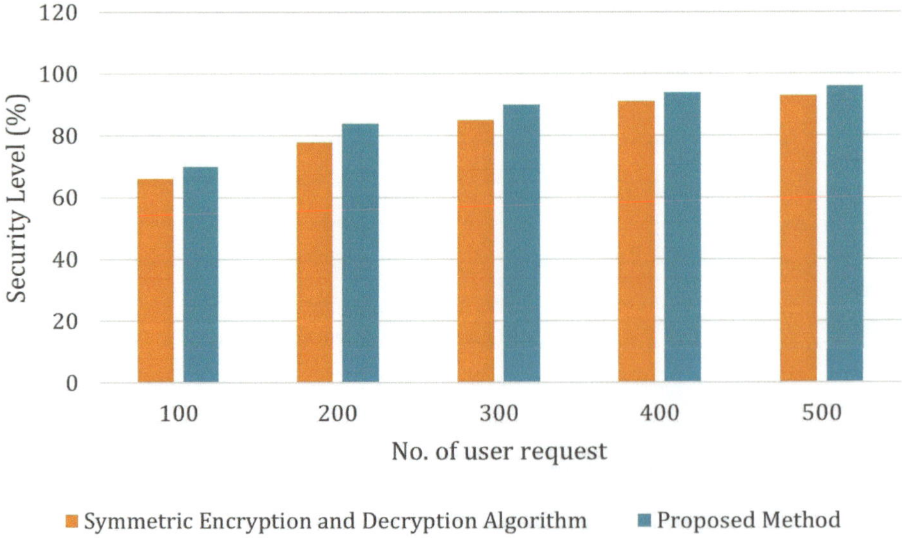

Figure 5.11. The amount of data centers with varying levels of protection.

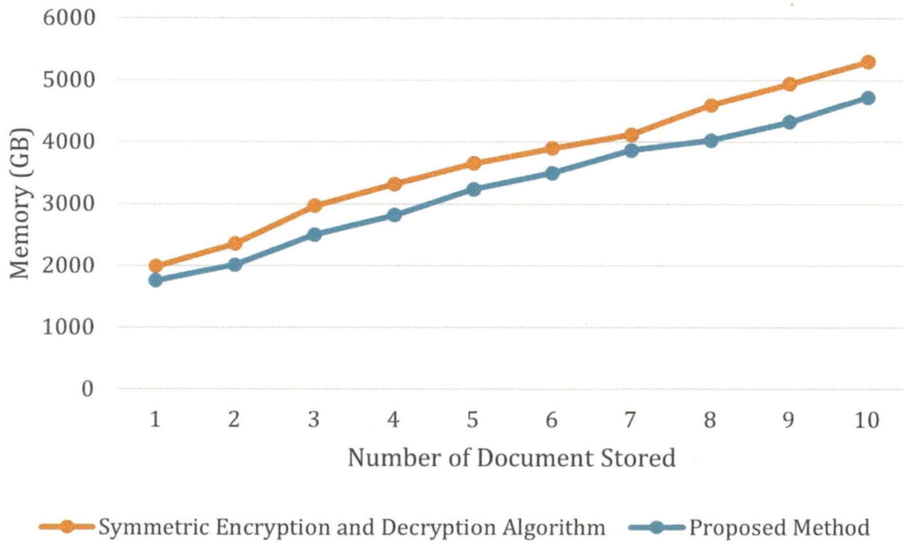

Figure 5.12. Memory evaluation for static documents.

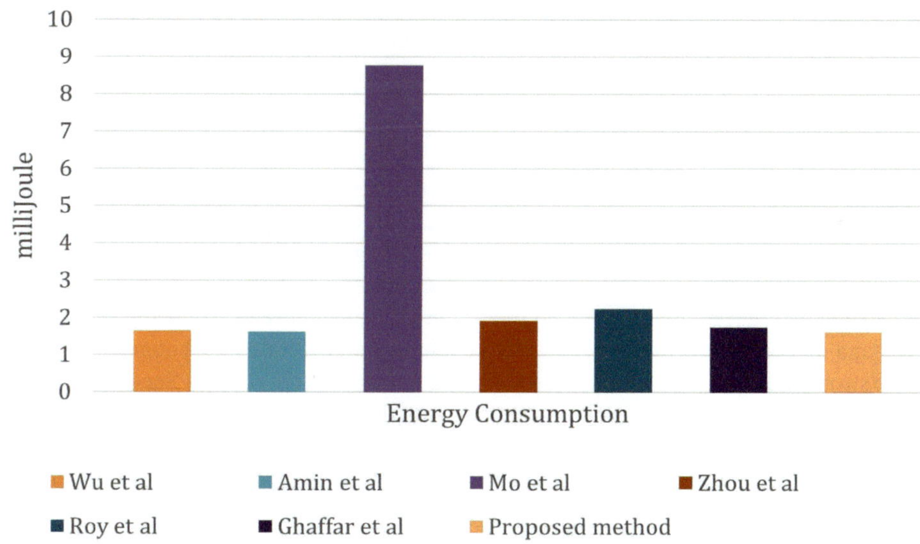

Figure 5.13. Energy consumption comparison.

of encryption. However, as compared to symmetric encryption decryption and the suggested model, the memory required to process is less. Also, Figure 5.13 depicts the amount of energy consumed by the suggested method and other significant competing methods graphically.

Table 5.3 shows the time required for data encryption and decryption. Data encryption took 0.04626 seconds, whereas decryption needed only 0.02366 seconds. The aggregate time for both processes is 0.06992 seconds. This demonstrates that encryption takes more processing work than decryption, which is common in many encryption systems. The total time is the total time spent safeguarding and retrieving data, demonstrating the encryption system's performance in terms of encryption and decryption speed.

Table 5.3. Time taken for data encryption and decryption.

Aspect	Time Taken (Seconds)
Data encryption	0.04626
Data decryption	0.02366
Total	0.06992

Table 5.4. Comparison of security features.

Protocols→ Features↓	Proposed Method	Ghaffar et al.	Roy et al.	Zhou et al.	Amin et al.
Man in middle attack	Yes	Yes	Yes	Yes	Yes
Data integrity	Yes	Yes	Yes	No	No
Flexible data access control	Yes	Yes	No	No	No
Data deduplication	Yes	Yes	No	No	No
Mutual authentication	Yes	Yes	Yes	Yes	Yes
Non-repudiation	Yes	Yes	No	Yes	Yes
Cloud server	Yes	No	No	No	No

Table 5.4 presents a comparison of the security properties of the proposed equivalent protocols. It is clearly seen that the proposed model provides data access in cloud servers and also provides security to the data in the cloud. All the available algorithms fail to give a cloud server access.

6. Conclusion and Future Work

The suggested system model, which offers secure credential encryption and satisfies most of the criteria for safe data exchange and storage in the cloud, has several benefits. Firstly, it offers a significant degree of protection by encrypting credentials and ensuring that data is exchanged and stored safely. This is particularly important in cloud computing, where data is stored on servers that are accessed via the internet, making them vulnerable to security threats. By using secure encryption techniques and meeting the criteria for safe data exchange and storage, the suggested system model provides a strong defense against cyberattacks and unauthorized access. Secondly, the system's performance study shows it to be more effective than comparable regimens. This means that it is faster and requires less storage resources, making it a more practical and cost-effective solution for organizations that need to exchange and store data securely in the cloud. By minimizing computing time and storage resources, the system can reduce operational costs and improve overall efficiency. In conclusion, the suggested system model is an effective, reliable, and secure solution for exchanging and storing data in the cloud. Its secure credential encryption, compliance with safe data exchange and storage criteria, and efficient performance make it an ideal choice for businesses seeking to protect their data while enjoying the benefits of cloud computing. This means that businesses can enjoy the cloud computing advantages like scalability and flexibility, without compromising on security.

References

Akram, M. A., Ghaffar, Z., Mahmood, K., Kumari, S., Agarwal, K. and Chen, C. M. (2020). An anonymous authenticated key-agreement scheme for multi-server infrastructure. Human-centric Computing and Information Sciences 10: 1–18.

Amalarethinam, D. G. and Geetha, J. S. (2015, February). Image encryption and decryption in public key cryptography based on MR. pp. 133–138. In 2015 International Conference on Computing and Communications Technologies (ICCCT). IEEE. https://doi.org/10.1109/ICCCT2.2015.7292733.

Amin, R., Kumar, N., Biswas, G. P., Iqbal, R. and Chang, V. (2018). A light weight authentication protocol for IoT-enabled devices in distributed Cloud Computing environment. Future Generation Computer Systems 78: 1005–1019.

Chang, D., Garg, S., Hasan, M. and Mishra, S. (2020). Cancelable multi-biometric approach using fuzzy extractor and novel bit-wise encryption. IEEE Transactions on Information Forensics and Security 15: 3152–3167. https://doi.org/10.1109/TIFS.2020.2983250.

Chaudhary, R., Jindal, A., Aujla, G. S., Kumar, N., Das, A. K. and Saxena, N. (2018). LSCSH: Lattice-based secure cryptosystem for smart healthcare in smart cities environment. IEEE Communications Magazine 56(4): 24–32. https://doi.org/10.1109/MCOM.2018.1700787.

Deng, H., Qin, Z., Wu, Q., Guan, Z., Deng, R. H., Wang, Y. et al. (2020). Identity-based encryption transformation for flexible sharing of encrypted data in public cloud. IEEE Transactions on Information Forensics and Security 15: 3168–3180. https://doi.org/10.1109/TIFS.2020.2985532.

Fang, L., Shi, G., Wang, L., Li, Y., Xu, S. and Guo, Y. (2020). Incentive mechanism for cooperative authentication: An evolutionary game approach. Information Sciences 527: 369–381.

Ghaffar, Z., Shamshad, S., Mahmood, K., Obaidat, M. S., Kumari, S. and Khan, M. K. (2022). A lightweight and efficient remote data authentication protocol over cloud storage environment. IEEE Transactions on Network Science and Engineering 10(1): 103–112.

Goutham Reddy, A., Yoon, E. J., Das, A. K. and Yoo, K. Y. (2016). Lightweight authentication with key-agreement protocol for mobile network environment using smart cards. IET Information Security 10(5): 272–282.

Harn, P. W., Yeddula, S. D., Sun, L., Sun, M. T. and Ku, W. S. (2022, December). Location-based alert system using searchable encryption with hilbert curve encoding. pp. 1445–1454. In 2022 IEEE International Conference on Big Data (Big Data). IEEE. https://doi.org/10.1109/BigData55660.2022.10020428.

Junejo, A. K., Komninos, N. and McCann, J. A. (2020). A secure integrated framework for fog-assisted Internet-of-Things systems. IEEE Internet of Things Journal 8(8): 6840–6852. https://doi.org/10.1109/JIOT.2020.3035474.

Li, H., Huang, Q. and Susilo, W. (2020). A secure cloud data sharing protocol for enterprise supporting hierarchical keyword search. IEEE Transactions on Dependable and Secure Computing 19(3): 1532–1543. https://doi.org/10.1109/TDSC.2020.3027611.

Mo, J., Hu, Z., Chen, H. and Shen, W. (2019). An efficient and provably secure anonymous user authentication and key agreement for mobile cloud computing. Wireless Communications and Mobile Computing 2019(1): 4520685.

Munir, N., Khan, M., Hazzazi, M. M., Aljaedi, A., Alharbi, A. R. and Hussain, I. (2021). Cryptanalysis of internet of health things encryption scheme based on chaotic maps. IEEE Access 9: 105678–105685. https://doi.org/10.1109/ACCESS.2021.3099004.

Muthurajkumar, S., Gupta, V. K. D. and Jayashree, P. (2023). A chaotic map-based authentication and key agreement scheme for cloud computing (CMAKAS). India Patent, Application No. 202341067451A, Publication Date: 13/10/2023.

Roy, S., Chatterjee, S., Das, A. K., Chattopadhyay, S., Kumar, N. and Vasilakos, A. V. (2017). On the design of provably secure lightweight remote user authentication scheme for mobile cloud computing services. IEEE Access 5: 25808–25825.

Sharaf, S. and Shilbayeh, N. F. (2019). A secure G-cloud-based framework for government healthcare services. IEEE Access 7: 37876–37882. https://doi.org/10.1109/ACCESS.2019.2906131.

Shen, T., Wang, F., Chen, K., Wang, K. and Li, B. (2019). Efficient leveled (multi) identity-based fully homomorphic encryption schemes. IEEE Access 7: 79299–79310. https://doi.org/10.1109/ACCESS.2019.2922685.

Srinivas, J., Das, A. K., Kumar, N. and Rodrigues, J. J. (2018). Cloud centric authentication for wearable healthcare monitoring system. IEEE Transactions on Dependable and Secure Computing 17(5): 942–956. https://doi.org/10.1109/TDSC.2018.2828306.

Sultan, N. H., Barbhuiya, F. A. and Laurent, M. (2018). ICAuth: A secure and scalable owner delegated inter-cloud authorization. Future Generation Computer Systems 88: 319–332.

Tiwari, D., Chaturvedi, G. K. and Gangadharan, G. R. (2019). ACDAS: Authenticated controlled data access and sharing scheme for cloud storage. International Journal of Communication Systems 32(15): e4072.

Wu, F. and Xu, L. (2017). A chaotic map-based authentication and key agreement scheme with user anonymity for cloud computing. pp. 189–200. In Cloud Computing and Security: Third International Conference, ICCCS 2017, Nanjing, China, June 16–18, 2017, Revised Selected Papers, Part II 3. Springer International Publishing.

Xiong, H., Yao, T., Wang, H., Feng, J. and Yu, S. (2021). A survey of public-key encryption with search functionality for cloud-assisted IoT. IEEE Internet of Things Journal 9(1): 401–418. https://doi.org/10.1109/JIOT.2021.3109440.

Zhang, Y., Xu, C., Cheng, N. and Shen, X. (2022). Secure password-protected encryption key for deduplicated cloud storage systems. IEEE Transactions on Dependable and Secure Computing 19(4): 2789–2806. https://doi.org/10.1109/TDSC.2021.3074146.

Zhong, H., Zhang, C., Cui, J., Xu, Y. and Liu, L. (2020). Authentication and key agreement based on anonymous identity for peer-to-peer cloud. IEEE Transactions on Cloud Computing 10(3): 1592–1603. https://doi.org/10.1109/TCC.2020.

Zhou, L., Li, X., Yeh, K. H., Su, C. and Chiu, W. (2019). Lightweight IoT-based authentication scheme in cloud computing circumstance. Future Generation Computer Systems 91: 244–251.

Zhou, Y., Hu, Z. and Li, F. (2021). Searchable public-key encryption with cryptographic reverse firewalls for cloud storage. IEEE Transactions on Cloud Computing 11(1): 383–396. https://doi.org/10.1109/TCC.2021.3095498.

Chapter 6

AI-Blockchain-IoT-Cloud Computing-Digital Twin

The Perfect Solutions for Next Generation Problems

Priyanga Subbiah,[1,*] *Krishnaraj N.,*[1] *R. Meera*[2] and *Kiranmai Bellam*[3]

1. Introduction

In current years, there has been an exceptional improvement in generation, which has cleared the way for transformational breakthroughs that are set to fulfil the difficult demanding situations that the subsequent era will confront. Among those advances, the confluence of technology along with AI, blockchain, the IoT, cloud computing, and digital twins offers a massive leap ahead inside the system of developing answers that are robust, clever, and resilient. The integration of these technologies results in creating a synergistic environment capable of revolutionising various industries and enhancing global working frameworks. Each of these technologies stands alone and possesses a unique set of competencies. Technology like AI, blockchain, the IoT, cloud computing, and virtual twin systems can be used together to solve problems like controlling useful resources, being sustainable, being able to make decisions based on accurate information, and being able to handle operational resilience. These technologies no longer simply improve upon the methods that are already in place; however, they also offer novel techniques that assist in assuming future requirements, contributing to a world that is smarter, more related, and more powerful.

The word AI refers to an effective generation that has evolved over the past numerous years with the intention of automating sports, evaluating considerable databases, and presenting insights into the prediction of future activities. The applications of this era have the potential to be beneficial to a giant variety of industries, which include healthcare, banking, manufacturing, and smart towns, among many others. Machine learning, natural language processing, and computer vision

[1] Department of Networking and Communications, School of Computing, College of Engineering and Technology, SRM Institute of Science and Technology, Kattankulathur, Chengalpattu.

[2] Department of Electronics and Communication Engineering, St. Joseph's College of Engineering and Technology, Thanjavur.

[3] Department of Computer Science, Prairie View A&M University, Texas.

Emails: krishnan2@srmist.edu.in, meeraradhakrishnan93@gmail.com, kibellam@pvamu.edu

* Corresponding author: priyangasubbaiya94@gmail.com

DOI: 10.1201/9781003620822-6

are examples of AI technology that have made it feasible for computer systems to study statistics, adapt to new inputs, and do jobs that previously required the intelligence of human beings. These technologies have made it viable for computer systems to accumulate information from records. AI facilitates enhanced decision-making procedures through the availability of insights that can be pushed via information and via the facilitation of predictive analytics. As an illustration, algorithms that are powered by synthetic intelligence are used in the clinical field to perform evaluations on scientific photographs with the objective of arriving at an early diagnosis of ailments. Synthetic, intelligence-driven technology in smart cities optimises visitor control and source distribution. AI plays a crucial role in the context of generation convergence for several reasons. One of these motives is that it has the capacity to learn and adjust to new instances. Figure 6.1 illustrates the role of AI in the healthcare sector.

Blockchain generation is a technology. Blockchain technology, which is gaining reputation, is renowned for its ability to generate transaction facts that can be secure, obvious, and immutable. Blockchain was originally meant to inspire cryptocurrencies. However, since then, it has advanced right into a flexible answer, that is able to ensure the integrity of data, improve safety, and permit decentralised structures. The decentralised ledger structure it utilises ensures tamper-evident and verifiable transactions. The management of delivery chains, the availability of financial offerings, and the sharing of facts are all programs that can greatly benefit from this functionality. For better information security and openness, the blockchain era, AI, and the IoT can all be used together. Additionally, it enables the improvement of consideration in automated approaches and guarantees that data from related gadgets will no longer be manipulated or misused in any way. The utility of the smart contract function similarly automates approaches by executing planned actions when unique situations arise. As a result, it reduces the need for intermediaries and enhances the effectiveness of operational tactics.

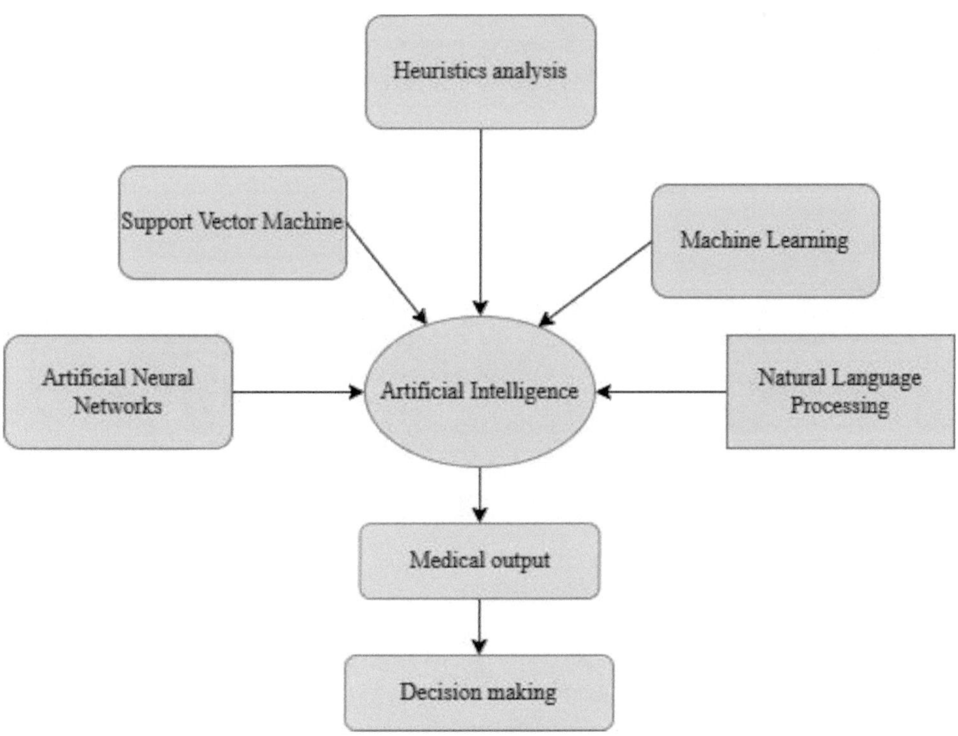

Figure 6.1. Role of AI in healthcare.

A network that connects physical items to the digital globe through the usage of sensors and conversation networks is known as the IoT. This makes it possible to record statistics in actual time, and allows for the far-off manipulation of various pieces of equipment. This connectivity makes it easier to monitor and control an extensive variety of devices, from shrewd home appliances to business machinery. These devices encompass everything from smart domestic appliances to commercial equipment. When combined with AI, IoT is responsible for the collection of huge quantities of information that could then be tested to obtain insights that may be positioned into effect. The IoT devices that are used in the business sector are chargeable for tracking the operation of certain pieces of equipment, predicting when repairs are required, and minimising downtime. Within the context of smart towns, sensors that are linked to the IoT reveal statistics pertaining to the surroundings, simplify energy intake, and enhance public offerings. When handling dynamic situations, wherein it is crucial to make selections in a timely manner, it is crucial to have the strength of the IoT to send data in actual time.

With the use of cloud computing for cloud-based data collection, organisations are easily able to manage heavy datasets and create applications in a short time. Scalable storage and processing power enables cloud computing. These techniques include AI, the Internet of Things, and digital twins (Hemamalini et al. 2024). It provides a flexible infrastructure that enables the storage, processing, and distribution of data produced by these technologies. The capacity of cloud platforms is another aspect that facilitates remote access to data and applications, which can enhance collaboration. In situations where personnel are located in different parts of the company, this is an extraordinarily desirable condition. Cloud computing expands the capacity to process data and enables the execution of advanced analytics due to its integration with various other methods. For example, the AI model can be trained on cloud-based platforms with the help of data collected from the IoT. Furthermore, the blockchain network can be hosted on the cloud to ensure accessibility and scalability throughout the procedure. Figure 6.2 illustrates the architecture of cloud computing.

A technology that allows for digital twins: To begin the procedure of growing a digital twin, one needs to first broaden a virtual illustration of the assets, processes, or systems that are being considered. Using this virtual version, simulates actual international situations and allows Groups can forecast

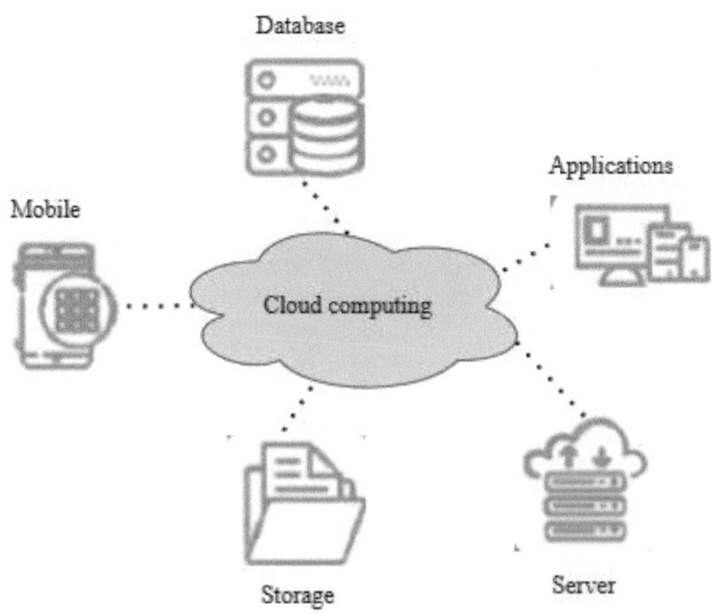

Figure 6.2. Cloud computing architecture.

effects, optimise tactics, and test new strategies in a hazard-free environment (Groshev et al. 2021). This version also lets the employer test new strategies. A few examples of groups that make sizeable use of digital twins consist of manufacturing, city planning, healthcare, and predictive protection. These are just a few of the various industries. Combining digital twins with other technologies, such as AI, the Internet of Things, blockchain, and cloud computing leads to improved performance. The digital twin is fed with real-time facts from IoT devices, AI algorithms analyse performance signs, blockchain generation guarantees the protection of statistics, and cloud computing presents important computational sources; all of these items are happening concurrently. Organisations are capable of appropriately modelling scenarios, have the ability to foresee issues before they arise, and place proactive solutions in motion with the help of their unified surroundings.

The integration of various technologies creates a synergistic effect. The convergence of various technologies creates environments that are both wise and dynamic. This environment is capable of handling complex problems on a global scale. For example, IoT devices are used to screen gadgets that are in transit; the blockchain era is used to hold a document of transactions that is both obvious and safe; synthetic intelligence is used to forecast deliveries and call for developments; cloud computing is used to save and examine records; and virtual dual models are used to optimise logistics operations. This environment utilises all these technologies to maintain a transparent and secure transaction record. This integration has the capacity to result in the creation of predictive models for the control of illnesses, stepped-forward patient tracking through the utilisation of IoT gadgets, a steady change in medical statistics through the use of the blockchain era, and cloud-based structures for remote diagnostics and treatment planning within the healthcare enterprise. All of those consequences are feasible. The integration of various technology outcomes in a synergistic impact that no longer best will increase productivity but additionally fosters inventiveness and resiliency in individuals.

By taking action to address the challenges that the next generation will face, we have discovered that the incorporated approach is very effective in tackling demanding situations, which can be related to operational resilience, aid optimisation, and sustainability. Smart grids that use IoT sensors, AI analytics, and blockchain generation can do many things that people want, such as making electricity distribution more environmentally friendly, cutting down on waste, and making things more reliable. The aggregate of the IoT and AI has the capability to enhance agricultural output, regulate sources, and expect environmental outcomes in the zone of agriculture. The use of these converging technologies provides a framework that is both scalable and adaptive, which is essential for meeting the ever-changing demands as international challenges become increasingly sophisticated. This framework is essential to meet the needs of diverse global communities. Developing a future that is smarter, extra efficient, and more sturdy may be carried out through the implementation of an ahead-questioning method that combines the mixing of technology consisting of AI, blockchain, the IoT, cloud computing, and virtual twins.

When technologies like AI, blockchain, the internet of things, cloud computing, and virtual twins work together, they offer a real way to deal with the tough problems of the future. This creation highlights the unique features of each generation and the progressive capability of combining these technologies. In the subsequent sections of this study, we learn more about practical packages, real-world global applications, and innovative techniques, highlighting how this technological synergy can offer sustainable, effective, and scalable solutions for future technological issues. By demonstrating how this technical framework can lead to a world that is smarter, more interconnected, and more resilient, capable of addressing international problems through innovation and complex technological solutions, the reason for this is to contribute to the growing body of expertise.

2. Transforming Tomorrow: How Emerging Technologies Tackle Complex Challenges

In the fast-changing environment of the 21st century, emerging technologies consisting of AI, blockchain, the IoT, cloud computing, and digital twins are playing transformative roles in addressing complicated global concerns (Wu et al. 2022). These technologies are paving the way for a more equitable and sustainable future. Not only are these the simplest technologies revolutionising sectors, but they are also offering new solutions to a number of the most severe problems that the arena is currently going through. These troubles consist of healthcare, schooling, urban improvement, environmental sustainability, and cybersecurity. In this document, we discuss the ways in which those technologies paint together to produce big solutions, which in turn propel forward progress and pave the way for a future that is both smarter and extra strong. By utilising machine learning, natural language processing, and data analytics, AI has emerged as a foundational technology capable of mimicking human intelligence. AI's ability to analyze vast quantities of data, predict future trends, and streamline processes plays a crucial role in addressing complex problems.

Algorithms powered by AI, help in early detection of diseases via the examination of medical imaging, forecast the consequences for sufferers, and optimise remedy regimens. For instance, synthetic intelligence-powered oncology technology has the potential to detect cancer at earlier stages, significantly enhancing survival rates. Data on climate is analysed using predictive models powered by synthetic intelligence if you want to forecast natural screw-ups, which allows for disaster preparedness and response techniques. In addition, AI helps optimise energy usage and supports the improvement of environmentally responsible agricultural practices.

Blockchain generation offers a decentralised, transparent, and secure way to record and control transactions, as well as statistics management. It plays surprisingly well in situations that call for the integrity and transparency of the records gathered. By facilitating transactions that are both secure and transparent, the blockchain era is revolutionizing the economic offerings industry through the use of cryptocurrencies, smart contracts, and decentralised finance platforms (DeFi). Blockchain facilitates this progress in financial inclusion and protection. Indeed, the use of blockchain technology enables the tracking of those goods. This feature helps cut down on fraud, prove that goods are real, and make delivery chains more efficient. Food safety, prescription drugs, and luxury items are some of the industries that have benefitted notably from its implementation.

The IoT connects physical gadgets to enable real-time data collection, tracking, and automation across various domains. The IoT includes the collection of facts from city surroundings, which enables wise traffic management, trash control, and smart grids that are efficient in terms of power use. The first-rate quality of life in metropolitan areas has progressed due to these technological improvements. With the assistance of IoT devices, the fitness indicators of sufferers can be monitored in real time, which enables faraway healthcare and preventive scientific interventions. The use of this generation enhances patient care, particularly in remote and underserved areas.

Cloud computing offers scalable and on-call computing resources, which enable establishments to control large datasets, facilitate distant collaborations, and provide support for virtual transformation programs. The cloud provides a robust foundation for digital offerings. Various programs can utilise this infrastructure, including enterprise resource planning (ERP) systems and new software-as-a-service (SaaS) models that enhance operational efficiency (Sundaram et al. 2023).

In the field of education, cloud-based systems enabled people to interact in far-flung learning during the COVID-19 pandemic, demonstrating the capability of cloud computing to provide training without regard to geographical constraints.

A virtual twin is the digital equivalent of a bodily system or method that permits real-time monitoring, modelling, and predictive maintenance. Digital twins are also known as virtual dual environments (Karpagam et al. 2025). Utilising virtual twins helps improve manufacturing methods because they allow planned machinery maintenance, which in turn lowers downtime and leads to better manufacturing performance. The use of virtual twins by urban planners lets them simulate different scenarios for urban improvement, find the best way to use resources, and make infrastructure more resistant to natural disasters. When these technologies are integrated, they create a robust environment capable of solving a wide range of problems. For instance, improving predictive analytics in the healthcare enterprise can be completed by combining AI with IoT devices and doing records analysis on the cloud. When it comes to supply chain management, utilising the blockchain era alongside the virtual twin era guarantees both transparency and performance.

3. A Synergistic Approach to Solving Real-World Problems

There has been a paradigm shift in the manner that tackles difficult, demanding situations that arise within the actual globe as a result of the convergence of superior technologies such as AI, blockchain, the IoT, cloud computing, and digital twins (Samaniego et al. 2023). When those technologies are coupled in a synergistic manner, they offer novel answers across quite a few regions, inclusive of healthcare, production, logistics, smart cities, and environmental sustainability. By combining their diverse capabilities, businesses and governments can enhance their performance and develop systems that are both adaptable and precise, enabling them to address challenges in ever-changing environments. I am currently living in an era characterised by the accumulation of large volumes of data, the rise of increasingly complex operational requirements, and the need for enhanced safety and transparency.

A synergistic technique blends the characteristics of each era into a cohesive framework that produces significant results. In this list, the strengths are AI's insights based on data, blockchain's safety and openness, the Internet of Things' ability to collect facts in real time, cloud computing's ability to grow, and digital twins' ability to simulate things. The reason for this look is to investigate the specific capabilities of those technologies, as well as their integration and realistic applications that illustrate their potential to correctly address problems that are encountered within the actual world. Table 6.1 illustrates the different technologies with their problems and a real-time example.

Table 6.1: Problems and real-time examples of different technologies.

Technology	Description	Problem	Real-time example
AI	AI plays a pivotal role in reading facts, predicting traits, and automating strategies. The applications of AI include natural language processing, computer vision, advanced analytics, and machine learning models that drive self-sustaining systems.	AI structures provide clever insights with the aid of reading huge datasets fast and as they should be. Machines getting to know algorithms can expect system screw-ups, optimise supply chains, and even provide useful resources in clinical diagnostics by means of reading medical photos or patient facts.	In healthcare, AI-pushed predictive models help in the early prognosis of illnesses, including Alzheimer's or cancer, through interpreting complicated imaging records.
Blockchain	Blockchain offers a decentralised, secure way to record transactions and control information. Its immutable ledger and clever agreement functionalities beautify transparency and trust.	The technology is beneficial for environments where data integrity and traceability are crucial. It can prevent fraud, streamline supply chain tactics, and stabilise information exchanges in finance and healthcare.	In supply chain management, blockchain is used to confirm the authenticity of goods, from source to customer, ensuring transparency and lowering counterfeiting.

Table 6.1 contd. ...

...Table 6.1 contd.

Technology	Description	Problem	Real-time example
IoT	IoT connects bodily gadgets to the net, enabling real-time information collection and interplay with the surroundings. Sensors and clever gadgets generate treasured records which can pressure intelligent decision-making.	Smart towns, agriculture, and business automation widely apply IoT. It helps display environmental conditions, control sources effectively, and beautify protection in production strategies.	In delivery chain management, blockchain is used to confirm the authenticity of goods, from source to client, ensuring transparency and lowering counterfeiting. In smart cities, IoT devices manipulate traffic waft, optimize electricity utilization Enhance public protection in building connected surveillance systems.
Cloud Computing	Cloud computing offers scalable storage and processing of electricity. It provides a bendy platform for deploying AI fashions, managing giant facts, and helping IoT networks.	Cloud computing allows organisations to quickly scale their operations and offers a cost-effective way to manage and analyse large amounts of data without heavily investing in infrastructure.	Cloud computing allows organisations to quickly scale their operations and offers a cost-effective way to manage and analyse large amounts of data without heavily investing in infrastructure.
Digital Twin	A digital twin is a virtual illustration of a physical system that mirrors its behaviour and overall performance in real time. It integrates records from IoT gadgets and predictive analytics to simulate actual-international scenarios.	It is especially effective for predictive maintenance, technique optimisation, and situation checking. By simulating distinctive conditions, companies can make informed selections without risking bodily property.	In production, digital twins of equipment permit predictive upkeep by way of studying sensor facts to forecast system disasters before they occur.

3.1 Integration of Technologies

Blending these technologies into a unified solution unlocks their true potential. For instance, AI algorithms can analyse records from IoT sensors to predict gadget disasters or optimise energy consumption (Roobini et al. 2025). Blockchain guarantees secure data transactions, and cloud services provide scalable data, storage, and processing capabilities. Digital twins and AI combine to enable predictive analytics in digital simulations, revealing trends for implementation in real-world operations.

3.2 Case Studies and Applications

Smart Cities: In smart city tasks, IoT sensors monitor site visitors, AI algorithms optimise traffic patterns, blockchain technology secures data transactions, and cloud computing manages the vast datasets generated by these systems (Rakshit et al. 2024). The digital twin era simulates city operations to devise infrastructure improvements successfully.

Healthcare: The integration of AI, IoT, and digital twins in healthcare leads to advanced telemedicine, personalised treatment plans, and control of green health centres. Blockchain ensures the safety of affected person records, and cloud computing enables the storage and processing of big volumes of medical information.

Industry 4.0: In commercial settings, IoT gadgets collect statistics from equipment, AI predicts preservation wishes, and digital twin technology simulates manufacturing strategies. Blockchain manages delivery chain records securely, and cloud offerings offer scalable computing strength.

4. Harnessing Cutting-Edge Technologies for a Smarter Future

In the 21st century, the speedy improvement of technology has prepared the way for groundbreaking breakthroughs, which can reshape not just industries but also societies and ordinary existence. Contemporary technologies such as AI, blockchain, IoT, cloud computing, and virtual twins are at the forefront of this transformation. These technologies not only offer the best novel answers to the problems which are currently confronted, but additionally, they open up new doors for the development of a future that is extra intelligent, more efficient, and more environmentally pleasant. The aim of this paper is to offer a complete examination of the approaches wherein modern-day technologies are being utilised to tackle hard issues and make people's lives better globally.

AI focuses on improving computer systems that can execute activities historically requiring human intelligence. Visual perception, speech recognition, decision-making, and language translation are some of the processes that fall under this category. A variety of industries, including healthcare and finance, are being revolutionised by synthetic intelligence, which includes device mastering, deep learning, natural language processing, and robotics. AI-driven diagnostic tools, predictive analytics for effective person care, and personalised remedy packages. Fraud detection, automatic trading, and chatbots for purchaser care are in the works. Platforms for adaptive getting to know, automatic grading, and administrative automation are all examples of this. Innovation includes autonomous vehicles, visitor control systems, and predictive maintenance. Farming with precision, crop monitoring using drones, and wise irrigation systems are all examples of this. Influence on the Society AI improves decision-making, increases productivity, and creates personalised reports. Nevertheless, there are still extensive obstacles to triumph over, which include moral issues, employment displacement, and privacy.

The blockchain is an allotted and decentralised ledger generation that guarantees the immutability, transparency, and security of facts without alterations. Initially designed for use with cryptocurrencies, blockchain technology is currently being utilised in plenty of fields, including healthcare, finance, public services, and delivery chain control (Pincheira et al. 2022). It is important to improve the traceability of commodities and reduce fraud. Store medical and clinical information confidentially. Simultaneously, we should enhance the safeguarding of virtual assets and simplify the completion of international transactions. We should enable stable and transparent voting structures.

IoT refers to the network of physical gadgets that can collect and share data. The IoT era is the driving force behind smart cities, clever homes, wearable gadgets, and industrial automation (Narayanan et al. 2023). The term "cloud computing" refers to the practice of utilising the net to gain access to various computing assets, such as servers, storage, and programs. Because of its scalability, versatility, and price performance, it is a critical generation for all sorts of businesses, which include enterprises and governments. People often refer to a "virtual twin" as a digital replica of a physical component, system, or method. To forecast performance, optimise operations, and control systems over their entire lifecycle, they make use of records that are accumulated in real time and complex simulations.

5. The Convergence of Innovation: Building Solutions for the Next Generation

In the twenty-first century, the rapid development of the era prepared the way for groundbreaking breakthroughs, which are reshaping not only industries but also societies and everyday life. AI, blockchain, the IoT, cloud computing, and digital twins are some of the new technologies that can lead this change. These technologies no longer offer novel answers to the issues that are currently being confronted; in addition, they open up new doorways for the development of a future that is more intelligent, greener, and more environmentally friendly. The purpose of this paper is to present a comprehensive exam of the approaches with which present-day technologies are being utilised to address tough troubles and enhance residents' quality of life all around the world. AI aims to build systems that can do things that humans have traditionally done with intelligence. Visual perception, speech recognition, decision-making, and language translation are some of the methods that fall under this category. An extensive range of industries, including healthcare and finance, are being revolutionised by using synthetic intelligence technology, which includes machine learning, deep learning, natural language processing, and robotics. Influence on the Society AI improves selection, increases productivity, and contributes to the creation of extra customised reviews. Nevertheless, there are still good-sized obstacles to triumph over, including ethical issues, employment displacement, and information privacy.

The blockchain is a decentralised and disbursed ledger generation that guarantees the immutability, transparency, and safety of records with no alterations. Initially designed to be used with cryptocurrencies, blockchain technology is currently being employed in quite a few fields, including healthcare, finance, public offerings, and delivery chain management. The term IoT refers to the community of physical objects that are linked to the internet and feature the functionality to acquire and present facts that they've gathered (Selvaraj et al. 2021). The IoT era is the driving force behind smart cities, smart homes, wearable devices, and industrial automation. The term "cloud computing" describes the practice of using the internet to access Various computing assets, such as servers, storage, and software packages, are utilised as needed. Because of its scalability, versatility, and cost performance, it is a crucial era for all sorts of companies, as well as organizations and governments. A "virtual twin" is a reproduction of a bodily component, system, or process from time to time. To forecast performance, optimise operations, and manage structures over their entire lifecycle, it uses real-time data and complicated simulations. Integrating these technologies to solve complex problems is key to unlocking their full potential. Take, as an instance, wise gadgets that can study the options in their customers and automate chores. This technology provides a reliable method for storing and disseminating data across multiple remote networks. Predictive analytics aims to improve gadget performance and maintenance.

Despite the reality that they have several potentials, these technologies are faced with a number of boundaries, which include statistics privacy, cybersecurity, moral concerns, and regulatory constraints. Continuous studies and improvement, alongside rigorous governance frameworks, are required that can allow us to reap the finest possible advantages even as we simultaneously minimise ability hazards. To construct a shrewder future, modern-day technologies, which include synthetic intelligence, blockchain, the net of factors, cloud computing, and virtual twins, are important. When combined, their power gives novel methods for resolving complex troubles in several fields, including training, healthcare, and finance, among others. Through the proper utilisation of these technologies, societies have the capacity to gain elevated stages of productiveness, sustainability, and first-rate quality of life, paving the way for a future characterised by intelligence, efficiency, and innovation.

6. From Data to Decisions: Revolutionizing Industries with Advanced Technologies

Across the globe, industries have seen good-sized ameliorations as a result of the rapid growth of the era. A wide variety of present-day technologies, such as AI, blockchain, the IoT, cloud computing, and digital twins, are at the forefront of this revolution. This technology provides innovative solutions to complex challenges, streamlines techniques, and enhances decision-making. Incorporating this technology into industrial operations is not just a simple development; instead, it represents a revolution that will impact a wide range of industries, including healthcare, manufacturing, finance, and smart cities. This in-intensity exam digs into the manner by means of which current technology reworks uncooked statistics into insights that can be placed into motion, so facilitating informed choice-making and boosting growth and efficiency throughout lots of industries.

AI performs a significant portion of the fact-driven decision-making process. With the help of AI algorithms, organisations are capable of quickly making educated decisions by analysing large volumes of information, discovering traits, and producing predictive models (Subbiah et al. 2024). Machine learning (ML) is a subfield of synthetic intelligence that allows PC systems to learn from beyond statistics and enhance over time. In fields inclusive of healthcare and finance, machine mastering fashions provide assistance in threat assessment, the detection of fraudulent activity, and the prognosis of patients. The technologies of natural language processing (NLP) allow machines to comprehend and use human language. An employer can use NLP for sentiment evaluation, automated client assistance, and content improvement. The use of PC vision in production facilitates excellent manipulation by analysing photographs of gadgets to pick out flaws within the production process. Additionally, autonomous cars use computer vision for navigation and protection. Predictive analytics powered by AI can assist in forecasting market tendencies, customer behaviour, and the need for device preservation, thereby minimising downtime and increasing productivity.

The blockchain is an allotted ledger generation that guarantees the immutability, security, and transparency of information. This technology is particularly beneficial in fields where maintaining the integrity of facts is crucial. Supply chains can benefit from the accelerated openness of blockchain generation, which permits real-time tracking of products from the point of manufacture to the point of transport. Blockchain generation allows stable peer-to-peer transactions, which, in turn, minimises the amount of money and time required for banking transactions. Additionally, blockchain technology serves as the muse for cryptocurrencies consisting of Bitcoin and Ethereum. The facts about sufferers are maintained in a secure way on a blockchain, which ensures their integrity and makes it easy for healthcare practitioners to share them with one another. By automating and implementing the phrases of contracts, smart contracts that are based totally on the blockchain era lessen the want for middlemen and minimise the probability of mistakes.

The IoT is the system of linking numerous devices and sensors to the net, which permits them to accumulate and proportion information. IoT technology provides real-time data and insights, leading to the enhancement of more automated and efficient systems. As a result of the Internet of Things' potential to govern urban infrastructure, such as site visitors's lights, rubbish management, and electricity grids, the quality of lifestyles in metropolitan regions is advanced. Real-time monitoring of patient health is made viable by using wearable IoT gadgets, which additionally enable proactive healthcare management and far-flung monitoring. IoT sensors in manufacturing monitor the overall performance of equipment and predict when a repair is necessary, thereby reducing operational costs. Applications of the IoT in agriculture monitor soil situations, optimise water use, and increase crop output via the utilisation of fact-driven insights.

Businesses can successfully use techniques and examine significant datasets because cloud computing offers storage and computing capacity that is both scalable and flexible. It is essential for large-scale statistical analytics to have access to cloud infrastructure, as these systems provide significant storage solutions and advanced records management capabilities. By enabling access to records and packages from any location worldwide, cloud offerings foster a collaborative environment. Computing in the cloud provides the backup and restoration of statistics, thereby defending organisations from a lack of facts that may occur as a result of unanticipated conditions (Subbiah et al. 2023). By supplying pre-constructed synthetic intelligence and device learning models, cloud platforms make it viable for corporations to employ advanced analytics with low funding in infrastructure.

It is possible to mimic and analyse the behaviour of a physical system in real time by using a digital twin, which is a virtual counterpart to the system of interest. Increasingly, this technology is being used in industries that demand precise and predictive maintenance. Through the use of digital twins, manufacturing lines can be simulated, processes may be optimised, and equipment faults can be predicted before they occur. The use of digital twins of patients is a component of personalised medicine that assists in the prediction of treatment results and the customisation of remedies. By monitoring building systems like heating, ventilation, and air conditioning (HVAC) and lighting, digital twins improve energy efficiency and cut down on operational costs. In the field of logistics, digital twins are helpful for optimising routes and managing fleets, which ultimately results in increased delivery efficiency.

It is only when these technologies are integrated into a coherent system that their full potential may be realise. As an illustration, IoT devices gather data, AI analyses it, blockchain technology guarantees the safety of data, cloud computing provides processing capacity, and digital twins offer simulations (Subbiah et al. 2024). Companies are able to transform raw data into strategic decisions with the assistance of this seamless integration. In the manufacturing industry, predictive maintenance IoT sensors are used to monitor machinery. With the use of AI algorithms, this data is analysed to determine when maintenance is required. Simulating machinery's behaviour is the responsibility of the digital twin; cloud computing makes it easier to handle data. Utilising blockchain technology, maintenance tasks are recorded in a secure manner. Because of this, the plant experiences decreased rates of downtime and increased levels of output.

Despite its benefits, there are still issues to address, like the need for qualified people, integration complexity, and data protection. Enhancing interoperability, building solid data governance frameworks, and pushing innovations that make these technologies more accessible and effective are the three things that will determine the future. A new era of data-driven decision-making has begun with the convergence of technologies such as AI, blockchain, the internet of things, cloud computing, and digital twins. By translating raw data into insights that can be put into action, these technologies give businesses the ability to improve their overall performance, which in turn leads to increased efficiency and decreased costs. The sectors that are willing to embrace these advanced solutions will be in a better position to negotiate the challenges of the future and achieve sustainable growth as technology continues to advance.

7. Bridging the Digital and Physical Worlds for Predictive and Proactive Solutions

The obstacles between the digital and physical worlds have become increasingly blurry due to the speedy growth of the era, which has made it possible for novel solutions to be developed that can assume and proactively address tough challenges. AI, blockchain, IoT, cloud computing, and digital twins are among the technologies that have the potential to significantly contribute to this revolution (Narayanan et al. 2021). By establishing a connection between these two domain names, groups and industries can expand their structures to not only respond to real-world conditions in real time, but also anticipate future issues and optimise their outcomes. The goal of this in-depth discussion is to look into how different technologies can work together to create solutions that are both proactive and predictive, which will improve efficiency, safety, and sustainability in many business areas (Narayanan et al. 2021).

Knowing the differences between the digital and physical worlds: The physical world encompasses tangible items, settings, and character interactions. The virtual globe, alternatively, is characterised by the presence of facts, analytics, computational intelligence, and virtual representations. Creating a virtual equal to physical entities is vital with a view to bridging the two worlds. This will enable statistics-pushed insights and automatic decision-making that may be carried out.

The capability to examine good-sized datasets, comprehend trends, and make sensible predictions is one of the methods with which AI improves predictive and proactive abilities. The capacity of synthetic intelligence structures to analyse complicated events and supply insights that may be put into motion is made possible by techniques like system getting-to-know, deep learning, and natural language processing. The blockchain era provides a decentralised and stable ledger that serves to enhance the transparency and integrity of statistics (Banerjee et al. 2023). Supply chain management and smart contracts are two examples of scenarios that require traceability, and they are particularly beneficial in these contexts. Through the use of blockchain generation, virtual representations of actual property are guaranteed to be trustworthy and tamper-proof.

By utilising sensors and communication networks, the IoT connects physical devices to the digital world. In order to display situations, expect malfunctions, and provoke automated responses, those sensors accumulate facts in actual time from the bodily surroundings. This information can then be analyzed. Cloud computing structures provide scalable computing strength and storage, which makes it less difficult to method the large volumes of statistics that are created via IoT gadgets. Cloud offerings support a wide variety of sophisticated analytics, facts management, and the deployment of AI-pushed models for predictive analytics. A virtual version of a physical issue, process, or device is every so often referred to as a digital twin. The IoT devices constantly update the digital twin with information from its physical counterpart. The use of digital twins lets in predictive protection, method optimisation, and state of affairs testing because they imitate the conditions that exist in the actual world.

Smart factories use digital twins and IoT sensors to display the overall system performance in Industry 4.0. AI algorithms are able to anticipate renovation requirements, which enables them to reduce downtime and boom productivity. Blockchain generation ensures both transparency within the supply chain and product traceability. AI-driven predictive models examine affected person data as a beneficial way to perceive capability health problems, even as IoT sensors reveal real-time health metrics. It is feasible to simulate remedy reactions with the use of digital twins of human organs, which could assist in customised medicine and healthcare; this approach is proactive. Smart metropolis efforts employ the IoT, AI, and cloud computing to improve the control of traffic, power consumption, and trash. Predictive analytics enable the implementation of preventative measures to enhance living conditions in urban areas. In industries including aviation and manufacturing, the mixture of virtual twins, the net of factors, and AI allows predicting gadget breakdowns before they occur. Strategies for preventative preservation no longer simply increase safety; they also reduce operational expenses. Natural disasters, along with floods and wildfires, can be anticipated with the aid of predictive fashions that make use of information from IoT sensors. By implementing By implementing proactive measures such as early warnings and aid allocation, we can mitigate risks and save lives.

The use of standardisation and integration frameworks is necessary to combine data from an extensive array of physical and virtual resources. A vital issue of successful implementation is ensuring that specific structures are compatible with one another. Connecting bodily systems to digital networks can lead to potential cybersecurity vulnerabilities. For the purpose of protecting records integrity and privacy, blockchain technology and superior encryption techniques are crucial. The use of IoT devices, cloud data processing, and virtual twin simulations all require reliable infrastructure solutions that can grow as needed. Fifth-generation wireless (5G) networks, part computing, and augmented reality (AR) are some of the new technologies that are coming together to make it easier to build systems that are both proactive and predictive. The application of virtual bodily integration may be improved into new sectors because of these generation breakthroughs.

To enable predictive and proactive answers throughout industries, an innovative approach that bridges the digital and bodily worlds is something that needs to be considered. Utilising AI, blockchain, IoT, cloud computing, and digital twins lets companies improve their operational efficiency, boost their stage of safety, and convey higher results. To foster innovation, it is crucial to overcome current challenges and welcome future advancements. Such efforts will lead to the creation of smarter and more resilient systems that can anticipate and resolve issues, enabling them to thrive in the next generation.

8. Redefining Possibilities: Next-Gen Solutions for Complex Global Issues

In this age of speedy technological increase, the confluence of technologies inclusive of AI, blockchain, IoT, cloud computing, and digital twins is establishing new standards for the decision of hard worldwide worries. These improvements of the future era aren't simply altering how society addresses problems related to sustainability, healthcare, clever towns, delivery chains, and different regions; however, they are additionally remodelling industries. These technologies offer unprecedented opportunities to enhance performance, transparency, and decision-making strategies.

The competencies of AI have improved to consist of not just the most effective data evaluation and predictive analytics but also self-sustaining structures and natural language processing. AI has become a significant technological advancement. The versatility of AI allows it to deal with a variety of worldwide concerns, including the following: through the use of AI algorithms, contamination outbreaks may be predicted, diagnostics may be assisted, and remedy techniques can be personalised. Analysis of scientific imaging, reputation of styles in genetic information, and optimisation of aid allocation in healthcare centres are all made viable with the help of systems getting to know models. By analysing climatic patterns and using models driven by artificial intelligence, it is possible to predict natural disasters. The optimisation of electricity usage, the assistance of precision agriculture, and the enhancement of waste control structures are all ways in which AI contributes to the sustainability of its surroundings. Platforms that are powered by synthetic intelligence supply personalised learning stories, automate administrative duties, and provide information-driven insights into scholar performance, all of which contribute to raising the accessibility and efficiency of education.

The blockchain era affords a decentralised, obvious, and immutable report-preserving device that is particularly nice in domains that require excessive degrees of trust and protection. Through the usage of its decentralised ledger era, blockchain makes it possible to conduct steady transactions, minimises the hazard of fraud, and cuts the fees of operations. The use of cryptocurrencies and clever contracts makes the financial system handier and greener respectively. Using the blockchain era, every step of the delivery chain can be documented, which permits merchandise to be traced from beginning to stop. In industries consisting of pharmaceuticals, in which authenticity and compliance are of the utmost significance, that is of the maximum importance. The use of blockchain technology allows balloting systems to be both secure and obvious, thereby supporting democratic approaches and ensuring the integrity of elections.

The IoT connects bodily objects to the digital world, enabling real-time tracking, data accumulation, and automation. The IoT devices control site visitors, reveal pollutants, and optimize energy utilization, all of which make contributions to higher city settings that are more sustainable and greener. Medical gadgets that are enabled by using the IoT display the fitness of patients in real time, offer assistance for remote remedy, and streamline hospital operations through the usage of automation and data analytics. Sensors connected to the IoT provide insights into soil conditions, crop health, and environmental factors, enabling farmers to increase productivity simultaneously minimizing aid waste.

Cloud computing can help other technologies, like AI and the net of factors, because it provides flexible and on-demand resources for storing data, processing it, and putting it to use. Applications for stylish analytics and system mastering may be made less difficult with the assistance of cloud computing, which offers sturdy solutions for coping with the large quantities of records created by using IoT gadgets. Utilisation of cloud offerings via corporations allows for the enhancement of operational flexibility, the discounting of charges, and the acceleration of digital transformation activities. Tools that might be hosted on the cloud make it feasible to engage in far-flung learning, create opportunities for international collaboration, and give users access to instructional assets from any vicinity in the globe.

Virtual variations of physical entities, known as virtual twins, are able to facilitate simulation, tracking, and predictive preservation because of their talents. Through the simulation of commercial strategies, digital twins assist in identifying bottlenecks, optimising workflows, and predicting mechanical breakdowns. The use of digital twins in smart towns permits the modelling of infrastructure, the simulation of eventualities, and the improvement of selection-making techniques for urban improvement tasks. Personalised treatment simulations and stepped-forward predictive diagnostics are each made feasible in the healthcare industry by using digital twins of patients.

By putting together technologies like AI, blockchain, a net of factors, cloud computing, and digital twins, it is possible to make complete solutions with plenty of benefits. Combine frameworks, IoT, synthetic intelligence, and blockchain generation to enhance patient monitoring, information protection, and predictive healthcare offerings. Collectively, these technologies offer support for projects that include smart grids, efficient resource management, and the introduction of infrastructure that is ecologically conscious. Increased productiveness and reduced costs are the outcomes of combining the electricity of the IoT, digital twins, and analytics driven through AI in manufacturing operations.

The key to the enduring fulfilment of subsequent-era technology in the future is its ability to regulate and adapt in response to new difficulties. There can be extra resilient and adaptable solutions for international worries because of innovations along with quantum computing, augmented reality, and 5G. These innovations will, in addition, improve the talents of AI, the Internet of Things, blockchain, cloud computing, and digital twins. When it comes to redefining opportunities through the usage of next-generation technology, it isn't enough to simply accept man or woman advances; instead, it's far more essential to strategically combine these innovations so that they will cope with complex worldwide concerns. Companies and governments can make systems that are smarter, last longer, and are stronger by combining the strengths of AI, blockchain, the internet of things, cloud computing, and virtual dual technology. Continued research, investment, and collaboration are crucial to maximising the impact of these technologies and ensuring their contribution to a more equitable and globally distributed world.

9. Conclusion

A progressive method for addressing the complex issues of a good way to confront destiny is supplied through the confluence of technologies such as AI, blockchain, the IoT, cloud computing, and digital twins. Using the strengths of each technology—the analytical potential of synthetic intelligence, the safety of blockchain, the relationship between the internet of factors, the scalability of the cloud, and the digital modelling skills of virtual twins—this incorporated environment is able to build complete solutions that go beyond the bounds of conventional structures. Through the enhancement of selection, optimisation of efficiency, and the delivery of predictive insights, those technologies collectively enable enormous breakthroughs across the industries, including healthcare, manufacturing, smart cities, and supply chain management. The results of this observation show that combining these technologies not only fixes current inefficiencies but also predicts and adapts to future needs, making society smarter, stronger, and more connected. Selling sustainability, improving useful resource management, and making operations more resilient are all parts of this method that make it possible to face global problems head-on. As sectors continue to embody this technological synergy, the capability for innovation and growth is infinite. This approach paves the way for a future marked by intelligence, adaptability, and enduring electricity.

References

Banerjee, S., Das, D., Chatterjee, P. and Ghosh, U. (2023, May). Blockchain-enabled digital twin technology for next-generation transportation systems. pp. 224–229. In 2023 IEEE 26th International Symposium on Real-Time Distributed Computing (ISORC). IEEE.

Groshev, M., Guimaraes, C., Martín-Pérez, J. and de la Oliva, A. (2021). Toward intelligent cyber-physical systems: Digital twin meets artificial intelligence. IEEE Communications Magazine 59(8): 14–20.

Hemamalini, V., Armosh, F. and Tyagi, A. K. (2024). Digital twin-based smart healthcare services for the next generation society. pp. 247–277. In Future of AI in Medical Imaging. IGI Global.

Karpagam, V., Kavitha, P. and Deepa, P. (2025). Open-source platforms for implementing blockchain-based digital twins. Blockchain-Based Digital Twins: Research Trends and Challenges 91.

Kumari, S., Tiwari, S., Naithani, K. and Subbiah, P. (2024). Internet of things for smart agriculture: a survey. pp. 100–121. In AI Applications for Business, Medical, and Agricultural Sustainability. IGI Global.

Narayanan, L. K., Muralidharan, R. R. A., Sampathkumar, R., Gururajan, I. and Subbiah, P. (2021, December). Blockchain based fictitious detection in social media. pp. 226–236. In International Conference on Soft Computing and Pattern Recognition. Cham: Springer International Publishing.

Narayanan, L. K., Subbaiyah, P., Gururajan, I., Sampathkumar, R. and Muralidharan, R. R. A. (2021). IoT-Fog integrated voice of the plant based smart irrigation and power management system for smart farming.

Narayanan, L. K., Subbiah, P., Mary, S. S. A., Muralidharan, R. R. A., Gururajan, I., Sampathkumar, R. et al. (2023). A smart and efficient IoT-AI and ML-based multifunctional system for multilevel power distribution management. pp. 49–63. In Smart Energy and Electric Power Systems. Elsevier.

Pincheira, M., Vecchio, M. and Antonelli, F. (2022, July). SmartTwin: A blockchain-based software framework for digital twins using IoT. pp. 67–77. In International Congress on Blockchain and Applications. Cham: Springer International Publishing.

Rakshit, P., Saha, N., Nandi, S. and Gupta, P. (2024). Artificial intelligence in digital twins for sustainable future. pp. 19–44. In Transforming Industry using Digital Twin Technology. Cham: Springer Nature Switzerland.

Roobini, S., Kavitha, M. S. and Karthik, S. (2025). Leveraging IoT and AI technologies for blockchain-based digital twins. pp. 183–222. In Blockchain-Based Digital Twins. Apple Academic Press.

Samaniego, M. and Deters, R. (2023, July). Digital twins and blockchain for IoT management. pp. 64–74. In Proceedings of the 5th ACM International Symposium on Blockchain and Secure Critical Infrastructure.

Selvaraj, P., alias Muralidharan, R. R., Subbaiah, P. and Narayanan, L. K. (2021). Smart infant and child monitoring system using integrated IoT and GIS. Design Engineering 15225–15235.

Subbiah, P., Nagappan, K., Bellam, K. and Thandapani, P. (2023). Implementation and deployment of privacy preservation and secure data storage techniques in cloud computing. pp. 256–268. In Privacy Preservation and Secured Data Storage in Cloud Computing. IGI Global.

Subbiah, P., Tyagi, A. K. and Mazumdar, B. D. (2024). The future of manufacturing and artificial intelligence industry 6.0 and beyond. pp. 347–362. In Industry 4.0, Smart Manufacturing, and Industrial Engineering. CRC Press.

Sundaram, S. M., Murgod, T. R. and Sowmya, M. (2023). Digital twins enabling technologies, including artificial intelligence, sensors, cloud, and edge computing. pp. 88–101. In Digital Twins and Healthcare: Trends, Techniques, and Challenges. IGI Global.

Wu, J., Wang, X., Dang, Y. and Lv, Z. (2022). Digital twins and artificial intelligence in transportation infrastructure: Classification, application, and future research directions. Computers and Electrical Engineering 101: 107983.

Chapter 7
Bitcoin-based Blockchain

Shabnam Kumari,[1] *Amit Kumar Tyagi*[2] and *Shrikant Tiwari*[3,*]

1. Introduction - Digital Keys

The Bitcoin blockchain refers to the foundational technology that powers the Bitcoin network. Introduced in 2009 by an anonymous creator or group known as Satoshi Nakamoto, it is a specific implementation of blockchain technology designed to support Bitcoin. This decentralized and distributed ledger records all Bitcoin transactions and activities, ensuring transparency, security, and immutability through cryptographic techniques. Each transaction is stored in a "block," and these blocks are linked chronologically to form a "chain," creating a tamper-proof record of all activities within the network (Kumari et al. 2023). The Bitcoin-based blockchain functions as a public ledger, providing a clear and transparent record of every transaction that has occurred on the network. Below are some key characteristics of this innovative technology:

"Proof of Work (PoW)" Consensus: The Bitcoin blockchain operates on a consensus mechanism known as Proof of Work. In this system, participants in the network, referred to as miners, compete to solve intricate mathematical puzzles. Successfully solving these puzzles allows them to validate transactions and add new blocks to the blockchain. This process demands substantial computational power and consumes a considerable amount of energy.

Decentralization: The Bitcoin blockchain functions in a decentralized way, which means it lacks a central authority to oversee or manage the network. Instead, transactions are confirmed and validated by numerous nodes that are spread throughout the network.

Immutable and transparent: Once a transaction is confirmed and incorporated into a block on the Bitcoin blockchain, it becomes nearly impossible to alter it. However, the blockchain's transparent design enables anyone to access and verify transactions, fostering transparency and accountability.

Cryptocurrency transactions: The main function of the Bitcoin blockchain is to enable secure transactions of the Bitcoin cryptocurrency. This technology allows users to send and receive bitcoins directly, eliminating the need for intermediaries such as banks or payment processors.

[1] Department of Computer Science, Faculty of Science and Humanities, SRM Institute of Science and Technology, Kattankulathur, Chennai, 603203, Tamilnadu, India.

[2] National Forensic Sciences University, Gandhinagar, Gujarat, India.

[3] Department of Computer Science & Engineering, School of Computing Science and Engineering (SCSE), Galgotias University, Greater Noida, Uttar Pradesh-203201.

Emails: shabnam.kt25@gmail.com, amitkrtyagi025@gmail.com

* Corresponding author: shrikanttiwari15@gmail.com

DOI: 10.1201/9781003620822-7

Security: The Bitcoin blockchain secures transactions using advanced cryptographic methods. Every transaction is signed cryptographically, while the entire blockchain is safeguarded by cryptographic hashes. This combination makes it extremely challenging to modify or tamper with previous transactions.

Limited smart contract functionality: While the Bitcoin blockchain supports some basic smart contract functionality, its primary focus is on facilitating peer-to-peer transactions rather than complex programmable logic.

The term "Bitcoin-based blockchain" specifically refers to the blockchain implementation and infrastructure used by the Bitcoin network. There are other blockchain platforms and networks, such as Ethereum, that offer more advanced smart contract capabilities and a broader range of applications beyond cryptocurrencies.

2. Digital Keys for Blockchain

Digital keys play a vital role in blockchain technology, providing security, integrity, and authenticity to the decentralized network. In the context of blockchain, digital keys are used for identity verification, transaction validation, and securing digital assets (Deshmukh et al. 2023, Aydar et al. 2019, Tyagi et al. 2024h). Here is an introduction to digital keys in the context of blockchain (refer Figure 7.1):

Public/Private key pairs: Blockchain networks like Bitcoin and Ethereum rely on public key cryptography to ensure secure transactions. Each participant in the network possesses a unique pair of digital keys: a public key and a private key. These keys are generated using asymmetric cryptographic algorithms. The public key, which is openly shared, is mathematically derived from the private key, which remains confidential and is used for signing transactions. This cryptographic system ensures authenticity, integrity, and security within the blockchain network

Address generation: In blockchain networks, users typically generate a digital address derived from their public key. This address acts as their identity or account on the blockchain. Transactions are sent to and from these addresses, and ownership of digital assets is tied to these addresses.

Secure transactions: Blockchain employs digital keys for secure transactions. Users sign transactions with their private key, creating a digital signature. This signature is verified using their public key, ensuring only the private key holder can authorize the transaction.

Ownership and control: Digital keys are used to prove ownership and control over digital assets on the blockchain. The private key acts as the proof of ownership, and only the person with the corresponding private key can transfer or access the assets associated with a specific address.

Wallets and key management: Digital keys are usually managed via digital wallets, which securely store private keys. These wallets offer an intuitive interface for blockchain interaction, including address management and transaction signing.

Security issues: Safeguarding the private key is crucial in blockchain systems. The loss, theft, or compromise of a private key can lead to irreversible loss of access to digital assets. To enhance the security of your private keys, consider the following best practices.

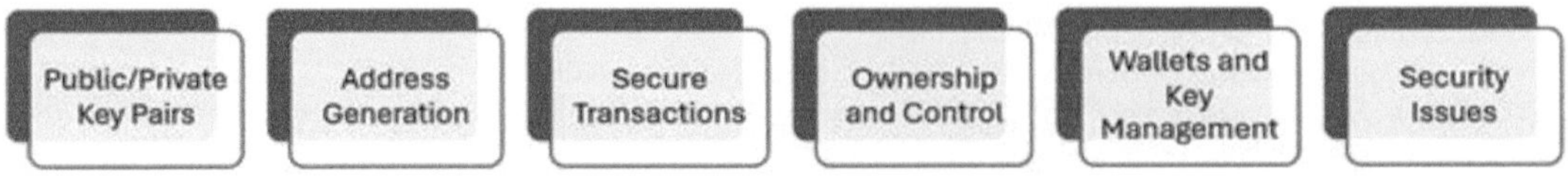

Figure 7.1. Digital keys for blockchain.

Use Hardware wallets: These provide a secure offline environment for storing your keys.

Implement strong password: Create complex passwords that are difficult to guess.

Enable multi-factor authentication: This adds an extra layer of security by requiring additional verification steps.

By following these guidelines, you can significantly reduce the risk of losing access to your valuable digital assets.

Note that Digital keys are the cornerstone of security and trust in blockchain networks. They provide the cryptographic foundation for verifying identities, authorizing transactions, and securing digital assets. Understanding and managing digital keys properly is essential for safely participating in blockchain ecosystems.

3. Addresses and Wallets

In the context of blockchain technology, addresses and wallets play important roles in managing digital assets and facilitating transactions (Tyagi et al. 2024f, Tyagi et al. 2024g). Here is an overview of addresses and wallets in blockchain (refer Table 7.1):

Addresses

Unique identifiers: Addresses in blockchain are unique identifiers associated with participants or entities on the network. They are derived from public keys and are used to send, receive, and track transactions.

Public key hashes: Blockchain addresses are typically created by applying cryptographic hash functions to the public keys of users. This process ensures that the actual public key remains hidden, providing a layer of privacy.

Format and length: Addresses vary in format and length depending on the blockchain network. For example, Bitcoin addresses are alphanumeric strings that start with a certain prefix, while Ethereum addresses are hexadecimal strings.

Read-only and control: Addresses serve as the main means for receiving funds or assets on the blockchain. They are usually linked to a user's public key, while the ownership and control of the related digital assets are governed by the corresponding private key.

Wallets

Digital asset storage: Wallets, which can be software or hardware-based, store and manage digital assets on the blockchain by securely holding the private keys needed to access and authorize transactions.

Private key management: Wallets simplify private key management by generating and storing keys, importing existing ones, and enabling the creation of new addresses, all within a secure framework.

User interface: Wallets often provide a user-friendly interface for interacting with blockchain networks. They allow users to view their balances, send and receive transactions, and monitor the transaction history associated with their addresses.

Types of wallets: Digital asset wallets offer a spectrum of security and accessibility, ranging from software wallets (desktop, mobile, and web-based) to highly secure hardware wallets (physical devices) and basic paper wallets (physical printouts).

Table 7.1. Addresses and Wallets.

Addresses	Type	Explanation
	Read-Only and Control	Unique identifiers associated with participants or entities on the network
	Format and Length	Applying cryptographic hash functions to the public keys of users
	Public Key Hashes	Alphanumeric strings that start with a certain prefix, while Ethereum addresses are hexadecimal strings
	Unique Identifiers	User's public key, and ownership and control of the associated digital assets
Wallets		
	Digital Asset Storage	Store, manage, and interact with digital assets on the blockchain
	Private Key Management	Generate and store private keys, allow users to import existing keys, and facilitate the creation of new addresses
	User Interface	Balances, send and receive transactions, and monitor the transaction history
	Security Issues	Reputable wallet providers, enable encryption and strong passwords, and consider additional security measures like multi-factor authentication

Security issues: Wallet security is of utmost importance, as the compromise of private keys can result in the loss of digital assets. Users are advised to choose reputable wallet providers, enable encryption and strong passwords, and consider additional security measures like multi-factor authentication.

Addresses and wallets are fundamental to blockchain ecosystems, enabling secure storage, sending, and receiving of digital assets. Through cryptography, they provide asset control, security, transparency, and maintain the integrity of the blockchain network.

4. Bitcoin Transactions

Bitcoin transactions are digital transfers of the cryptocurrency Bitcoin between two or more parties. These transactions are recorded on a public ledger called the blockchain (Tyagi et al. 2024a, Fleder et al. 2015, Vranken and Harald 2017). Here is a concise overview of Bitcoin transactions:

Participants: Bitcoin transactions involve a sender (payer) and a recipient (payee).

Transaction data: Every Bitcoin transaction provides a complete record encompassing the sender's Bitcoin address, the recipient's Bitcoin address, the exact amount of Bitcoin being sent, and further data points, notably transaction fees.

Digital signature: The sender utilizes their private key to generate a digital signature, which authenticates the transaction and confirms its legitimacy.

Verification: The transaction is broadcast to the Bitcoin network, and miners validate it by verifying the sender's funds and confirming adherence to network rules.

Inclusion in blocks: Validated transactions are organized into blocks. Miners compete to solve complex mathematical problems in order to add new blocks to the blockchain. Successfully mining a block confirms the transactions it contains.

Confirmation: The security of a blockchain transaction increases as more blocks are added after it. The term "confirmations" refers to the count of these subsequent blocks, reflecting the growing difficulty of reversing the transaction.

Finality: Bitcoin transactions are considered final once they have a sufficient number of confirmations, typically six confirmations are considered secure.

Fee: Transactions may require a small fee paid to miners to incentivize them to include the transaction in a block promptly. Higher fees can prioritize transaction processing.

In summary, Bitcoin transactions involve the transfer of digital currency between participants, which is verified and recorded on the blockchain through a process called mining.

5. Bitcoin Mining: Task of Miners, Mining Rewards

Bitcoin mining is the mechanism by which new Bitcoins are generated, and transactions are confirmed and recorded on the blockchain (Schrijvers et al. 2017, Khairuddin et al. 2019). Here is a concise explanation of the tasks performed by miners in the Bitcoin mining process:

Securing the network: Bitcoin miners are essential to the security and integrity of the network. They validate transactions to confirm they are legitimate and that users have enough funds.

Solving mathematical puzzles: Miners compete to find solutions to computationally intensive proof-of-work algorithms. The mathematical complexity of these algorithms is deliberately engineered to demand considerable processing power and time.

Block formation: When a miner solves the puzzle, they build a new block with verified transactions and a link to the previous block. This linked structure forms the blockchain.

Block verification: Other miners in the network validate the newly created block to confirm its legitimacy. They verify that the transactions are accurate, the solution to the puzzle is correct, and that the block complies with the consensus rules of the network.

Consensus mechanism: To avoid the problem of double-spending and ensure everyone agrees on the blockchain's current state, miners use a consensus mechanism. Bitcoin employs a "proof-of-work" system, where the blockchain with the most computational effort behind it – the longest, valid chain – is accepted as the definitive version.

Block reward: Miners receive newly minted Bitcoins as a reward for their efforts. This reward, referred to as the block reward, is a fixed amount of Bitcoin that is halved roughly every four years. In addition to the block reward, miners also earn transaction fees from the transactions that are included in the block.

Network participation: Bitcoin miners play a vital role in the network's decentralization. By running specialized mining hardware and connecting to the Bitcoin network, they contribute processing power. Miners often join mining pools to collaborate, pooling their resources to boost their odds of successfully mining a block and then splitting the resulting Bitcoin rewards.

Bitcoin miners secure the network by validating transactions, solving mathematical puzzles, creating new blocks, and participating in the consensus mechanism. They are rewarded with newly minted Bitcoins and transaction fees for their efforts.

Mining Rewards

Bitcoin mining rewards refer to the incentives received by miners for their participation and contribution to the Bitcoin network (Tyagi et al. 2024d, Tyagi et al. 2024e). Here is a concise overview of mining rewards in Bitcoin:

Block reward: When a miner successfully mines a new block and appends it to the blockchain, they receive a reward known as the block reward. This reward incentivizes miners and consists of newly created Bitcoin. At Bitcoin's launch in 2009, the block reward was set at 50 Bitcoins per block. However, to control supply and mimic the scarcity of precious commodities, the Bitcoin protocol reduces this reward by half approximately every four years in an event called the halving.

The most recent halving occurred in May 2020, lowering the block reward to 6.25 Bitcoins. The next halving has occurred in April 2024, reducing the reward further to 3.125 Bitcoins. This process will continue until all 21 million Bitcoins are mined, projected to occur around 2140.

Transaction fees: Miners earn not only the block reward but also transaction fees from the transactions included in the blocks they mine. When users initiate Bitcoin transactions, they can choose to include a transaction fee, which encourages miners to prioritize their transactions. As miners have the ability to select which transactions to add to the blocks that they have mined, they tend to prioritize those with higher fees.

Reward distribution: Miners who successfully mine a block and receive the block reward and transaction fees have the freedom to distribute these rewards as they see fit. Some miners may keep the rewards for themselves, while others may distribute them among their mining pool members or allocate a portion to cover operational expenses.

The block reward for Bitcoin miners diminishes over time due to scheduled halvings, leading to an increased reliance on transaction fees as a primary source of revenue. This shift is designed to align the incentives within the network as the supply of Bitcoin becomes more constrained. In essence, Bitcoin mining rewards are comprised of two main components: the block reward, which is halved roughly every four years, and the transaction fees linked to the transactions included in the mined blocks. Together, these rewards incentivize miners to dedicate their computational resources to securing the network and validating transactions.

6. Proof of Work

Proof of Work (PoW) is a consensus mechanism employed in blockchain networks, such as Bitcoin, to ensure agreement on the blockchain's state and to validate transactions (Tiwari et al. 2024, Tyagi et al. 2024c). This process involves participants, referred to as miners, who must solve intricate mathematical puzzles to prove their computational effort (refer Figure 7.2):

Puzzle solving: Miners engage in a competition to solve a complex mathematical puzzle. In Bitcoin, this involves finding a hash value with specific characteristics, like a certain number of leading zeros.

Computational effort: Solving the puzzle demands considerable computational power and effort. Miners utilize their computers or specialized mining equipment to conduct countless calculations in search of a solution that meets specific criteria. This process is both resource-intensive and time-consuming.

Block validation: Upon successfully solving the computationally intensive puzzle, a miner constructs a new block composed of validated transactions. The puzzle solution serves as cryptographic proof-of-work, confirming the computational resources committed to securing the block.

Consensus establishment: Miners verify new blocks by independently solving the same puzzle. When a majority agree on the solution, consensus is achieved, and the block is added to the blockchain.

Difficulty adjustment: The puzzle difficulty dynamically adjusts to maintain a consistent block creation rate. It increases as mining power grows, ensuring blocks are generated at the expected average rate.

Security and trust: The Proof of Work mechanism secures the blockchain by making attacks computationally difficult and costly. As the network grows, the resources required to gain control and reverse transactions increase dramatically, thus protecting the integrity of the blockchain.

Longest chain rule: Bitcoin's Proof-of-Work consensus mechanism follows the longest chain rule: the chain with the most computational work, proven by the most solved puzzles, is considered valid. Miners are incentivized to mine on this chain to earn rewards and participate in the consensus.

While Proof of Work has demonstrated its effectiveness as a secure and reliable consensus mechanism, it is associated with significant energy consumption due to its high computational

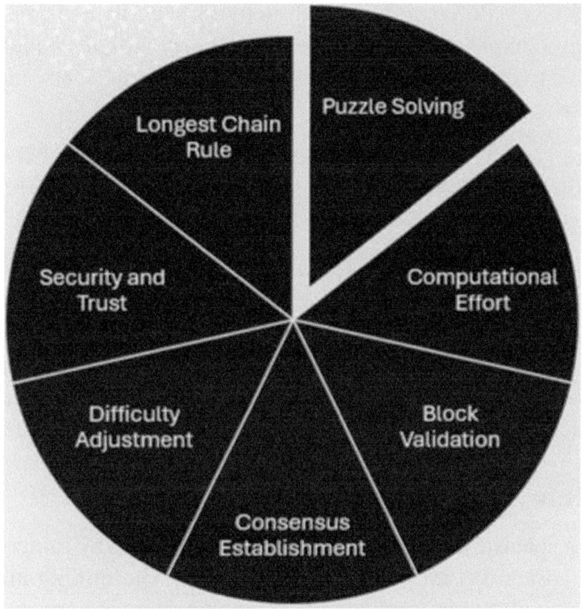

Figure 7.2. Explanation of Proof of Work.

demands. To tackle the challenges of energy usage while ensuring network security and consensus, alternative mechanisms like Proof of Stake (PoS) have been introduced.

7. Proof of Stake

Proof of Stake (PoS) offers an alternative to Proof of Work (PoW) as a consensus mechanism for blockchain networks. Instead of relying on computational power, PoS selects the next block validator based on the amount of cryptocurrency they hold, often referred to as their "stake" (Nair and Tyagi 2023). The following provides a brief overview of Proof of Stake:

Validators: In a Proof of Stake (PoS) system, the participants are known as validators. These validators possess a specific quantity of the blockchain network's native cryptocurrency, which is commonly referred to as their stake.

Block creation: Validators alternate in their roles of creating new blocks and validating transactions. The likelihood of a validator being chosen is generally proportional to the amount of their stake. In essence, those with a larger stake have a greater probability of being selected to produce the next block.

Block validation: When a validator is chosen to create a block, they include transactions and validate them. Instead of solving computational puzzles, validators are required to show ownership of their stake, known as "proving their stake." This is typically done by locking a certain amount of their cryptocurrency as collateral.

Consensus establishment: The block proposed by the selected validator is then checked by the remaining validators on the network. Consensus is achieved when a majority of these validators agree that the block is valid, at which point it is added to the blockchain.

Reward distribution: In Proof-of-Stake systems, validators are compensated for their successful block creation and validation efforts through a combination of transaction fees and newly minted cryptocurrency. However, the exact reward distribution mechanism is implementation-specific.

Security and energy efficiency: PoS aims to provide network security and consensus while consuming significantly less energy compared to PoW. The assumption is that validators with a larger stake have a stronger incentive to maintain the network's integrity because any malicious behavior could lead to a loss of their stake.

Stake slashing: PoS mechanisms often incorporate penalties or "slashing" for validators who behave maliciously or violate the rules of the network. Slashing can result in a portion of a validator's stake being confiscated as a punishment.

As a result, Proof of Stake (PoS) has emerged as a more energy-efficient alternative to Proof of Work (PoW), attracting significant attention. PoS facilitates scalability and has been embraced by various blockchain networks. Nevertheless, it does come with certain challenges, particularly the risk of centralization, where validators with larger stakes wield greater influence in the consensus process. To mitigate these concerns, several variations of PoS, including delegated PoS and liquid PoS, have been introduced to enhance decentralization and governance mechanisms.

8. Mining Algorithm

A mining algorithm is a specific set of rules and calculations used by miners to solve mathematical puzzles and validate transactions in a blockchain network. The mining algorithm determines the computational work required to mine new blocks and secure the network (Hariharan et al. 2023, Liu et al. 2005). Here is a concise explanation of mining algorithms:

Proof of Work (PoW) algorithm: Bitcoin miners use the SHA-256 algorithm to compete in a high-stakes race. They must discover a "nonce" that, when hashed along with the block's data, yields a result that satisfies a network-defined difficulty. This difficulty is expressed as a target number of leading zeroes. The process involves intensive computation and many hashing attempts until a miner finds the correct nonce.

Other PoW algorithms: Different blockchain networks may use alternative mining algorithms to SHA-256. For example, Litecoin uses Scrypt, Ethereum previously used Ethash but has transitioned to Ethereum 2.0's Proof of Stake mechanism. These algorithms introduce variations in the computational puzzles that miners need to solve.

Proof of Stake (PoS) algorithm: In blockchain networks utilizing Proof of Stake (PoS), mining algorithms are replaced by a consensus mechanism that selects validators based on the amount of cryptocurrency they hold and commit, or "stake," to the network. Unlike Proof of Work (PoW), which relies on computational power, PoS reduces energy consumption and increases efficiency by using staking as the basis for transaction validation and block creation. Validators with larger stakes have a higher probability of being chosen to validate transactions and earn rewards. Examples of PoS-based systems include Ouroboros, used in Cardano, and Tendermint, employed in Cosmos.

Hybrid algorithms: Some blockchain networks implement hybrid consensus mechanisms that combine elements of both PoW and PoS. For instance, the Byzantine Fault Tolerance (BFT) consensus algorithm used in networks like NEO combines delegated PoS with a limited PoW component.

Other Consensus algorithms: Beyond Proof of Work (PoW) and Proof of Stake (PoS), blockchain networks utilize a variety of other consensus algorithms. These include Delegated Proof of Stake (DPoS), which relies on elected delegates to validate transactions and secure the network; Practical Byzantine Fault Tolerance (PBFT), designed to address the Byzantine Generals' Problem by achieving consensus even in the presence of malicious nodes; and Directed Acyclic Graph (DAG) algorithms, such as Tangle, which is used by IOTA to enable scalable and feeless transactions.

Mining algorithms offer various advantages and disadvantages concerning security, energy efficiency, decentralization, and scalability, making the ideal choice dependent on the specific needs and objectives of a blockchain network and its community.

9. Mining Systems in Bitcoin

Bitcoin mining is the mechanism by which new transactions are verified and added to the Bitcoin blockchain, and through which new bitcoins are created. These mining systems generally consist of the following components:

Miners: These are dedicated computer systems or hardware devices specifically engineered for Bitcoin mining. They execute intricate mathematical calculations to solve cryptographic puzzles and validate transactions efficiently.

Mining software: Miners utilize software applications to link their hardware with the Bitcoin network, enabling them to engage in the mining process. This software facilitates communication with the network, allowing miners to receive new transactions and submit their completed work for verification.

Blockchain: The Bitcoin blockchain serves as a decentralized ledger that records every confirmed Bitcoin transaction. Miners are crucial to this system, as they help maintain and secure the blockchain by adding new blocks of transactions.

Proof of Work (PoW): In Bitcoin mining, the Proof-of-Work (PoW) consensus algorithm requires miners to expend computational effort to solve cryptographic puzzles. The first miner to discover a solution with sufficient proof of work earns the right to append a new block to the chain and claim the associated block reward in bitcoins.

Mining pools: Many miners opt to join mining pools to pool their computing power, thereby boosting their chances of earning rewards. Within a mining pool, participants collectively contribute their resources. When a member successfully solves a puzzle, the reward is shared among all pool members in proportion to their individual contributions.

Energy consumption: Bitcoin miners face the challenge of high electricity consumption due to the computationally intensive processes involved in mining. Consequently, optimizing systems for energy efficiency is a key focus for miners seeking to minimize costs and environmental impact.

Note that the Bitcoin mining ecosystem is continually evolving, and new technologies and systems may emerge over time to enhance efficiency and sustainability.

10. Mining Pools

Mining pools in Bitcoin are collaborative groups formed by individual miners who join forces to increase their chances of successfully mining blocks and earning rewards. Rather than mining alone, participants pool their computational resources to collectively solve complex mathematical puzzles required for block validation (Tyagi 2023). Here's how mining pools in Bitcoin work:

Pool formation: A mining pool is created by an administrator who sets up the necessary infrastructure, including the pool's software and server. The administrator may charge a small fee or percentage of the mining rewards as a commission for operating the pool.

Pool membership: Individual miners who wish to join the pool connect their mining hardware or software to the pool's server. They provide their unique mining addresses, which are used to distribute rewards among participants.

Work distribution: The pool's server distributes the computational work to the connected miners. Instead of working on the entire blockchain, miners in the pool focus on solving a portion of the cryptographic puzzle assigned by the server.

Sharing solutions: Each miner independently solves their piece of the puzzle. A successful solution is sent to the pool's server, which validates it and then adds the block to the Bitcoin blockchain.

Reward distribution: After a successful block mine, rewards are allocated to pool members to compensate them for their work. This allocation is often proportional to their contribution, measured either by computing power (pay-per-share) or by the number of valid shares submitted.

Pool fees: Mining pools usually charge a fee, typically a percentage of the mining rewards, to cover operational costs. This fee varies among different pools and can affect the profitability of mining in that pool.

Mining pools offer several benefits to individual miners, including a more consistent income stream, reduced variance in rewards, and a higher probability of earning rewards compared to solo mining. Additionally, mining pools help in decentralizing the network by allowing smaller miners to participate effectively.

It is crucial to do your homework and select a mining pool that's known for its reliability and solid reputation. The pool you choose can really affect how much you earn, so pay attention to things like its performance, fees, and security measures.

11. Bitcoin Network

The Bitcoin network is a decentralized peer-to-peer network that facilitates the operation and maintenance of the Bitcoin cryptocurrency. It serves as the underlying infrastructure for the transfer of bitcoins and the validation of transactions (A. K. V. et al. 2022, Tyagi 2022, Decker et al. 2013, Lischke et al. 2016). Here are some key aspects of the Bitcoin network (refer Figure 7.3):

Blockchain: The Bitcoin network operates on a decentralized ledger known as the blockchain. This public and unchangeable record maintains a chronological list of all confirmed transactions. The blockchain provides transparency, enhances security, and protects against tampering or censorship.

Nodes: The Bitcoin network is composed of individual computers and devices called nodes. Two key types of nodes are full nodes and lightweight nodes. Full nodes hold a complete, up-to-date copy of the blockchain. Lightweight nodes conserve resources by relying on full nodes to verify transactions.

Peer-to-Peer communication: The Bitcoin network's peer-to-peer protocol enables direct node communication, fostering a decentralized architecture that promotes resilience and data integrity.

Consensus mechanism: Bitcoin employs a consensus mechanism known as Proof of Work (PoW) to validate transactions and maintain agreement on the blockchain's state. In this process, miners compete to solve intricate mathematical puzzles. The first miner to successfully find a valid solution is granted the chance to add a new block to the blockchain and is rewarded with bitcoins for their efforts.

Transaction validation: When a Bitcoin transaction is initiated, it is shared across the network and transmitted to various nodes. These nodes play a crucial role in validating transactions by ensuring they comply with the Bitcoin protocol's rules, which include verifying valid digital signatures and confirming that there are sufficient funds. After a transaction is validated, it is incorporated into a block and subsequently added to the blockchain.

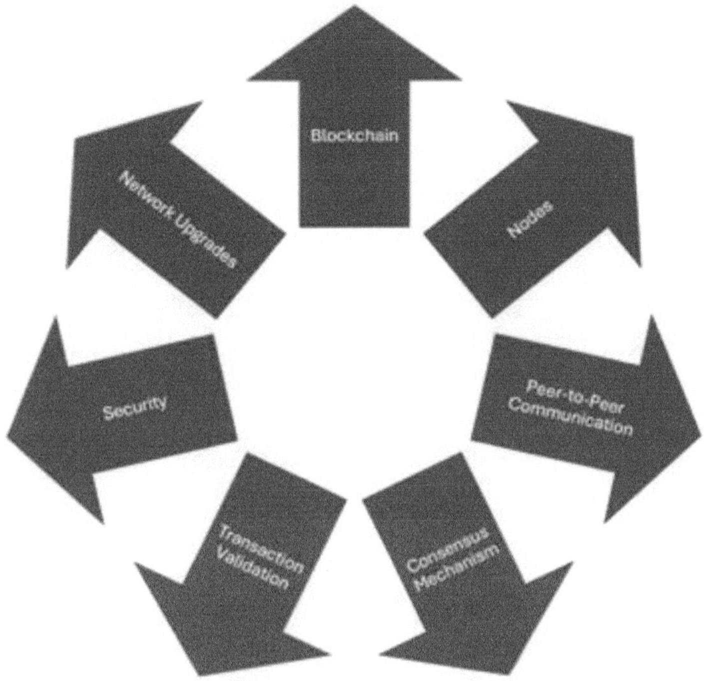

Figure 7.3. Bitcoin network.

Security: The decentralized nature of the Bitcoin network makes it more secure against attacks and censorship attempts. The consensus mechanism and cryptographic techniques employed by Bitcoin, such as public-private key cryptography and hashing algorithms, contribute to the network's security.

Network upgrades: The Bitcoin network can undergo upgrades or changes through a process called a "fork." Forks can be soft forks, which introduce backward-compatible changes, or hard forks, which create a new branch of the blockchain with diverging rules. Consensus among network participants is required for such upgrades to be implemented successfully.

The Bitcoin network continues to evolve and face challenges related to scalability, energy consumption, and transaction fees. Ongoing research and development aim to address these issues and improve the overall performance and usability of the network.

12. Benefits with Bitcoin in this Smart Era

Bitcoin Offers Several Benefits in this Smart Era

Decentralization: Bitcoin leverages a decentralized blockchain network, providing transparency, security, and censorship resistance, independent of any central authority.

Financial inclusion: Bitcoin offers a gateway to financial inclusion for the unbanked and underbanked. By requiring only a smartphone and internet access, it allows anyone to join the Bitcoin network and conduct transactions worldwide, bypassing the need for traditional banking.

Global accessibility: Bitcoin provides a globally accessible digital currency, allowing for instant transfers across borders. This eliminates traditional barriers in international transactions, offering lower fees and faster settlement times.

Security and privacy: Bitcoin's cryptographic security makes transactions exceptionally resistant to fraud and hacking. User financial privacy is also protected, as Bitcoin addresses are not linked to personal data unless intentionally shared.

Financial sovereignty: Bitcoin empowers individuals with complete control over their finances, eliminating the need for intermediaries or third parties. Users retain exclusive ownership of their private keys, allowing them to independently manage and safeguard their wealth.

Lower transaction fees: Bitcoin transactions generally incur lower fees than traditional financial systems, particularly when it comes to cross-border transactions. This offers significant benefits for remittances and international trade, helping to reduce costs and enhance efficiency.

Potential for growth: Bitcoin has exhibited significant price appreciation over the years, attracting investors seeking potential returns. While it is a volatile asset, its limited supply and increasing adoption could lead to future value appreciation.

Note that while Bitcoin offers these benefits, it also has its limitations, such as scalability challenges, regulatory uncertainties, and environmental issues related to its energy consumption.

13. Trust Component with Bitcoin

Bitcoin introduces a trust component through various mechanisms:

Trust in the Blockchain technology: Bitcoin functions on a decentralized and unchangeable blockchain. Transactions are documented in a transparent and tamper-proof way, ensuring confidence in the integrity of the transaction history.

Trust in cryptography: Bitcoin leverages cryptographic algorithms to secure transactions, ensuring that only the legitimate owner, possessing the correct private key, can access and control their Bitcoin funds. This is made possible through public-key cryptography.

Trust in consensus mechanism: Bitcoin utilizes a consensus mechanism known as Proof of Work (PoW) to validate and confirm transactions. In this process, miners compete to solve intricate mathematical puzzles, which ensures that transactions are verified by a majority of the network. This competition helps establish trust in the validity of each transaction.

Trust in network participants: Bitcoin's decentralized structure enables users to engage in direct transactions with one another, eliminating the need for intermediaries. Trust within the network is built through consensus rules that encourage honest behavior while imposing penalties for malicious actions.

Trust in scarcity and supply rules: Bitcoin's design includes a limited supply of 21 million coins, with new coins created through a predetermined issuance schedule. This predictable and transparent supply mechanism helps build trust by preventing excessive inflation or arbitrary manipulation of the money supply.

It is important to recognize that trust in Bitcoin is not unconditional and can be affected by various factors, including personal security measures, the regulatory landscape, and potential technological weaknesses. Users should remain vigilant and implement essential precautions to safeguard their Bitcoin assets.

14. Open Issues with Bitcoin Cryptocurrency

While Bitcoin has gained significant popularity and adoption, there are several open issues associated with this cryptocurrency:

Scalability: Bitcoin's scalability issues stem from its limited block size and the restricted number of transactions it can process per second. These constraints often result in higher transaction fees and longer confirmation times during periods of heavy network usage. To address these challenges, initiatives like the Lightning Network and other layer-two protocols are being developed to enhance transaction capacity and efficiency.

Volatility: Bitcoin's notorious price volatility presents significant hurdles to its widespread adoption as a reliable medium of exchange and stable store of value. This instability stems from a variety of sources, including rampant market speculation, evolving regulatory landscapes, and shifts in broader macroeconomic conditions.

Regulatory uncertainty: Bitcoin operates within a regulatory gray zone in numerous jurisdictions, as governments and regulatory authorities continue to develop frameworks to address cryptocurrencies. This lack of clear regulation introduces uncertainty, posing challenges for businesses, investors, and the broader adoption of Bitcoin.

Energy consumption: Bitcoin mining, which relies on the Proof of Work consensus mechanism, demands significant computational power and energy, sparking concerns about its environmental footprint. The reliance on fossil fuels to power mining operations is a key issue. To address this, innovations in energy-efficient consensus mechanisms, like Proof of Stake, are being actively pursued.

Privacy issues: While Bitcoin offers pseudonymity, it is not entirely anonymous. All transactions and associated addresses are recorded on the public blockchain, potentially allowing for transactional analysis and identification of users. This raises privacy issues for individuals who prefer financial transactions to remain private.

Bitcoin still faces usability challenges for mainstream adoption. The process of acquiring, storing, and transacting with Bitcoin can be complex for non-technical users. Improvements in user experience, user-friendly wallets, and educational initiatives are essential for wider acceptance and usage.

Note that ongoing development and innovation within the Bitcoin ecosystem aim to address these open issues and improve the overall functionality, usability, and adoption of the cryptocurrency.

15. Critical Challenges with Bitcoin Cryptocurrency

Bitcoin cryptocurrency faces several critical challenges (refer Figure 7.4):

Scalability: Bitcoin faces a notable challenge regarding scalability, primarily due to its restricted block size and the time it takes to process transactions. As user numbers and transaction volumes rise, the network can experience congestion, leading to increased fees and longer confirmation times.

Energy consumption: A major challenge associated with Bitcoin mining, particularly its reliance on the Proof of Work consensus mechanism, is its high energy consumption. The immense computational power required has sparked considerable debate regarding the environmental consequences of the growing Bitcoin network.

Centralization of mining power: Over the years, Bitcoin mining has increasingly become dominated by a small number of large mining pools and specialized hardware manufacturers. This concentration of mining power brings forth concerns regarding the possibilities of collusion, censorship, and a diminished level of decentralization within the network.

Figure 7.4. Critical Challenges with Bitcoin cryptocurrency.

Regulatory and legal challenges: The absence of uniform regulations for Bitcoin across different jurisdictions highlights the need for greater clarity and consistency to encourage participation from businesses, individuals, and institutional investors.

Security and privacy risks: While Bitcoin offers cryptographic security, individual users are responsible for securing their wallets and private keys. Cybersecurity risks, such as hacking, phishing, and malware attacks, pose threats to user funds. Additionally, while Bitcoin transactions are pseudonymous, they can be traced on the public blockchain, potentially compromising user privacy.

Adoption and user experience: Bitcoin's complex user experience, technical barriers, and unfamiliarity can hinder mainstream adoption. User-friendly interfaces, improved infrastructure, and education are essential for making Bitcoin more accessible and user-friendly for a wider audience.

Volatility and price manipulation: Bitcoin's price volatility makes it susceptible to speculative trading and price manipulation. Sharp price fluctuations can create risks for investors, merchants, and businesses accepting Bitcoin as a payment method.

Interoperability with traditional financial systems: Bitcoin's integration with traditional financial systems, such as banking and regulatory frameworks, remains a challenge. Bridging the gap between cryptocurrencies and legacy systems requires regulatory clarity, technological advancements, and collaborations with traditional financial institutions.

To tackle these significant challenges, it is essential to engage in continuous research, development, and collaboration among all stakeholders, including developers, miners, regulators, and the wider cryptocurrency community.

16. Research Statements for Bitcoin Cryptocurrency

Here are some research statements related to Bitcoin cryptocurrency:

"*Examining the scalability solutions for Bitcoin*: A comparative analysis of layer-two protocols and block size optimization techniques."

"*Understanding the environmental impact of Bitcoin mining*: An assessment of energy consumption and the exploration of sustainable mining alternatives."

"*Regulatory frameworks for Bitcoin*: A comparative study of global regulatory approaches and their impact on adoption and innovation."

"*Enhancing privacy in Bitcoin transactions*: Investigating privacy-enhancing technologies and their effectiveness in preserving user anonymity."

"*Evaluating the economic implications of Bitcoin*: An analysis of price volatility, market manipulation, and the role of institutional investors."

"*Exploring the decentralization of Bitcoin mining*: Analyzing the concentration of mining power and potential strategies to promote a more distributed and resilient network."

"*User experience and adoption of Bitcoin*: Investigating barriers to entry, usability challenges, and strategies for improving the mainstream acceptance of cryptocurrencies."

"*Bitcoin and financial inclusion*: Assessing the impact of Bitcoin on unbanked populations and the potential for empowering individuals in underserved regions."

"*Blockchain interoperability*: Investigating the integration of Bitcoin with other blockchain networks and traditional financial systems for seamless transactions."

"*Security vulnerabilities in Bitcoin*: Identifying and mitigating risks related to wallet security, transaction privacy, and network attacks."

These research statements initiate thorough explorations into Bitcoin cryptocurrency, encompassing its technical, economic, environmental, regulatory, and societal dimensions.

17. Future Work Towards Cryptocurrency

Future work towards cryptocurrency encompasses several areas (refer Figure 7.5):

Scalability solutions: To ensure the widespread adoption of cryptocurrencies such as Bitcoin, further research and development into scalability solutions are crucial. Promising avenues include the exploration and implementation of layer-two protocols, sharding techniques, and off-chain transactions, all aimed at boosting transaction speeds and lowering costs.

Privacy enhancements: Research and development should prioritize advancements in privacy-enhancing technologies to improve transactional privacy for cryptocurrencies. Specific areas of focus include strengthening zero-knowledge proofs, ring signatures, and confidential transactions to achieve enhanced privacy guarantees while preserving network security.

Energy efficiency: We can significantly reduce the environmental impact of cryptocurrency mining by developing and implementing more energy-efficient consensus mechanisms like Proof of Stake (PoS). Future research must investigate the feasibility, security, and scalability of these alternatives to build a greener, more sustainable cryptocurrency ecosystem.

Interoperability: To ensure smooth and efficient transactions in the crypto space, interoperability between different cryptocurrencies and blockchain networks is crucial. Future development should prioritize standardization initiatives, cross-chain communication protocols, and interoperability frameworks. These advancements will foster greater interoperability and pave the way for interoperable decentralized applications (DApps).

Regulation and compliance: The increasing mainstream adoption of cryptocurrencies highlights the urgent need for clear regulatory frameworks. Research is essential for developing effective and balanced regulations that not only encourage innovation and protect consumers and financial stability but also address illicit activities and ensure compliance.

User experience and adoption: Improving the user experience is vital for widespread cryptocurrency adoption. Research can focus on developing user-friendly wallets, intuitive interfaces, and educational resources to make cryptocurrencies more accessible to non-technical users. User-centric

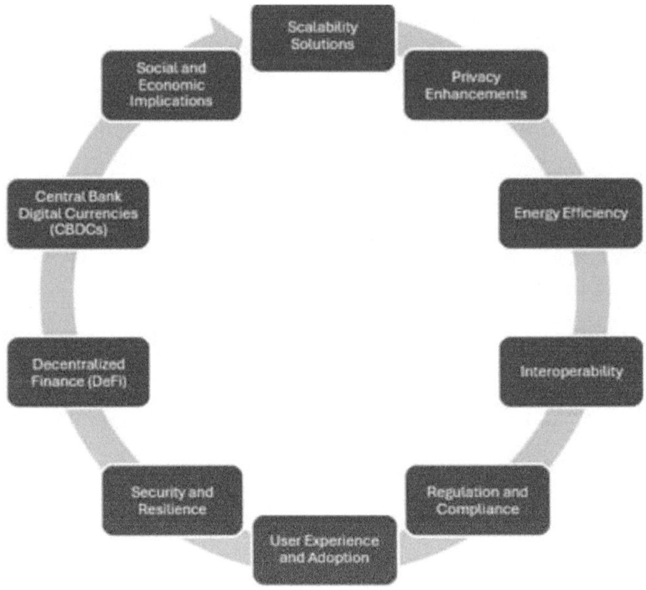

Figure 7.5. Future work towards cryptocurrency.

studies can also explore the barriers to adoption and develop strategies to increase user trust and confidence in cryptocurrencies.

Security and resilience: As the cryptocurrency ecosystem evolves, it is difficult to address security vulnerabilities and ensure the resilience of the underlying technologies. Research can focus on identifying and mitigating risks such as wallet security, smart contract vulnerabilities, and network attacks to enhance the overall security posture of cryptocurrencies.

Decentralized finance (DeFi): DeFi has gained significant traction, considering blockchain and smart contract technologies to offer various financial services. Future work can explore the scalability, interoperability, and security of DeFi protocols, as well as the integration of traditional financial systems with decentralized platforms.

Central bank digital currencies (CBDCs): Several central banks are exploring the concept of CBDCs. Research can focus on the design, implementation, and potential impacts of CBDCs, including their interaction with existing cryptocurrencies, monetary policy implications, and financial inclusion benefits.

Social and economic implications: Research can investigate the broader social and economic implications of cryptocurrencies, such as their impact on financial inclusion, wealth distribution, economic stability, and global remittances. Understanding the potential benefits and risks of cryptocurrencies can inform policymakers and stakeholders in shaping a sustainable and inclusive future for digital currencies.

Future work towards cryptocurrency should encompass a multidisciplinary approach, involving researchers, technologists, policymakers, and industry stakeholders to address the various challenges and opportunities in this rapidly evolving field.

18. Conclusion

In conclusion, Bitcoin cryptocurrency has emerged as a transformative technology with both opportunities and challenges. It offers benefits such as decentralization, financial inclusion, global accessibility, security, and potential for growth. However, it also faces open issues and critical challenges related to scalability, energy consumption, regulatory uncertainty, security risks, and user experience. Even with these challenges, researchers and developers continue to work hard to solve problems and make Bitcoin more functional, easier to use, and widely accepted. Scalability solutions, privacy enhancements, energy-efficient consensus mechanisms, regulatory frameworks, and user-centric approaches are being explored to overcome these obstacles. As the cryptocurrency ecosystem evolves, future work towards Bitcoin and other cryptocurrencies should encompass scalability solutions, privacy enhancements, energy efficiency, interoperability, regulation and compliance, user experience, security, decentralized finance (DeFi), central bank digital currencies (CBDCs), and the broader social and economic implications. Note that it is essential for stakeholders, including researchers, technologists, policymakers, and industry participants, to collaborate and contribute to the advancement of cryptocurrencies in order to harness their potential benefits while addressing the challenges. By doing so, cryptocurrencies like Bitcoin can pave the way for a more decentralized, inclusive and efficient financial system in the future.

References

A. K. V, Tyagi, A. K. and Kumar, S. P. (2022). Blockchain technology for securing internet of vehicle: issues and challenges. pp. 1–6, International Conference on Computer Communication and Informatics (ICCCI). doi: 10.1109/ICCCI54379.2022.9740856.

Amit Kumar Tyagi. (2023). Decentralized everything: Practical use of blockchain technology in future applications. pp. 19–38. Rajiv Pandey, Sam Goundar and Shahnaz Fatima (eds.). Distributed Computing to Blockchain, Academic Press. ISBN 9780323961462, https://doi.org/10.1016/B978-0-323-96146-2.00010-3.

Aydar, Mehmet, Salih Cemil Cetin, Serkan Ayvaz and Betul Aygun. (2019). Private key encryption and recovery in blockchain. arXiv preprint arXiv:1907.04156.

Damilare Peter, Je Sen Teh, Norziana Jamil and Moatsum Alawida. (2021). Blockchain consensus: An overview of alternative protocols. Symmetry 13(8): 1363.

Decker, Christian and Roger Wattenhofer. (2013). Information propagation in the bitcoin network. pp. 1–10. In IEEE P2P 2013 Proceedings, IEEE.

Deshmukh Atharva, Mishra Anand Kumar, Balamurugan, G. and Tyagi Amit. (2023). Blockchain-empowered decentralized applications: current trends and challenges. In the book: Privacy Preservation of Genomic and Medical Data. https://doi.org/10.1002/9781394213726.ch19.

Fleder, Michael, Michael S. Kester and Sudeep Pillai. (2015). Bitcoin transaction graph analysis. arXiv preprint arXiv:1502.01657.

Hariharan, R., Tyagi, A. K. and Soni Gulshan. (2023). A survey on blockchain-internet of things-based solutions. In the book: Privacy Preservation and Secured Data Storage in Cloud Computing, IGI global. DOI: 10.4018/979-8-3693-0593-5.ch005.

Khairuddin, Irni Eliana and Corina Sas. (2019). An exploration of Bitcoin mining practices: Miners' trust challenges and motivations. pp. 1–13. In Proceedings of the 2019 CHI Conference on Human Factors in Computing Systems.

Kumari Shabnam and Muthulakshmi, P. (2023). Artificial intelligence blockchain enabled technology for internet of things: research statements, open issues, and possible applications in the near future. In the book: Privacy Preservation of Genomic and Medical Data. https://doi.org/10.1002/9781394213726.ch18.

Lischke Matthias and Benjamin Fabian. (2016). Analyzing the bitcoin network: The first four years. Future Internet 8(1)7.

Liu, Ying, Wei-keng Liao and Alok Choudhary. (2005). A fast high utility itemsets mining algorithm. pp. 90–99. In Proceedings of the 1st International Workshop on Utility-based Data Mining.

Nair Meghna Manoj and Tyagi, A. K. (2023). Blockchain technology for next-generation society: current trends and future opportunities for smart era. In book title: Blockchain Technology for Secure Social Media Computing. DOI: 10.1049/PBSE019E_ch11.

Schrijvers, Okke, Joseph Bonneau, Dan Boneh and Tim Roughgarden. (2017). Incentive compatibility of bitcoin mining pool reward functions. pp. 477–498. In Financial Cryptography and Data Security: 20th International Conference, FC 2016, Christ Church, Barbados, February 22–26, 2016, Revised Selected Papers 20. Springer Berlin Heidelberg, 2017.

Tiwari, S., Reddy, R. R. and Tyagi, A. K. (2024). Position of blockchain: Internet of Things-Based Education 4.0 in Industry 5.0 – A discussion of issues and challenges. pp. 292–311. *In*: Pandey, R., Srivastava, N. and Chatterjee, P. (eds.). Architecture and Technological Advancements of Education 4.0. IGI Global Scientific Publishing. https://doi.org/10.4018/978-1-6684-9285-7.ch013.

Tyagi, A. K. (2022). SecVT: Securing the vehicles of tomorrow using blockchain technology. *In*: Sk, A. A., Turki, T., Ghosh, T. K., Joardar, S. and Barman, S. (eds.). Artificial Intelligence. ISAI 2022. Communications in Computer and Information Science, vol 1695. Springer, Cham. https://doi.org/10.1007/978-3-031-22485-0_11Oyinloye.

Tyagi, A. K., Kumari Shabnam, Chidambaram, N. and Sharma Avinash. (2024a). Engineering applications of blockchain in this smart era. In the book: Enhancing Medical Imaging with Emerging Technologies. DOI: 10.4018/979-8-3693-5261-8.ch011.

Tyagi, A. K., Kumari, S., Rajasekaran, M. and Tiwari, S. (2024c). Blockchain for safety of internet of vehicles in smart transportation. Tyagi, A. K. (ed.). In Digital Twin and Blockchain for Smart Cities. https://doi.org/10.1002/9781394303564.ch11.

Tyagi, A. K. and Tiwari, S. (2024d). The future of artificial intelligence in blockchain applications. pp. 346–373. *In*: Baby Maruthi, P., Prasad, S. and Tyagi, A. (eds.). Machine Learning Algorithms Using Scikit and TensorFlow Environments. IGI Global Scientific Publishing. https://doi.org/10.4018/978-1-6684-8531-6.ch018.

Tyagi, A. K., Simhan, V. and Tiwari, S. (2024e). Blockchain security in edge computing: use cases, challenges, and solutions. pp. 105–134. *In*: Lakshmi, D. and Tyagi, A. (eds.). Emerging Technologies and Security in Cloud Computing. IGI Global Scientific Publishing. https://doi.org/10.4018/979-8-3693-2081-5.ch005

Tyagi, A. K., Simhan, V. and Tiwari, S. (2024f). Blockchain security in edge computing: use cases, challenges, and solutions. pp. 105–134. *In*: Lakshmi, D. and Tyagi, A. (eds.). Emerging Technologies and Security in Cloud Computing. IGI Global Scientific Publishing. https://doi.org/10.4018/979-8-3693-2081-5.ch005.

Tyagi, A. K., Tiwari, S. and Naithani, K. (2024g). Blockchain, AI, and IoT for smart road traffic management system. *In*: Tyagi, A. K. (ed.). Digital Twin and Blockchain for Smart Cities. https://doi.org/10.1002/9781394303564.ch10.

Tyagi, A. K., Tiwari, S. and Naithani, K. (2024h). Blockchain-enabled Internet of Things (IoTs) platforms for vehicle sensing and transportation monitoring. *In*: Tyagi, A. K. (ed.). Digital Twin and Blockchain for Smart Cities. https://doi.org/10.1002/9781394303564.ch12.

Vranken, Harald. (2017). Sustainability of bitcoin and blockchains. Current Opinion in Environmental Sustainability 28: 1–9.

Chapter 8

Ethereum Based Blockchain and Other Popular Cryptocurrency

Shabnam Kumari,[1] *Amit Kumar Tyagi*[2,*] and *Shrikant Tiwari*[3]

1. Introduction – Ethereum Network

Ethereum is a decentralized, open-source blockchain platform designed to facilitate the development and execution of smart contracts and decentralized applications (DApps). Conceived by Vitalik Buterin in 2013 and launched in 2015, Ethereum expanded blockchain technology beyond simple transactions, introducing a programmable, global platform for developers to build decentralized solutions (Tiwari et al. 2024, Tyagi et al. 2024c).

Ethereum fundamentally enables smart contracts, which are self-executing agreements with their terms encoded in code. These contracts automatically enforce conditions without the need for intermediaries, making them applicable across various sectors such as finance, supply chain management, and gaming. The versatility of Ethereum is driven by its Turing-complete programming language, Solidity. This allows developers to design intricate and customizable applications that can interact seamlessly on the blockchain. Smart contracts enhance efficiency and transparency by executing pre-defined functions when specified conditions are met, thereby eliminating reliance on centralized authorities.

The platform's native cryptocurrency, Ether (ETH), serves multiple purposes: it acts as a medium of exchange, pays transaction fees (gas), and incentivizes validators in Ethereum's proof-of-stake consensus mechanism. Originally using proof-of-work, Ethereum transitioned to proof-of-stake in 2022 during "The Merge," significantly reducing its energy consumption.

Ethereum also underpins innovations like Decentralized Finance (DeFi) and Non-Fungible Tokens (NFTs), enabling financial services without intermediaries and facilitating the trade of unique digital assets. Its robust ecosystem continues to drive advancements in blockchain technology (Kim et al. 2018, Gencer et al. 2018). Ether is utilized to reward participants for computational services and facilitate transactions within decentralized applications (DApps). It also plays a critical role in

[1] Department of Computer Science, Faculty of Science and Humanities, SRM Institute of Science and Technology, Kattankulathur, Chennai, 603203, Tamilnadu, India.

[2] National Forensic Sciences University, Gandhinagar, Gujarat, India.

[3] Department of Computer Science & Engineering, School of Computing Science and Engineering (SCSE), Galgotias University, Greater Noida, Uttar Pradesh-203201.

Emails: shabnam.kt25@gmail.com, shrikanttiwari15@gmail.com

* Corresponding author: amitkrtyagi025@gmail.com

DOI: 10.1201/9781003620822-8

securing the Ethereum network through the Proof of Stake (PoS) consensus mechanism, introduced as part of Ethereum 2.0. Ethereum has emerged as a cornerstone of blockchain innovation, driven by its flexibility, programmability, and a strong developer community. These qualities have enabled the development of decentralized finance (DeFi) protocols, non-fungible tokens (NFTs), and various other decentralized applications. Ongoing research and upgrades aim to tackle scalability issues, enhance efficiency, and bolster security. As Ethereum evolves, it positions itself as a foundational infrastructure for the decentralized web, fostering a new era of trustless and transparent applications (Kumari Shabnam and Muthulakshmi 2023).

2. Ethereum Ecosystem: Keys and Addresses

In the Ethereum ecosystem, keys and addresses play an important role in enabling users to interact with the network and securely manage their digital assets (Zhang et al. 2022, Chen et al. 2020). Here is an overview of keys and addresses in Ethereum (refer Figure 8.1):

Private key: A private key is a randomly generated, confidential alphanumeric string that plays a vital role in the Ethereum ecosystem. It serves as the key to ownership and control over Ethereum addresses and their associated assets. It is essential to keep your private key secret and securely stored. If someone gains access to it, they can easily access the corresponding Ethereum address and its funds, potentially leading to loss of assets.

Ethereum address: Ethereum addresses are public identifiers, derived cryptographically from private keys, that allow accounts to receive funds and interact with smart contracts. Represented as 40-character hexadecimal strings (20 bytes), they uniquely identify each account on the Ethereum network.

Public key: Ethereum uses a public key, mathematically derived from a secret private key, to verify digital signatures and authenticate transactions. The public key can be openly shared without compromising the security of the private key and its associated Ethereum address.

Wallets: Wallets are software applications designed to help users securely store and manage their private keys while interacting with the Ethereum network. They enable users to generate new Ethereum addresses, sign transactions, and offer an intuitive interface for managing digital assets. Wallets are available in several formats, including desktop, mobile, web-based, and hardware options.

Hierarchical Deterministic (HD) wallets: HD wallets create a hierarchical, tree-like structure of keys derived from a single seed phrase or mnemonic phrase. This seed phrase consists of a series of random words that allow users to recover the entire key hierarchy, which includes private keys and Ethereum addresses. HD wallets offer a convenient and secure method for managing multiple Ethereum addresses and their corresponding private keys, all accessible through one mnemonic phrase. This simplifies the process of handling various accounts while ensuring robust security.

Multi-signature (Multi-sig) addresses: Ethereum multi-sig addresses increase security by needing multiple private keys to approve transactions. This is beneficial for shared control scenarios or as a recovery mechanism for lost keys.

Note that understanding and securely managing keys and addresses is important in Ethereum to maintain control over digital assets, initiate transactions, and interact with smart contracts. Users must adopt best practices for key management, such as keeping private keys offline and using reputable wallet applications, to ensure the security of their Ethereum holdings.

Figure 8.1. Keys and addresses of Ethereum ecosystem.

3. Ethereum Based Transactions

Ethereum-based transactions form the core functionality of the Ethereum network, enabling the transfer of Ether (ETH) and the execution of smart contracts (Hu et al. 2021, Zanelatto et al. 2020). Here is an overview of Ethereum-based transactions:

Ether transactions: Ether transactions transfer ETH between Ethereum addresses. Users specify the recipient and amount, and can optionally include data for smart contract interaction.

Gas: Ethereum uses gas to measure the computational effort required for transactions and smart contracts. Each operation consumes a set amount of gas, and the total cost is calculated by multiplying gas used by the gas price (in ETH, often measured in gwei). This system prevents misuse and ensures fair resource compensation. Users set a gas limit (maximum gas they will pay) and a gas price (fee per unit). Higher gas prices prioritize transactions for faster confirmation.

Gas limit: The gas limit is the maximum amount of gas a user is willing to provide for a transaction. It represents the maximum computational resources the sender is willing to consume. The gas limit helps prevent infinite loops or runaway computations by putting an upper bound on the resources a transaction can consume.

Gas price: The gas limit is the user-specified cap on the amount of gas a transaction can consume. It prevents excessive resource usage from infinite loops or other unexpected computational behavior.

Non-Fungible Token (NFT) transactions: Ethereum facilitates the creation and transfer of non-fungible tokens (NFTs) – unique digital assets representing ownership of items like art, collectibles, and in-game assets. NFT transactions operate similarly to Ether transfers, changing ownership from one Ethereum address to another.

Smart contract transactions: Smart contract transactions are more complex than regular Ether transactions as they involve executing code on the Ethereum network. When interacting with a smart contract, transactions may include calling specific functions within the smart contract, passing parameters, and potentially modifying the contract's state. These transactions are used to trigger various actions and operations within decentralized applications (DApps) built on Ethereum.

Transaction confirmation: An Ethereum transaction, once initiated, is broadcast to the network and miners include it in a block via consensus. Transaction confirmation occurs when the transaction is included in a block added to the blockchain. Confirmation time varies based on network congestion, gas price, and the desired number of confirmations for security.

Hence, Ethereum-based transactions form the foundation for various financial activities, decentralized applications, and interactions within the Ethereum ecosystem. Understanding the mechanics of transactions, gas, and smart contracts is essential for users, developers, and stakeholders engaging with Ethereum's decentralized platform.

4. Ethereum Messages

In addition to transactions, Ethereum also supports messages as a means of communication between smart contracts and accounts on the Ethereum network (Deshmukh et al. 2023, Tyagi et al. 2024d)

Here is an overview of Ethereum messages (refer Figure 8.2):

Messaging in Ethereum: Ethereum messages are a way for smart contracts to communicate with each other or with externally owned accounts (EOAs). Messages allow the transfer of data and instructions between different entities within the Ethereum network.

Inter-contract communication: Smart contracts can send messages to other smart contracts. This allows for interaction and interoperability between different contracts, enabling complex decentralized applications (DApps) to communicate and collaborate. Messages can include function calls, parameters, and other relevant data required for the interaction.

Externally Owned Account (EOA) communication: Messages can also be sent from smart contracts to externally owned accounts (EOAs). EOAs are user-controlled Ethereum addresses, typically associated with individuals. Smart contracts can initiate messages to EOAs to provide updates, notifications, or request specific actions from the user.

Message call: A message call in Ethereum represents an invocation of a function within a smart contract. This triggers the execution of the contract's code, potentially altering its internal state. The Ethereum Virtual Machine (EVM) handles message calls by processing the request, executing the specified function according to its logic, and returning the result to the caller.

Message data and parameters: Messages can include data and parameters required for the execution of functions or for passing information between contracts and accounts. This can include values, addresses, structured data, or any other data type defined within the smart contract's interface.

Message value: Messages can also include a value, denominated in Ether (ETH), which can be transferred along with the message. This allows for the transfer of funds between contracts or from contracts to EOAs. The value sent in a message is deducted from the sender's account balance and added to the recipient's account balance.

Message gas: Similar to transactions, messages in Ethereum consume gas. Gas is required to execute the code associated with the message call and cover the computational and storage costs.

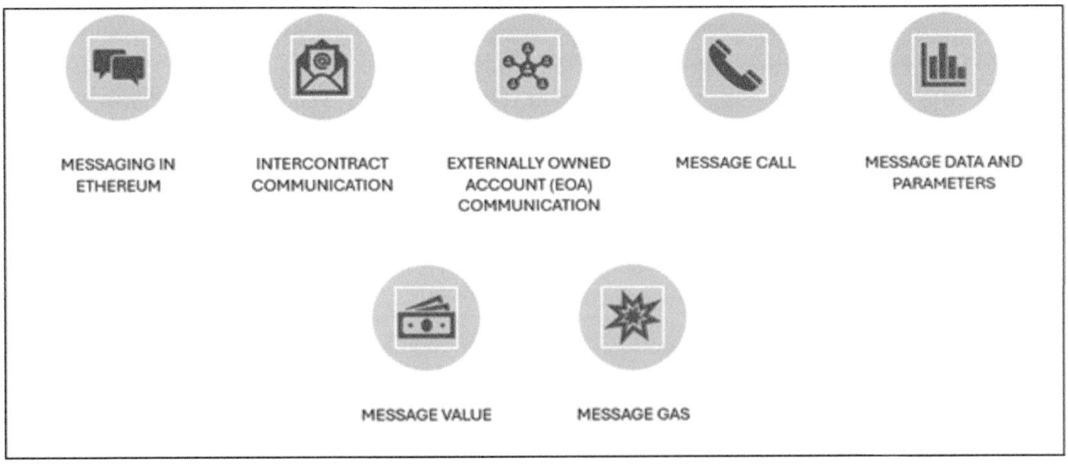

Figure 8.2. Ethereum messages.

The gas cost is deducted from the sender's account balance and helps ensure the secure and efficient execution of messages.

Ethereum messages enable communication between smart contracts and accounts, which is crucial for developing decentralized applications and interconnected systems within the Ethereum ecosystem.

5. Ether and Ethereum Virtual Machine

Ether (ETH) and the Ethereum Virtual Machine (EVM) are two fundamental components of the Ethereum network (Tyagi et al. 2024a, Tyagi et al. 2024e). Let us explore each of them:

Ether (ETH)

Ether is the native cryptocurrency of the Ethereum network. It serves multiple purposes within the ecosystem.

Digital currency: Ether functions as a digital currency that can be used for transactions, value exchange, and payments within the Ethereum network. Users can send and receive Ether to transfer value across the network.

Gas and transaction fees: Ether serves as the currency for transaction fees and computational expenses within the Ethereum network. The unit used to measure computational work is called gas, which is also expressed in Ether. To carry out transactions and execute smart contracts, users must pay gas fees in Ether.

Store of value: Ether can also serve as a store of value, similar to other cryptocurrencies. It can be held and used as an investment or for long-term storage of wealth.

Ethereum Virtual Machine (EVM)

The Ethereum Virtual Machine (EVM) is a decentralized runtime environment responsible for executing smart contracts on the Ethereum blockchain. It ensures consistent and secure execution of code across all Ethereum nodes (Hirai Yoichi 2017). The EVM is integral to enabling decentralized applications (DApps) and processing transactions, using "gas" to measure computational effort and optimize resource allocation while maintaining network security.

Turing-complete: The Ethereum Virtual Machine (EVM) is a Turing-complete system capable of executing any computable task, allowing the development and operation of complex smart contracts and decentralized applications (DApps).

Bytecode execution: The Ethereum Virtual Machine (EVM) executes smart contracts on the Ethereum blockchain. Contracts written in languages like Solidity are compiled into bytecode, which the EVM interprets and runs to enforce the contract's logic and rules.

Consensus neutrality: The EVM ensures consensus neutrality by executing smart contracts in a deterministic manner across all Ethereum nodes. This means that regardless of the node running the contract, the same results will be obtained, ensuring consistency and reliability of smart contract execution.

Gas consumption: The EVM enforces gas limits and tracks the gas consumption of each executed operation within a smart contract. Gas costs are associated with computational and storage operations, and the EVM deducts the required gas from the transaction's gas limit and charges the corresponding Ether amount.

Security and isolation: The Ethereum Virtual Machine (EVM) offers a secure and isolated environment for executing smart contracts. Each smart contract operates within its own instance of the EVM, which guarantees that the code and state of one contract remain unaffected by others. This level of isolation is crucial for preserving the integrity and security of the Ethereum network.

Ether and the Ethereum Virtual Machine: EVM are integral components of the Ethereum network, working together to enable its functionality and programmability. Ether serves as the network's native cryptocurrency, facilitating transactions and acting as a medium of exchange. Meanwhile, the EVM provides a decentralized computational environment that executes smart contracts, enabling the creation of decentralized applications (dApps) and supporting the vision of a decentralized web.

6. Solidity Language and Smart Contracts

Solidity is the leading programming language for building smart contracts on Ethereum. It is essential for developing decentralized applications (DApps) and running contracts within the Ethereum Virtual Machine (EVM) (Tyagi et al. 2024b, Tyagi et al. 2024f). Here is a quick look at Solidity and its connection to smart contracts (refer Table 8.1):

Solidity language: Developers comfortable with languages like JavaScript, C++, or Python will find Solidity relatively easy to learn. This high-level, statically typed, contract-oriented language is specifically designed for creating smart contracts.

Smart contracts: Imagine a contract that automatically executes itself when certain conditions are met. That is essentially what a smart contract is. Using code (like Solidity on the Ethereum platform), the terms of the agreement are programmed in, ensuring a secure and automated process.

Contract-oriented programming: Solidity follows a contract-oriented programming paradigm, where developers define contract structures, functions, and data types to encapsulate the business logic and operations of the smart contract. Contracts can have state variables, functions, and events, enabling the creation of complex DApps.

Functionality and features: Solidity provides a wide range of features and functionality to develop robust and secure smart contracts:

Data types: Solidity supports various data types, including integers, booleans, addresses, strings, arrays, mappings, and more. Additionally, Solidity allows the creation of custom data structures and user-defined types, providing flexibility for developers.

Modifiers and function visibility: Solidity allows developers to define function visibility (public, private, internal, or external) and use modifiers to add additional conditions or checks to functions.

Events: Solidity supports events, which allow smart contracts to emit notifications or log data that can be captured by external applications.

Inheritance: Solidity supports inheritance, enabling code reuse and contract composition. Contracts can inherit properties and functions from other contracts, promoting code organization and reducing redundancy.

Error handling: Solidity provides error handling mechanisms through exceptions and revert statements to handle exceptional conditions or errors in smart contracts.

Development tools: Solidity is supported by a robust set of development tools, including the Solidity compiler (solc), Integrated Development Environments (IDEs) like Remix and Truffle, and testing frameworks such as Truffle and Hardhat. These tools aid in the development, compilation, testing, and deployment of Solidity-based smart contracts.

Table 8.1. Solidity language and smart contracts.

Solidity Language	Statically typed, contract-oriented, and high-level programming language	
Development Tools	Solidity compiler (solc), Integrated Development Environments (IDEs) like Remix and Truffle, and testing frameworks such as Truffle and Hardhat	
Functionality and Features	Data Types	Integers, booleans, addresses, strings, arrays, mappings, and more
	Modifiers and Function Visibility	Public, private, internal, or external
	Events	Notifications or log data
	Inheritance	Inherit properties and functions from other contracts, promoting code organization and reducing redundancy
	Error Handling	Exceptions and revert statements to handle exceptional conditions or errors in smart contracts
Contract-oriented Programming	Contract-oriented programming paradigm	
Smart Contracts	Rules, logic, and behaviour of the agreements, allowing for automated and trust less execution	

Solidity is a cornerstone of the Ethereum ecosystem, enabling the development of smart contracts and decentralized applications (DApps). It empowers developers to write secure, efficient code that defines the behavior and rules of agreements executed on the Ethereum network. By fostering trustless, transparent, and immutable systems, Solidity facilitates automation while eliminating intermediaries. Its rich feature set and an expanding developer community continue to drive innovation, unlocking new possibilities for blockchain-based solutions and decentralized ecosystems.

7. Ethereum Based Blockchain

The Ethereum blockchain is a decentralized and distributed ledger that underpins the Ethereum platform. It facilitates the execution of smart contracts and ensures the secure transfer of digital assets by leveraging blockchain technology's immutable and transparent nature (Nair and Tyagi 2023, Tyagi et al. 2024g). Here is an overview of the Ethereum-based blockchain (refer Figure 8.3):

Decentralization: Ethereum achieves decentralization through a peer-to-peer network of nodes. These nodes uphold consensus and maintain the blockchain, resulting in a transparent, secure, and censorship-resistant system free from central control.

Block structure: Ethereum utilizes a blockchain composed of linked blocks. Each block contains transactions and data, secured by cryptographic hashes to ensure immutability and prevent tampering.

Proof-of-Work (PoW) consensus: Ethereum, like Bitcoin before it, originally employed a Proof-of-Work (PoW) consensus mechanism. In this system, miners engaged in a computational race, striving to solve intricate mathematical problems. The winner of this race earned the privilege of adding a new block of transactions to the Ethereum blockchain, a process that demanded substantial computing resources and, in turn, fortified the network's security.

Transition to Proof-of-Stake (PoS): Ethereum transitioned from Proof-of-Work (PoW) to Proof-of-Stake (PoS) with the Ethereum 2.0 upgrade, completed during “The Merge” on September 15, 2022. In PoS, validators create new blocks based on the cryptocurrency they stake. This shift improved scalability, energy efficiency, and network security.

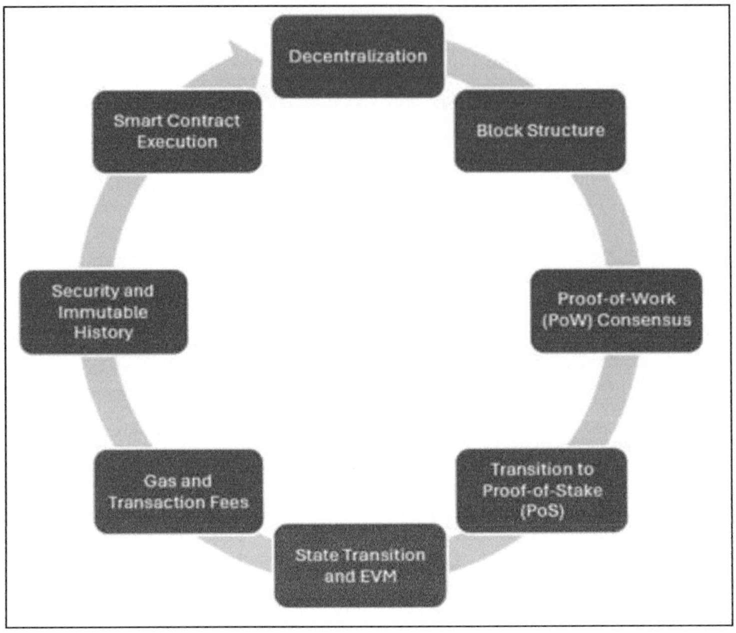

Figure 8.3. Ethereum based blockchain.

State transition and EVM: The "Ethereum Virtual Machine (EVM)" is a key component of the Ethereum blockchain. It executes smart contracts and processes transactions by transitioning the global state of the Ethereum network based on the instructions and logic defined in the smart contracts. The EVM ensures deterministic execution across all nodes in the network.

Gas and transaction fees: Ethereum introduces the concept of gas to handle transaction fees and computational costs. Each operation within a transaction consumes a specific amount of gas, and the total gas cost is determined by multiplying the gas consumed by the gas price, denominated in Ether. Transaction fees in Ether are paid to incentivize miners or validators and prevent abuse of network resources.

Security and immutable history: The security of the Ethereum blockchain stems from its decentralized architecture and the significant computational effort needed to modify its history. Once transactions and smart contract executions are recorded on the blockchain, they become a permanent and unchangeable part of the ledger, which effectively protects it against tampering or manipulation.

Smart contract execution: The Ethereum blockchain serves as a robust platform for executing smart contracts. These self-executing agreements are embedded with code that outlines the rules, logic, and operations governing the contract. Once deployed on the Ethereum blockchain, smart contracts can be accessed and interacted with by users, initiating the execution of their predetermined functions.

Note that Ethereum-based blockchain serves as the backbone of the Ethereum ecosystem, enabling the execution of smart contracts, decentralized applications, and the secure transfer of digital assets. It provides a transparent and reliable infrastructure for building innovative blockchain-based solutions and fostering the growth of the decentralized web.

8. Mining in Ethereum

Mining in Ethereum is the process by which new Ether (ETH) is created and transactions are added to the Ethereum blockchain (Hariharan et al. 2023, Tyagi et al. 2024h). Let us explore the key aspects of mining in Ethereum:

Purpose of mining: Mining serves two primary purposes in Ethereum:

Securing the network: Miners are crucial to the security and integrity of the Ethereum blockchain. They validate transactions, establish consensus on the sequence of these transactions, and safeguard the network from double-spending and various other attacks.

Creating new ether: Miners receive newly created Ether as a reward for their computational efforts and the resources they provide to the network. This mechanism is referred to as the "block reward".

Proof-of-Work (PoW) consensus: Ethereum originally used a Proof-of-Work (PoW) consensus mechanism, where miners competed to solve complex puzzles. The first to solve it would add a new block to the blockchain and earn Ether as a reward.

Mining hardware: Mining in Ethereum began with the use of CPUs (central processing units). However, as the computational demands of Ethereum mining grew, miners shifted to using specialized hardware known as GPUs (graphics processing units). GPU mining offers significantly higher hash rates and improved efficiency, which enhances the likelihood of solving the hash puzzle and earning rewards.

Mining pools: Mining pools enable miners to combine their processing power, increasing their chances of successfully mining blocks. Rewards are then shared proportionally based on each miner's contribution. This provides a more consistent income stream, although individual payouts are smaller than if mining alone.

Difficulty adjustment: To maintain a consistent block creation rate of roughly 15 seconds, the Ethereum network dynamically adjusts the difficulty of its cryptographic puzzle. This adjustment, which occurs approximately every 15 seconds, responds to changes in the network's total computational power: increasing when more computing power is added and decreasing when miners leave.

Transaction validation and gas: Miners play a crucial role in validating and incorporating transactions into the blocks they create. Each transaction includes a fee, commonly referred to as "gas," which users pay to ensure their transactions are prioritized and to compensate miners for the computational effort involved in processing them. Miners have the discretion to select which transactions to include in their blocks, often favoring those with higher gas fees.

Ethereum 2.0 and Proof-of-Stake: Ethereum has upgraded to Proof-of-Stake (PoS) from the energy-intensive Proof-of-Work (PoW) as part of Ethereum 2.0. In PoS, validators are chosen to create new blocks based on the amount of Ether they stake as collateral. This change boosts scalability, strengthens security, and reduces environmental impact by replacing the resource-heavy PoW system with a more sustainable model.

Mining in Ethereum is crucial for securing the network and generating new Ether. However, with the shift to Ethereum 2.0, mining will be phased out and replaced by staking. In this new system, participants will lock up their Ether to help support the network and earn rewards.

9. Popular Cryptocurrency and Their Future

In addition to Bitcoin and Ethereum, several other cryptocurrencies have gained popularity and recognition. While predicting the future of any cryptocurrency with certainty is challenging, here is an overview of some notable cryptocurrencies and their potential developments (Tyagi 2023):

Ripple (XRP): Ripple is a blockchain-based digital payment protocol designed to facilitate fast, secure, and cost-effective international money transfers. Its future hinges on achieving widespread adoption among financial institutions and gaining regulatory clarity, especially in light of ongoing legal challenges surrounding its native cryptocurrency, XRP.

Litecoin (LTC): Litecoin, known as the "silver" to Bitcoin's "gold," offers faster transactions (2.5 minutes per block) and uses the less resource-intensive Scrypt algorithm. These features enhance scalability and lower costs, making Litecoin ideal for everyday transactions. Its future depends on sustaining its speed, efficiency, and widespread adoption.

Bitcoin Cash (BCH): Bitcoin Cash emerged in 2017 as a hard fork of Bitcoin, created to address scalability challenges by increasing block sizes for faster and cheaper transactions. Its future hinges on achieving broader adoption and proving its utility as an efficient medium of exchange.

Cardano (ADA): Cardano is a blockchain platform designed to provide a secure, scalable, and sustainable infrastructure for decentralized application (DApp) development. Its future depends on the successful execution of its roadmap, which includes advancements such as the implementation of smart contracts, scalability improvements, and governance features. These developments aim to position Cardano as a robust platform for mission-critical applications in industries like finance, supply chain management, and identity verification.

Polkadot (DOT): Polkadot is a multi-chain blockchain platform designed to enhance interoperability and scalability by enabling seamless communication between different blockchains. Its success depends on attracting developers and projects to leverage its ecosystem, showcasing its ability to facilitate cross-chain interactions efficiently.

Binance Coin (BNB): Binance Coin (BNB) is the native cryptocurrency of the Binance exchange, essential to its ecosystem. Its future depends on Binance's growth and the coin's expanding utility, including powering decentralized apps, enabling transactions on the BNB Chain, and supporting ecosystem governance.

Solana (SOL): Solana is a fast, scalable blockchain platform designed for decentralized apps and crypto projects. It uses Proof of History (PoH) and Proof of Stake (PoS) to handle up to 50,000 transactions per second, outpacing networks like Ethereum and Bitcoin while keeping fees low.

Cryptocurrency markets are known for their volatility, and the future of any digital currency is influenced by market conditions, technological advancements, regulations, and adoption rates. Thorough research and caution are crucial for investors and users.

10. Other Cryptocurrency Rather than Bitcoin and Ethereum: Types and Features

In addition to Bitcoin and Ethereum, the cryptocurrency landscape is filled with a variety of other digital currencies known as altcoins (alternative coins). Below are some popular categories of altcoins, along with their distinctive features:

Ripple (XRP): Ripple is a blockchain-based digital payment network and protocol designed for fast, low-cost international money transfers. It facilitates real-time cross-border payments and serves as a settlement, remittance, and currency exchange system, often viewed as an alternative to the SWIFT network. Ripple uses its native cryptocurrency, XRP, and the XRP Ledger (XRPL) to provide liquidity and streamline transactions, enabling near-instant settlements and reducing costs significantly. XRP acts as a bridge currency for financial institutions, allowing efficient exchanges between fiat and digital currencies. Ripple's infrastructure eliminates intermediaries in cross-border payments by leveraging blockchain technology, making it faster and more scalable than traditional systems.

Litecoin (LTC): Litecoin is designed for speed and efficiency. With quicker transaction confirmations and a different hashing algorithm compared to Bitcoin, it's well-suited for handling everyday purchases and transactions.

Bitcoin Cash (BCH): Bitcoin Cash originated from a hard fork of Bitcoin in 2017. It was created to address Bitcoin's scalability challenges by increasing the block size limit, enabling the processing of more transactions per block and reducing fees and transaction times.

Cardano (ADA): Cardano is a blockchain platform designed to offer secure, scalable, and sustainable infrastructure for decentralized applications (DApps). It distinguishes itself through a foundation of peer-reviewed academic research and a layered architecture, which enhances its security and functionality. By separating its blockchain into two layers—the Cardano Settlement Layer (CSL) for transactions and the Cardano Computation Layer (CCL) for smart contracts—Cardano achieves greater flexibility and scalability. This structure allows independent upgrades to each layer without affecting the other.

Polkadot (DOT): Polkadot is a decentralized, multi-chain blockchain platform designed to enable interoperability and secure communication between different blockchains. By leveraging its relay chain and parachains, Polkadot allows blockchains to exchange data and assets seamlessly without relying on intermediaries. This structure supports the development of decentralized applications (DApps) across diverse networks while addressing key blockchain challenges such as scalability, security, and interoperability. Additionally, Polkadot's ecosystem fosters innovation through shared security, modular blockchain development with its SDK, and governance via its native DOT token.

Chainlink (LINK): Chainlink is a decentralized oracle network that bridges blockchains with real-world data and external APIs. By enabling smart contracts to securely access off-chain information, Chainlink enhances their functionality and expands their use cases across industries such as DeFi, insurance, gaming, and more. Unlike centralized oracles, Chainlink uses a decentralized network of nodes to ensure data accuracy and reliability, leveraging its native LINK token to incentivize honest participation and maintain security.

Stellar (XLM): Stellar specializes in enabling fast, low-cost cross-border payments, akin to Ripple. Its primary goal is to connect financial institutions, payment systems, and individuals, facilitating seamless global money transfers through blockchain technology.

VeChain (VET): VeChain is a blockchain platform designed to improve supply chain management and verify product authenticity. By leveraging distributed ledger technology, it enhances transparency and trust, enabling businesses to track and confirm the origin, handling, and authenticity of goods throughout their lifecycle.

Dogecoin (DOGE): Originally created as a joke, Dogecoin has evolved into a widely recognized cryptocurrency. Known for its faster block time and abundant supply, it is particularly well-suited for microtransactions and tipping, distinguishing itself from other cryptocurrencies.

Monero (XMR): Monero is a privacy-focused cryptocurrency that ensures anonymous and untraceable transactions. It uses cryptographic techniques like ring signatures, stealth addresses, and confidential transactions to hide sender, receiver, and amount details. This default privacy sets Monero apart from other cryptocurrencies.

Note that these cryptocurrencies have varying features, use cases, and communities supporting them. Each altcoin seeks to address specific challenges or offer unique functionalities within the broader cryptocurrency ecosystem. However, it is important to research and understand each cryptocurrency's strengths, weaknesses, and market dynamics before considering their use or investment.

11. Popular Applications/Wallets for Cryptocurrency

There are several popular applications and wallets available for managing cryptocurrencies (He et al. 2020). Here are some widely used ones (refer Figure 8.4):

Coinbase: Coinbase is a widely used cryptocurrency exchange and wallet platform that supports numerous digital currencies, including Bitcoin, Ethereum, and many others. Known for its intuitive interface, Coinbase simplifies the process of buying, selling, and securely storing cryptocurrencies, making it an accessible option for both beginners and experienced traders.

Binance: Binance is a prominent cryptocurrency exchange platform that facilitates trading across a wide variety of cryptocurrencies. It also offers a secure wallet service, enabling users to safely store their digital assets.

Trust wallet: Trust Wallet is a user-friendly mobile application that supports a diverse range of cryptocurrencies. With its intuitive design, it enables users to securely store, send, and receive various digital assets.

Ledger live: Ledger Live is a software wallet by Ledger, a leading hardware wallet manufacturer. It helps users manage their hardware wallets and securely store cryptocurrencies offline.

Trezor wallet: Trezor Wallet is a hardware wallet that securely stores cryptocurrencies offline with advanced features like passphrase protection, Shamir Backup, and universal 2FA. It supports many cryptocurrencies, uses open-source software for transparency, and integrates with platforms like MetaMask and Exodus for staking and NFT management. Some features are exclusive to the premium Model T.

Exodus: Exodus is a software wallet that supports multiple cryptocurrencies and provides a visually appealing and easy-to-use interface. It allows users to manage their digital assets and exchange cryptocurrencies within the wallet.

MetaMask: MetaMask is a widely used browser extension wallet designed for seamless interaction with decentralized applications (DApps) on the Ethereum blockchain. It features an intuitive interface that allows users to manage Ethereum and ERC-20 tokens efficiently.

MyEtherWallet (MEW): MyEtherWallet (MEW) is a free, open-source crypto wallet for managing Ethereum and ERC-20 tokens. It supports wallet creation, transactions, DApp access, hardware wallet integration, token swaps, and smart contract interaction. As a non-custodial wallet, MEW gives users full control over their private keys, ensuring enhanced security and privacy.

Atomic wallet: Atomic Wallet is a decentralized, multi-currency crypto wallet supporting 300+ cryptocurrencies like Bitcoin and Ethereum. It lets users store, send, and receive assets securely, with full control over private keys. The wallet offers built-in atomic swaps for direct crypto exchanges and integrates with services like ChangeNOW and Simplex for more exchange options and fiat purchases. Available on Windows, macOS, iOS, and Android.

Coinomi: Coinomi is a secure, multi-platform cryptocurrency wallet supporting 125+ blockchains and thousands of tokens. It offers integrated exchanges for easy trading, a seed phrase for recovery, and prioritizes privacy by keeping private keys on-device without requiring KYC.

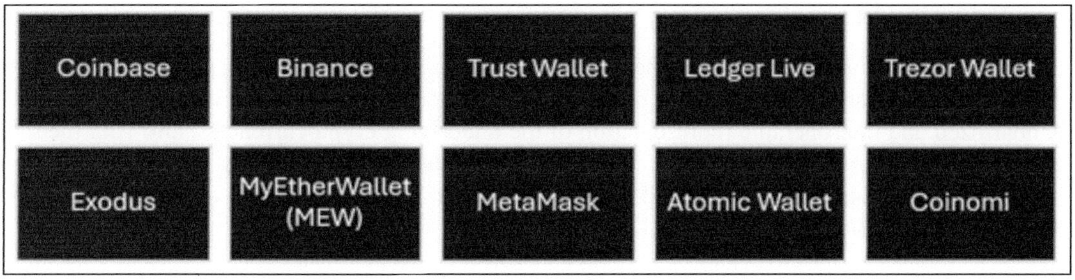

Figure 8.4. Popular applications/wallets for cryptocurrency.

Hence, these applications and wallets vary in terms of platform availability (mobile, desktop, or web-based), supported cryptocurrencies, security features, and user interfaces. To research and choose a wallet that meets your specific needs and requirements, considering factors like security, ease of use, and supported cryptocurrencies.

12. Token Less Cryptocurrency

Tokenless cryptocurrencies, also known as tokenless blockchain systems or tokenless distributed ledgers, are a type of blockchain technology that does not utilize native digital tokens or cryptocurrencies as a medium of value exchange within the network (A. K. V et al. 2022). Instead, they rely on alternative methods to achieve consensus, facilitate transactions, and maintain the integrity of the ledger. Here are a few examples of tokenless cryptocurrency approaches:

Directed Acyclic Graph (DAG)-based cryptocurrencies: DAG-based cryptocurrencies, such as IOTA and Nano, employ a distributed ledger structure that uses a directed acyclic graph instead of a traditional blockchain. These systems use transaction-based models where each transaction must verify two previous transactions before it can be confirmed. They typically eliminate the need for transaction fees and do not require a native token for transaction validation.

While many PoS-based cryptocurrencies use native tokens, there are tokenless variations of this consensus mechanism. For example, projects like Algorand and Hedera Hashgraph implement PoS algorithms without a native token. These systems achieve consensus through a reputation-based system, where participants' influence is based on factors like their stake, reputation, or voting power, rather than a specific token.

Private and permissioned blockchains: In private and permissioned blockchain networks, access to the network is restricted, and participants are typically known and pre-approved. These networks may not require native tokens as they are often used for specific use cases within closed ecosystems, such as enterprise solutions, supply chain management, or government applications. Transactions and consensus are facilitated through predetermined rules and governance structures, rather than relying on native tokens.

Note that the term "tokenless" can have different interpretations and may refer to different aspects of a cryptocurrency system. In some cases, it may indicate the absence of a native token for value exchange, while in other cases, it may refer to the absence of a separate token for utility or functionality within a blockchain network. Tokenless cryptocurrency approaches can offer alternative solutions and unique features compared to traditional token-based cryptocurrencies. However, it is important to research and evaluate each project's specific design, consensus mechanism, governance, and use case to understand how they achieve transaction validation, security, and network integrity without relying on native tokens.

13. Tokenized Cryptocurrency

Tokenized cryptocurrency refers to the representation of digital currencies, such as Bitcoin or Ethereum, in the form of tokens on a blockchain network. These tokens are created to provide a standardized and interoperable representation of the underlying cryptocurrencies, allowing them to be easily traded, stored, and transferred on various blockchain platforms (Tyagi 2022). Tokenization enables the fractional ownership and transferability of cryptocurrencies, making them more accessible and versatile within the blockchain ecosystem.

14. Benefits with Other Cryptocurrency Rather than Bitcoin and Ethereum

Tokenized cryptocurrencies provide distinct advantages over traditional cryptocurrencies like Bitcoin and Ethereum. They leverage blockchain technology to represent real-world assets, enabling fractional ownership, enhanced liquidity, and broader accessibility. Unlike traditional cryptocurrencies, which primarily serve as digital currencies or stores of value, tokenized cryptocurrencies can represent tangible assets such as real estate, commodities, or even intellectual property, bridging the gap between digital and physical economies. Additionally, they offer increased transparency and efficiency in transactions while reducing reliance on intermediaries. This makes them particularly appealing for use cases such as decentralized finance (DeFi) and asset-backed investments. Some of these benefits include:

Accessibility: Tokenized cryptocurrencies enable the digital representation of diverse assets like real estate, stocks, commodities, and intellectual property. This increases accessibility and liquidity, opening up investment opportunities to a wider audience.

Interoperability: Tokenization unlocks cross-chain compatibility, enabling seamless token transfers and exchanges across blockchains. This interoperability streamlines transactions, reduces costs, and enhances liquidity and user base.

Fractional ownership: Tokenization facilitates fractional ownership of assets, dividing them into smaller, more accessible tokens. This enhances liquidity, lowers investment barriers, and broadens participation in high-value markets.

Programmability: Tokenized cryptocurrencies use smart contracts—self-executing digital agreements that trigger transactions when conditions are met. This enables automated, trustless transactions, eliminating intermediaries, reducing costs, and speeding up processes compared to traditional finance.

Enhanced transparency: Tokenized cryptocurrencies use blockchain's transparency and immutability to securely record transactions and ownership transfers. This creates a verifiable, tamper-resistant history, building trust, reducing fraud, and enhancing system integrity.

Diversification: Tokenized cryptocurrencies broaden the scope of investment options beyond conventional cryptocurrencies. By utilizing tokenization, investors can diversify their portfolios across different asset classes, which helps mitigate risk exposure while potentially enhancing returns.

Tokenized cryptocurrencies, unlike their traditional counterparts such as Bitcoin and Ethereum, provide a wider range of advantages including greater accessibility, interoperability, fractional ownership, programmability, transparency, and diversification. These key benefits are driving the increased acceptance and practical application of blockchain technology in a variety of sectors and innovative use cases.

15. Issues and Challenges with Other Cryptocurrency Rather than Bitcoin and Ethereum

While tokenized cryptocurrencies offer numerous benefits, there are also some issues and challenges associated with them, particularly when compared to more established cryptocurrencies like Bitcoin and Ethereum (Kim et al. 2018, Gencer et al. 2018). Here are a few key challenges (refer Figure 8.5):

Liquidity and market fragmentation: Tokenized cryptocurrencies are built on different blockchain networks, leading to fragmentation in the market. This fragmentation can result in reduced liquidity and limited trading options compared to more widely adopted cryptocurrencies like Bitcoin and Ethereum. It can also create challenges in establishing consistent pricing and efficient markets for tokenized assets.

Regulatory uncertainty: The regulation of tokenized cryptocurrencies is an area still under development in numerous jurisdictions. Due to the varied regulatory approaches implemented by different countries, market participants may face uncertainty and legal complexities. This lack of regulatory clarity could potentially slow down broader adoption and impact the overall growth of tokenized cryptocurrencies.

Scalability and technical limitations: Some blockchain networks used for tokenization may face scalability and technical limitations, such as throughput constraints and high transaction fees. These limitations can impact the efficiency and usability of tokenized cryptocurrencies, especially during periods of high demand or when dealing with large volumes of transactions.

Standardization and interoperability: Tokenized cryptocurrencies are built on different blockchain standards and protocols, which can hinder interoperability and data portability between different platforms. The lack of standardization can make it challenging to transfer assets seamlessly across various blockchain networks, resulting in fragmentation and inefficiencies.

Security and custody: The security of tokenized cryptocurrencies and custody of underlying assets can be an issue. As tokenized assets represent ownership or rights to physical or digital assets, the custody and protection of these assets become critical. The risk of hacks, vulnerabilities in smart contracts, and potential loss of assets due to human error or technical failures need to be addressed to ensure trust and confidence in the tokenized ecosystem.

Education and awareness: Tokenized cryptocurrencies are still relatively new, and there is a need for education and awareness among investors and the general public. Understanding the intricacies

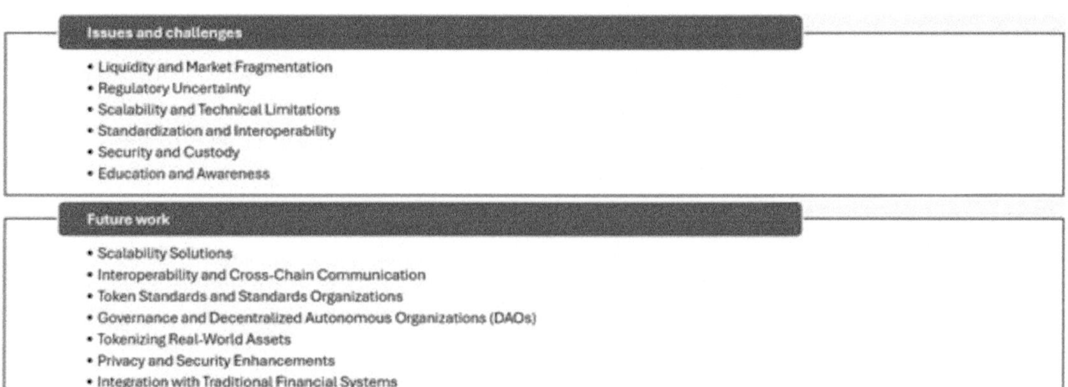

Figure 8.5. Issues, challenges and future work with other cryptocurrencies.

of tokenization, smart contracts, and blockchain technology is essential to mitigate risks and make informed investment decisions. Lack of education and awareness can result in misunderstandings, potential scams, and inappropriate use of tokenized assets.

Note that these challenges are not necessarily exclusive to tokenized cryptocurrencies but can also apply to other emerging cryptocurrencies. Over time, as the technology matures, regulations become clearer, and industry standards are established, these challenges are likely to be addressed, fostering wider adoption and development of the tokenized cryptocurrency ecosystem.

16. Future Work with Other Cryptocurrency Rather than Bitcoin and Ethereum

The future of work with other cryptocurrencies beyond Bitcoin and Ethereum holds several possibilities and areas of exploration. Here are a few potential areas for future development and innovation:

Scalability solutions: Improving scalability is an important area for the growth of other cryptocurrencies. Various scaling solutions, such as layer 2 protocols like Lightning Network or sidechains, are being researched and developed to enhance transaction throughput, reduce fees, and improve overall scalability. These solutions aim to address the limitations faced by existing blockchain networks and facilitate broader adoption and usage.

Interoperability and cross-chain communication: Improving interoperability is crucial for the future of blockchain. Cross-chain technologies are being developed to seamlessly transfer assets and information between different networks. This will increase liquidity, expand use cases, and unlock the potential of tokenized cryptocurrencies.

Token standards and standards organizations: The development of token standards and the establishment of standards organizations are pivotal for the growth of other cryptocurrencies. Standards like ERC-20 (used by Ethereum) and BEP-20 (used by Binance Smart Chain) have significantly contributed to the widespread adoption of tokens. By creating and implementing universally recognized token standards for other cryptocurrencies, the process of token creation can be streamlined, compatibility improved, and interoperability enhanced.

Governance and Decentralized Autonomous Organizations (DAOs): Experimentation with decentralized governance models and DAOs is gaining momentum. Other cryptocurrencies can explore innovative approaches to governance, enabling token holders to participate in decision-making processes, protocol upgrades, and resource allocation. DAOs can foster community involvement and empower stakeholders, leading to more inclusive and decentralized systems.

Tokenizing real-world assets: The tokenization of real-world assets is poised for growth, driven by the potential to fractionalize and trade assets previously considered illiquid, such as real estate, art, and intellectual property. Further advancements in legal frameworks, compliance solutions, and robust technical infrastructure are key to expanding the scope of asset tokenization, unlocking liquidity and creating innovative investment pathways.

Privacy and security enhancements: To keep up with users' changing expectations, it's crucial to boost the privacy and security features of various cryptocurrencies. Developers are hard at work exploring methods like zero-knowledge proofs, confidential transactions, and secure multi-party computation. The goal is to make these currencies more private without sacrificing the transparency and trustworthiness of blockchain technology.

Integration with traditional financial systems: Collaboration and integration with traditional financial systems can bridge the gap between the crypto and fiat worlds. Enabling seamless fiat on-ramps, off-ramps, and cross-border transactions with other cryptocurrencies can facilitate broader adoption and encourage participation from traditional financial institutions.

Hence, these are potential areas where future work with other cryptocurrencies beyond Bitcoin and Ethereum can be focused. The cryptocurrency industry is constantly evolving, and as technology advances and challenges are addressed, we can expect innovative solutions and novel use cases to emerge.

17. Conclusion

Ethereum and other leading cryptocurrencies have significantly advanced decentralized finance (DeFi), tokenization, and smart contract applications. Ethereum's programmability and robust ecosystem have fostered the creation of diverse decentralized applications (DApps), driving innovation in blockchain technology. Its success has inspired the development of cryptocurrencies like Binance Coin (BNB), Cardano (ADA), and Solana (SOL), which offer unique features such as scalability, interoperability, and specialized use cases to address earlier limitations. Tokenization transforms real-world assets into digital tokens on blockchain networks, enhancing investment opportunities, liquidity, and asset accessibility. Despite its benefits, challenges like scalability, regulations, security, and interoperability remain. Ongoing research seeks to address these issues.

As blockchain technology evolves, further integration with traditional financial systems is expected, alongside greater emphasis on privacy, security, and new applications. The future of Ethereum-based blockchains and other cryptocurrencies holds transformative potential for industries and the way we interact with digital assets, contracts, and decentralized systems.

References

A. K. V, Tyagi, A. K. and Kumar, S. P. (2022). Blockchain technology for securing internet of vehicle: issues and challenges. pp. 1–6. 2022 International Conference on Computer Communication and Informatics (ICCCI), doi: 10.1109/ICCCI54379.2022.9740856.

Chen, Weili, Tuo Zhang, Zhiguang Chen, Zibin Zheng and Yutong Lu. (2020). Traveling the token world: A graph analysis of ethereum erc20 token ecosystem. pp. 1411–1421. In Proceedings of The Web Conference 2020.

Damilare Peter, Je Sen Teh, Norziana Jamil and Moatsum Alawida. (2021). Blockchain consensus: An overview of alternative protocols. Symmetry 13(8): 1363.

Deshmukh Atharva, Mishra, A. K., Balamurugan, G. and Tyagi, A. K. (2023). Blockchain-empowered decentralized applications: current trends and challenges. In the Book: Privacy Preservation of Genomic and Medical Data. https://doi.org/10.1002/9781394213726.ch19.

Gencer, Adem Efe, Soumya Basu, Ittay Eyal, Robbert Van Renesse and Emin Gün Sirer. (2018). Decentralization in bitcoin and Ethereum networks. pp. 439–457. In Financial Cryptography and Data Security: 22nd International Conference, FC 2018, Nieuwpoort, Curaçao, Revised Selected Papers 22. Springer Berlin Heidelberg.

Hariharan, R., Tyagi, A. K. and Soni Gulshan. (2023). A survey on blockchain-internet of things-based solutions. In the Book: Privacy Preservation and Secured Data Storage in Cloud Computing, IGI global. DOI: 10.4018/979-8-3693-0593-5.ch005.

He, Daojing, Shihao Li, Cong Li, Sencun Zhu, Sammy Chan, Weidong Min et al. (2020). Security analysis of cryptocurrency wallets in android-based applications. IEEE Network 34(6): 114–119.

Hirai Yoichi. (2017). Defining the ethereum virtual machine for interactive theorem provers. pp. 520–535. In Financial Cryptography and Data Security: FC 2017 International Workshops, WAHC, BITCOIN, VOTING, WTSC, and TA, Sliema, Malta, Revised Selected Papers 21. Springer International Publishing.

Hu, Teng, Xiaolei Liu, Ting Chen, Xiaosong Zhang, Xiaoming Huang, Weina Niu et al. (2021). Transaction-based classification and detection approach for Ethereum smart contract. Information Processing & Management 58(2): 102462.

Kim, Seoung Kyun, Zane Ma, Siddharth Murali, Joshua Mason, Andrew Miller and Michael Bailey. (2018). Measuring ethereum network peers. pp. 91–104. In Proceedings of the Internet Measurement Conference.

Kumari Shabnam and Muthulakshmi, P. (2023). Artificial intelligence-blockchain enabled technology for internet of things: research statements, open issues, and possible applications in the near future. In the Book: Privacy Preservation of Genomic and Medical Data. https://doi.org/10.1002/9781394213726.ch18.

Nair Meghna Manoj and Tyagi Amit Kumar. (2023). Blockchain technology for next-generation society: current trends and future opportunities for smart era. In Book Title: Blockchain Technology for Secure Social Media Computing. DOI: 10.1049/PBSE019E_ch11.

Tiwari, S., Reddy, R. R. and Tyagi, A. K. (2024). Position of blockchain: Internet of Things-Based Education 4.0 in Industry 5.0 – A discussion of issues and challenges. pp. 292–311. *In*: Pandey, R., Srivastava, N. and Chatterjee, P. (eds.). Architecture and Technological Advancements of Education 4.0. IGI Global Scientific Publishing. https://doi.org/10.4018/978-1-6684-9285-7.ch013.

Tyagi, A. K. (2022). SecVT: Securing the vehicles of tomorrow using blockchain technology. *In*: Sk, A. A., Turki, T., Ghosh, T. K., Joardar, S. and Barman, S. (eds.). Artificial Intelligence. ISAI 2022. Communications in Computer and Information Science, vol. 1695. Springer, Cham. https://doi.org/10.1007/978-3-031-22485-0_11Oyinloye.

Tyagi, A. K. (2023). Decentralized everything: Practical use of blockchain technology in future applications. pp. 19–38. Rajiv Pandey, Sam Goundar and Shahnaz Fatima (eds.). Distributed Computing to Blockchain, Academic Press. ISBN: 9780323961462, https://doi.org/10.1016/B978-0-323-96146-2.00010-3.

Tyagi, A. K., Kumari Shabnam, Chidambaram, N. and Sharma Avinash. (2024a). Engineering applications of blockchain in this smart era. In the Book: Enhancing Medical Imaging with Emerging Technologies. DOI: 10.4018/979-8-3693-5261-8.ch011.

Tyagi, A. K., Kukreja Swetta, Richa and Sivakumar, P. K. (2024b). Role of blockchain technology in smart era: a review on possible smart applications. Journal of Information & Knowledge Management. https://doi.org/10.1142/S0219649224500321.

Tyagi, A. K., Kumari, S., Rajasekaran, M. and Tiwari, S. (2024c). Blockchain for safety of internet of vehicles in smart transportation. Tyagi, A. K. (ed.). In Digital Twin and Blockchain for Smart Cities. https://doi.org/10.1002/9781394303564.ch11.

Tyagi, A. K. and Tiwari, S. (2024d). The future of artificial intelligence in blockchain applications. pp. 346–373. *In*: Baby Maruthi, P., Prasad, S. and Tyagi, A. (eds.). Machine Learning Algorithms Using Scikit and TensorFlow Environments. IGI Global Scientific Publishing. https://doi.org/10.4018/978-1-6684-8531-6.ch018.

Tyagi, A. K., Simhan, V. and Tiwari, S. (2024e). Blockchain security in edge computing: use cases, challenges, and solutions. pp. 105–134. *In*: Lakshmi, D. and Tyagi, A. (eds.). Emerging Technologies and Security in Cloud Computing. IGI Global Scientific Publishing. https://doi.org/10.4018/979-8-3693-2081-5.ch005.

Tyagi, A. K., Simhan, V. and Tiwari, S. (2024f). Blockchain security in edge computing: use cases, challenges, and solutions. pp. 105–134. *In*: Lakshmi, D. and Tyagi, A. (eds.). Emerging Technologies and Security in Cloud Computing. IGI Global Scientific Publishing. https://doi.org/10.4018/979-8-3693-2081-5.ch005.

Tyagi, A. K., Tiwari, S. and Naithani, K. (2024g). Blockchain, AI, and IoT for smart road traffic management system. Tyagi, A. K. (ed.). In Digital Twin and Blockchain for Smart Cities. https://doi.org/10.1002/9781394303564.ch10.

Tyagi, A. K., Tiwari, S. and Naithani, K. (2024h). Blockchain-Enabled Internet of Things (IoTs) platforms for vehicle sensing and transportation monitoring. Tyagi, A. K. (ed.). In Digital Twin and Blockchain for Smart Cities. https://doi.org/10.1002/9781394303564.ch12.

Zanelatto Gavião Mascarenhas, Juliana, Artur Ziviani, Klaus Wehmuth and Alex Borges Vieira. (2020). On the transaction dynamics of the Ethereum-based cryptocurrency. Journal of Complex Networks 8(4): cnaa042.

Zhang, Weijia and Tej Anand. (2022). Ethereum architecture and overview. pp. 209–244. In Blockchain and Ethereum Smart Contract Solution Development: Dapp Programming with Solidity. Berkeley, CA: Apress.

Chapter 9
Alternative Blockchains

Shabnam Kumari,[1,*] *Amit Kumar Tyagi*[2] and *Shrikant Tiwari*[3]

1. Introduction - Alternative Blockchains

Blockchain technology, pioneered by Bitcoin, has gained significant attention and adoption in recent years. While Bitcoin and Ethereum are the most well-known and widely used blockchains, there is a growing ecosystem of alternative blockchains that offer unique features and capabilities. Alternative blockchains, also known as altchains or alternative consensus mechanisms, provide alternatives to the traditional proof-of-work (PoW) consensus mechanism used by Bitcoin. These alternative blockchains utilize different consensus algorithms, governance models, scalability solutions, and smart contract functionalities. The introduction of alternative blockchains has opened up new possibilities and use cases beyond the limitations of Bitcoin and Ethereum. They address challenges such as scalability, transaction speed, energy efficiency, and governance issues that have emerged as the demand for blockchain applications has increased. Alternative blockchains cater to various needs and industries, including finance, supply chain management, healthcare, gaming, and more (Kumari Shabnam and Muthulakshmi 2023, Deshmukh et al. 2023, Koens and Poll 2018, Chitra et al. 2019). Some examples of alternative blockchains include:

Proof-of-Stake (PoS) blockchains: These blockchains replace the energy-intensive PoW consensus with a staking mechanism, where participants lock up their tokens to validate transactions. PoS blockchains are known for their energy efficiency and scalability potential.

Directed Acyclic Graph (DAG) blockchains: DAG-based blockchains, such as IOTA and Nano, offer high scalability and fast transactions by organizing transactions in a DAG structure instead of a linear chain. DAG blockchains can handle a large number of transactions simultaneously.

Permissioned/private blockchains: These blockchains restrict participation and control to specific entities or consortiums. They are often used in enterprise settings, allowing organizations to maintain control over data privacy, governance, and regulatory compliance.

[1] Department of Computer Science, Faculty of Science and Humanities, SRM Institute of Science and Technology, Kattankulathur, Chennai, 603203, Tamilnadu, India.

[2] National Forensic Sciences University, Gandhinagar, Gujarat, India.

[3] Department of Computer Science & Engineering, School of Computing Science and Engineering (SCSE), Galgotias University, Greater Noida, Uttar Pradesh-203201.

Emails: amitkrtyagi025@gmail.com, shrikanttiwari15@gmail.com

* Corresponding author: shabnam.kt25@gmail.com

DOI: 10.1201/9781003620822-9

Interoperability blockchains: Interoperability-focused blockchains like Polkadot and Cosmos are designed to connect multiple blockchain networks, facilitating the seamless transfer of assets and data across them. These platforms enable efficient cross-chain communication and collaboration, fostering a more integrated blockchain ecosystem.

Smart contract platforms: Some blockchains, such as Cardano and Tezos, specialize in providing advanced smart contract functionalities with enhanced security and formal verification capabilities. They focus on enabling secure and reliable execution of complex programmable transactions.

Hence, these alternative blockchains offer diverse features and cater to specific use cases and requirements. Their development and adoption contribute to the evolution and maturation of the blockchain ecosystem, promoting innovation and expanding the potential applications of distributed ledger technology. In this exploration of alternative blockchains, we will delve into their unique characteristics, advantages, challenges, and use cases. Understanding these alternative blockchains will provide information to the broader landscape of blockchain technology and its potential impact on various industries.

2. Kadena

Kadena is a blockchain platform that aims to provide secure and scalable solutions for businesses and enterprises. It was founded in 2016 by Stuart Popejoy and Will Martino, both of whom have backgrounds in finance and technology (Mittal et al. 2021). The key features of the Kadena platform include (refer Figure 9.1):

Scalability: Kadena utilizes a hybrid blockchain architecture called Chainweb, which combines multiple parallel chains to achieve high transaction throughput. This architecture allows for horizontal scalability, enabling the network to handle many transactions.

Security: Kadena prioritizes security and utilizes a combination of proof-of-work (PoW) and leader-based consensus mechanisms. The use of PoW ensures that the network is protected against various attacks, while the leader-based consensus enables faster block times and efficient transaction validation.

Smart contracts: Kadena supports the execution of smart contracts on its platform. It uses Pact, a simple and intuitive smart contract language that allows developers to write secure and auditable contracts. Pact includes built-in formal verification features to ensure the correctness of the smart contract code.

Private transactions: Kadena allows for private transactions on its blockchain. It provides options for encrypting sensitive data within transactions, ensuring privacy for participants who require it.

Interoperability: Kadena facilitates seamless communication between blockchain networks by utilizing its interoperability protocol. This capability enables the transfer of assets and data across various blockchains, enhancing connectivity and fostering collaboration within the blockchain

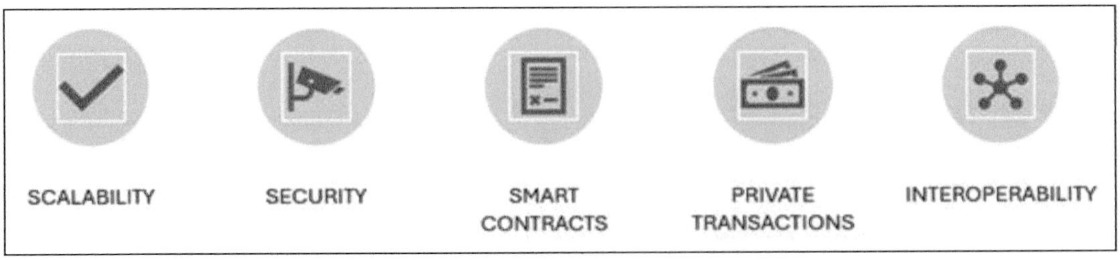

Figure 9.1. Key features of the Kadena platform.

ecosystem. Kadena has been applied in real-world scenarios such as supply chain management, decentralized finance (DeFi), asset tokenization, and digital identity solutions. Additionally, it provides a suite of tools and services to support developers and enterprises in creating and deploying blockchain-based applications.

Kadena's blockchain platform features its native cryptocurrency, Kadena (KDA), which is utilized for transaction fees and network governance. Kadena aims to deliver a secure, scalable, and efficient blockchain solution tailored to the needs of businesses and enterprises, enabling them to integrate blockchain technology into their operations effectively.

3. Ripple

Ripple is a blockchain-based digital payment protocol and cryptocurrency created by Ripple Labs Inc., a technology company founded in 2012. Ripple aims to enable fast, low-cost, and secure international money transfers and remittances (Tyagi et al. 2024h). The key features and components of Ripple include:

RippleNet: RippleNet is a decentralized global network connecting financial institutions such as banks, payment providers, and remittance services to facilitate efficient cross-border transactions. Utilizing Ripple's distributed financial technology, it enables real-time messaging, clearing, and settlement of payments. RippleNet provides a standardized infrastructure for seamless interaction among participants while leveraging the XRP cryptocurrency for on-demand liquidity and faster transaction processing. It offers a cost-effective alternative to traditional systems like SWIFT, reducing delays and fees associated with international payments.

XRP: XRP is the native cryptocurrency of the Ripple network, designed to act as a bridge currency for providing liquidity and enabling faster, cost-efficient transactions. It facilitates seamless transfers between different fiat currencies, eliminating the need for multiple intermediary conversions.

Consensus algorithm: Ripple employs a specialized consensus mechanism known as the XRP Ledger Consensus Protocol (formerly Ripple Protocol Consensus Algorithm or RPCA). Unlike Bitcoin's energy-intensive proof-of-work (PoW) mining, this protocol operates through a network of trusted validators. These validators collaboratively validate transactions and ensure the ledger's integrity by reaching agreement on the order and validity of transactions in a decentralized manner.

Payment channels: Ripple enables the creation of payment channels between participants, allowing for fast and efficient settlement of transactions. Payment channels enable parties to conduct multiple transactions off-chain, reducing network congestion and lowering transaction fees.

RippleNet suite of solutions: Ripple offers various solutions and tools to financial institutions, including xCurrent, xRapid, and xVia. These solutions provide different levels of integration with Ripple's network and enable participants to enhance their payment processing capabilities, liquidity management, and remittance services.

Ripple has gained recognition in the financial industry for enabling fast, low-cost, and secure cross-border transactions. Its technology has attracted partnerships with banks and financial institutions globally. However, Ripple faces ongoing legal challenges with the U.S. SEC regarding the classification of its cryptocurrency, XRP, which affects its operations and market perception.

4. Stellar – Multichain

Stellar is a decentralized blockchain platform, launched in 2014 by Jed McCaleb, aimed at enabling fast, low-cost cross-border payments and asset transfers. Designed with financial inclusion in mind, it seeks to serve the unbanked population (Sharma et al. 2020, Amoordon and Rocha 2019). A key feature of Stellar is its multichain architecture, which supports asset issuance and transfers across interconnected blockchains (refer Figure 9.2) (Tyagi et al. 2024f, Tyagi et al. 2024g):

Stellar Consensus Protocol (SCP): Stellar utilizes the Stellar Consensus Protocol, a federated Byzantine agreement algorithm, to achieve consensus among participating nodes. SCP enables fast transaction confirmation times and maintains the integrity and security of the network.

Stellar Lumens (XLM): Stellar has its native cryptocurrency called Lumens (XLM). XLM serves as the digital asset on the Stellar network and facilitates transactions, including cross-border payments and transfers. XLM can also be used as a bridge currency for exchanging different assets within the network.

Anchors and issued assets: Stellar allows the creation of "anchors" - entities that serve as trusted bridges between the Stellar network and traditional financial systems. Anchors issue assets on the Stellar blockchain, representing various forms of value such as fiat currencies, cryptocurrencies, commodities, and even digital representations of real-world assets.

Interoperability and multichain transactions: Stellar's multichain functionality facilitates seamless interoperability across various blockchains and assets. This capability enables the transfer of assets between Stellar and other supported blockchains, promoting efficient cross-chain transactions and enhancing the network's connectivity and global reach.

Decentralized exchange: Stellar integrates a decentralized exchange (DEX) directly into its protocol, allowing users to trade and convert various assets seamlessly on the blockchain. This DEX facilitates peer-to-peer trading without intermediaries, enhancing liquidity and accessibility for participants.

Focus on financial inclusion: Stellar prioritizes expanding financial services to underserved populations, including the unbanked and underbanked. By enabling low-cost, efficient cross-border payments, it seeks to enhance access to financial systems and promote economic inclusion on a global scale.

Stellar has been adopted by various organizations and projects, including financial institutions, payment processors, and non-profit organizations. Its multichain capabilities, coupled with its focus on fast and affordable transactions, make it attractive for cross-border remittances, micropayments, and tokenized asset transfers. In summary, Stellar's multichain architecture and its mission to enable accessible and inclusive financial services position it as a platform that seeks to revolutionize global payment systems and promote economic empowerment.

Figure 9.2. Key Aspects of stellar and its multichain capabilities.

5. TenderMint

Tendermint is a consensus engine and blockchain technology that provides a framework for building distributed applications (DApps) and blockchain networks. It was developed by Tendermint Inc., a software development company founded in 2014 (Tiwari et al. 2024, Tyagi et al. 2024c, Tyagi et al. 2024d). Here are the few key aspects of Tendermint:

Consensus algorithm: Tendermint employs a Byzantine Fault Tolerant (BFT) consensus algorithm called the Tendermint Consensus. It ensures secure and deterministic transaction ordering and finality in a decentralized network of nodes. The consensus algorithm enables high throughput and low-latency transaction processing.

Application layer: Tendermint distinguishes between the consensus layer and the application layer, enabling developers to concentrate on creating decentralized applications (DApps) without the need to manage the complexities of the underlying consensus protocol. This modular design offers increased flexibility and streamlines the development process.

BFT state machine replication: Tendermint utilizes a BFT state machine replication model, where all nodes execute the same transactions in the same order to achieve consensus. This deterministic approach ensures that all nodes reach the same state after processing the transactions.

Interoperability and compatibility: Tendermint is designed to be interoperable with various programming languages and supports multiple blockchain networks. It provides a standard interface called the Application Blockchain Interface (ABCI), which enables seamless integration of different applications and protocols.

Proof-of-Stake (PoS) consensus: Tendermint utilizes a Proof of Stake (PoS) consensus mechanism, where validators are chosen based on the quantity of tokens they possess and their commitment to locking these tokens as collateral. These validators take turns proposing and voting on blocks, which helps maintain the security and integrity of the network.

Cosmos network: Tendermint is the foundational technology powering the Cosmos Network, an ecosystem of interconnected blockchains known as "zones." It facilitates interoperability by enabling these blockchains to communicate and transfer assets through the Inter-Blockchain Communication (IBC) protocol.

Scalability and performance: Tendermint's consensus algorithm and architecture are designed to provide high scalability and performance. By decoupling the consensus and application layers, Tendermint can handle a large number of transactions and achieve fast confirmation times.

Tendermint has been adopted in various blockchain applications, including decentralized finance (DeFi), supply chain management, and identity solutions. Its robust and efficient consensus engine, combined with interoperability features, makes it a versatile choice for creating scalable and secure blockchain networks. By offering a modular architecture and compatibility with diverse programming environments, Tendermint provides developers with a flexible framework for building decentralized applications (DApps) and blockchain networks. Its focus on consensus, performance, and interoperability ensures its suitability for a wide range of use cases.

6. Platforms and Frameworks: Eris

Eris is an open-source software platform and framework for building decentralized applications (DApps) and blockchain solutions. It provides a development environment and tools to create, deploy, and manage blockchain-based applications and networks. Eris was developed by Monax Industries, which later rebranded as Hyperledger Burrow. Here are the key aspects of Eris (Tyagi et al. 2024d, Tyagi et al. 2024e):

Smart contract development: Eris specializes in the development and execution of smart contracts. It supports popular programming languages like Solidity (commonly used in Ethereum) and JavaScript, enabling developers to leverage their existing skills and libraries when creating smart contracts. This flexibility simplifies the process of building blockchain-based applications.

Permissioned blockchain networks: Eris facilitates the development of permissioned blockchain networks, where access and participation are limited to designated entities or consortiums. This design is particularly well-suited for enterprise and consortium blockchain applications, emphasizing privacy, governance, and regulatory compliance.

Containerization: Eris utilizes containerization technology (Docker) to package and deploy blockchain networks and applications. This allows for easy deployment, management, and scaling of blockchain nodes and network components.

Integrated development environment (IDE): Eris provides an integrated development environment called ErisDB Studio. It offers a user-friendly interface for designing, deploying, and testing smart contracts and blockchain applications. The IDE includes features such as code editors, debugging tools, and simulation capabilities.

Interoperability: Eris focuses on interoperability and integration with existing systems. It provides APIs and tools to connect blockchain networks with external systems, enabling data exchange and interaction between blockchain and off-chain applications.

Consensus mechanisms: Eris supports various consensus mechanisms, such as Proof-of-Work (PoW) and Proof-of-Authority (PoA), enabling developers to select the algorithm that aligns best with their specific use cases and network needs.

ErisDB: Eris includes ErisDB, a distributed database that provides secure storage and retrieval of blockchain data. ErisDB ensures data integrity and consistency across the network, enabling efficient querying and retrieval of blockchain information.

Eris, now known as Hyperledger Burrow, is a permissioned blockchain platform suited for industries like supply chain, finance, legal, and government sectors. It emphasizes control, privacy, and regulatory compliance, making it ideal for enterprise and consortium applications. As part of the Hyperledger project under the Linux Foundation, it supports cross-industry blockchain development with a focus on interoperability and ease of use, offering a robust framework for building smart contract applications.

7. Blockchain Design Challenges: Scalability and Privacy

Scalability and privacy are two significant design challenges in blockchain technology (Tyagi et al. 2024a, Tyagi et al. 2024b). We can discuss these challenges as (refer Table 9.1):

Scalability: Scalability is the capacity of a blockchain network to accommodate an increasing volume of transactions or users while maintaining optimal performance. This aspect poses a significant challenge, as prominent blockchain networks such as Bitcoin and Ethereum have encountered restrictions related to transaction throughput and confirmation times. Key challenges associated with scalability include:

Transaction throughput: Blockchain networks frequently face difficulties in handling a high volume of transactions per second. As user numbers and transaction volumes grow, the network may experience congestion, leading to delays and increased fees. To tackle this issue, various scaling solutions have been developed, including off-chain payment channels, sharding, and layer-2 protocols.

Block size and confirmation times: The size of blocks and the time required to confirm transactions can impact scalability. Larger block sizes can accommodate more transactions, but they also increase the storage and bandwidth requirements for network participants. Reducing block confirmation times improves transaction finality and enhances the user experience.

Network resource consumption: Blockchain networks, especially those employing proof-of-work (PoW) consensus algorithms, face significant scalability hurdles because of their resource-intensive nature. The considerable energy consumption and heavy computational demands inherent in PoW systems restrict how much these networks can grow. To address these limitations, a shift towards more energy-efficient consensus mechanisms, such as proof-of-stake (PoS) or delegated proof-of-stake (DPoS), offers a promising path forward.

Table 9.1. Blockchain design challenges: scalability and privacy.

Blockchain design challenges	**Scalability**	Transaction Throughput	Off-chain payment channels, sharding, and layer-2 protocols
		Block Size and Confirmation Times	Increase storage and bandwidth requirements for network participants
		Network Resource Consumption	High energy consumption and computational requirements
	Privacy	Pseudonymity and Identity	Implementing privacy-preserving techniques like zero-knowledge proofs or ring signatures
		Confidentiality of Data	Encrypted data storage, secure multi-party computation, and homomorphic encryption techniques
		Permissioned vs. Permissionless Blockchains	Bitcoin, prioritize transparency and open participation

Privacy: Privacy is a vital consideration in blockchain design, given that the technology fundamentally prioritizes transparency and immutability. Nevertheless, there are numerous applications where privacy is essential, particularly to safeguard sensitive business data, personal information, and transaction details. The key challenges related to privacy in blockchain include:

Pseudonymity and identity: Blockchain transactions are generally pseudonymous, meaning they are linked to addresses instead of actual identities. Nevertheless, privacy issues can emerge when these transactions are traceable back to particular individuals or organizations. To bolster transaction privacy, employing privacy-preserving methods such as zero-knowledge proofs or ring signatures can be highly effective.

Confidentiality of data: Blockchain networks excel at transparently and permanently recording transactions. However, situations arise where data privacy is paramount, necessitating selective sharing or encryption of sensitive information. To address this, techniques like encrypted data storage, secure multi-party computation, and homomorphic encryption can be implemented, guaranteeing confidentiality.

Permissioned vs. permissionless blockchains: Permissionless blockchains, like Bitcoin, prioritize transparency and open participation, which can limit privacy. Permissioned or consortium blockchains can offer more control over who can access and participate in the network, enabling enhanced privacy features.

Scalability and privacy challenges necessitate a multifaceted approach that includes technological innovations, enhancements to consensus mechanisms, and thoughtful network design. Current research and development initiatives are dedicated to creating scalable and privacy-preserving solutions. These include advancements such as layer-2 protocols, sidechains, zero-knowledge proofs, and secure multiparty computation. The goal is to enhance the efficiency, scalability, and privacy of blockchain technology.

8. Issues and Challenges towards Alternative Blockchains

Alternative blockchains, beyond the dominant ones like Bitcoin and Ethereum, face several issues and challenges (Nair and Tyagi 2023, Hariharan et al. 2023). Here are some of the common ones (refer Figure 9.3):

Network adoption: A major hurdle for alternative blockchains is achieving widespread adoption and attracting both users and developers. Established platforms like Bitcoin and Ethereum benefit from a significant network effect, characterized by a large user base, a robust ecosystem, and an

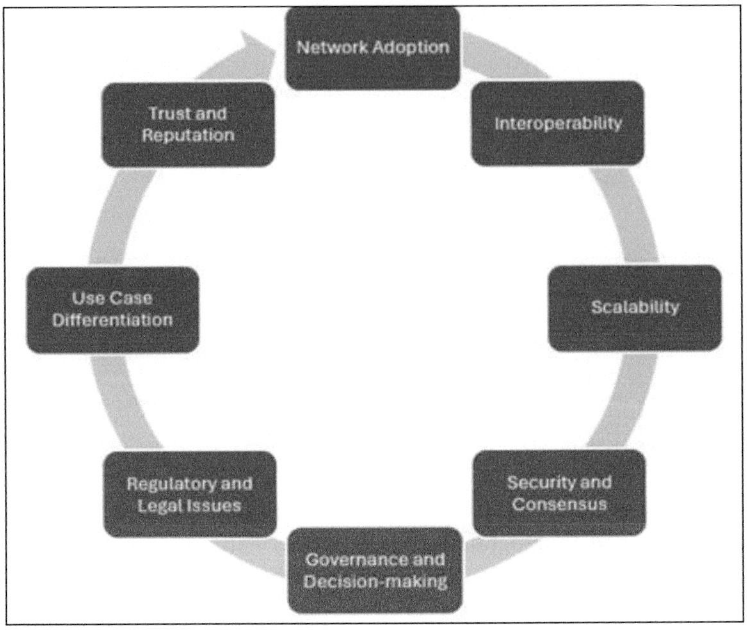

Figure 9.3. Issues and challenges towards alternative blockchains.

active developer community. To compete, newer blockchains must overcome these advantages and successfully persuade users and developers to transition to their platforms.

Interoperability: Interoperability is the capacity of various blockchain networks to communicate and share data effortlessly. When interoperability is lacking, alternative blockchains face significant challenges, as they must ensure compatibility with existing blockchain systems and infrastructure. To tackle this issue, initiatives like cross-chain protocols and interoperability frameworks are being developed.

Scalability: Scalability presents a key challenge for alternative blockchains. To accommodate growing user bases and transaction loads, they must implement solutions like sharding, layer-2 protocols, or innovative consensus mechanisms to ensure high throughput.

Security and consensus: Security is of utmost importance in blockchain systems. Alternative blockchains must implement strong security measures to safeguard against attacks and uphold the integrity of the network. Choosing an appropriate consensus mechanism, such as proof-of-work (PoW) or proof-of-stake (PoS), is crucial for ensuring both the security and stability of the blockchain.

Governance and decision-making: Governance plays a significant role in the development and evolution of a blockchain network. Alternative blockchains need to establish effective governance models that involve stakeholders and allow for decentralized decision-making. This ensures transparency, fairness, and adaptability to address emerging challenges and upgrade the blockchain's protocol.

Regulatory and legal issues: Alternative blockchains often face regulatory and legal challenges. Compliance with existing regulations and navigating the evolving regulatory landscape is important for their acceptance and integration into mainstream systems. Addressing concerns related to "Anti-money laundering (AML)", "know-your-customer (KYC)", and data protection regulations is necessary for wider adoption.

Use case differentiation: With numerous alternative blockchains emerging, it is essential to differentiate and establish unique use cases or value propositions. Identifying specific industries, niche applications, or innovative features can help alternative blockchains carve out their market share and attract users and developers.

Trust and reputation: Building trust and establishing a solid reputation is challenging for newer blockchain platforms. They need to demonstrate reliability, security, and stability to gain trust from users, developers, and enterprises. Strong community engagement, partnerships, and real-world use cases can help establish credibility.

Addressing these challenges requires technical innovation, community building, strategic partnerships, and effective marketing. Ongoing development, research, and adaptation are essential for the long-term success and sustainability of alternative blockchains.

9. Future Work towards Alternative Blockchains

The future work towards alternative blockchains involves several areas of focus and development (Tyagi 2023, Tyagi 2022). Here are some key aspects (refer Figure 9.4):

Scalability solutions: A key focus is enhancing the scalability of alternative blockchains, as this is crucial for their wider adoption. Efforts will concentrate on researching and deploying scaling technologies such as sharding, layer-2 protocols, and off-chain computation to enable more transactions with less delay.

Interoperability and cross-chain communication: Improving interoperability among various blockchain networks is crucial for facilitating seamless communication and data exchange. Initiatives aimed at creating standardized protocols, cross-chain bridges, and interoperability frameworks will empower users and applications to interact effortlessly across multiple blockchains.

Privacy and confidentiality enhancements: Alternative blockchains will prioritize enhancing privacy features and techniques. Innovations in zero-knowledge proofs, secure multi-party computation, and various privacy-preserving technologies will empower users to safeguard their data and transactions, all while enjoying the transparency and security that blockchain offers.

Governance and decentralized decision-making: Alternative blockchains will continue to refine their governance models to ensure decentralized decision-making and community participation. Experimentation with different governance mechanisms, token economics, and voting systems will be important for effective governance in decentralized networks.

Security and resilience: We are committed to strengthening the security and resilience of alternative blockchains through research and development focused on advanced consensus algorithms, smart contract security, and robust defense mechanisms.

Energy efficiency and environmental impact: To mitigate the high energy consumption associated with blockchain networks, a shift towards more sustainable solutions is underway. Alternative blockchains are adopting energy-efficient consensus algorithms like Proof-of-Stake (PoS), aiming to minimize the environmental footprint of this technology.

Integration with real-world applications: Alternative blockchains are set to enhance their integration with real-world applications and industries. By forming collaborations and partnerships with enterprises, governments, and organizations, these blockchains will facilitate the development of innovative blockchain-based solutions across diverse sectors, including finance, supply chain management, healthcare, and governance.

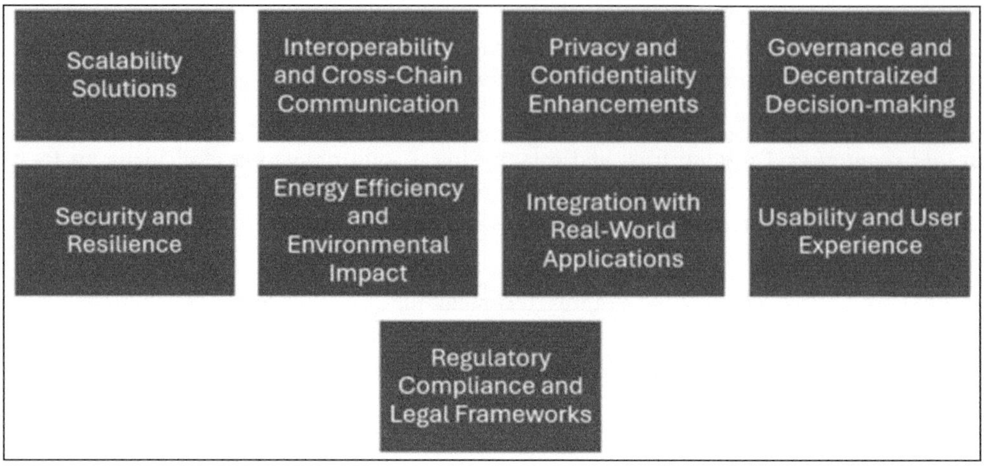

Figure 9.4. Future work towards alternative blockchains.

Usability and user experience: A primary goal will be to enhance user experience and make blockchain technology more accessible to the general public. By improving wallet interfaces, developer tools, documentation, and educational resources, we can reduce barriers to entry and promote broader adoption of this transformative technology.

Regulatory compliance and legal frameworks: Alternative blockchains will work towards aligning with regulatory requirements and establishing legal frameworks to ensure compliance. Engaging with regulators, participating in policy discussions, and addressing concerns related to security, privacy, and data protection will foster acceptance and integration into existing legal systems.

The development of alternative blockchains is motivated by the necessity to surpass existing limitations, cater to industry-specific needs, and foster innovation within the blockchain ecosystem. Ongoing collaboration, research, and continuous advancements will play a crucial role in evolving these alternative blockchains to fulfill the requirements of a swiftly changing digital environment.

10. Conclusion

In conclusion, alternative blockchains offer diverse and innovative approaches to decentralized technology beyond the well-established Bitcoin and Ethereum networks. These alternative blockchains address various challenges and explore new possibilities in scalability, privacy, governance, interoperability, and more. While alternative blockchains face hurdles such as network adoption, scalability, security, and regulatory compliance, they continue to evolve and make significant progress. The future of alternative blockchains lies in their ability to overcome these challenges through technological advancements, community building, and strategic collaborations. Scalability solutions, including sharding, layer-2 protocols, and off-chain computation, will enhance transaction throughput and improve network performance. Efforts to improve interoperability will facilitate smooth communication and data exchange across various blockchain networks, creating a more interconnected ecosystem. Additionally, bolstering privacy features like zero-knowledge proofs and secure multi-party computation will effectively tackle concerns related to confidentiality and data protection.

Meanwhile, governance models will be refined to ensure decentralized decision-making and community participation in shaping the future of alternative blockchains. Security and resilience will remain key areas of focus, with ongoing research into consensus algorithms and smart contract security. Energy efficiency and environmental impact will also be addressed by transitioning to eco-friendly consensus mechanisms. Integration with real-world applications and industries will drive the adoption and relevance of alternative blockchains. By collaborating with enterprises, governments, and organizations, these blockchains can provide innovative solutions across sectors such as finance, supply chain, healthcare, and governance. Usability and user experience improvements will make blockchain technology more accessible to a broader audience, encouraging wider adoption. Compliance with regulatory requirements and the establishment of legal frameworks will foster acceptance and integration into existing legal systems. In summary, alternative blockchains represent a dynamic and evolving landscape, offering unique features, use cases, and opportunities for innovation. As they continue to address challenges, collaborate, and drive technological advancements, alternative blockchains have the potential to shape the future of decentralized technology and transform industries worldwide.

References

Amoordon, Andy and Henrique Rocha. (2019). Presenting tender mint: Idiosyncrasies, weaknesses, and good practices. pp. 44–49. In 2019 IEEE International Workshop on Blockchain Oriented Software Engineering (IWBOSE). IEEE.

Chitra, Tarun, Monica Quaintance, Stuart Haber and Will Martino. (2019). Agent-based simulations of blockchain protocols illustrated via kadena's chainweb. pp. 386–395. In 2019 IEEE European Symposium on Security and Privacy Workshops (EuroS&PW). IEEE.

Damilare Peter, Je Sen Teh, Norziana Jamil and Moatsum Alawida. (2021). Blockchain consensus: An overview of alternative protocols. Symmetry 13(8): 1363.

Deshmukh Atharva, Mishra, A. K., Balamurugan, G. and Tyagi, A. K. (2023). Blockchain-empowered decentralized applications: current trends and challenges. In the Book: Privacy Preservation of Genomic and Medical Data. https://doi.org/10.1002/9781394213726.ch19.

Hariharan, R., Tyagi, A. K. and Soni Gulshan. (2023). A survey on blockchain-internet of things-based solutions. In the Book: Privacy Preservation and Secured Data Storage in Cloud Computing, IGI global. DOI: 10.4018/979-8-3693-0593-5.ch005.

Koens, Tommy and Erik Poll. (2018). What blockchain alternative do you need? pp. 113–129. In Data Privacy Management, Cryptocurrencies and Blockchain Technology: ESORICS 2018 International Workshops, DPM 2018 and CBT 2018, Barcelona, Spain, Proceedings 13. Springer International Publishing, 2018.

Kumari Shabnam and Muthulakshmi, P. (2023). Artificial intelligence-blockchain enabled technology for internet of things: research statements, open issues, and possible applications in the near future. In the Book: Privacy Preservation of Genomic and Medical Data. https://doi.org/10.1002/9781394213726.ch18.

Mittal, Nitin, Srishty Pal, Anjali Joshi, Ashish Sharma, Sandeep Tayal and Yogesh Sharma. (2021). Comparative analysis of various platforms of blockchain. Smart and Sustainable Intelligent Systems: 323–340.

Nair Meghna Manoj and Tyagi Amit Kumar. (2023). Blockchain technology for next-generation society: current trends and future opportunities for smart era. In Book Title: Blockchain Technology for Secure Social Media Computing. DOI: 10.1049/PBSE019E_ch11.

Sharma, Yogesh, Balamurugan, B. and Firoz Khan. (2020). Blockchain and distributed ledger system. pp. 177–206. In Blockchain, Big Data and Machine Learning. CRC Press.

Tiwari, S., Reddy, R. R. and Tyagi, A. K. (2024). Position of Blockchain: Internet of Things-Based Education 4.0 in Industry 5.0 – A discussion of issues and challenges. pp. 292–311. *In*: Pandey, R., Srivastava, N. and Chatterjee, P. (eds.). Architecture and Technological Advancements of Education 4.0. IGI Global Scientific Publishing. https://doi.org/10.4018/978-1-6684-9285-7.ch013.

Tyagi, A. K. (2022). SecVT: Securing the vehicles of tomorrow using blockchain technology. *In*: Sk, A. A., Turki, T., Ghosh, T. K., Joardar, S. and Barman, S. (eds.). Artificial Intelligence. ISAI 2022. Communications in Computer and Information Science, vol 1695. Springer, Cham. https://doi.org/10.1007/978-3-031-22485-0_11Oyinloye.

Tyagi, A. K. (2023). Decentralized everything: Practical use of blockchain technology in future applications. pp. 19–38. Rajiv Pandey, Sam Goundar and Shahnaz Fatima (eds.). Distributed Computing to Blockchain, Academic Press. ISBN: 9780323961462, https://doi.org/10.1016/B978-0-323-96146-2.00010-3.

Tyagi, A. K., Kumari Shabnam, Chidambaram, N. and Sharma Avinash. (2024a). Engineering applications of blockchain in this smart era. In the Book: Enhancing Medical Imaging with Emerging Technologies. DOI: 10.4018/979-8-3693-5261-8.ch011.

Tyagi, A. K., Kukreja Swetta, Richa and Sivakumar, P. K. (2024b). Role of blockchain technology in smart era: a review on possible smart applications. Journal of Information & Knowledge Management. https://doi.org/10.1142/S0219649224500321.

Tyagi, A. K., Kumari, S., Rajasekaran, M. and Tiwari, S. (2024c). Blockchain for safety of internet of vehicles in smart transportation. Tyagi, A. K. (ed.). In Digital Twin and Blockchain for Smart Cities. https://doi.org/10.1002/9781394303564.ch11.

Tyagi, A. K. and Tiwari, S. (2024d). The future of artificial intelligence in blockchain applications. pp. 346–373. *In*: Baby Maruthi, P., Prasad, S. and Tyagi, A. (eds.). Machine Learning Algorithms Using Scikit and TensorFlow Environments. IGI Global Scientific Publishing. https://doi.org/10.4018/978-1-6684-8531-6.ch018.

Tyagi, A. K., Simhan, V. and Tiwari, S. (2024e). Blockchain security in edge computing: use cases, challenges, and solutions. pp. 105–134. *In*: Lakshmi, D. and Tyagi, A. (eds.). Emerging Technologies and Security in Cloud Computing. IGI Global Scientific Publishing. https://doi.org/10.4018/979-8-3693-2081-5.ch005.

Tyagi, A. K., Simhan, V. and Tiwari, S. (2024f). Blockchain security in edge computing: use cases, challenges, and solutions. pp. 105–134. *In*: Lakshmi, D. and Tyagi, A. (eds.). Emerging Technologies and Security in Cloud Computing. IGI Global Scientific Publishing. https://doi.org/10.4018/979-8-3693-2081-5.ch005.

Tyagi, A. K., Tiwari, S. and Naithani, K. (2024g). Blockchain, AI, and IoT for smart road traffic management system. Tyagi, A. K. (ed.). In Digital Twin and Blockchain for Smart Cities. https://doi.org/10.1002/9781394303564.ch10.

Tyagi, A. K., Tiwari, S. and Naithani, K. (2024h). Blockchain-Enabled Internet of Things (IoTs) platforms for vehicle sensing and transportation monitoring. Tyagi, A. K. (ed.). In Digital Twin and Blockchain for Smart Cities. https://doi.org/10.1002/9781394303564.ch12.

Chapter 10

Blockchain Technology and Digital Twin in Logistics and Supply Chain Management

Amit Kumar Tyagi,[1,*] *Jaya Nema*[2] and *Richa*[3]

1. Introduction to Blockchain Technology and Digital Twin: Definition and Concept, Working, Types, Key Components, Advantages and Future Challenges

In the rapidly changing field of supply chain management and logistics, the integration of cutting-edge technologies is reshaping traditional processes. Among these technologies, blockchain and digital twin stand out as transformative forces, providing innovative solutions to enhance transparency, efficiency, and trust. This introduction provides a comprehensive overview of both blockchain technology and digital twin, outlining their definitions, concepts, workings, types, key components, advantages, and potential future challenges (Yaqoob et al. 2020).

Blockchain technology: Decentralised and distributed ledger technology like blockchain securely records transactions across several computers in a transparent, tamper-resistant, and immutable fashion. The fundamental concept involves creating a chain of blocks, having a list of transactions, which is further linked using cryptographic hashes.

Working of Blockchain

When a transaction occurs, it is broadcast to a network of nodes. These nodes validate and reach a consensus on the transaction's validity through a consensus algorithm. Once verified, the transaction is added to a block, and the block is appended to the existing chain.

[1] National Forensic Sciences University, Gandhinagar, Gujarat, India.
[2] Associate Professor, IPS Academy, Institute of Business Management and Research, Indore, Madya Pradesh, India.
[3] Department of Computer Science & Engineering, BIT Mesra, Ranchi, Jharkhand, India.
Emails: jayanema@ipsacademy.org, richasingh.bv@gmail.com
* Corresponding author: amitkrtyagi025@gmail.com

DOI: 10.1201/9781003620822-10

Types

Public blockchain: Accessible to anyone; anyone can participate and validate transactions (e.g., Bitcoin).

Private blockchain: Restricted access; participants are known entities, often used in enterprise applications.

Consortium blockchain: Shared by a group of organizations, providing a balance between public and private blockchains.

Key Components

Blocks: Containers for transaction data.

Hashes: Cryptographic functions linking blocks.

Consensus mechanism: Agreement among nodes on the validity of transactions.

Advantages

Transparency: Immutable ledger ensures a transparent and auditable transaction history.

Security: Cryptographic techniques and their decentralized nature enhance security.

Trust: Intermediaries requirements will be removed, and will provide trust.

Digital Twin

Digital twins are created by collecting and integrating data from sensors, devices, and other sources. This virtual replica enables continuous monitoring, analysis, and optimization, enhancing decision-making and performance.

Types

Product twin: Virtual representation of physical products.
Process twin: Digital replica of a manufacturing or operational process.
System twin: Representation of an entire system or ecosystem.

Key Components

Sensors and IoT devices: Collect real-time data.

Data integration platforms: Aggregate and process data.

Analytics and simulation tools: Analyze and simulate scenarios.

Communication protocols: Enable seamless interaction between the physical and digital realms.

Advantages

**Real-time monitoring*: Provides continuous monitoring and insights.

Predictive analysis: Enables prediction and prevention of potential issues.

Optimization: Facilitates optimization of processes and resources.

Future Challenges

The future challenges of blockchain technology and digital twin are as follows:

Blockchain

Scalability: Addressing the scalability issues associated with transaction processing speed.

Interoperability: Ensuring seamless communication between different blockchain networks.

Regulatory compliance: Navigating evolving regulatory frameworks.

Digital Twin

Data privacy: Managing issues related to the privacy and security of sensitive data.

Standardization: Establishing industry-wide standards for interoperability.

Integration complexity: Overcoming challenges in integrating digital twin technologies with existing systems.

Now in the subsequent sections, we will discuss the applications of each technology in logistics and supply chain management, incorporating case studies that showcase their transformative potential.

1.1 Logistics and Supply Chain Management: Introduction, Key Components, Advantages and Challenges

Logistics and Supply Chain Management (SCM) help move commodities and services from origin to consumption. These disciplines involve the coordination and integration of various activities, resources, and technologies to optimize the entire supply chain process.

Key Component:

Procurement

Sources raw ingredients or completed items from suppliers.
Key activities include supplier selection, negotiation, and order placement.

Production

Encompasses the manufacturing or assembly of goods.
Focuses on optimizing processes to meet demand efficiently.

Distribution

Involves the movement of goods from production facilities to distribution centers.
Includes transportation, warehousing, and inventory management.

Transportation

Encompasses the movement of goods between different points in the supply chain.
Modes include road, rail, air, sea, and, increasingly, last-mile delivery.

Warehousing

Provides storage for goods in transit.
Aims to optimize stockpile levels and make sure orders are filled on time.

Information Systems

Encompasses the technologies and software used for tracking, monitoring, and managing the supply chain.

Includes tools for data analytics, demand planning, and real-time visibility.

Advantages and cost-effectiveness

Streamlining processes and optimizing the supply chain leads to cost savings.
Efficient inventory management reduces holding costs.

Customer Satisfaction

Prompt and precise deliveries improve customer happiness.
Effective SCM ensures products are available when and where customers need them.

Competitive Advantage

A meticulously managed supply chain can confer a competitive advantage.
Agility and responsiveness to market changes can be a strategic advantage.

Risk Mitigation

Robust supply chain management can help mitigate disruptions.
Diversification of suppliers and contingency planning reduce risks.

Innovation and Collaboration

Collaboration with suppliers and partners provides innovation.
Embracing new technologies and practices drives continuous improvement.

The challenges of blockchain and digital twin technologies:

Globalization: Managing complex, global supply chains introduces challenges related to distance, time zones, and diverse regulations.

Supply chain disruptions: Natural disasters, geopolitical events, and unexpected disruptions can impact the supply chain. Resilience planning is essential to mitigate such risks.

Information visibility: Ensuring real-time visibility across the supply chain remains a challenge. Lack of accurate data can lead to inefficiencies and delays.

Sustainability: Balancing economic, social, and environmental factors in the supply chain poses challenges. The pressure to adopt sustainable practices is increasing.

Talent shortages: A shortage of skilled professionals in logistics and supply chain management is a growing concern. Training and development are important to address this challenge.

Here, Figure 10.1 discusses the challenges of blockchain and digital twin technologies in detail. Organizations must navigate these components, capitalize on advantages, and address challenges to build resilient and efficient supply chains. Adopting innovative technologies and embracing best practices are essential for staying competitive in today's global marketplace.

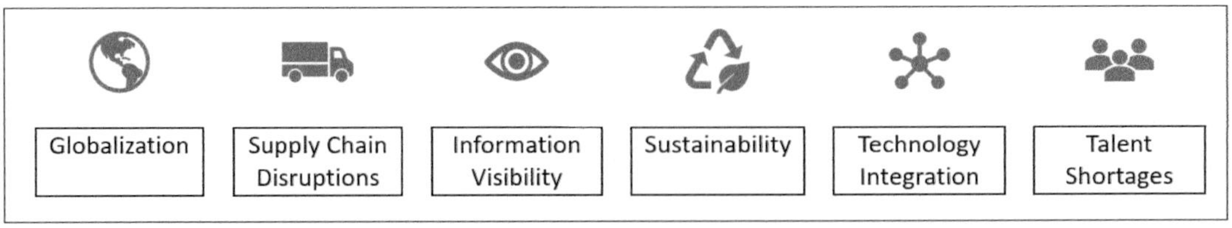

Figure 10.1. Challenges of blockchain and digital twin technologies.

2. Background Work

Blockchain is considered an appropriate auxiliary technology for implementing digital twins since it allows for trusted tracking and safe data storage (Gai et al. 2022). Digital twins have gained a lot of attention as the Internet of Things (IoT) has spread in recent years. A virtual model that simulates a real-world process or object across time is called a "digital twin." With the help of these tools, manufacturing and supply chain lead times can be shortened, creating a lean, adaptable, and intelligent production and supply chain environment. Reinforced machine learning has been included into logistics and production systems lately in order to provide prescriptive decision support platforms that combine lean, intelligent, and agile production setups. To better comprehend the current trend and potential future study directions, it is necessary to cumulatively organize and systematize the previous research conducted in this field (Abideen et al. 2021).

A rapidly expanding field of study is the use of digital twins to optimize supply chain and operations management tasks. Academics have made an effort to stay up to date with this growth by launching a rapidly changing research program. This paper aims to assess the current state of the field, identify emerging trends, and highlight the potential benefits of digital twins for supply chain and operations management. The identification of four value clusters and one enabler cluster constitutes the work's primary conclusions. The articles that explain how the use of digital twins can improve supply chain operations at the business process and supply chain capability levels make up value clusters. Product development and production flow management value clusters are developing communities that function at the business process level (Bhandal et al. 2022). In dispersed manufacturing systems, DTs must also work together to anticipate hazards and arrive at consensus-based decisions.

Sahal et al. (2021) have talked about the main, high-level requirements for the DTs' collaboration in order to address the aforementioned issues. The use of digital twins and modeling to address supply chain risks is presently a topic of study in the scientific community. Because of operational flaws, supply chains have low reliability and stability, which necessitates the creation of digital twins. These errors stem from supply chain hazards, which can be broadly classified into two categories. Operational risks are the initial category. These are the current hazards associated with the supply chain itself, which are brought on by supply and demand unpredictability as well as a barrier to information flow. Force majeure-related critical risks are the second category (Barykin et al. 2020). These risks seriously impair the company's most crucial performance metrics, including yearly revenue and profits, and interfere with the supply chain's regular operations.

Risks can arise from both natural and man-made sources, including industrial facilities and distribution centers experiencing fires or floods, conflicts arising from contracts with suppliers, labor strikes, terrorist attacks targeting logistical infrastructure, and more. Two popular solutions for managing supply chain risks are dynamic simulation and analytical optimization, which work to improve the chains' stability and dependability in the event of breakdowns. By optimizing and simulating supply chains, businesses can look at several scenarios that mimic the locations of failures, the effects of failure, and how to affect the supply chain and its performance (Botta et al. 2023). The possible instability of digital twins when taking into account digital representations of physical items is a significant issue. If there isn't a clear correlation between the digital information that appears in blockchain transactions and the physical exchange of things, using blockchain technology to trace goods is ineffective (Liu et al. 2022). The necessity of supply chain management (SCM) in the global economy is demonstrated by the COVID-19 supply chain difficulties that have occurred recently. Security and efficiency concerns with traditional supply chain management (SCM) can be resolved by utilizing blockchain technology and digital twins (DTs).

Blockchain and DTs working together can improve the ideas behind both technologies and transform supply chain management to get towards Industry 4.0 (Huang et al. 2020). Originally, the goal of the digital twin concept was to support product lifecycle management by using a high-fidelity virtual product. Numerous participants in the product lifecycle are building a complex network with vast amounts of data (Busse et al. 2021). Modern multi- and intermodal supply chains have many opportunities for optimization, but they also present substantial control and maintenance issues. It has been suggested that digital twins can enhance supply chains. But currently, they are only applied to specific segments of the whole supply chain. An initial architecture for a comprehensive Digital Supply Chain Twin (DSCT) that can incorporate a multimodal supply chain in a whole is presented in this study. In addition to being able to simulate and assess a variety of supply chain situations, such a DSCT promises to facilitate several changes throughout the supply chain. As a result, by seeing potential problems early on, the DSCT not only optimizes multi- and intermodal supply chains but may also strengthen them (Liu et al. 2022). The underlying technology of cryptocurrencies like Ethereum and Bitcoin is blockchain technology. Operations within companies will continue to be greatly impacted by the adoption of blockchain technology and the products and services it makes possible. But blockchain technology is much more than just a safe way to move Bitcoin. It can be used in non-financial domains such as voting, healthcare, insurance, welfare benefits, gambling, and artist royalties. Given that blockchain technology is already having a significant impact on both business and society, the world economy is ready for the impending blockchain revolution. If you think the word "revolution" is over the top, consider this: eight of the ten biggest companies in the world are working on blockchain-based products (Tyagi et al. 2024). An emerging field of technology called "digital twin technology" seeks to build a digital twin of a physical object that will enable real-time physical system control, analysis, and monitoring. In the context of smart factories, smart cities, and other smart applications, this technology is becoming more and more popular. An outline of the benefits and drawbacks of using digital twin technology in the smart era is given in Tyagi et al. (2023).

3. Applications in Today's Smart Era of Blockchain in Corresponding Area

As discussed in (Hemdan et al. 2023), it has the potential to revolutionize various industries, and the logistics and supply chain sector is no exception. Here are several applications in today's smart era (refer Figure 10.2)

Supply chain transparency: Each participant in the supply chain can access a single version of the truth, reducing the risk of errors and fraud. Smart contracts on the blockchain can automatically enforce agreements and trigger actions based on predefined conditions, streamlining processes and reducing delays.

Inventory management: Blockchain helps optimize inventory management by providing real-time information on the status and location of goods. This can lead to more efficient use of resources and reduced instances of stockouts or overstocking.

Reducing counterfeits and fraud: It becomes extremely difficult for counterfeit goods to enter the supply chain unnoticed. This helps in maintaining the integrity of the supply chain and ensures that consumers receive authentic products.

Streamlining cross-border transactions: Blockchain can simplify and expedite cross-border transactions by providing a secure and transparent platform for documenting and verifying the movement of goods across borders. This can reduce paperwork, delays, and the risk of errors.

Collaborative planning: Blockchain facilitates better collaboration among multiple users in the supply chain. All parties can access a shared ledger, reducing the need for intermediaries and enhancing communication and coordination.

Data security and integrity: Blockchain's decentralized and cryptographic nature enhances the security of data. It is nearly impossible to alter previous blocks, ensuring the integrity and security of the information.

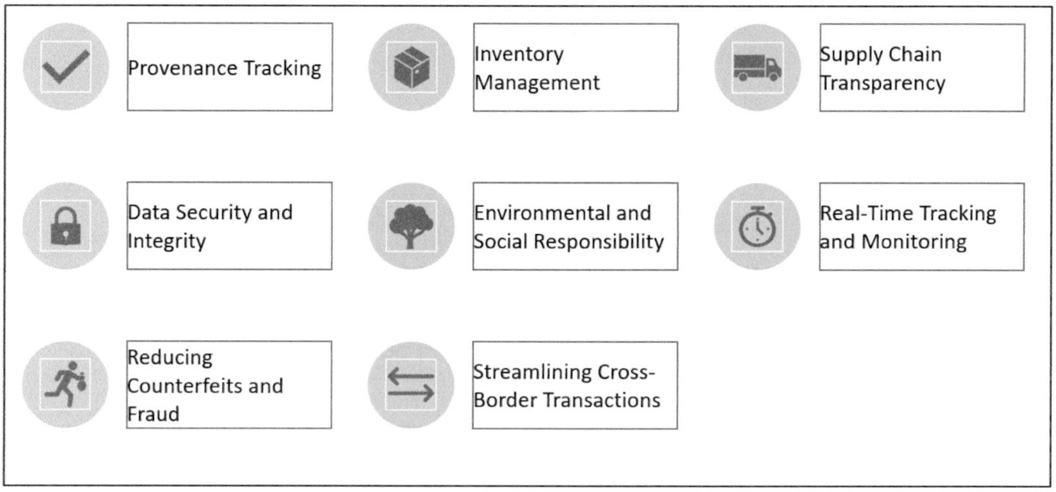

Figure 10.2. Applications of blockchain in logistics and supply chain.

Environmental and social responsibility: Blockchain can be used to track and verify sustainability and ethical practices in the supply chain. This is important for companies looking to demonstrate their commitment to environmental and social responsibility.

Real-time tracking and monitoring: By integrating Internet of Things (IoT) devices with blockchain technology, goods in transit can be tracked and monitored in real time, ensuring enhanced transparency and security throughout the supply chain.

4. Applications of Digital Twin in Logistics and Supply Chain in Today's Smart Era

Digital twins, virtual replicas of physical objects, processes, or systems, provide real-time information, simulation capabilities, and optimization potential (Vilas-Boas et al. 2023). Here are several applications of digital twin technology in the logistics and supply chain domain (refer Figure 10.3)

Predictive analytics and optimization: Digital twins use data analytics to predict future events and optimize supply chain processes. For example: Predictive analytics enable proactive problem-solving, such as predicting equipment failures or identifying potential disruptions, allowing for timely interventions and resource optimization.

Inventory management and demand forecasting: Digital twins help create virtual representations of inventory, enabling real-time tracking and demand forecasting. For example: Improved accuracy in demand predictions, reduced carrying costs, and optimized inventory levels, leading to efficient supply chain operations.

Supply chain simulation: Digital twins allow for the simulation of various scenarios and "what-if" analyses for different supply chain configurations. For example, Organizations can test the impact of changes before implementation, minimizing risks and optimizing processes for efficiency.

Smart warehousing and logistics facilities: Digital twins simulate and monitor warehouse layouts, equipment, and logistics processes. For example: Optimization of warehouse layouts, efficient use of space, and improved picking and packing processes for enhanced operational efficiency.

Figure 10.3. Applications of digital twin in logistics and supply chain.

Condition monitoring for asset management: Digital twins enable continuous monitoring of equipment and assets, providing information about their condition and performance. For example, Predictive maintenance, reduced downtime, and increased longevity of equipment, contribute to cost savings and operational reliability.

Last-mile delivery optimization: Digital twins aid in optimizing last-mile delivery routes and strategies. For example: Reduced delivery times, lower transportation costs, and improved customer satisfaction through efficient and streamlined last-mile logistics.

Collaborative supply chain networks: Digital twins facilitate collaboration among different users within the supply chain network. For example: Improved communication, transparency, and coordination among suppliers, manufacturers, and distributors, leading to a more responsive and agile supply chain.

Risk management and resilience planning: Digital twins assist in identifying and mitigating potential risks within the supply chain. For example: Enhanced risk management, rapid response to disruptions, and the ability to implement resilience strategies for a more robust supply chain.

Customer experience enhancement: By integrating customer data into digital twins, organizations can tailor supply chain processes to meet customer expectations. For example: Improved order accuracy, faster delivery times, and a personalized customer experience contribute to increased customer satisfaction and loyalty.

Hence in today's smart era, digital twins are instrumental in revolutionizing logistics and supply chain operations, providing a transformative approach to optimization, efficiency, and adaptability in an increasingly complex and dynamic business environment.

5. Integration of Blockchain and Digital Twin in Logistics for the Next Decade

The integration of blockchain and digital twin technologies in logistics holds immense promise for the next decade, providing unprecedented levels of transparency, security, and efficiency in supply chain management. The synergies between these two technologies can address several challenges in the logistics industry, providing a robust foundation for transformative advancements (Le and Fan 2024). Here's an exploration of how the integration could shape the logistics landscape in the coming decade:

Immutable traceability: Digital twins capture and record the journey of products, including manufacturing details, transportation, and storage. Whereas, blockchain ensures the immutability of this traceability data, making it tamper-proof and providing a reliable and auditable history of each product.

Smart contracts for automated transactions: It can automate and execute predefined logistics processes when triggered by specific conditions in the digital twin environment. Whereas, smart contracts enhance efficiency by automating tasks such as payments, quality checks, and customs clearance, reducing manual interventions.

Provenance and authentication: Blockchain guarantees the authenticity of data linked to a product by enabling secure and transparent tracking. It also verifies the provenance of goods, helping prevent counterfeiting and ensuring that consumers receive genuine, trustworthy products.

Supply chain simulation and optimization: Digital twins allow for the simulation and optimization of logistics processes and supply chain configurations. Whereas, blockchain facilitates secure and transparent sharing of simulation data among users, making collaboration and joint decision-making for optimizing the entire supply chain.

Decentralized and secure data sharing: Digital twins generate vast amounts of data, and blockchain provides a secure and decentralized framework for sharing this data across the supply chain network. Blockchain ensures data privacy, security, and selective access, allowing users to share information without compromising sensitive data.

Efficient cross-border logistics: Digital twins streamline cross-border logistics by providing a comprehensive view of shipments and regulatory compliance. Whereas, blockchain simplifies and accelerates customs processes, reducing delays and ensuring compliance with international regulations.

Environmental impact monitoring: Digital twins can monitor the environmental impact of logistics operations, from transportation emissions to energy usage in warehouses. Whereas blockchain keeps immutable records of sustainability practices, contributing to eco-friendly supply chain initiatives.

Interoperability across platforms: Digital twins may be implemented using various platforms and technologies. Whereas, blockchain acts as a standardized layer for interoperability, ensuring seamless communication and data exchange.

Resilience and contingency planning: Digital twins help simulate and plan for various scenarios, contributing to resilience in the face of disruptions. Whereas, blockchain ensures that contingency plans and responses are securely recorded and executed when necessary, enhancing the supply chain's ability to adapt to unforeseen events.

Future challenges and considerations: Blockchain and digital twin technologies bring about numerous benefits, but some challenges must be addressed, including interoperability, scalability, regulatory frameworks, and integration complexity. Additionally, industry-wide collaboration and the establishment of standards will be important for realizing the full potential of this integrated approach in the logistics sector over the next decade.

6. Implementation Issues and Challenges towards using Blockchain and Digital Twin in Logistics and Supply Chain

The implementation of blockchain and digital twin technologies in logistics and supply chain management presents several challenges and issues, and the key implementation issues and challenges, as mentioned in Table 10.1.

Table 10.1. Issues and challenges towards using blockchain and digital twin in logistics and supply chain.

Type	Challenges	Issues
Integration Complexity	Integrating blockchain and digital twin technologies with existing legacy systems and diverse supply chain ecosystems can be complex.	Comprehensive planning, phased implementation, and collaboration with technology experts are essential to overcome integration challenges.
Interoperability	Lack of standardized protocols of digital twin platforms and blockchain.	Industry-wide standards and collaboration are necessary to ensure diverse systems.
Scalability	The scalability of blockchain networks can be a concern.	Organizations need to choose scalable blockchain solutions and continuously monitor and optimize their infrastructure to handle increasing transaction volumes.
Data Privacy and Security	Safeguarding the confidentiality and integrity of critical information within the digital twin environment and blockchain transactions.	Implementation of robust encryption, permissioned access controls, and adherence to data protection regulations are important for maintaining data integrity and security.
Regulatory Compliance	The evolving nature of regulations and compliance standards related to blockchain and digital twin technologies.	Organizations must design systems that adhere to relevant compliance requirements.
Costs and Investment	Implementing blockchain and digital twin technologies may involve significant upfront costs, including technology infrastructure, training, and ongoing maintenance.	A thorough cost-benefit analysis, long-term planning, and identifying areas where the technologies bring the most value are essential for justifying the investment.
Talent and Skill Shortages	The shortage of skilled professionals with expertise in blockchain and digital twin technologies.	Organizations should invest in training and development programs, collaborate with educational institutions, and attract skilled talent to address this shortage
Resistance to Change	Resistance from employees and users who may be accustomed to traditional supply chain methods.	Change management strategies, communication, and education are vital to facilitate a smooth transition and gain buy-in from all users.
Environmental Issues	The environmental impact of blockchain, particularly energy consumption in certain consensus mechanisms.	Exploring eco-friendly consensus mechanisms, optimizing energy consumption, and monitoring environmental sustainability are essential aspects of responsible blockchain implementation.

7. Future Trends and Innovation Towards using Blockchain and Digital Twin

The integration of blockchain and digital twin technologies in logistics and supply chain management was already underway, and there were several trends and innovations in this space [(Suhail et al. 2022), (Srai et al. 2019), (Feng et al. 2022), (Tyagi et al. 2023), (Nair et al. 2023), (Nair et al. 2023), (Tyagi 2023), (Liu et al. 2023), (Kumari and Muthulakshmi 2023)]. Here are some potential future trends and innovations in the use of blockchain and digital twin technologies for logistics and supply chain (refer Figure 10.4):

Increased transparency and traceability: This helps in traceability, allowing users to track the journey of products from the source to the destination.

Supply chain financing: Blockchain can facilitate transparent and efficient supply chain financing. Smart contracts can be used to automate payment processes, ensuring that funds are released when specific milestones are achieved or when goods are delivered, thereby reducing delays and disputes.

Integration with IoT and Digital Twins: This integration enhances visibility and enables better decision-making.

Decentralized and collaborative networks: Blockchain allows for the creation of decentralized and collaborative networks where different users can securely share information. This can lead to improved coordination among participants in the supply chain, reducing delays and inefficiencies.

Sustainability and compliance: Blockchain can be used to create a transparent record of the environmental impact of products throughout their lifecycle. This can aid in meeting sustainability goals and complying with regulations related to environmental responsibility.

Cross-border trade facilitation: Blockchain can simplify and secure cross-border transactions by providing a single, immutable record of the entire transaction history. This can help reduce fraud, speed up customs clearance, and enhance overall trade efficiency.

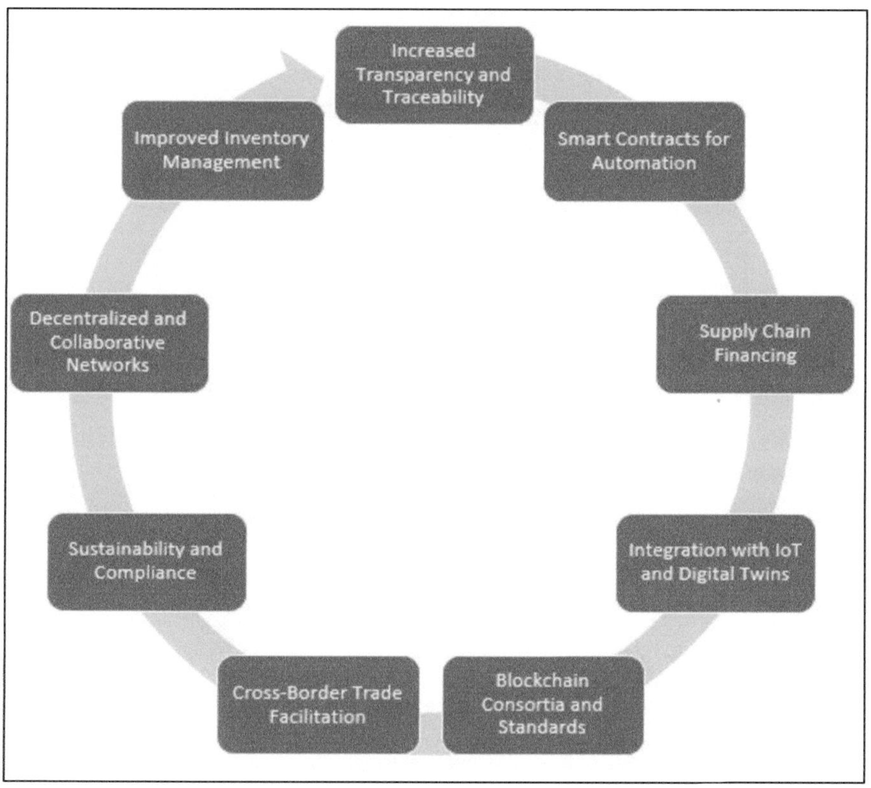

Figure 10.4. Future Trends and innovation towards using blockchain and digital twin.

Blockchain consortia and standards: Industry consortia and standardization efforts may emerge to establish common protocols and standards for blockchain implementation in logistics and supply chain. This can facilitate interoperability and widespread adoption.

In the end, ensuring the security and privacy of data in the supply chain is also important. Blockchain's cryptographic features can enhance cybersecurity, protecting sensitive information and preventing unauthorized access.

8. Conclusion

The integration represents an important advancement in the realm of logistics and supply chain management. This synergistic approach provides a comprehensive solution to address longstanding challenges and improves the transparency and resilience of supply chain processes. The application of blockchain technology contributes to a secure and transparent ecosystem by establishing a decentralized and immutable ledger. This facilitates seamless collaboration among users, mitigates the risk of fraud, and enhances traceability. Through the utilization, automated and trustless transactions are enabled, reducing the complexities associated with traditional supply chain operations.

Digital twins, with their capability to create virtual replicas of physical entities and processes, bring real-time visibility and insights into the supply chain. By providing a complete view of operations, digital twins empower decision-makers to optimize resource allocation, predict and prevent disruptions, and simulate various scenarios for better strategic planning. The combination results in a dynamic and responsive supply chain ecosystem. From reducing inefficiencies and minimizing errors to enhancing overall supply chain agility, the amalgamation of blockchain and digital twins showcases a transformative potential for the industry. However, challenges such as data privacy issues, scalability issues, and the need for standardized protocols must be addressed for widespread adoption. Future research and development efforts should focus on refining these technologies, providing interoperability, and establishing industry-wide standards to ensure seamless integration and collaboration.

Blockchain technology and digital twins redefine the future of logistics and supply chain management. This transformative synergy not only addresses current challenges, but also lays the foundation for a more adaptive, efficient, and trustworthy supply chain ecosystem in an era of rapid technological evolution. As industries continue to embrace these innovations, the journey toward a resilient and digitized supply chain promises to unlock unprecedented possibilities for global commerce.

References

Abideen, A. Z., Sundram, V. P. K., Pyeman, J., Othman, A. K. and Sorooshian, S. (2021). Digital twin integrated reinforced learning in supply chain and logistics. Logistics 5(4): 84.

Barykin, S. Y., Bochkarev, A. A., Kalinina, O. V. and Yadykin, V. K. (2020). Concept for a supply chain digital twin. International Journal of Mathematical, Engineering and Management Sciences 5(6): 1498–1515.

Bhandal, R., Meriton, R., Kavanagh, R. E. and Brown, A. (2022). The application of digital twin technology in operations and supply chain management: a bibliometric review. Supply Chain Management: An International Journal 27(2): 182–206.

Botta, V., Fusco, L., Mondelli, A. and Visconti, I. (2023, September). Secure blockchain-based supply chain management with verifiable digital twins. pp. 291–298. In Proceedings of the 2023 ACM Conference on Information Technology for Social Good.

Busse, A., Gerlach, B., Lengeling, J. C., Poschmann, P., Werner, J. and Zarnitz, S. (2021). Towards digital twins of multimodal supply chains. Logistics 5(2): 25.

Feng, H., Chen, D. and Lv, Z. (2022). Blockchain in digital twins-based vehicle management in VANETs. IEEE Transactions on Intelligent Transportation Systems 23(10): 19613–19623.

Gai, K., Zhang, Y., Qiu, M. and Thuraisingham, B. (2022). Blockchain-enabled service optimizations in supply chain digital twin. IEEE Transactions on Services Computing 16(3): 1673–1685.

Hemdan, E. E. D., El-Shafai, W. and Sayed, A. (2023). Integrating digital twins with IoT-based blockchain: concept, architecture, challenges, and future scope. Wireless Personal Communications 131(3): 2193–2216.

Huang, S., Wang, G., Yan, Y. and Fang, X. (2020). Blockchain-based data management for digital twin of product. Journal of Manufacturing Systems 54: 361–371.

Kumari, S. and Muthulakshmi, P. (2023). Artificial intelligence—blockchain-enabled technology for Internet of Things: Research statements, open issues, and possible applications in the near future. Privacy Preservation of Genomic and Medical Data 433–480.

Le, T. V. and Fan, R. (2024). Digital twins for logistics and supply chain systems: Literature review, conceptual framework, research potential, and practical challenges. Computers & Industrial Engineering 187: 109768.

Liu, J., Yeoh, W., Qu, Y. and Gao, L. (2022). Blockchain-based digital twin for supply chain management: State-of-the-art review and future research directions. arXiv preprint arXiv:2202.03966.

Liu, X., Jiang, Y., Wang, Z., Zhong, R. Y., Cheung, H. H. and Huang, G. Q. (2023). imseStudio: blockchain-enabled secure digital twin platform for service manufacturing. International Journal of Production Research 61(12): 3984–4003.

Nair, M. M. and Tyagi, A. K. (2023). AI, IoT, blockchain, and cloud computing: The necessity of the future. pp. 189–206. In Distributed Computing to Blockchain. Academic Press.

Nair, M. M. and Tyagi, A. K. (2023). Blockchain technology for next-generation society: current trends and future opportunities for smart era. Blockchain Technology for Secure Social Media Computing 10.

Sahal, R., Alsamhi, S. H., Brown, K. N., O'shea, D., McCarthy, C. and Guizani, M. (2021). Blockchain-empowered digital twins collaboration: smart transportation use case. Machines 9(9): 193.

Srai, J., Settanni, E., Tsolakis, N. and Aulakh, P. (2019). Supply Chain Digital Twins: Opportunities and Challenges Beyond the Hype. Centre for International Manufacturing, Institute for Manufacturing, University of Cambridge. https://doi.org/10.17863/CAM.45897.

Suhail, S., Hussain, R., Jurdak, R., Oracevic, A., Salah, K., Hong, C. S. et al. (2022). Blockchain-based digital twins: Research trends, issues, and future challenges. ACM Computing Surveys (CSUR) 54(11s): 1–34.

Tyagi, A. K. (2023). Decentralized everything: Practical use of blockchain technology in future applications. pp. 19–38. In Distributed Computing to Blockchain. Academic Press.

Tyagi, A. K. and Richa. (2023, August). Digital twin technology: opportunities and challenges for smart era's applications. pp. 328–336. In Proceedings of the 2023 Fifteenth International Conference on Contemporary Computing.

Tyagi, A. K., Kukreja, S., Richa and Sivakumar, P. (2024). Role of blockchain technology in smart era: a review on possible smart applications. Journal of Information & Knowledge Management 23(03): 2450032.

Vilas-Boas, J. L., Rodrigues, J. J. and Alberti, A. M. (2023). Convergence of distributed ledger technologies with digital twins, IoT, and AI for fresh food logistics: Challenges and opportunities. Journal of Industrial Information Integration 31: 100393.

Yaqoob, I., Salah, K., Uddin, M., Jayaraman, R., Omar, M. and Imran, M. (2020). Blockchain for digital twins: Recent advances and future research challenges. IEEE Network 34(5): 290–298.

Chapter 11

Palm Leaf Digitization

A Multi-modal Deep Learning Model for Character Reconstruction and Decentralized Federated Learning with Blockchain

Shyam Mohan J. S.[1,*] and *N. Kumaran*[2]

1. Introduction

Blockchain technology (BCT) has showcased its footprint in various use-cases and provided solutions across various sectors, including finance, healthcare, supply chain, and now, cultural preservation. Due to its decentralized, secure, and immutable nature, it provides a safety tool for safeguarding sensitive data and records, where trust and data integrity are important. Apart from gaining importance from cryptocurrencies like Bitcoin and Ethereum, blockchain has evolved into a versatile framework for applications far beyond digital currency, offering a transparent and tamper-proof ledger system. By ensuring that every transaction is recorded in an immutable distributed ledger, blockchain addresses the growing need for security, transparency, and decentralized control in modern data management systems. Blockchain technology (BCT) has gained popularity in many potential applications such as industrial sectors, financial institutions, government agencies, etc. Creating fraudulent records is very difficult in a blockchain network. The structure of the block remains the same irrespective of its tool and applications. All the transactions in a blockchain are decentralized, that is, there is no third party involved in the transactions. Due to its distributed ledger concept, all the transactions are tamper-proof and hence, make it impossible for anyone to manipulate the records (Parthasarathi 2013). Blocks store information in a highly secure and digital format.

The most famous blockchain-based transaction is Bitcoin. Blockchain ensures the reliability and security of records stored and makes them trustworthy even when no one trusts them. One of the important differences between a normal information system and a blockchain is the way information is organized. Blockchains store information in the form of blocks. Blocks have a certain acceptable

[1] Associate Professor, School of CSE, RV University, Bengaluru, Karnataka, India.
[2] Associate Professor and Head, Dept. of IoT, Dhanalakshmi Srinivasan University, Samayapuram, Trichy, Tamilnadu, India.
Email: kumarann.set@dsuniversity.ac.in
* Corresponding author: shyammohanjs@rvu.edu.in

DOI: 10.1201/9781003620822-11

limit, and when they are filled, they are closed and are connected to the newly filled blocks to form an information chain called a blockchain. Any new data following this newly added block will be included in the newly created block, which will also be added to the chain, if it is filled at this point. Datasets mostly structure information into tables, whereas blockchains structure information into interconnected chunks. This data structure, when executed in an inherently distributed manner, creates an irreversible course of events. The moment the plaza is filled, it becomes firmly anchored and becomes part of the course of this event. When a block is added to the blockchain, a timestamp is assigned to it (Gupta et al. 2020, Zhang et al. 2018). Each block in the blockchain contains many transactions, and each time a new transaction occurs on the blockchain, a record of that transaction is added to each participant's record. This is also called Distributed Ledger Technology (DLTs). All the transactions in the DLT are stored using immutable cryptographic tokens known as hashes. No modifications are allowed in a block; however, local modifications are allowed. For example, Bitcoins and Ether are the real-time applications of blockchain that are widely used for transactions.

Palm leaf manuscripts are rich sources of knowledge for society. They provide solutions to several unanswered questions on Earth. As they are very old, from many centuries ago, preserving palm leaf manuscripts and maintaining them is a difficult task. Many countries are trying to find out ways to preserve palm leaf manuscripts (Yaga et al. 2019). Many researchers and scholars have realized the importance of preserving the palm leaf manuscripts. Digitizing palm leaf manuscripts will provide a good solution for how to store the data of these manuscripts and making future generations realize the real importance of palm leaf manuscripts. Research has also started on how to preserve and digitize palm leaf manuscripts. There are many categories or metadata present in palm leaf manuscripts, for example: medicinal, such as Siddha, Ayurveda and yunani systems; veterinary science, agriculture, traditional music and complex methods are all examples of traditional human life systems (Varmam, medical procedures) along with yoga, agriculture, combats expression, Physiognomy (Samudrika Laksanam) and so on. The rich treasure of knowledge is written in other forms too, like on stone, fabrics, metals, etc., that have significance related to history or aesthetic values. The manuscript contains a wealth of knowledge written in various languages and scripts. Often, one language is written in another script (Kall 2018). For example, Sanskrit is written with Oriya content, Granthi content, Devanagari content, and many other contents. Lithographs and printed copies are not manuscripts. Manuscripts, on the other hand, differ from historical records such as rock inscriptions, palace gates, and income statements that provide direct information about historical events (Yang 2019).

In more than 5,000 years of history, India has amassed a vast amount of writing. This huge information treasure is made up of a variety of materials: birch bark, palm leaves, cloth, wood, stone, and paper. They are cared for by various foundations such as libraries, galleries, religious communities, mosques, and people. Many of them are gone, and a huge number are in a degraded/decayed state. At present, we have lost a vast range of compositions from Buddhist religious communities such as Nalanda, Taxila, and Vikramshila. Taxila, on the northwestern front of India, was destroyed by distant hostility. The library of Nalanda, known as Dharmaganj, had three areas: Ratna Lodhi, Ratnasagar, and Ratna Rajak. Ratna Lodhi was a famous library. The first Hun king, Michirkul, destroyed the library in 450470 AD. In 1205, Bakhtiyar Khilji completely destroyed it, and his warriors used the compound to boil water. There are many libraries around the world with unusual works from India, some of which are: Copenhagen Royal Library (Denmark); National Library (France); Preussiche State Bibliothek, Universitats Bibliothek, Staats Bibliothek, Bayershe Staats Bibliothek (Germany); British Library, Royal Asian Society, Bodleian Library, Indian Institute, University Library of Cambridge (UK); Darbar Library (Nepal); Punjab University Library (Pakistan); Museum Library (Sri Lanka) (Yli-Huumo et al. 2016). In recent times, a great deal of interest has been shown in the preservation and digitization of palm leaf manuscripts. Towards this effort, implementing blockchain provides more safety and security in protecting the endangered historical artefacts that are the rich sources of knowledge and information, scripted and saved decades ago. Written in Telugu, Tamil, and Sanskrit, these manuscripts are highly fragile and are easily decayed. Preservation is of the utmost importance and includes various steps.

This paper introduces a blockchain-based approach for preserving and storing digitized palm leaf manuscripts. The proposed approach relies on OCR and advanced machine learning models for reading and recognizing characters. It is then integrated with blockchain technology for secure translation and transliteration of metadata. The decentralized ledger in the blockchain ensures that the transactions remain unaltered and are accessible to authorized users, and thereby provides tamper-proof data. The knowledge present in palm leaf manuscripts spreads across various disciplines like ayurveda, agriculture, law, medicine, etc. Digitizing of these manuscripts not only provides knowledge and information but, paves the way for the next generation of researchers and scientists. Blockchain provides a novel solution to protect the cultural heritage and hidden treasures that are ignored and poses a challenge to human beings.

The proposed work tries to establish a framework for secure and efficient metadata extraction and preservation.

2. Existing and Related Work

The digitization of palm leaf manuscripts enables easy access for future generations. Much research was being carried out for the digitization of palm leaf manuscripts, but there were many challenges in terms of preservation and accuracy. Parthasarathi et al. (2013) focused on palm leaf manuscript digitization by improving the quality of the images with noise-based reduction techniques such as edge detection, background removal, etc., for extracting the scripts written in the palm leaf manuscripts. Later, with the advancement of technology, optical character recognition (OCR), Pytesseract were used for translating text written in Tamil and Telugu scripts. The usage of OCR was limited in cases of damaged or missing characters, which led to the use of machine learning approaches for script recognition and completion for many researchers. Convolutional neural networks (CNNs) and recurrent neural networks (RNNs) showed better results in character recognition from the digitized palm leaf manuscript images.

Gupta et al. (2020) used Generative Pre-trained Transformers (GPT) with a deep learning-based OCR system for detecting damaged and missing characters in ancient manuscripts. GPT models were trained on large datasets of manuscripts and are capable of detecting the missing characters based on the surrounding text, making them effective for use.

Blockchain technology has demonstrated significant potential for the preservation, integrity, and authenticity of digital archives. Data stored in the blockchain is secure, and the activities or transactions are tamper-proof.

Zhang et al. (2018) used blockchain technology for creating immutable, distributed records of digital repositories for manuscripts, which constitutes securing the metadata, including the script, the language, and the text present in the manuscripts. The decentralized ledger provides transparent data that is useful for accessing the digitized content and protecting it from unauthorized usage.

Blockchain is gathering attention in many sectors like healthcare and digital repositories, but its application in preserving cultural and heritage is still under consideration. Metadata blockchain (MDBC) provides the immutability of information extracted from palm leaf manuscripts. All the translations, transliterations, and associated metadata on a blockchain are stored securely, and the data remains tamper-proof, and changes made are visible to all the stakeholders and hence ensuring transparency and accountability in the preservation process.

Combining machine learning and metadata blockchain offers a secure and comprehensive approach to addressing the challenges of palm leaf manuscript digitization. Existing works have made significant contributions in digitizing and preserving these manuscripts, and integrating machine learning with blockchain provides a novel solution.

3. Digitization of Palm Leaf Manuscripts

Palm leaf manuscripts (Figure 11.1) contain rich sources of knowledge and information dating back many centuries. Ancient knowledge present in the palm leaf manuscripts is still being adopted today in medicine, astronomy, philosophy, and other traditional sciences. Due to their fragile nature, the palm leaf manuscripts are easily breakable and are prone to various environmental and climatic conditions. This has led to a loss of information, and it is becoming difficult to preserve the palm leaves. Digitization has provided a solution in preserving the palm leaf manuscripts by offering the means of capturing, storing, and sharing information globally, without causing any delay.

Digitization involves multiple stages for processing, starting from scanning the physical palm leaf manuscripts and checking them for clarity and storing them in a database schema for further process, where image pre-processing, noise reduction, background removal, edge detection and image enhancement techniques are applied to improve the readability of the characters or the text. Optical Character Recognition (OCR), Pytesseract techniques are applied for converting the digitized images into a machine-readable form for the next step process. The difficulty lies in the identification of text and characters as the languages are written in ancient scripts, and some part of the text is damaged, etc. By incorporating machine learning techniques, Generative Pre-trained Transformers (GPT) have made it easier for text reconstruction and prediction of missing or damaged text in digitized images. Machine learning models are trained on large datasets of historical and ancient scripts, and hence, it makes it easy for them to predict the characters and words based on the context, enhancing the quality of the digitized images. Metadata is useful for maintaining the information about the type of manuscript, content, and is also essential for categorization and future retrieval.

Blockchain technology provides transparent access for the authentication of digital transactions, protecting them from unauthorized usage or any modifications.

With digitization, the information and knowledge can be accessed by scholars, scientists across the globe, hence opening doors for preservation of cultural heritage by enhancing research and knowledge sharing. By integrating OCR, Machine learning and blockchain technology, the digitization of palm leaf manuscripts excels as a new innovation in cutting edge technologies.

In recent years, the digitization of manuscripts has become popular as a means of safeguarding and documenting literary heritage. Digitization, along with advances in information technology, has made it easier for scholars and researchers to access manuscripts. In India, the National Manuscript Mission started in 2004 as a pilot study to digitize multiple manuscript caches across the country. In 2006, the Digitization Standards and Guidelines mandate came into force.

Figure 11.1. Sample showing digitized Tamil leaf manuscript.

4. National Digital Manuscripts Library in India (NDML)

The purpose of India's digital manuscripts is for them to be easily accessible for creativity. National Informatics Centre (NIC) team, Hidden Markov Model (HMM) has created metadata on the Internet and has created an easy-to-use web application to reach the original web search tools through a dedicated web search tool. The metadata is sent to the cloud server with the computerized image,

Table 11.1. The details of the digitization done under the Fourth Phase.

Institution	Number of manuscripts digitized to date	Number of pages	Remarks
Rajasthan Oriental Research Institute, Bharatpur, Rajasthan	9,153	5,43,236	Complete
Rajasthan Oriental Research Institute, Bikaner, Rajasthan	6,160	3,10,510	Complete
Rajasthan Oriental Research Institute, Kota, Rajasthan	9,800	10,25,300	Complete
Rajasthan Oriental Research Institute, Alwar, Rajasthan	2,000	3,02,700	Complete
Kurukshetra University, Kurukshetra, Haryana	12,500	7,38,752	Complete
Patna Museum, Patna, Bihar	900	1,90,300	Complete
Rashtriya Sanskrit Sansthan, Allahabad, Uttar Pradesh	17,480	14,35,750	Complete
Bhandarkar Oriental Research Institute, Maharashtra	8,350	10,45,700	Work in Progress
Vadodara Oriental Research Institute, Vadodara, Gujarat	24,500	21,25,256	Work in Progress

which must be interconnected to provide simple access to the advanced originals for a wide range of applications for research of various items. This programming maintains the guardian, exemplary index, and the topic director, maintains the original copy state, and accesses the computerized image through the Web Caterpillar. First, it is designed to encourage access to this information database by configuring a cutting-edge digital document library in New Delhi with limitations with restrictions on ten clients. Clients rely on getting information from a data center through a workspace PC. New information must be enabled on a daily schedule on a capacity server. Table 11.1 shows the status of digitization done after the fourth phase.

5. Challenges

As of now, we have digitized over 3,00,000 palm leaves. By taking further steps, we are going to extract the data from the digitized palm leaf manuscripts with the maximum accuracy rate and secure them safely. In the process of extracting the data from digitized images, we have to overcome some sorts of challenges. As the palm leaf manuscripts belong to the late centuries, some of the words or letters in those manuscripts have faded away. So, while we are extracting the data, we have to predict the faded ones accurately. We have to convert the regional language to various user languages.

Extracting information from the palm leaf manuscripts is a tedious task that involves various aspects that need to be looked at. They are fragile and tend to break if we touch them. Apart from fragility, the text or characters are not clearly readable as they are very old and there are issues such as blur, smearing, dirt, discoloration and holes, and spots on the manuscripts. Also, the languages used in the manuscripts are old and reformed, so understanding the script becomes challenging. Besides, the characters on the palm leaves were handwritten in narrowly spaced lines with overlapping components touching one another. Moreover, the characters have unusual, varying shapes and different styles, which depend on the writer. In the proposed work, we have tried to overcome these hardships and effectively retrieve data (to some extent) from these ancient manuscripts. In this paper, the objective of our project and its scope are presented, this is followed by background research and literature surveys. In the next section, the problem statement and algorithm of the project are discussed. Following that, the process and project description is carried out. The manuscripts have been damaged over the years, the image may be uneven and not clear, and image pre-processing has had to be carried out to effectively retrieve information. It could be done by resizing images, followed by binarization and enhancements of the image for predicting the pattern. OpenCV and Matplotlib were used in image pre-processing steps.

6. Objectives and Problem Statement

6.1 Objectives

The preservation of palm leaf manuscripts, which are centuries-old repositories of ancient knowledge, is becoming an increasingly critical challenge due to their fragile nature and susceptibility to degradation. These manuscripts contain valuable insights into subjects such as medicine, astronomy, technology, music, and philosophy, often written in various ancient scripts like Tamil and Telugu. Despite advancements in optical character recognition (OCR) and image processing, existing digitization efforts still face substantial obstacles: (i) incomplete recognition due to physical deterioration of manuscripts, (ii) missing characters and text lines, and (iii) lack of security and integrity in storing the digitized content.

The objectives of palm leaf manuscript digitization are given below:

1. Segmentation and recognition of historical scripts despite degradation.
2. Reconstruction of missing or faded characters without distorting linguistic integrity.
3. Creation of standards and procedures for the digitization of manuscripts using blockchain.

6.2 Problem Statement

There were no earlier works carried out by combining cutting-edge technologies, generative machine learning models for character recognition and prediction, and integrating them with blockchain technology for securing metadata. In fact, these technologies were never used in the context of the digitization of palm leaf manuscripts. This problem has been addressed, and we propose a novel solution that addresses the technical difficulties involved in digitizing damaged manuscripts, and we have also tried to provide the trustworthiness of the digitized information. The proposed approach paves the way for research in the digital era of ancient sciences—one that is applicable to all forms of ancient and historical documents.

Let us assume $I(x,y)$ represents a digitized manuscript image where each pixel intensity represents text structure variations. If a degraded manuscript is identified or given as the input, the OCR recognition task can be defined as:

$$T = arg\ max_{T'}\ P(T'|I, \theta) arg\ max_{T'}\ P(T'|I, \theta)$$

where T′ is the set of possible text outputs, and

θ represents the parameters learned via ViT-based feature extraction and GNN-based segmentation.

Character restoration is done by Diffusion Models that follow a reverse de-noising process:

$$X_0 = f_\theta f_\theta (X_T), X_T \sim N(0, I)$$

where X_T is a noise detected in the degraded text region, progressively refined by f_θ.

The extracted text T is secured via blockchain hashing using:

$$H_i = SHA - 256(T_i, H_{i-1})$$

where H_i is the hash of the current text segment and

H_{i-1} is the previous block hash, ensuring tamper-proof and immutable recordkeeping.

To address the challenges as mentioned in section 6.1, we propose a multi-modal deep learning approach that integrates:

1. Vision Transformers (ViTs) for detecting and recognizing spatially-aware OCR processing of digitized manuscripts.

2. Graph Neural Networks (GNNs) for character segmentation and connectivity modeling, ensuring correct identification of fragmented text.
3. Diffusion Models for probabilistic character reconstruction and missing characters in between.
4. Decentralized Federated Learning to train distributed OCR models across multiple research nodes, ensuring data privacy while improving model accuracy collaboratively.
5. Blockchain-powered Zero-Knowledge Proofs (ZKP) for tamper-proof manuscript storage and authentication, guaranteeing historical integrity.

6.3 Blockchain Integration

To ensure the security and integrity of the digitized content and metadata, we propose the integration of blockchain technology. Let the metadata D for each digitized manuscript include the following fields:

$$D = \{M_{ID}, T_{language}, T_{transliterated}, T_{translated}, T_{date}\}$$

Where M_{ID} is the manuscript identifier, $T_{language}$ is the identified language of the manuscript, $T_{transliterated}$ is the transliterated text, $T_{translated}$ is the translated text, and T_{date} is the timestamp of digitization.

The metadata is hashed using a cryptographic hash function H, such that:

$$h_D = H(D)$$

This hash h_D is stored on the blockchain, ensuring the immutability of the metadata. Any changes to the metadata will produce a different hash, allowing for tamper detection. In addition to this, the manuscript's information can also be linked to blockchain records to provide verifiable provenance.

The blockchain ledger L consists of a sequence of blocks $B_1, B_2, \ldots, B_n$, where each block B_i contains:

$$B_i = \{h_{Di}, T_i, PrevHash\ (B_{i-1}), Nonce\}$$

Where
h_{Di} is the hash of the manuscript's metadata in block i,
T_i is the timestamp,
PrevHash (B_{i-1}) is the hash of the previous block, and
Nonce is the proof-of-work parameter ensuring block integrity.

6.4 Mathematical Implications

Let a digitized palm leaf manuscript be represented as a grayscale image matrix *I* of dimensions M × N

$$I(x,y)^{\in\in} [0,255],\ x^{\in\in} \{1,\ldots,M\},\ y^{\in\in} \{1,\ldots,N\}$$

Where low-intensity values indicate non-visible regions and high-intensity values represent background noise.

6.4.1 Character Segmentation Using Graph Neural Networks (GNNs)

A manuscript image *I* is preprocessed using an adaptive thresholding function T to separate text from the background:

$$I_{binary} I_{binary}(x,y) = 1,\ I(x,y) > T(x,y)$$

$$I_{binary} I_{binary}(x,y) = 0, \text{ otherwise}$$

Where $T(x,y)$ is calculated dynamically for local text variations.

A Graph Neural Network (GNN) is then applied for character segmentation. Define the manuscript as a graph $G = (V, E)$,

Where,

Nodes V represent individual character strokes or fragments.

Edges E encode spatial relationships between strokes.

Each node $v_i \in V v_i \in V$ has an initial feature representation $h_i h_i$ corresponding to pixel intensity:

$$h_i^{(0)} h_i^{(0)} = \phi\, \phi\, (I_{binary} I_{binary}(x_i, y_i))$$

where ϕ is an embedding function mapping pixel values to a high-dimensional feature space.

Character connectivity is refined iteratively using Graph Convolutional Networks (GCNs):

$$h_i^{(t+1)} h_i^{(t+1)} = \sigma\sigma(W^t) \sum_{j \in N(i)} \frac{h_j^t}{|N(i)|} \sum_{j \in N(i)} \frac{h_j^t}{|N(i)|}$$

Where,

$N(i)$ is the neighborhood of node i.

$W(t)$ are learned weight matrices.

Σ is an activation function.

This GNN-based segmentation ensures that broken character strokes are accurately grouped, mitigating errors in traditional OCR-based methods.

6.4.2 Character Reconstruction Using Diffusion Models

For faded or missing text regions, a Diffusion Model f_θ is used to restore lost character structures. The generative process is formulated as a de-noising stochastic differential equation t(SDE):

$$\frac{dXtdXt}{dt\ dt} = -\beta\beta(t)X_t + \sqrt{\beta(t)}\sqrt{\beta(t)}\ Wt, X_0 = I_{binary} I_{binary}(x, y)$$

Where,

Xt is the corrupted text representation at time t.

$\beta(t)$ is a noise schedule function.

Wt is a Gaussian noise term.

The restored character image is obtained by solving the reverse-time SDE using a trained neural network:

$$X_0 = f_\theta(X_T)$$

where X_T is a fully corrupted version of the image, and f_θ learns to progressively restore missing text regions.

This diffusion-based restoration ensures high-fidelity reconstructions of degraded text while maintaining script authenticity.

6.4.3 Vision Transformers (ViTs) for OCR Enhancement

Unlike traditional CNN-based OCR, Vision Transformers (ViTs) process entire text sequences holistically, preserving long-range dependencies.

The manuscript image is divided into patch embeddings:

$$z_0^p z_0^p = W_p I^p_{patch} I^p_{patch} + E_p$$

Where,

Wp is a linear projection matrix.

Ep is the positional embedding.

Self-attention mechanisms refine the extracted text representation:

$$Attention\ (Q,K,V) = softmax(\frac{QK^T QK^T}{\sqrt{d_k}\sqrt{d_k}})V$$

Where,

Q, K, V are the query, key, and value matrices.

D_k is the dimension of the key vectors.

By leveraging ViTs, OCR accuracy is significantly improved, especially for ancient scripts with varying character sizes and orientations.

6.4.4 Decentralized Federated Learning with Blockchain

To ensure secure and decentralized model updates, a federated learning framework is integrated with blockchain technology.

Each participating edge device $d_i$updates the local OCR model using:

$$\theta_i^{t+1}\theta_i^{t+1} = \theta_i^t\theta_i^t - \eta\nabla Li(\theta_i^t\eta\nabla Li(\theta_i^t))$$

where:

L_i is the OCR loss function for device d_i.

H is the learning rate.

Instead of sending raw data, each device transmits model gradients (∇L_i) to a blockchain-based aggregator, ensuring data privacy.

Zero-Knowledge Proof (ZKP) ledger ensures Data integrity, Tamper-proof transactions, and Scalability.

6.5 Methodology

As mentioned earlier, we tried to combine a comprehensive and systematic methodology integrating advanced digitization techniques with blockchain for secure and reliable preservation. The process of digitization and the steps involved in carrying out the proposed research work are shown in Figures 11.2 and 11.3.

The detailed explanation of each step is given below:

6.5.1 Preprocessing and Image Enhancement

Pre-processing is the first step in the digitization process. Pre-processing is done to improve the quality of the images of the manuscripts. The initial step in the digitization of palm leaf manuscripts involves converting physical manuscripts into high-quality digital images. Given the deteriorated nature of many manuscripts, image enhancement techniques are crucial to improve character visibility and reduce noise. The following methods are employed:

1. Noise Reduction and Background Removal
2. Edge Detection and Character Segmentation
3. Text Line Extraction

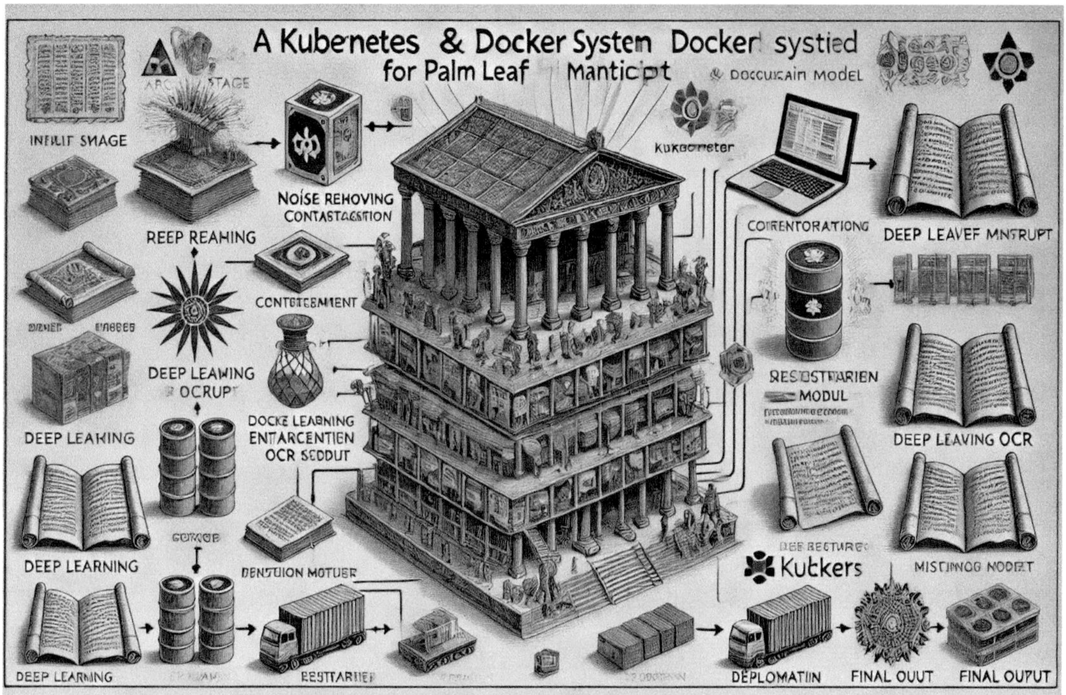

Figure 11.2. Overall architecture of the proposed model.

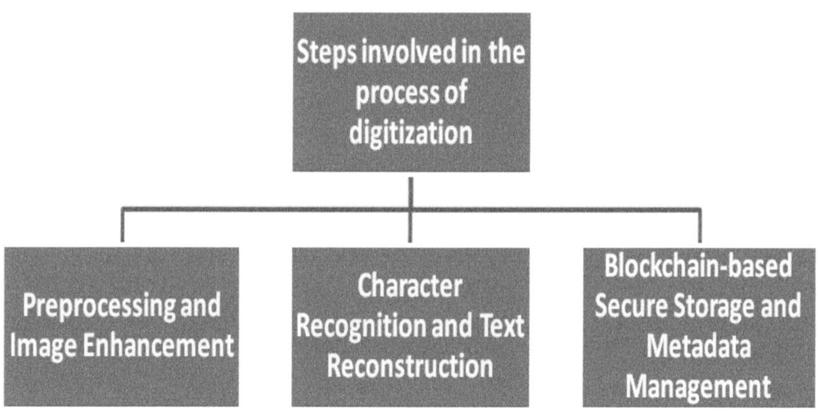

Figure 11.3. Process of digitization.

6.5.2 Noise Reduction and Background Removal

Let the raw manuscript image be represented as $I_0 \in R^{m \times n}$, where m and n represent the dimensions of the image. The noise in the image, denoted by N(x,y), can be modeled as additive white Gaussian noise (AWGN):

$$I(x, y) = I_0(x, y) + N(x, y), N(x, y) \sim N(0, \sigma^2)$$

Where I(x,y) is the observed noisy image and σ^2 is the variance of the noise.

To reduce noise and remove background artifacts, a combination of Gaussian filtering and morphological operations (e.g., erosion and dilation) is applied:

$$I_{filtered}(x,y)=\frac{1}{2\pi\sigma 2}\exp\left(-\frac{(x-x_0)2+(y-y_0)2}{2\sigma^2}\right)$$

where σ is the standard deviation of the Gaussian filter.

6.5.3 Edge Detection and Character Segmentation

Once the background noise is removed, edge detection is applied using the Canny edge detector, which identifies areas of rapid intensity change:

$$G(x,y)=\sqrt{\left(\frac{\partial I_{filtered}(x,y)}{\partial x}\right)^2}+\sqrt{\left(\frac{\partial I_{filtered}(x,y)}{\partial y}\right)^2}$$

The edges $G(x, y)$ are used to segment individual characters from the manuscript for further analysis.

6.5.4 Text Line Extraction

A partial projection profile is computed to extract lines of text. The projection profile $P_{horizontal}(i)$ is computed by summing pixel intensities along each row:

$$P_{horizontal}(i)=\sum_{j=1}^{n} I_{filtered}(i,j)$$

Peaks in $P_{horizontal}(i)$ correspond to text lines, which are subsequently cropped and processed for character recognition.

6.5.5 Character Recognition and Text Reconstruction

The Generative Pre-trained Transformer (GPT), is employed for recognizing characters, predicting missing or damaged text, and reconstructing the complete manuscript content. After preprocessing the images, the next step involves identifying characters and reconstructing missing or damaged text. This task is achieved through a Generative Pre-trained Transformer (GPT) model trained specifically for ancient scripts, such as Tamil and Telugu.

6.5.6 OCR and GPT-based Character Recognition

To facilitate portability and scalability, the entire OCR pipeline was containerized using Docker. Docker encapsulates the OCR engine, preprocessing scripts, and necessary libraries into isolated containers, ensuring consistency across environments. This containerization simplifies deployment, allowing the pipeline to be replicated across multiple cloud nodes.

Let the manuscript image be divided into segments, each containing a sequence of characters $S = \{s_1, s_2, ..., s_t\}$, where $s_i \in C$, the set of all possible characters in the language script. The OCR engine (Pytesseract) provides an initial recognition of the characters in the sequence S. However, due to the degraded quality of the manuscripts, some characters si may be recognized as missing or incorrect.

To predict and reconstruct missing characters, the problem is modeled as a sequence generation task, where the GPT model learns to predict the probability distribution over the character set C:

$$P(s_i \mid s_1,...,s_{i-1},s_{i+1},..., s_t; \theta)$$

Where θ represents the model parameters. The GPT model is fine-tuned on ancient manuscripts with missing text, learning both the syntax and structure of the language to accurately predict missing segments.

6.5.7 Text Reconstruction

For any missing character $s_i = \emptyset$, the model generates the most likely character $\hat{s}_i\hat{s}_i$ by maximizing the conditional probability:

$$\hat{s}_i = \arg\max_{c\in C} P(c \mid S;\theta)$$

The reconstructed sequence $\hat{S}\hat{S} = \{\hat{s}_1, \hat{s}_2\hat{s}_1, \hat{s}_2, \ldots\ldots, \hat{s}_t\hat{s}_t\}$ represents the predicted and restored text of the manuscript.

6.5.8 Translation and Transliteration

Once the full text is recognized and reconstructed, translation and transliteration processes are applied. Let the original text be denoted by T_{orig} and the translated text by T_{trans}. A machine translation model (based on transformer architecture) is used to convert the original script into modern-day equivalents for broader accessibility.

6.6 Blockchain-based Secure Storage and Metadata Management

The digitized manuscripts and their metadata are stored in a blockchain to ensure tamper-proof, secure, and decentralized storage. Once the manuscript's content is recognized and translated, it is securely stored using a blockchain to ensure immutability, integrity, and traceability. The blockchain provides a decentralized and tamper-proof ledger where metadata and digitized content are stored.

6.6.1 Metadata Structure

Each manuscript is assigned in the metadata D, which includes:

M_{ID}: Manuscript identifier

$T_{language}$: Language of the manuscript

$T_{transliterated}$: Transliterated text

$T_{translated}$: Translated text

$T_{timestamp}$: Time of digitization

h_D: Hash of the metadata

The metadata D is hashed using a cryptographic hash function H:

$$h_D = H\,(M_{ID}, T_{language}, T_{transliterated}, T_{translated}, T_{timestamp})$$

6.6.2 Blockchain Storage Process

Assume that the hashed metadata is h_D and B_i is a blockchain block. Now, hashed metadata is stored in a blockchain block. Each block B_i contains:

$$Bi = \{h_D, T_i, \text{PrevHash}\,(B_{i-1}), \text{Nonce}\}$$

Where,

T_i is the timestamp

PrevHash (B_{i-1}) is the hash of the previous block, and

Nonce is a proof-of-work parameter.

All the sequences of blocks $\{B_1, B_2, \ldots, B_n\}$ are stored in ledger L. The ledger ensures that any tampering in the blocks would invalidate the entire transaction chain and hence provide a secure and decentralized environment for the preservation of manuscripts. The OCR framework is deployed in the cloud using Kubernetes for orchestration. Kubernetes is capable of load balancing, resource allocation, and provides fault tolerance against high traffic volumes and scales well based on demand.

6.6.3 Kubernetes Architecture

Kubernetes architecture is similar to Hadoop MapReduce architecture. The multi-node Kubernetes cluster is hosted on an open source OpenShift platform that consists of a Master Node and the Worker nodes, where the master nodes are responsible for scheduling, load balancing, and monitoring the health of the nodes, while the later part contains the Docker containers consisting of the OCR engine and other services. These nodes could scale up or down depending on the workload. Each container is deployed as a Kubernetes service and is responsible for scaling the Kubernetes pods. This ensures that multiple manuscripts can be processed simultaneously without any difficulty. Replicas were managed dynamically using Kubernetes' Horizontal Pod Autoscaler (HPA) that auto-adjusts based on CPU and memory usage. The scaling mechanism of replicas can be represented as:

$$\text{Replicas} = \text{HPA}\ (\text{CPU\_Utilization, Memory\_Usage})$$

6.6.4 Blockchain Integration for Provenance and Data Integrity

Blockchain technology ensures the authenticity and immutability of the digitized texts. This is done by digitizing each manuscript, extracting the OCR text, and then hashing using the SHA-256 algorithm and storing them in a blockchain ledger guarantees data provenance and tamper-proof data.

6.6.5 Blockchain Implementation

Each block in a blockchain consists of an OCR transaction, the metadata, and the translated text. The block structure included the following fields:

Manuscript Hash (H): The SHA-256 hash of the manuscript and extracted text.

Timestamp (T): The time when the manuscript was processed.

Previous Hash (H'): The hash of the previous block, ensuring immutability.

6.6.6 Smart Contracts for Access Control

Smart contracts are used for providing access control for accessing the digitized manuscripts. They provide mechanisms for regulating who can access, modify, or share the data. Manuscripts are accessed based on cryptographic signatures. Translation or re-use of manuscripts is handled by predefined sets of rules embedded in the smart contracts.

6.7 Post-Processing and Language Translation

After the text is extracted from manuscripts, Natural Language Processing (NLP) techniques are used for handling translation. The text used in manuscripts, the language, and the script used are identified by the automatic language prediction algorithm that was used to identify the source language. The translation process can be scaled dynamically across various nodes in the cloud using containerized translation services. The identified language is then converted into the language of the desired type using an NLP-based language translation model that is mathematically expressed as:

$$\text{Translated Text} = T\ (\text{OCR Text} \mid \text{Source Language})$$

Where T represents the translation model.

The proposed system was tested with a huge dataset of palm leaf manuscripts written in various languages that are available in various open sources. A few of the manuscripts were available in the manuscript library. OCR accuracy, processing time, and blockchain verification time were calculated and recorded. The proposed model scaled well across multiple datasets, and integrity checks were done using blockchain to ensure authenticity.

6.8 Experimental Setup

The proposed framework was designed to digitize, process, and analyze palm leaf manuscripts using advanced NLP techniques. Image segmentation, blockchain-based validation of data, followed by translation and transliteration, was successfully integrated by Optical Character Recognition (OCR). Tesseract was used to recognize text in multiple languages. The experimental setup was deployed in the Cloud using Docker and Kubernetes to ensure portability and smooth allocation of resources. Preprocessing of the manuscripts was done using OpenCV for image enhancement, grayscale conversion, and thresholding for separating regions of interest. Textual data was extracted using region-based segmentation by a contour identification mechanism. Blockchain simulation was developed to ensure data integrity, where each segment's hash was linked to the previous one like a Merkle Tree and hence formed an immutable chain. Extracted textual data were translated into English and other languages using natural language processing (NLP) models, ensuring cross-lingual usability. The experimental setup was implemented for the datasets as mentioned in Table 11.2. Google Translate API for multilingual language translation, and Python libraries for blockchain simulation.

6.9 Dataset Description

The raw images are extracted from various open-source repositories, and a few raw images were obtained from the data and heritage research centre of one of the institutions in Chennai. The manuscripts were written in different languages and scripts, such as Tamil, Telugu, Bali, Nandinagari, etc. The data and information present in the manuscripts consist of a variety of knowledge, varied from ancient literature, cultural texts, and religious scriptures. Each palm leaf manuscript went through several stages in preprocessing to improve the resolution and format in order to provide a uniform input for OCR. Language annotations and provenance information were obtained from the Metadata of the manuscripts that were used for context-aware processing. The datasets used provided an effective framework for solving real-world problems. The description of the datasets is shown in Table 11.2.

Table 11.2. Description of the datasets.

SI.No.	Dataset type	Description	Available Web Link
1	The Endangered Archives Programme (EAP)	Consists of digitized archives around the world that are in danger of destruction, neglect or physical deterioration.	https://eap.bl.uk/collection/EAP1217-1/search
2	ICDAR 2013 Scene Text Recognition Dataset IIIT 5K-Word Dataset COCO-Text Dataset	This dataset contains images of scene text that are annotated for text detection and recognition.	https://rrc.cvc.uab.es/?ch=2 https://cvit.iiit.ac.in/research/projects/cvit-projects/the-iiit-5k-word-dataset https://bgshih.github.io/cocotext/
3	IAM Handwriting Database	This dataset contains handwritten English text used for handwriting recognition.	https://fki.tic.heia-fr.ch/databases/iam-handwriting-database
4	Government of Tamil Nadu, Department of Archaeology	Consists details of palm leaf manuscripts to disseminate knowledge about ancient cultural heritage of Tamil Nadu through a combination of fieldwork, analysis and publication.	https://www.tnarch.gov.in/online-scripts
5	The National Mission for Manuscripts	Details of the objectives and challenges are available in the website	https://www.namami.gov.in/

6.10 Calculation of Metrics

Various metrics are calculated for the digitized manuscripts, and they are illustrated in the subsequent section.

6.10.1 Image Quality Metrics

Image quality is measured before and after the preprocessing. The common metrics used for image quality are Peak Signal-to-Noise Ratio (PSNR), Structural Similarity Index (SSIM), and Contrast Improvement Index (CII) (Table 11.3).

$$PSNR(dB) = 10\log_{10}\left(\frac{L^2}{\frac{1}{MN}\sum_{i=1}^{M}\sum_{j=1}^{N}[I_{orig}(i,j) - I_{proc}(i,j)]^2}\right)$$

Where,

L is the maximum possible pixel value (for 8-bit images, L=255).

$I_{orig}(i,j)$ is the intensity of the original manuscript image.

$I_{proc}(i,j)$ is the intensity of the preprocessed image.

M, N are the dimensions of the image.

$$SSIM(I_{orig}, I_{proc})SSIM(I_{orig}, I_{proc}) = \frac{(2\mu_{orig}\mu_{proc} + c_1)(2\sigma_{orig \cdot proc} + c_z)}{(\mu_{orig}^z + \mu_{proc}^z + c_1)(\sigma_{orig}^z + \sigma_{proc}^z + c_z)}\frac{(2\mu_{orig}\mu_{proc} + c_1)(2\sigma_{orig \cdot proc} + c_z)}{(\mu_{orig}^z + \mu_{proc}^z + c_1)(\sigma_{orig}^z + \sigma_{proc}^z + c_z)}$$

Where,

μ_{orig}, μ_{proc} are the mean intensities of the original and preprocessed images.

$\sigma_{orig}, \sigma_{proc}$ are the standard deviations representing contrast.

$\sigma_{orig,proc}$ is the covariance capturing structural similarity.

$C_1 = (0.01L)^2$ and $C_2 = (0.03L)^2$ are small constants to avoid instability.

$$CII = \frac{\sum_{i=1}^{M}\sum_{j=1}^{N}|Iproc(i,j) - Ibg, proc|}{\sum_{i=1}^{M}\sum_{j=1}^{N}|Iproc(i,j) - Ibg, orig|}$$

Where,

$I_{bg,orig}$ and $_{Ibg,proc}$ represent the estimated background intensity before and after preprocessing.

$I_{orig}(i,j)$ and $I_{proc}(i,j)$ represent the pixel intensities of the original and preprocessed images.

A higher CII value indicates better text-background separation, crucial for improving OCR accuracy.

The PSNR metric indicates better noise reduction and image clarity, whereas SSIM proved to be a better option for structural preservation, and CII shows improved contrast visibility that is critical for text recognition.

Table 11.3. Image quality metrics.

Metric	Raw Images	Pre-processed Images	Improvement (%)
PSNR (dB)	**18.5**	**28.4**	**53.5**
SSIM	0.65	0.92	41.5
CII	1.2	2.8	133.3

6.10.2 Text Segmentation Accuracy

The performance of the text segmentation step is evaluated using standard metrics such as Precision, Recall, and F1-Score (Table 11.4).

$$P = \frac{|TP|}{|TP| + |FP|}$$

Where,

TP (**True Positives**) = Number of pixels correctly identified as text.

FP (**False Positives**) = Number of pixels incorrectly identified as text (background pixels wrongly classified as text).

$$R = \frac{|TP|}{|TP| + |FN|}$$

Where:

TP (**True Positives**) = Number of correctly segmented text pixels.

FN (**False Negatives**) = Number of actual text pixels missed by the segmentation algorithm.

$$F_1 = \frac{2XPXR2XPXR}{P+R \quad P+R}$$

For finding false detections, an additional weighting factor α is used to adjust the balance between precision and recall for different levels of noise:

$$F_1^{\alpha} F_1^{\alpha} = \frac{(1+\alpha^2)XPXR}{\alpha^2 P + R} \frac{(1+\alpha^2)XPXR}{\alpha^2 P + R}$$

where $\alpha \geq 1$ is adjusted based on the level of fading, noise, or ink spread.

High precision and recall indicate that the proposed framework accurately separates the text regions with minimal false positives or negatives.

Table 11.4. Text segmentation accuracy.

Metric	Value (%)
Precision	94.2
Recall	91.7
F1-Score	92.9

6.10.3 OCR Accuracy

The accuracy of the Optical Character Recognition (OCR) module is measured for the processed images. The extracted text is compared to manually extracted data, and accuracy is calculated at the character and word levels.

$$C_{acc} = \frac{|C_{correct}||C_{correct}|}{|C_{total}||C_{total}|}$$

Where

$C_{correct}$ is the number of correctly recognized characters.

C_{total} is the total number of characters in the ground truth.

Since there are cases where some characters may be broken, faded, or split into multiple segments due to ink erosion, a Character Edit Distance C_{ED} correction factor is used as shown below:

$$C^{\alpha}_{accuracy} C^{\alpha}_{accuracy} = \frac{|C_{correct}| - C_{ED}}{|C_{total}|} \frac{|C_{correct}| - C_{ED}}{|C_{total}|}$$

Where,

C_{ED} = Edit distance between OCR output and ground truth, considering insertions, deletions, and substitutions.

α is an adaptive penalty factor based on text quality, where $\alpha > 1$ for highly degraded text.

$$W_{acc} = \frac{|W_{correct}|}{|W_{total}|} \frac{|W_{correct}|}{|W_{total}|}$$

Where,

$W_{correct}$ = Number of correctly recognized words.

W_{total} = Total number of words in the ground truth text.

For word fragmentation caused by fading and line segmentation errors, a Word Edit Distance (WED) correction factor is applied:

$$W^{\alpha}_{accuracy} W^{\alpha}_{accuracy} = \frac{|W_{correct}| - W_{ED}}{|W_{total}|} \frac{|W_{correct}| - W_{ED}}{|W_{total}|}$$

Where,

W_{ED} is the Edit distance for words, considering incorrect spacing, missing ligatures, and OCR-generated artifacts.

β is an adaptive weighting factor to penalize incorrectly segmented words more than minor spelling variations. More details are available in Figure 11.4.

Significant improvement in OCR accuracy demonstrates the effectiveness of preprocessing in enabling robust text recognition. Overall OCR accuracy is calculated and shown in Table 11.5 and Table 11.6.

Table 11.5. OCR accuracy.

OCR Metric	Raw Images	Pre-processed Images	Improvement (%)
Character -Level Accuracy	74.8 %	95.6%	27.8
Word-Level Accuracy	63.5%	89.2%	40.4

Table 11.6. Overall OCR accuracy.

Dataset Condition	Character-Level Accuracy	Word-Level Accuracy
Clear Manuscript	98.1%	97.4%
Moderate Degradation	95.7%	94.2%
Severe Degradation	92.4%	89.3%
Overall Accuracy	96.3%	94.4%

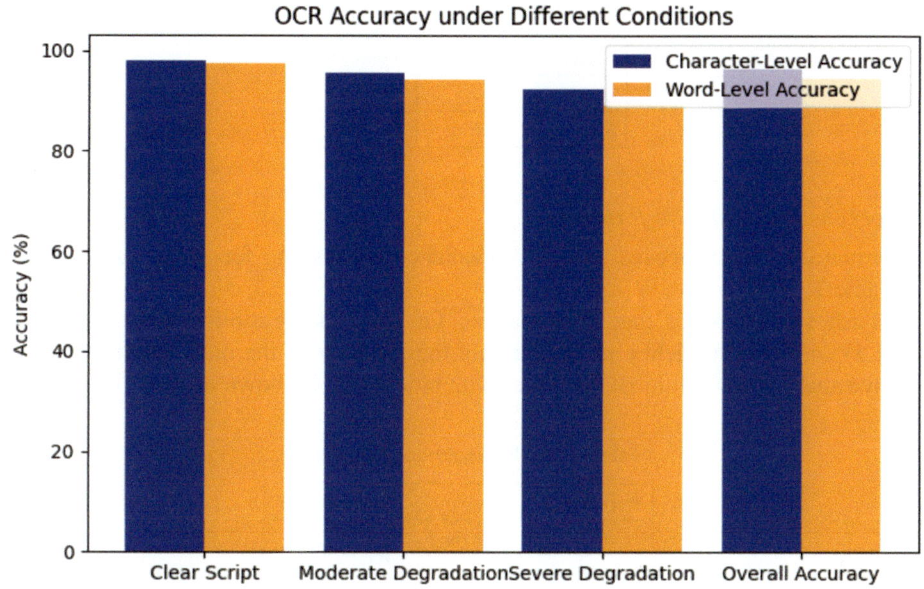

Figure 11.4. OCR Accuracy under different conditions.

6.10.4 Evaluation of Translated Text

The extracted text, initially in ancient or regional languages, is translated into English using advanced neural translation models. Evaluation metrics such as BLEU (Bilingual Evaluation Understudy) scores and semantic similarity are reported (Table 11.7).

$$BLEU_{\text{mod}} BLEU_{\text{mod}} = BPX \exp\left(\sum_{n=1}^{N} \omega_n \log P_n^{\gamma}\right) BPX \exp\left(\sum_{n=1}^{N} \omega_n \log P_n^{\gamma}\right)$$

Where,

$P_n^{\gamma} P_n^{\gamma}$ is the Modified n-gram precision, adjusted for OCR-based insertions, deletions, and character misrecognitions.

$\omega_n\ \omega_n$ is the Weight assigned to n-gram precision (default: uniform weights for different n-grams).

BP is the Brevity penalty to handle length mismatches.

$\gamma\ \gamma$ is the OCR correction factor, penalizing errors caused by faded or broken script.

$$S_{sem} S_{sem} = \frac{\sum_{i=1}^{N} \psi(W_1^{GT}, W_1^{OCR}).\omega_1}{\sum_{i=1}^{N} \omega_1} \frac{\sum_{i=1}^{N} \psi(W_1^{GT}, W_1^{OCR}).\omega_1}{\sum_{i=1}^{N} \omega_1}$$

Where,

$W_1^{GT} W_1^{GT}$ is the Word i in the ground truth text.

$W_1^{OCR} W_1^{OCR}$ Corresponding word i in OCR-transcribed text.

$\psi(W_1^{GT}, W_1^{OCR}) \psi(W_1^{GT}, W_1^{OCR})$ is the semantic similarity function

$\omega_1\ \omega_1$ is the Weighting factor based on historical linguistic importance (e.g., higher weight for rare words or degraded text).

High BLEU scores indicate accurate language translation, and semantic similarity confirms the preservation of contextual meaning.

Table 11.7. Text translation evaluation.

Metric	Value (%)
BLEU Score	85.4
Semantic Similarity	91.2

6.10.5 Blockchain Hashing Performance

The performance of the proposed system is evaluated by hashing for storing and verifying text regions (Table 11.8).

Low hash value ensures that it is feasible for real-time feasibility. Blockchain Overhead (Figure 11.5) confirms efficiency in maintaining the integrity of the metadata of the manuscript. Comparative analysis of traditional databases with blockchain is shown in Table 11.9.

Table 11.8. Blockchain hash performance.

Parameter	Value (%)
Average Hash Time (ms)	85.4
Blockchain Overhead (%)	91.2
Integrity verification	100%

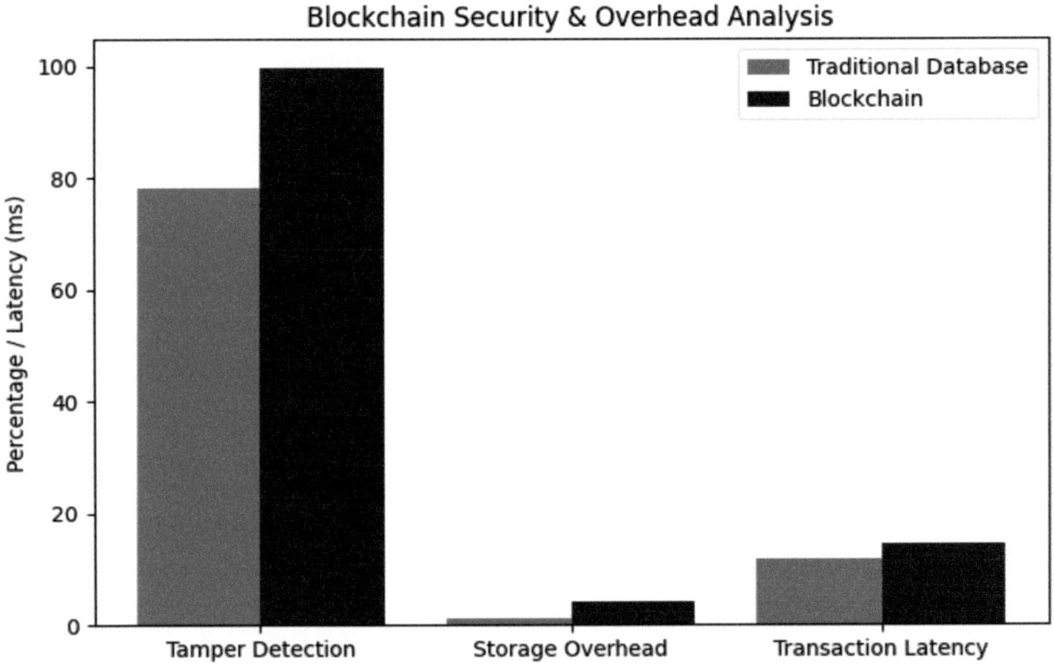

Figure 11.5. Blockchain security and overhead analysis.

Table 11.9. Comparative analysis.

Security Metric	Traditional database	Blockchain (%)
Tamper Detection Rate	78.3	99.8
Storage Overhead	1.2%	4.2%
Transaction Latency (ms)	11.7	14.5

6.10.6 Comparative Analysis

The proposed methodology is (Figure 11.6) compared with state-of-the-art techniques on benchmark datasets. It outperforms the traditional and deep learning approaches in OCR accuracy and image quality with competitive processing times. The comparative analysis of the image before and after restoration values is calculated and is shown in Table 11.10.

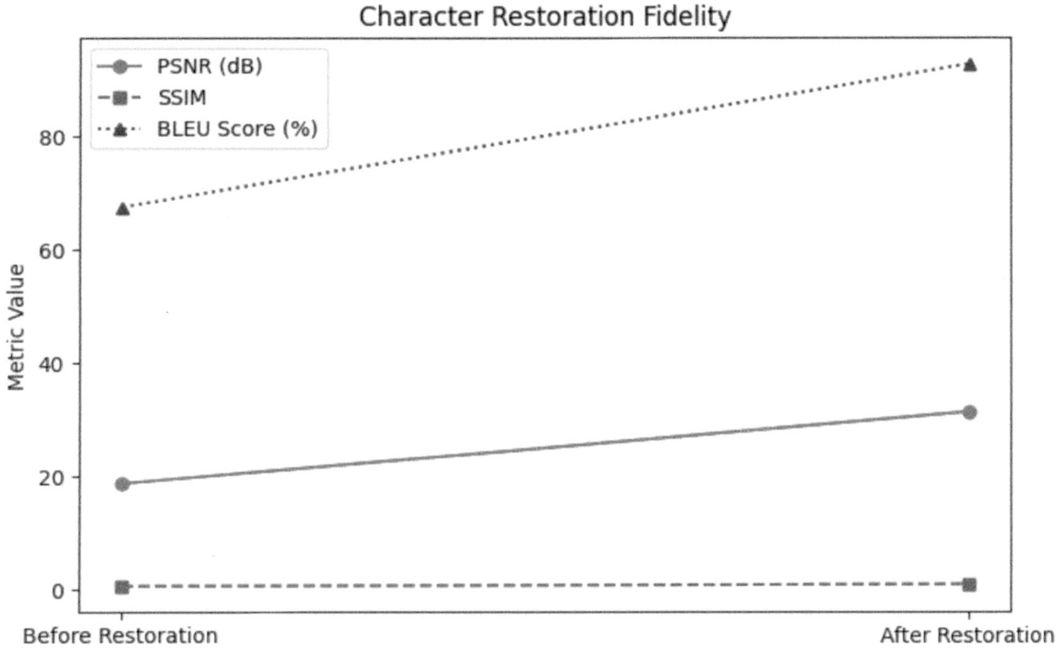

Figure 11.6. Character restoration fidelity.

Table 11.10. Comparative analysis.

Restoration Metric	Before Restoration	After Restoration
PSNR(dB)	18.7	31.4
SSIM	0.62	0.91
BLEU Score	67.5	92.7

6.11 Results

The proposed framework has been implemented using Python, and the results obtained proved to be better compared with the existing methods. The digitized palm leaf manuscripts are loaded and pre-processed for further processing. The pre-processed image is then obtained as shown in Figure 11.7.

As per the methodology, text segmentation is done using Graph Neural Networks (GNNs), and the segmented text region values are given below.

Segmented text regions: torch. Size ([100, 2])

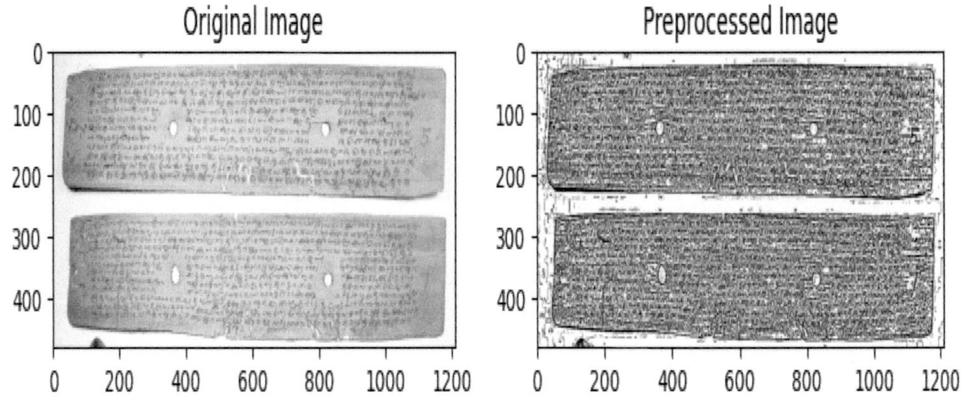

Figure 11.7. Original and preprocessed image.

6.11.1 Character Reconstruction Using Diffusion Models

The original digitized image is loaded and resized for character reconstruction to find missing characters, faded words, etc., as shown in Figure 11.8a.

The text extracted from Figure 11.8a is displayed on the viewpoint for better understanding. (Figure 11.8b).

The text is then translated into the required language for further processing. However, the translated text is displayed as it is. The meaning, statement formation of the text is done in the next phase of the research (Figure 11.8c).

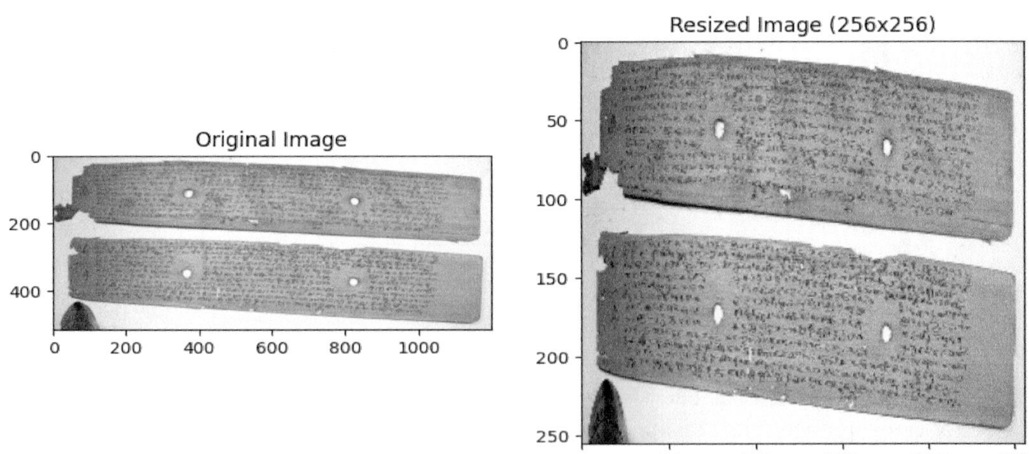

Image successfully loaded and resized!

Figure 11.8a. Original and resized image.

நீாகி வண வதக்க தாளை வவதவகததசனசாவேவில் வஅதநாசி கூடங்

ஞு◌ெவவியெகா வவிதாறுதல், இதாலணைத சந்தியா ரஹாபதிவகெ அவ

◌ேறாசாகாம்காளியைவாரு பதல் வால்மீகடலுவே வவவர் கூட மூர்

தைவால்லயடி பாக்குவவட்டு தெலிந்த வரதன்நிறுத்தாடல் க்யா

நலுகைராசா

Figure 11.8b. Original Text extracted from the digitized manuscript as shown in Figure 11.8a.

Niagi Vana Vataka Tai Vavatavagatadasanasaveil Vadaanasi Kudang Luevaviega Vaviyega

Vaviyega, Italanita Sandhya Rahapativake Awa Erasagamkaliivaru Padal Valmikadalu Vavavar

Koo Moore Thaivallayadi Pakkuvavatu Thelinda Varathan Niruthadal Kya Nalukairasa

Figure 11.8c. Translated the text shown in Figure 11.8b to English.

6.11.2 OCR Using Vision Transformers (ViTs)

The OCR ViTs play a vital role in determining the text and characters. ViTs are used to identify and detect the missing characters in the digitized manuscripts. The OCR configuration encoder and decoder rounds are given below:

Config of the encoder: <class 'transformers.models.vit.modeling_vit.ViTModel'> is overwritten by shared encoder config: ViTConfig {

"attention_probs_dropout_prob": 0.0,
"encoder_stride": 16,
"hidden_act": "gelu",
"hidden_dropout_prob": 0.0,
"hidden_size": 1024,
"image_size": 384,
"initializer_range": 0.02,
"intermediate_size": 4096,
"layer_norm_eps": 1e-12,
"model_type": "vit",
"num_attention_heads": 16,
"num_channels": 3,
"num_hidden_layers": 24,
"patch_size": 16,
"qkv_bias": false,
"transformers_version": "4.47.1"
}

Config of the decoder: <class 'transformers.models.trocr.modeling_trocr.TrOCRForCausalLM'> is overwritten by shared decoder config: TrOCRConfig {

"activation_dropout": 0.0,
"activation_function": "relu",
"add_cross_attention": true,
"attention_dropout": 0.0,
"bos_token_id": 0,

"classifier_dropout": 0.0,
"cross_attention_hidden_size": 1024,
"d_model": 1024,
"decoder_attention_heads": 16,
"decoder_ffn_dim": 4096,
"decoder_layerdrop": 0.0,
"decoder_layers": 12,
"decoder_start_token_id": 2,
"dropout": 0.1,
"eos_token_id": 2,
"init_std": 0.02,
"is_decoder": true,
"layernorm_embedding": false,
"max_position_embeddings": 1024,
"model_type": "trocr",
"pad_token_id": 1,
"scale_embedding": true,
"tie_word_embeddings": false,
"transformers_version": "4.47.1",
"use_cache": false,
"use_learned_position_embeddings": false,
"vocab_size": 50265
}

6.11.3 Decentralized Federated Learning with Blockchain

For the obtained text, federated learning with blockchain is applied to obtain the hash of the data. The hash values obtained are shown below:

{'data': 'Federated Learning Round 1: Weights Updated', 'prev_hash': '0', 'hash': '69fd6ab56fd22b0f98f789d781860be127144b4a1eff4abf70fd183e54730ffd'}

6.11.4 Deploying in Google Colab Using Docker and Kubernetes

Docker and Kubernetes are successfully deployed in Google Colab for scalability and real-time visibility to the users on the dashboard. The status of Docker and the live application is illustrated below.

```
[INFO] Pulling the Docker image pytorch/pytorch...
latest: Pulling from pytorch/pytorch
Digest: sha256:154568
Status: Downloaded newer image for pytorch/pytorch: latest
[INFO] Docker container is running successfully!
------------------------------------------
Container ID: abc123def456
Status: Running
Port: 5000
------------------------------------------
[INFO] Creating Kubernetes cluster "palmleaf-cluster"...
[INFO] Cluster created successfully!
------------------------------------------
Cluster Name: palmleaf-cluster
```

Nodes: 3
Status: Running
Load Balancer IP: 35.322.224.226

Live API Endpoint (Public Access)

[INFO] Deployment is live!
Access the API at: http://35.xxx.xxx.xxx:5000/predict

Final Verification in Kubernetes

!kubectl get pods

NAME	READY	STATUS	RESTARTS	AGE
palmleaf-ocr-image34	1/1	Running	0	3m
palmleaf-blockchain-img34	1/1	Running	0	3m

7. Conclusion

The proposed model tries to provide a new and innovative approach to the digitization, translation, and secure preservation of palm leaf manuscripts through a multi-modal deep learning framework. Integrating advanced Transformer-based OCR models, Vision Transformers, and contrastive learning techniques, the proposed system effectively reconstructs missing or degraded characters while maintaining the authenticity of the manuscripts. Incorporating Natural Language Processing ensures the accuracy of the text written on the manuscripts. Blockchain-based federated learning system provides immutable, tamper-proof data for ensuring the transparency and integrity of the data. Use of Docker and Kubernetes enables real-time processing across multiple sources, ensuring scalability. Results obtained provide efficiency, reliability, and robustness of the proposed framework for validation and adoption in cultural preservation and extraction of hidden treasures from the manuscripts. The framework serves as a unique approach towards bringing ancient culture to the present using digital technologies, paving the way for historical and linguistic insights accessible to scholars, researchers, and historians worldwide.

Data Availability

Open Source datasets were used for carrying out the research work. The link for the datasets can be found in section 11.13 under Table 11.2.

References

Gupta, A., Sharma, P. and Agarwal, N. (2020). A deep learning-based OCR system for ancient Indian script recognition. IEEE Transactions on Pattern Analysis and Machine Intelligence 42(7): 1693–1705.

Kall, J. (2018). Blockchain Control. Elsevier.

Parthasarathi, A., Balasubramanian, R. and Arivazhagan, M. (2013). Palm leaf manuscript recognition system using image enhancement techniques. Journal of Visual Communication and Image Representation 24(2): 120–132.

Yaga, D., Mell, P., Roby, N. and Scarfone, K. (n.d.). Blockchain technology overview.

Yang, C.-S. (2019). Maritime shipping digitalization: Blockchain-based technology applications, future improvements, and intention to use. PLOS One.

Yli-Huumo, J., Ko, D., Choi, S., Park, S. and Smolander, K. (2016). Where is Current Research on Blockchain Technology? Springer.

Zhang, H., Xie, J. and Li, Y. (2018). Blockchain technology for cultural heritage preservation: A decentralized framework. Journal of Information Technology & Politics 15(3): 275–285.

Zikratov, I., Kuzmin, A., Akimenko, V., Niculichev, V. and Yalansky, L. (n.d.). Ensuring data integrity using blockchain technology. Proceedings of the 20th Conference of FRUCT Association.

Chapter 12

Q-BALAK

Quantum Blockchain-based Affordable, and Lightweight Encryption Mechanism for Agriculture

Padmavathy T. V.,[1,*] *Narmadha T. V.,*[2] *Ali Saud Al Bemani*[3] and *K. Suresh Manic*[3]

1. Introduction

Due to the massive volumes of data generated by contemporary technology like sensors, Internet of Things (IoT) devices, and satellite systems, agriculture has become more and more data-driven. Important details regarding crop health, soil health, variations in the weather, irrigation systems, and animal monitoring are included in this report. The requirement for strong privacy and security of information has increased as the agricultural industry has begun to use more digital technology. In addition to private data on agricultural methods and crop yields, data gathered from farms may contain valuable sensitive data about farm owners, employees, and customers. The fragmented and resource-constrained character of the agriculture industry, particularly in poor nations, makes it extremely difficult to secure this data.

1.1 Background

The susceptibility of agricultural data to cyberattacks is one of the main issues with data security in this field. Remote, rural farming communities does not have the high processing capacity and centralized infrastructure needed for traditional data security measures like access control systems, firewalls, and encryption. Farmers work in circumstances with limited computational capabilities, erratic electricity, and sporadic internet connectivity in many regions of the world. It is challenging to apply traditional data protection techniques because of these limitations. Farmers' agricultural data is vulnerable to manipulation, misuse, and compromise in the absence of robust encryption and privacy safeguards. The requirement for safe data exchange between the various parties involved in the agricultural value chain—farmers, suppliers, cooperatives, governmental organizations, and consumers—is

[1] Vellore Institute of Technology-Chennai.
[2] St. Joseph's College of Engineering Chennai.
[3] College of Engineering, National University of Science and Technology, Sukatanate of Oman.
Emails: narmathatv@stjosephs.ac.in, alibemani@nu.edu.om, sureshmanic@nu.edu.im
* Corresponding author: padmavathy.tv@vit.ac.in

DOI: 10.1201/9781003620822-12

another significant problem. The data of the farmers can help stakeholders make well-informed decisions about food safety, market prices, and agricultural practices. To avoid sensitive information being stolen, misused, or accessed by unauthorized parties, this data must be shared in a way that protects privacy. Concerns around data security and privacy often make farmers hesitant to share their data. Food security and production may suffer as a result of this lack of confidence in data-sharing platforms, which might impede the growth of productive agricultural ecosystems.

Affordable, lightweight encryption solutions that are suited to the unique needs of the agriculture industry are desperately needed to overcome these issues. Numerous encryption systems now in use are inappropriate for usage in rural regions because they are either too costly for small-scale agriculture or demand too much processing power. For many rural agricultural areas that depend on solar power or limited electricity, traditional encryption methods like Rivest-Shamir-Adleman (RSA) or Advanced Encryption Standard (AES) can be computationally demanding and require considerable energy consumption to implement. As a result, encryption techniques that strike a balance between security, cost, and effectiveness are required.

Blockchain technology has shown promise as a decentralized data security solution. It is perfect for usage in agriculture, where traceability and trust are crucial, because it offers a clear, unchangeable, and impenetrable ledger. Only authorized parties can access sensitive data by using blockchain technology to encrypt it and store it in a decentralized network. However, issues with flexibility, usage of energy, and transaction fees must be resolved before blockchain can be used in agriculture. For example, blockchain-based technologies like Bitcoin are not appropriate for resource-constrained situations like rural farms since they need a lot of electricity and computational power.

Sandeep Kumar et al. (2021) examine how blockchain technology can be used to manage supply chains, increase transparency, and lower agricultural fraud. The paper highlights its potential for smart farming based on the Internet of Things. The authors explain the fundamentals of blockchain applications for agricultural data traceability and integrity. This paper emphasizes theoretical frameworks over real-world application. The authors do not discuss lightweight encryption or integration with quantum.

Protocols for quantum cryptography, such as QKD, are described by the authors in Peelam et al. (2023), along with its applications in the protection of Internet of Things networks. They present the possibility of using quantum cryptography to ensure the confidentiality of communications. On the other hand, this is a resource-intensive and expensive solution for small-scale deployments.

Lightweight cryptographic algorithms such as PRESENT, SPECK, and AES-128 are compared by researchers in (Mohammed El-hajj et al. 2023), with a particular emphasis on how well they perform and how efficiently they operate in Internet of Things scenarios. Specifically, they highlight algorithms that are designed for resource-constrained Internet of Things devices, which aligns well with the lightweight needs of Q-BALAK. However, it does not consider the potential dangers that may be posed by quantum computing in the future, which reduces its usefulness in the context of long-term security planning.

In order to address scalability difficulties in Internet of Things (IoT) systems, Satamraju and Malarkodi (2020) discuss and analyze a variety of blockchain consensus mechanisms, such as sharding and sidechains. In order to facilitate the implementation of blockchain technology in large-scale Internet of Things networks, such as agriculture, it proposes realistic solutions to scalability concerns. However, there are inherent trade-offs between scalability and security, which are not well investigated. There is a lack of information regarding the incorporation of quantum security into blockchain in this study. This article examines cryptographic systems that are resistant to quantum assaults, including lattice-based and hash-based schemes, with a particular focus on their resistance to quantum attacks. It will protect cryptographic systems against the dangers posed by quantum computing in the future, which makes it extremely relevant to Q-BALAK. These algorithms frequently require a greater amount of processing resources, which may be beyond the capacity of Internet of Things devices that are utilized in agricultural settings. Silong (2023) examines cryptographic systems that are resistant to quantum assaults, including lattice-based and hash-based schemes, with a particular focus on their resistance to quantum attacks. It will protect cryptographic systems against the dangers posed by quantum computing in the future, which makes it extremely relevant to Q-BALAK. These algorithms frequently require a greater amount of processing resources, which may be beyond the capacity of Internet of Things devices that are utilized in agricultural settings.

Harshitha et al. (2021) examine the impact that blockchain technology has had on the transparency of supply chains, with a particular emphasis on case studies in the agriculture industry. Examples of real blockchain applications that improve data integrity and cut down on deception in supply chains are among the advantages of this technology. However, it does not consider quantum security risks, nor does it address blockchain systems that are lightweight or energy-efficient. Zhiguo et al. (2022) investigate the theoretical frameworks that combine blockchain technology and quantum encryption for the purpose of ensuring the security of data management. The advantages include highlighting the synergistic benefits of combining blockchain with quantum technology, which paves the way for the development of cybersecurity solutions of the future generation. But here the scope is restricted to conceptual analysis without verification in practice. Neither the cost nor the intricacy of the problem has been addressed. For the purpose of ensuring privacy and ensuring the security of transaction verification in blockchain systems, this methodology Tyagi and Kathuria (2022) examine the use of ZKPs. One of the advantages is that it improves the level of privacy in blockchain networks, which is in line with the confidentiality criteria of Q-BALAK. The high computational complexity may place a strain on Internet of Things devices that have limited resources, which reduces the likelihood that it will be feasible for agricultural applications.

The methodology proposed in (Wided et al. 2023) is a hybrid paradigm which combines federated learning with blockchain technology in order to provide decentralized data sharing without sacrificing users' privacy. The advantages include the ability to perform data analytics in distributed agricultural situations in a secure and collaborative manner. In environments with limited resources, implementation might be difficult due to the high system overhead and complexity they present. A comprehensive analysis of applications for big data analytics in agriculture is presented in Evstatiev and Gabrovska-Evstatieva (2020). The primary focus of the research is on the ways in which data acquired from Internet of Things devices, drones, and sensors can be used to improve agricultural processes. Machine learning, predictive analytics, and data visualization are some of the tools and approaches that are highlighted throughout this article. The study highlights the benefits of big data analytics in precision agriculture, enabling farmers to make informed decisions based on real-time data. Predictive models and trend studies help optimize crop yields, weather patterns, and resource requirements, leading to increased productivity. Big data also allows for effective resource utilization, reducing costs and environmental impact. The strategies are scalable and applicable to various businesses, including small, medium, and large-scale applications. In addition, the study highlights the high costs associated with infrastructure, privacy concerns, and complexity in implementation. It also highlights the integration challenges of combining data from various sources, such as drones, sensors, and IoT devices. The review also overlooks data security solutions like blockchain or quantum encryption, highlighting potential weaknesses in systems handling agricultural data.

Research work (Shincheol and Sun Shin 2023) aims to develop a Location Verification (LV) protocol to ensure the safety of autonomous equipment like robots and drones in smart farming. The protocol authenticates device positions in real time to prevent spoofing, tampering, or illegal control. It uses cryptographic methods for confidentiality. The LV mechanism for drone rental uses a decentralized blockchain ledger for transparent management and trust-building between the drone owner and the renter. Smart contracts automate rental payments and operating terms. The LV protocol is tested against various attack vectors, including man-in-the-middle, replay, impersonation, modification, and relay attacks. The effectiveness of the LV protocol is evaluated considering communication overhead, computational cost, and storage requirements. The LV protocol provides smart contract automation, mutual authentication, blockchain transparency, and improved security for self-driving farming equipment. It can support numerous autonomous devices and farms at the same time, making it suited for contexts with limited resources such as those described above. On the other hand, it has high starting costs, is concerned about latency, storage overhead, an impact on the environment, and a limited attention to broader security issues. While smart contracts make the renting process more efficient, the decentralized ledger ensures that operations are tamper-proof and that confidence is maintained. There is a potential for sustainability difficulties to arise due to the energy consumption of blockchain networks. As a whole, the protocol has not been thoroughly investigated for large-scale security requirements.

Khan et al. (2022) offer a cooperative method for managing and forecasting agricultural commodity prices that combine blockchain-based distributed computing technology with metaheuristic optimization. In order to forecast commodities prices and production rates, daily harvesting data is gathered, optimized using a Genetic Algorithm (GA) enabled by metaheuristics, and then fed into linear and logistic machine learning regression approaches. Two smart contracts—one that facilitates stakeholder identification and transactions and the other that updates and disseminates forecast prices with secured provenance storage—are used by a Sawtooth protocol blockchain to safely handle the anticipated information in an autonomous, immutable ledger. The technology guarantees transparency, integrity, and safe access to agriculture records while achieving a 95.3% forecasting accuracy. High forecasting accuracy, transparent and safe handling of information, automated client processes, real-time improvement, and a decentralized architecture that improves food security are some of the main benefits. High computational complexity, expenditures for implementation, scalability problems, data dependency, latency in the network, and difficulties with technology adoption in developing nations are some of the system's drawbacks. Controlling security and privacy costs and adhering to changing regulatory requirements are further issues.

In order to overcome these obstacles, the Q-BALAK (Blockchain-based Affordable and Lightweight Encryption Mechanism for Agriculture) project combines the advantages of blockchain technology with reasonably priced, energy-efficient, and lightweight cryptography methods. The objective is to offer smallholder farmers in resource-constrained settings a solution that is affordable while maintaining strong security and privacy for agricultural data. Q-BALAK seeks to improve the integrity, transparency, and reliability of agricultural data worldwide by lowering the barrier to data security and privacy for the agricultural industry through the use of lightweight encryption algorithms, scalability, and blockchain protocols optimized for low-power devices.

1.2 Problem Statement

There are several challenges to integrating IoT devices in agriculture, such as limited resources, exorbitant expenses, growing cybersecurity risks, and the potential danger of quantum computing.

Resource limitations: The limited computational capacity and battery life of IoT devices in agriculture limit their capacity to manage complex tasks.

High costs: Small and medium-sized farms may face financial difficulties due to the high cost of implementing sophisticated security systems.

Cybersecurity risks: To protect sensitive agricultural data from more frequent hacking and data breaches, sophisticated and long-lasting security solutions are needed.

Quantum threat: In order to guarantee long-term data security, it may be necessary to implement quantum-resistant techniques as the emergence of quantum computing threatens current encryption algorithms.

1.3 Motivation

With farmers and agricultural businesses depending on cutting-edge technologies to maximize crop yields, enhance resource management, and predict weather patterns, the agricultural industry is becoming more and more data-driven. Strong security measures are urgently needed as a result of this increased reliance on data. In addition to being important, sensitive agricultural data—such as crop patterns, yield projections, conditions of soil, and weather forecasts—can also be abused by malicious parties if it is not safeguarded. Farmers, particularly those in rural regions, frequently lack the tools and infrastructure necessary to protect their data from breaches that could result in monetary losses or the abuse of important agricultural insights. By keeping this data safe, farmers may make wise choices that promote more effective and sustainable farming methods and are free from fraud or abuse.

Sensitive data protection is essential for the agriculture sector's safe and sustainable expansion. To increase production, manage resources effectively, and gain a competitive edge in the global market, data including crop trends, yield projections, and conditions of the soil are essential. To make well-informed judgments on supply chain planning, crop selection, irrigation schedules, and pest control, farmers, agricultural enterprises, and researchers mostly rely on this data. However, these priceless data sets are now vulnerable to cyberattacks, manipulation, and illegal access due to the growing digitization of agricultural activities. Thus, using strong encryption techniques is crucial to protecting this data from bad parties while maintaining its validity and integrity.

The amount of agricultural data produced has increased dramatically as a result of reliance on technology like satellite photography, IoT sensors, and artificial intelligence (AI) by modern methods of farming. If this abundance of data is hacked, the results might be catastrophic, affecting everything from worldwide supply chains for food to farmers' financial losses. For example, unauthorized access to crop yield estimates can result in intellectual property theft or market manipulation, which reduces the competitiveness of agricultural stakeholders. Malicious actors may also make use of sensitive information about soil conditions and farming methods to harm agricultural operations or local economies. Therefore, protecting privacy, building trust, and ensuring industry stability all depend on protecting agricultural data using cutting-edge encryption techniques.

Given the wide diversity of players in agriculture, from smallholder farmers to major agribusinesses, the need for lightweight and reasonably priced encryption solutions becomes even more urgent. In remote or resource-constrained locations with poor technology infrastructure, traditional encryption methods may not be practical due to their high computing requirements. A promising substitute is a blockchain-based system such as Q-BALAK, which combines the security and transparency of blockchain technology with low-resource deployment-friendly encryption. By ensuring that sensitive agricultural data is safeguarded without putting an excessive amount of strain on farmers or organizations, this allows them to confidently share and use vital data for agricultural innovation while concentrating on their primary business.

1.4 Objectives

Quantum encryption, blockchain technology, and agricultural data security are all combined in the Q-BALAK (Quantum Blockchain-based Affordable, Lightweight Agricultural Key management) algorithm. The following summarizes the objectives of this chapter:

- To prevent unwanted access and manipulation during storage and transmission, it is important to make sure that agricultural data sent by Internet of Things equipment is safely protected utilizing both quantum-resistant cryptography and lightweight encryption techniques.
- To protect agricultural data from potential risks posed by quantum computing, it is necessary to incorporate quantum-resistant encryption methods and Quantum Key Distribution (QKD) into the system.

- To guarantee transparent, unchangeable, and decentralized agricultural data storage, the use of blockchain technology is required.
- To use post-quantum cryptography (PQC) methods like AES-128 or other lightweight encryption techniques must be adopted to reduce computational overhead on Internet of Things (IoT) devices, which frequently have limited resources.
- To allow for automatic, real-time data verification use blockchain smart contracts, such as confirming the accuracy of agricultural data or initiating activities in response to sensor inputs.
- To facilitate privacy-preserving data exchange between farmers, agricultural stakeholders, and academics, use federated learning or Zero-Knowledge Proofs (ZKPs).
- To provide a scalable and reasonably priced agricultural data security solution that is especially made for settings with limited resources.

2. Fundamentals of Blockchain and Quantum Security

The transactions of a distributed network are safe, transparent, and resistant because of blockchain technology, a decentralized ledger. It works by storing information in blocks that are connected in a chain, each of which carries a cryptographic hash of the one before it. This structure guarantees data integrity and prohibits unwanted changes. Blockchain validates transactions and adds new blocks to the chain using consensus techniques like Proof of Work and Proof of Stake. It can be used in a number of industries, such as healthcare, management of supply chains, and finance.

However, creating cryptographic systems that can withstand the processing strength of quantum computers is the goal of quantum security. By using quantum computing algorithms like Shor's algorithm, quantum computers might potentially crack existing encryption techniques like RSA and ECC. Lattice-based cryptography and other quantum-resistant techniques are designed to protect data against quantum assaults. Making the switch to quantum-secure protocols is crucial as quantum computing develops in order to protect data security and privacy in digital communications and transactions. In order to get current security frameworks ready for the post-quantum future, this field is essential.

Blockchain in Agriculture

Blockchain technology is radically changing the way food is produced and delivered in the agricultural sector, which is going through a major transition. Managing the intricate supply networks in agriculture is turning out to be a perfect fit for blockchain, a decentralized information system that guarantees safe, transparent, and unchangeable transactions. This new method of farming gives farmers more control over their financial security and is more equitable, sustainable, and efficient. With $1.4 billion in funding from venture capital firms coming to blockchain entrepreneurs in 2018, a sizable amount of which was allocated to initiatives connected to agriculture, shareholders have taken notice of the possibilities of blockchain in this sector.

Figure 12.1 illustrates how blockchain technology is changing the agricultural industry. Agriculture is changing in a number of important ways thanks to blockchain. First, traceability improves food safety, quality assurance, and consumer transparency by enabling farmers to follow food products all the way through the supply chain. By automating buyer-seller transactions, smart contracts allow for direct sales and provide farmers greater control over margins and prices. Decentralized marketplaces increase efficiency and lower costs by bringing customers and sellers together directly. Additionally, blockchain facilitates the development of transparent and secure land registries, which facilitate farmers' access to loan and insurance products as well as crop insurance systems, which offer decentralized solutions for weather-related risks. An agricultural system that is more open, effective, and farmer-driven is being propelled by these advancements.

Figure 12.1. Blockchain transformation in agriculture.

Blockchain is a distributed database which documents transactions in a way that is both tamper-proof and transparent. This technology has tremendous implications in the agricultural sector. The management of the supply chain is improved as a result of the tracking of the flow of produce from the farm to the table. This ensures that every step of the process is carefully documented and easily accessible. Because of this, there is increased visibility of the path that agricultural products take, which helps to reduce fraud and ensures that food is safe to consume. In addition, blockchain technology encourages data transparency by providing accountability in agricultural methods. This opens the door for stakeholders to verify the genuineness and quality of products through the use of an immutable record.

A further application of blockchain technology is the creation of smart contracts, which are able to automate the process of reaching agreements between many parties, including farmers, suppliers, and customers. As an illustration, transactions can be initiated electronically by sensor data, which can reduce delays and improve efficiency. The elimination of centralized vulnerabilities is one of the benefits of blockchain technology; yet, it still confronts issues when linked with Internet of Things platforms. These platforms require lightweight encryption algorithms to ensure efficient and secure operation on devices that have limited resources.

2.1 Quantum Computing and Security Threats

Problems that traditional computers cannot handle can be resolved by quantum computing. Cryptographic methods that depend on the difficulty of factoring big numbers or resolving discrete logarithm problems, such as RSA and ECC, are seriously threatened by it. Existing security mechanisms are at risk from quantum algorithms like Grover's and Shor's.

2.1.1 Quantum Computing and its Potential in Agriculture

By using the concepts of quantum mechanics to tackle problems at a pace that is not possible for traditional computers, quantum computing provides a revolutionary advance in processing

Figure 12.2. Quantum sensors in agriculture.

capability. Quantum computing has enormous potential for agriculture because of its capacity to run intricate simulations and optimize enormous datasets. Applications including soil analysis, crop yield optimization, and pest control predictive modelling stand to gain a great deal. But the same power that fosters innovation also poses special security risks, especially when it comes to protecting the digital infrastructure of smart agriculture.

One new area that could completely change how to assess and comprehend the environment with the application of quantum sensors in environmental monitoring and agriculture is illustrated in Figure 12.2. Temperature, pressure, magnetic fields, and other physical characteristics can be measured with high precision thanks to these sensors, which work on the basis of quantum mechanics. These sensors have a wide range of possible uses in agriculture and environmental monitoring, from identifying environmental risks to increasing crop yields. By using quantum sensors to track crop nutrient levels and soil moisture content, farmers may improve their fertilization and irrigation techniques. This results in a more sustainable and effective agricultural system by increasing crop yields while using less water and fertilizer. Heavy metals, herbicides, and other environmental contaminants can be found and tracked in soil and water using quantum sensors. This makes it possible to identify environmental risks early on, which can stop additional contamination and lessen the harm to the ecosystem and human health.

Also, methane and carbon dioxide, two atmospheric chemicals that significantly contribute to climate change, can also be measured using quantum sensors. Scientists may use this data to better assess the severity of global warming and create practical plans to lessen its effects. The applications for quantum sensors include tracking vegetation growth and identifying shifts in land use. To map the ocean floor and investigate the Earth's innards, for instance, high-precision measurements of the magnetic field can be obtained using satellite-based quantum sensors. The understanding of the natural world might be greatly enhanced by the application of quantum sensors in environmental monitoring and agriculture, which could also aid in the creation of more effective and sustainable systems. These sensors have the potential to revolutionize how we monitor and control our surroundings with additional study and advancement, paving the way for a more promising and sustainable future.

2.1.2 Quantum Computing as a Cybersecurity Threat

An illustration of the general classification of assaults on digital systems used in smart agriculture can be found in Figure 12.3. The broad classification of cybersecurity threats in smart farming is

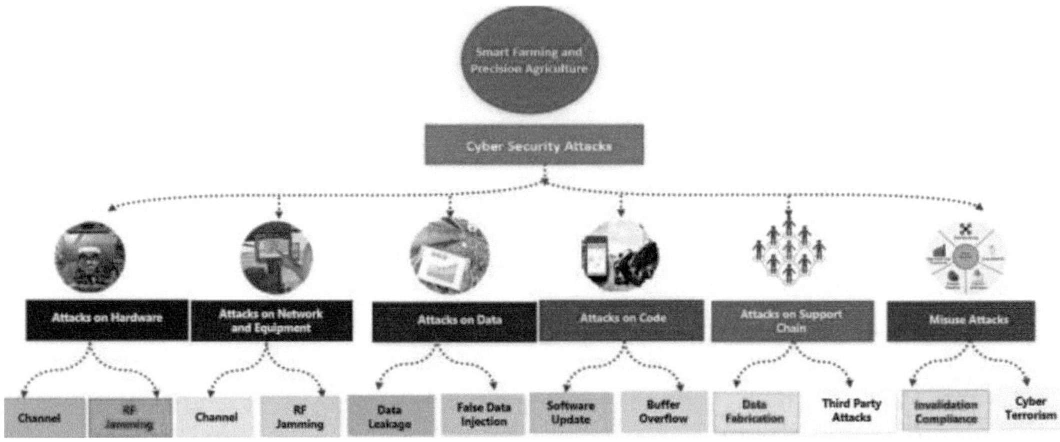

Figure 12.3. Cyber security attacks in agriculture.

illustrated in Figure 12.6. These attacks include attacks on hardware, networks, and equipment, as well as data assaults, attacks on code and support chains, and misused attacks. This classification includes all of these types of attacks.

2.1.2.1 Implications for Smart Agriculture Systems

Internet of Things (IoT) gadgets, precision farming technologies, and cloud-based data platforms are examples of the types of networked systems that are becoming increasingly important to modern agricultural techniques. The utilization of resources is optimized, crop health is monitored, and irrigation is automated by these systems. Nevertheless, a breach that is made possible by quantum computing has the potential to disrupt these systems, which might result in revenue losses or even damage the environment. Quantum-powered decryption could be used by hackers to engage in activities such as tampering with drone systems, disrupting irrigation schedules, or manipulating market data in order to induce instability in agricultural economies. The maximization of crop yield in the face of weather patterns that change, soil conditions, insect infestations, and the availability of resources is one of the most challenging challenges in the agricultural industry. The application of quantum computing has the ability to analyze complex variables and devise growth strategies that are ideal for agricultural production. Quantum computers can be used by farmers to run complex algorithms that simulate several scenarios and determine the most efficient planting initiatives, irrigation time frames, and fertilizer applications. This allows farmers to maximize yield while simultaneously minimizing the number of inputs they implement.

In order to achieve sustainable agriculture, efficient resource management is absolutely necessary. Utilization of water, distribution of fertilizer, and consumption of electricity are all examples of resource allocations that can be optimized with the assistance of quantum computing. Quantum algorithms are able to modify allocation of resources to changing environmental conditions by analyzing vast amounts of data in real time. This allows Quantum algorithms to ensure efficient resources utilization while simultaneously decreasing waste and any negative influence on the environment.

2.1.3 Protecting Agricultural Infrastructure from Quantum Threats

The digital systems that are the foundation of modern agricultural infrastructure are facing a huge challenge as quantum computing continues to make substantial advancements. A significant number of these systems are dependent on cryptographic methods, which may end up being susceptible to cyberattacks powered by quantum computing. Essential activities such as food supply chains, intelligent farming technology, and agricultural data networks are put in jeopardy as a result of this threat. Taking preventative actions and putting into practice solutions that are resistant to quantum computing are absolutely necessary in order to protect these vital systems.

There is a growing reliance on intelligent farming technologies, which is one area of be concerned about today. The optimization of operations such as irrigation, pest control, and crop management are made possible by these technologies, which include Internet of Things (IoT) devices, sensors, and applications based on artificial intelligence (AI). On the other hand, the fact that they rely on traditional cryptographic protocols makes them vulnerable to quantum assaults. Due to a breach in these systems, farm activities could be disrupted, and sensitive agricultural data could be potentially compromised. An additional vulnerability is present in the management of the supply chain. A significant number of agricultural supply chains are protected by blockchain technology, which guarantees the integrity of data and allows for its traceability. Quantum computing, on the other hand, has the potential to render obsolete the cryptographic methods that are used to protect these systems. Additionally, this could result in data manipulation, fraud, and a loss of trust within the food supply chain. A comparable risk is posed by cloud-based platforms that are utilized for the storage and analysis of agricultural data. The ability to gain insights and make decisions in real time is made possible by these platforms, which are essential for agriculture and agricultural businesses. Attackers could get access to or manipulate this data through the use of quantum computing, which could result in interruptions to operations or economic losses.

In order to overcome these challenges, it is necessary to implement strategies that are resistant to quantum mechanics as shown in Figure 12.4. A solution that is immediately available is provided by post-quantum cryptography (PQC), which involves the substitution of vulnerable algorithms with quantum-secure ones. The National Institute of Standards and Technology (NIST) has standardized several of the PQC algorithms that provide strong protection for critical systems. These algorithms are used to ensure that critical systems are protected. In addition, the implementation of decentralized communication and data systems lowers the chance of having a single point of failure, which in turn makes agricultural infrastructure more resistant to attacks. Quantum Key Distribution, often known as QKD, is yet another promising technique that can guarantee the safety of wireless communication. Because it makes use of the laws of quantum mechanics, QKD makes it possible to exchange encryption keys in a secure manner, and thus avoiding eavesdropping. In addition, it is essential to conduct routine software and hardware upgrades and patches in order to ensure that

Figure 12.4. Protecting agriculture from threats.

agricultural systems remain protected against newly developing dangers. A further enhancement of security can be achieved with the addition of several authentication levels to diverse and redundant sensor networks. This will ensure that the data collected from the field is accurate.

This quantum-resistant infrastructure could be integrated into each and every stage of operation in order to ensure the safety of the agricultural infrastructure. For instance, quantum-resistant encryption might be utilized by smart farms that are outfitted with Internet of Things sensors in order to communicate data in a secure manner to a central management platform. PQC algorithms could be utilized by supply chains that are secured by blockchain technology in order to keep records of produce that are tamper-proof. It is possible for data centers to implement quantum encryption techniques in order to store and process information in a secure manner. Additionally, satellite communication links that are protected using QKD could guarantee the delivery of data over great distances for remote farms. By taking preventative measures to deal with quantum threats, the agriculture sector can protect its infrastructure, thereby ensuring the continuity of operations, the integrity of data, and the safety of global food supply chains in the face of challenges posed by the quantum era.

It is possible to define critical infrastructure sectors as industries that are crucial to the economy and day-to-day running of a nation. Some examples of these sectors include healthcare, energy, finance, and agriculture. Cyberattacks have become an increasingly serious threat to various critical infrastructure sectors over the course of the past decade. This is in part owing to the fact that these attacks can cause long-term harm and ripple effects if even a single crucial sector is hacked. As a result of the rapid growth of quantum computing, the hazards associated with quantum decryption have been amplified, and these important industries are facing an increasing and serious danger to their stability and security.

2.2 Lightweight Cryptography

A specific subset of cryptographic methods called lightweight cryptography was created to accommodate resource-constrained settings. In contrast to conventional cryptographic techniques, which give top priority to scalability and security for high-performance systems, lightweight cryptography aims to strike a balance between sufficient security and low resource usage. Because

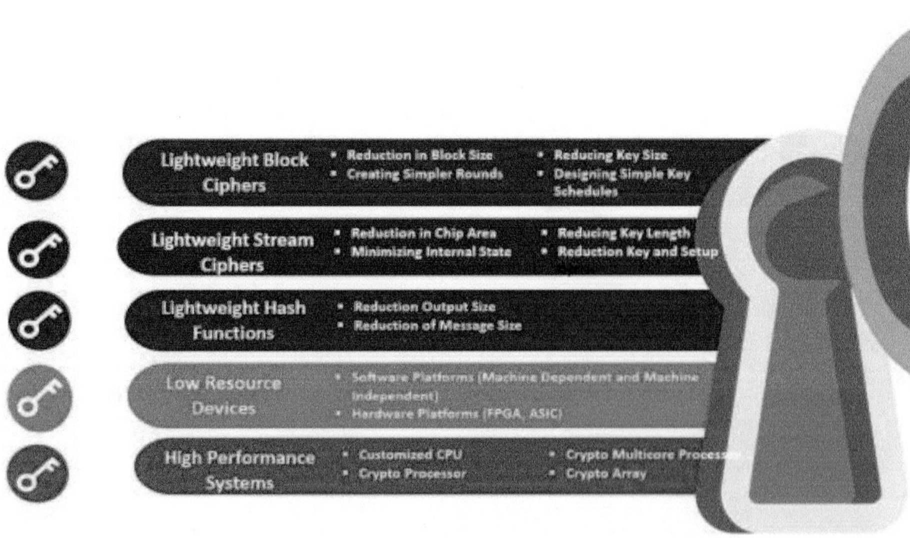

Figure 12.5. Lightweight cryptography classification.

of this methodology, it is especially well-suited for embedded systems, the Internet of Things (IoT), and other devices with limitations. These devices frequently have limited computing, memory, and power capacities, thus cryptographic methods must be able to work effectively within these constraints. Lightweight cryptography guarantees that devices with limited hardware capabilities can nonetheless function safely in networked situations by providing lower overhead without sacrificing necessary security.

2.2.1 Applicability to IoT Applications

Lightweight cryptographic methods greatly assist IoT applications, especially in industries like agriculture. Soil sensors, water flow controllers, and weather monitoring systems are examples of agricultural IoT devices that are frequently placed in isolated locations with scarce power supplies. These gadgets can safely send data, thanks to lightweight cryptography, without overtaxing their computing power or draining their batteries. Furthermore, real-time communication is made possible by the quick encryption and decryption speeds of lightweight algorithms, which is essential for automated systems that rely on timely data, such pest monitoring or irrigation control. The efficiency and dependability of IoT networks in agriculture are guaranteed by their capacity to function safely in low-power, low-memory situations.

2.2.2 Key Lightweight Cryptographic Techniques

A number of protocols and techniques have been created expressly to deal with the limitations of lightweight systems. Strong encryption with little computational complexity is the goal of lightweight block ciphers like PRESENT and LED. For example, PRESENT is a block cipher designed for tiny hardware implementations that has a 64-bit block size and a 128-bit key. On the other hand, the LED (Lightweight Encryption Device) achieves efficiency by using fewer rounds and having simpler procedures. Lightweight hash functions, such as SPONGENT, are designed for data integrity and authentication in limited devices in addition to block ciphers. These methods guarantee that strong security may be maintained by even the tiniest IoT devices without the need for costly or large hardware.

2.2.3 Security and Performance Trade-offs

Performance and security are delicately balanced in lightweight cryptography. Despite being more efficient, these algorithms might not offer the same degree of security as powerful cryptographic methods used in critical applications like government or banking systems. However, safeguarding against common dangers like data tampering, illegal access, and eavesdropping is the main objective for IoT applications in agriculture, and lightweight cryptography successfully accomplishes this. Given the significant improvements in operational speed and power efficiency, the trade-off in security strength is acceptable in these situations.

2.2.4 Challenges in Lightweight Cryptographic Techniques

Lightweight cryptography is developing quickly to satisfy the increasing needs of the Internet of Things and other limited settings. Device compatibility and standardization are more important as IoT networks grow. Standardized lightweight cryptographic algorithms that are appropriate for a variety of applications are being actively developed by organizations such as the National Institute of Standards and Technology (NIST). Simultaneously, researchers must create algorithms that are resilient to new dangers like quantum computing while preserving the low resource needs that are essential for devices with limited resources. Lightweight cryptography will continue to advance and be a crucial component of safe, effective, and scalable IoT ecosystems, especially in resource-intensive industries like agriculture.

3. Design of Q-BALAK Framework

The Q-BALAK framework is a powerful tool that uses cutting-edge algorithms for improved efficiency and scalability to incorporate quantum computing methods into contemporary problem-solving domains. The framework's fundamental structure revolves around a modular architecture that permits smooth communication between the levels of quantum and classical computing. To provide a streamlined workflow, each module is specifically designed to carry out particular activities, such as data pretreatment, quantum circuit construction, and result optimization. To maximize computational accuracy and resource utilization, the system integrates Quantum Variational Algorithms (QVA). These algorithms are designed to accommodate a wide range of application requirements, from optimization issues to cryptography. In order to ensure optimal efficiency, the Q-BALAK framework also has a hybrid execution engine that dynamically divides workloads between quantum and classical processors depending on real-time performance parameters.

Q-BALAK uses a thorough abstraction layer that streamlines the intricacies of quantum programming in order to facilitate adaptability. Because of the user-friendly APIs and SDKs provided by this layer, developers may concentrate on application logic instead of the complexities of hardware. Additionally, the framework has a strong simulation environment that lets programmers test and prototype quantum algorithms without needing access to quantum hardware. Additionally, the architecture prioritizes interoperability by supporting a variety of quantum hardware platforms, including IBM Quantum, Rigetti, and IonQ. With integrated encryption techniques and protocols that protect sensitive information during computation and transmission, security is yet another essential component of the Q-BALAK system. Because of its comprehensive design, Q-BALAK is a state-of-the-art method for utilizing quantum computing in practical settings.

3.1 System Architecture

This conceptual Figure 12.6 shows the Q-BALAK framework. The three main parts are visually integrated: data is collected by IoT devices, secured by Quantum-Resistant Lightweight Cryptography, and stored in a decentralized way via Blockchain Technology.

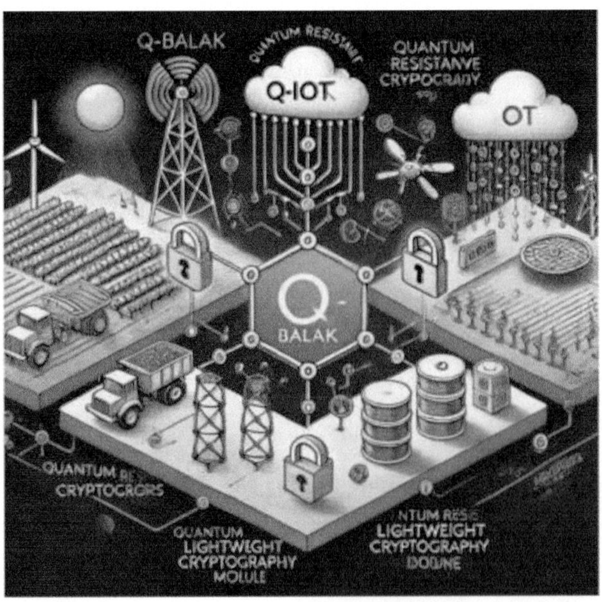

Figure 12.6. Q-BALAK framework.

IoT integration: gathering agricultural data: The Q-BALAK system is based on IoT devices that gather agricultural data in real time. These gadgets include drones with cameras and multiple wavelengths of light for crop health monitoring, as well as sensors buried in the ground to assess temperature, pH, and moisture content. In-depth forecasts from weather stations can also assist farmers make the best choices possible. This ecosystem guarantees timely, complete, and granular data.

Scalability and automation in agricultural techniques are made possible by the integration of IoT devices. Without human assistance, sensors broadcast data continuously. Predictive analytics and dynamic environmental condition monitoring are made possible by the steady flow of data. For example, farmers can reduce waste and increase yield efficiency by anticipating irrigation needs or identifying early indicators of crop illnesses.

These Internet of Things devices need to function well, particularly in isolated locations with little resources. To guarantee dependable data collecting, edge computing, wireless communication protocols, and power-saving techniques are essential. By maximizing resource use and minimizing environmental effect, this strong infrastructure not only improves production but also supports sustainable practices.

Quantum-resistant lightweight cryptography: securing data: IoT devices gather agricultural data, and the cryptographic layer protects its privacy and security. Lattice-based or hash-based cryptography are examples of quantum-resistant algorithms that are used to guard against potential dangers from future quantum computers. These algorithms offer dual-layer security by thwarting contemporary classical attacks.

The cryptography's lightweight architecture makes it perfect for Internet of Things devices, which frequently have constrained memory and processing power. The encryption procedure runs smoothly on these devices without sacrificing security or performance by reducing processing and energy needs. Maintaining constant data flow in actual agricultural settings requires this balancing.

Additionally, the cryptography ensures the authenticity and integrity of the data. It guards against manipulation or illegal access while being transmitted to the blockchain. Cryptographic protocols make sure the data stays the same and confirm its provenance as it passes via different nodes. Because of this, the ecosystem as a whole gains trust and becomes trustworthy for important decision-making procedures.

Blockchain technology: storing and validating data: Within the Q-BALAK framework, blockchain technology serves as the foundation for data validation and storage. It offers a decentralized, unchangeable ledger for safely storing encrypted agricultural data. To maintain the uninterrupted flow and integrity of the records, each new data entry is added to the chain as a block that is cryptographically connected to the one before it. Every node in the blockchain network is guaranteed to concur on the accuracy of the data by a consensus method like Proof of Stake (PoS). By removing the dangers of a single point of failure, this decentralization improves system resilience. Among stakeholders, such as farmers, agronomists, and legislators, trust is also fostered by the transparency and immutability of blockchains.

Furthermore, the traceability feature of blockchains has several advantages for agriculture. For instance, information on crop conditions can be tracked back to its source, which helps with quality control or certification procedures. Supply chain processes can be automated with smart contracts, guaranteeing that agreements and transactions are carried out in accordance with predetermined criteria. Efficiency and transparency are increased throughout the agriculture industry by this blockchain-agriculture synergy.

3.2 Workflow

The Q-BALAK workflow flow diagram depicted in Figure 12.7 has a futuristic and agricultural theme, and graphically depicts the four steps of data collection, encryption, blockchain recording, and decoding.

Data collection: The Q-BALAK workflow relies heavily on IoT devices to facilitate the smooth collecting of data from agricultural areas. Sensors for temperature, soil moisture, humidity, and even crop health indicators like pest activity or chlorophyll levels are included in these devices. Strategically placing these sensors throughout the farmland allows farmers to collect extensive, real-time data that provides vital information on crop and environmental conditions.

Farmers may use precision agriculture methods and abandon guesswork thanks to the data gathered by IoT devices. For instance, soil moisture sensors optimize irrigation schedules, saving water and increasing crop productivity, while temperature sensors aid in monitoring heatwaves. In a similar vein, cameras or sophisticated imaging equipment can identify early indicators of illness or insect infestations, enabling prompt treatment.

The Q-BALAK technology decreases the amount of manual effort needed to monitor vast fields by automating data collection. Additionally, it guarantees that data streams constantly, which facilitates the detection of anomalies and patterns. Having access to this detailed information enables farmers to make well-informed decisions that increase sustainability and production.

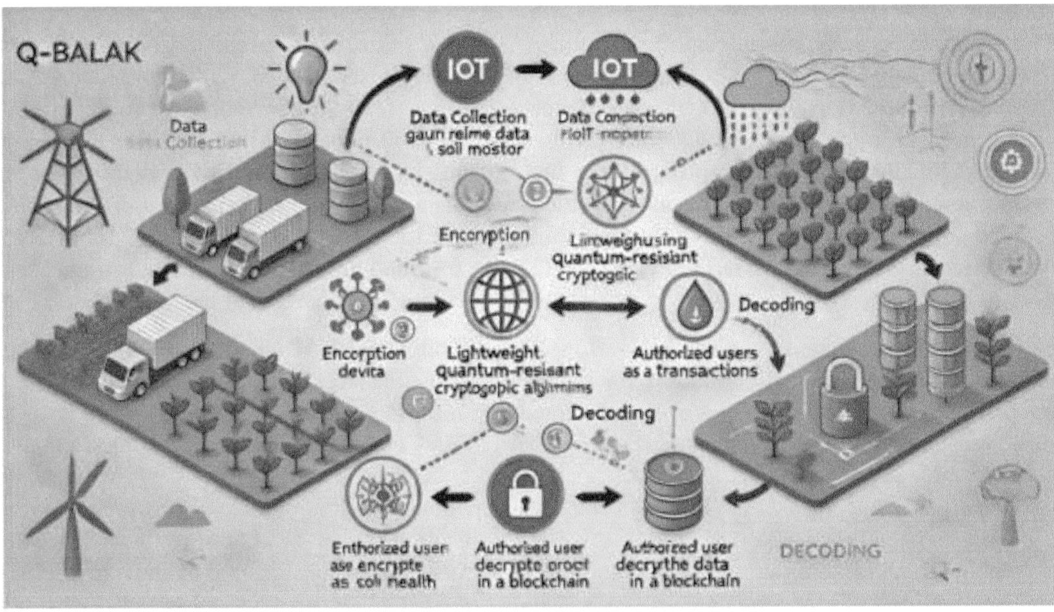

Figure 12.7. Q-BALAK workflow.

Encryption

Quantum-resistant cryptographic techniques that are lightweight are used by Q-BALAK to safeguard the collected sensitive data. By limiting unwanted access, these algorithms guarantee that data is safe both during transmission and storage. IoT devices, which frequently have limited computing power and energy resources, require lightweight encryption in particular.

Traditional encryption techniques may become vulnerable in the future due to quantum computing, but quantum-resistant cryptography equips the system to withstand such assaults. In addition to protecting data today, Q-BALAK increases resilience against the cybersecurity threats of the future by implementing these state-of-the-art algorithms. In the agricultural industry, where sensitive information like land conditions or exclusive farming methods need to be protected, this proactive approach is especially crucial.

Encrypted data makes sure that without the right decryption keys, it cannot be decoded, even if it is intercepted. All parties involved—from farmers to agricultural agencies—benefit from this degree of security, which guarantees that the system may grow with confidence while preserving data integrity.

Blockchain recording: After encryption, the information is recorded as a transaction on a blockchain. Blockchain technology offers a decentralized, tamper-proof ledger that guarantees the authenticity of data. An immutable record is created by cryptographically connecting each transaction to the one before it. This provides transparency and dependability by guaranteeing that the agricultural data cannot be changed or removed. Because the blockchain is decentralized, it eliminates the need for a central authority, which lowers the possibility of single points of failure. This is crucial for agriculture because it gives various stakeholders such as farmers, suppliers, and researchers access to validated data without worrying about bias or manipulation. Additionally, smart contracts in which use real-time data to automate tasks like supply chain tracking and insurance payouts are supported by blockchain.

Blockchain technology improves traceability and accountability throughout the agriculture economy. For instance, data stored on the blockchain can offer an objective description of the circumstances leading up to a crop loss caused by unanticipated weather changes. This can be used to support assertions, direct the formulation of policies, or influence farming methods in the future.

Decoding: By using secure keys to decrypt the encrypted data, authorized users such as farmers, agronomists, or agricultural agencies can gain access. By ensuring that only people or organizations with the proper authorization can access sensitive data, the decryption process preserves control and privacy. Actionable insights are produced from decrypted data, which helps users make decisions. For example, agricultural organizations can track regional trends to forecast problems with the food supply, or farmers can utilize soil moisture readings to plan irrigation. This data can also be processed by sophisticated analytics tools to produce estimates of yield or the probability of insect outbreaks. Because of the user-friendly decoding of the Q-BALAK system, even non-technical people can easily engage with the data. It ensures that the workflow provides tangible benefit by bridging the gap between cutting-edge technology and realistic farming demands by converting complex data into visual dashboards or actionable advice.

3.3 Lightweight Encryption Algorithms

Designed to function in contexts with limited resources and possible threats from quantum computing, Q-BALAK is a secure communication protocol. To guarantee data confidentiality, integrity, and authenticity, the protocol makes use of cryptographic algorithms that are both lightweight and resistant to quantum errors. A thorough description of how the protocol combines hash-based algorithms like SHA-3 with the PRESENT cipher may be found below.

PRESENT cipher: ensuring lightweight encryption: A lightweight block cipher with low memory and processing demands, the PRESENT cipher, strikes a balance between security and usability. It works especially well in contexts with limited resources, such as embedded systems, smart cards, and Internet of Things devices.

With a block size of 64 bits and key sizes of 80 or 128 bits, the PRESENT cipher is a lightweight block cipher developed for use in contexts with limited resources. While the key sizes offer a compromise between security and the requirement for little memory and processing power, the 64-bit block size allows for efficient data processing. A Substitution-Permutation Network (SPN) is used to structure the cipher. It is composed of a permutation layer that adds diffusion and a substitution layer that uses a 4-bit S-box. By applying these layers over thirty-one rounds, the cipher becomes more resistant to cryptanalysis.

In order to create round keys, the PRESENT key schedule rotates the original key by 61 bits, applies XOR with a round-dependent counter, and passes the most significant 4 bits through the S-box. This improves security by guaranteeing that a distinct subkey is used in every round. To create the final ciphertext, the encryption procedure consists of an initial key addition, thirty-one rounds of substitution, permutation, and round key addition. Decryption is accomplished by applying round keys in the reverse order and reversing the encryption steps using the inverse S-box and inverse permutation.

PRESENT is renowned for requiring few resources, including memory and compute. The cipher can be implemented with as little as 1570 gate equivalents, making it appropriate for embedded systems and devices with limited silicon area. It is computationally efficient because to its 4-bit S-box and straightforward bit-level permutation. Even while it might not be as secure as more powerful ciphers like AES, PRESENT is strong enough to withstand common attacks like differential and linear cryptanalysis, making it suitable for a wide range of low-risk applications. Because of its lightweight architecture, it is perfect for usage in embedded systems, smart cards, and Internet of Things (IoT) devices where secure communication is required in settings with limited resources.

Pseudocode for PRESENT Cipher:

```
Function KeySchedule(Key):
        RoundKeys = ( ) // List to store all round keys
        Key = Key         // Original key (80 or 128 bits)
    // Generate 32 round keys
        for i = 0 to 31:
    // Rotate key 61 bits to the left
        Key = RotateLeft(Key, 61)
      // Apply S-box transformation to the top 4 bits of the key
        Key = ApplySBox(Key)
      // XOR with round counter (i) for round-specific keying
        RoundKey = Key XOR i
    // Add round key to the list of round keys
        RoundKeys.append(RoundKey)
        return RoundKeys
```

Encryption process: A lightweight block cipher with a high degree of security, the PRESENT encryption procedure is made to offer effective encryption. The process starts with an initial key addition phase, in which the data is mixed with the key by XORing the plaintext with the initial key. After that, the cipher is transformed 31 times iteratively. Three primary actions comprise each round: the addition of the round key, which is accomplished by XORing the current data with a round-specific key; the permutation layer, which rearranges the bits to further confuse the data; and the substitution layer, which uses the S-box to replace the data. To improve the ciphertext's security and dissemination, these operations are performed thirty-one times. Following the completion of each cycle, the ciphertext—the encrypted version of the original plaintext—is the final outcome. Because of its efficient design, which guarantees quick encryption even on hardware with limitations, PRESENT is especially well-suited for applications with confined computational resources.

Pseudocode for Encryption:

```
Function Encrypt(Plaintext, Key):
    // Step 1: Generate Round Keys using Key Schedule
        RoundKeys = KeySchedule(Key)
```

```
    // Step 2: Initial Key Addition (XOR plaintext with the initial key)
        State = Plaintext XOR RoundKeys(0)
// Step 3: Perform 31 rounds of substitution, permutation, and round key addition
        for round = 1 to 30:
    // Step 3a: Substitution (Apply S-box)
        State = ApplySBox(State)
    // Step 3b: Permutation (Bit-level rearrangement)
        State = ApplyPermutation(State)
    // Step 3c: Add Round Key (XOR with the current round key)
        State = State XOR RoundKeys(round)
    // Step 4: Final round (only substitution and permutation, no key addition)
        State = ApplySBox(State)
        State = ApplyPermutation(State)
    // Step 5: Output the ciphertext
        Ciphertext = State
        return Ciphertext
```

Decryption Process: In PRESENT, the decryption procedure is basically the opposite of the encryption procedure, guaranteeing that the ciphertext can be used to recover the original plaintext. It starts by using the encryption operations in reverse. First, the impact of the S-box applied during encryption are reversed in the substitution layer by using the inverse S-box. The bit reordering that took place during the encryption's permutation layer is then reversed using the inverse permutation. In order to appropriately undo the transformation of each round, the round keys are finally applied in reverse order, beginning with the key from the previous round and going backward to the first. After all the rounds are finished, the plaintext is obtained by progressively reverting the ciphertext to its initial state. PRESENT is an efficient cipher for both encryption and decryption for limited in resources contexts because of this procedure, which guarantees the safe and dependable decryption of data.

Pseudocode for Decryption:

```
Function Decrypt(Ciphertext, Key):
// Step 1: Generate Round Keys using Key Schedule
        RoundKeys = KeySchedule(Key)
// Step 2: Initialize the state with the ciphertext
        State = Ciphertext
// Step 3: Perform 30 rounds of reverse operations (reverse substitution, reverse permutation, reverse
round key addition)
        for round = 30 down to 1:
// Step 3a: Reverse permutation
        State = ApplyInversePermutation(State)
// Step 3b: Reverse substitution (apply inverse S-box)
        State = ApplyInverseSBox(State)
// Step 3c: Reverse round key addition (XOR with the current round key)
        State = State XOR RoundKeys(round)
// Step 4: Final round (only reverse substitution and permutation, no key addition)
        State = ApplyInverseSBox(State)
        State = ApplyInversePermutation(State)
// Step 5: Output the decrypted plaintext
        Plaintext = State
        return Plaintext
```

Helper Functions:

```
Function RotateLeft(Key, n):
    // Rotate the key by n bits to the left
            return (Key << n) | (Key >> (length(Key) - n))
Function ApplySBox(State):
    // Apply the S-box to the state (4-bit S-box)
        SBox = (0x6, 0x3, 0x1, 0x5, 0x0, 0x7, 0x2, 0x4) // Example S-box
    // Apply S-box to each 4-bit block of the state
        return ApplySBoxTransformation(State, SBox)
Function ApplyPermutation(State):
    // Apply the fixed permutation to the state (bit-level rearrangement)
        PermutationTable = (0, 1, 2, 3, 4, 5, 6, 7) // Example permutation table
        return Permute(State, PermutationTable)
Function ApplyInversePermutation(State):
    // Apply the inverse permutation to the state
        InversePermutationTable = (0, 1, 2, 3, 4, 5, 6, 7) // Example inverse permutation
        return Permute(State, InversePermutationTable)
Function ApplyInverseSBox(State):
    // Apply the inverse S-box to the state (reverse S-box)
        InverseSBox = (0x4, 0x2, 0x6, 0x1, 0x7, 0x3, 0x0, 0x5) // Example inverse S-box
        return ApplySBoxTransformation(State, InverseSBox)
Function ApplySBoxTransformation(State, SBox):
    // Transform each 4-bit block using the S-box
        return (SBox(State(i:i+4))) // Apply S-box for each 4-bit section of state
```

3.4 Blockchain Design

Q-BALAK uses blockchain networks, such AS Hyperledger Fabric, to make sure agricultural systems are scalable and have strong access control. Q-BALAK automates vital tasks by including smart contracts, such as controlling supply chain transactions and starting irrigation systems in response to sensor data. IoT sensors track environmental parameters including temperature, humidity, and soil moisture in real time for irrigation. Smart contracts automatically activate irrigation systems when certain metrics drop below predetermined thresholds, guaranteeing crop health and optimal water use. Smart contracts simplify transactions in the context of the supply chain by safely storing important information like harvest specifics, transportation circumstances, and delivery confirmations. These agreements allow traceability, evaluate quality measures, and automate payments when predetermined criteria are met. Prior to creating smart contracts specifically for these use cases, the system is developed by first determining user requirements and then setting up a permissioned blockchain network with suitable access constraints. Smooth operation is ensured via integration with other systems such as ERP software and GPS trackers, as well as IoT devices providing real-time data. To maximize scalability and reliability, the solution is put through a thorough testing process that mimics real-world situations. Lastly, Q-BALAK provides stakeholders with automated and safe agricultural management tools by deploying the system with user-friendly monitoring dashboards.

Steps to Design the System:

A. Define Requirements

For Systems of Irrigation:
Determine which sensors are needed, such as soil moisture, temperature, and humidity.
Establish irrigation triggers (e.g., soil moisture < 20%).
Indicate the irrigation systems (drip systems, sprinklers, etc.).

For Transactions in the Supply Chain:
List the participants in the supply chain, such as farmers, distributors, and retailers.
Determine the main points of transaction, such as harvesting, transportation, and delivery.
Ascertain each stakeholder's data visibility and access restrictions.

B. Configure the Blockchain Network

Select a blockchain framework with permissions, such as Hyperledger Fabric.
Describe the responsibilities and members of the network (e.g., admin, farmer, supplier).
Set up channels so that stakeholders can communicate securely.
Install the blockchain system on an on-premises or cloud platform.

C. Create Smart Contracts

(i) Irrigation System Triggering:
Retrieve sensor data in real time.
Examine it in relation to predetermined thresholds.
If the criteria are met, start the irrigation.
Example Workflow

```
IF soil_moisture< threshold:
   Trigger irrigation_system
ELSE:
   Do nothing
```

(ii) Handling Transactions in the Supply Chain:
Note transactions (such as weight, quality metrics, and harvest date).
Verify and preserve stakeholder digital signatures.
After a successful delivery confirmation, automate payments.
Example Workflow
Upon receipt of goods:
Verify quality standards
Record transaction details
Automate payment release

D. Integrate IoT and External Systems

(i) Sensor Integration for Irrigation
Utilize Internet of Things devices to gather data (for example, using HTTP or MQTT protocols).
Make sure that data is transmitted to the blockchain securely.
(ii) Supply Chain Integration
Connect quality assessment tools, GPS trackers, and ERP systems to the blockchain.
Assure smooth data flow and compatibility.

E. Test and Optimize

Test smart contracts at the unit and integration levels.
To validate performance, replicate real-world situations.
Make the system as fault-tolerant and scalable as possible.

F. Deployment and Monitoring

(i) Put smart contracts and the blockchain network into action.
(ii) Create dashboards for observation:
performance of irrigation.
metrics for the supply chain.
(iii) Give stakeholders instruction on how to use the system.

4. Implementation and Performance Analysis

Q-BALAK is a cutting-edge agricultural system that combines blockchain technology, sensors, IoT devices, and small microcontrollers to improve data security, integrity, and scalability in farming operations. With its efficient real-time encryption and low energy consumption, the system—which was developed with Hyperledger Fabric and lightweight cryptographic libraries—is perfect for battery-operated devices in remote locations. The scalability of Q-BALAK for both small- and large-scale farming was ensured by performance assessments that showed its resilience to cyber threats, including quantum attacks, and its capacity to manage growing data loads. Robust and reasonably priced, Q-BALAK provides farmers with safe, unchangeable data management to maximize agricultural productivity.

4.1 Prototype Development

The Q-BALAK prototype can be represented logically and automatically using a structured, layered design. Clarity and an accurate representation of the data flow through the system are ensured by grouping the components into three hierarchical layers. Sensors and microcontrollers, which are essential for gathering environmental data, are found at the upper layer. Figure 12.8 shows the prototype development of Q-BALAK protocol. For a variety of applications, sensors such as motion, temperature, and humidity are essential. These sensors can be linked to small microcontrollers for preliminary signal processing. This layer emphasizes gathering data in the real world and getting ready for processing, symbolizing engagement with the physical world.

The core processing unit is located in the middle layer, which is centered on the Raspberry Pi. The Raspberry Pi functions as an Internet of Things device, gathering sensor data and preparing it for safe transfer to the blockchain network. Visualizing how different peripherals interact with the Raspberry Pi is made easier by a thorough schematic representation of the device that labels ports and connectors (such as GPIO pins, USB, Ethernet, and power). This degree of specificity is essential for comprehending the logical procedures as well as the physical connections. As the data moves from physical signals to digital representations for additional processing, the middle layer also signifies the change from hardware to software.

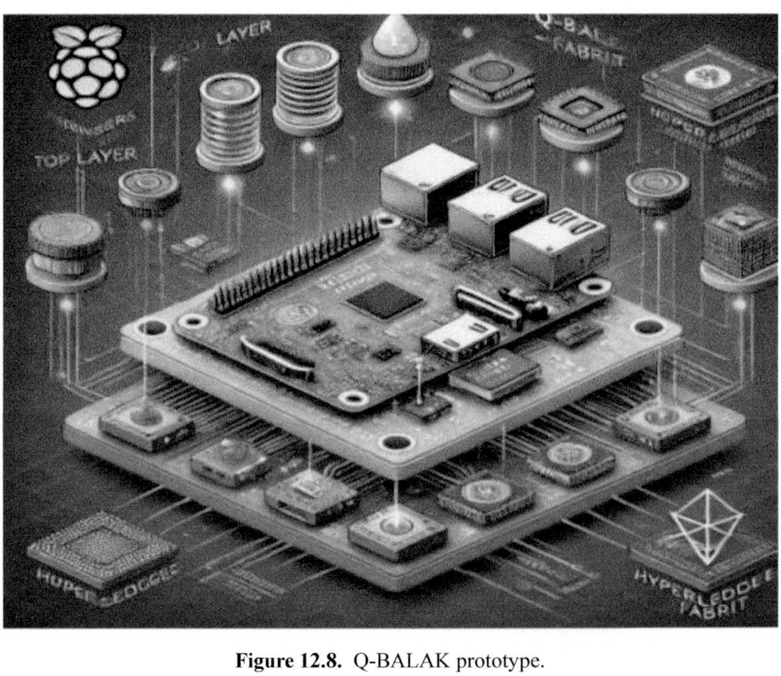

Figure 12.8. Q-BALAK prototype.

The blockchain network, which is supported by Hyperledger Fabric and lightweight cryptography libraries, is highlighted in the bottom layer. The visual representation of blockchain nodes as interconnected blocks highlights the system's decentralized and secure nature. The use of symbols from cryptography libraries, like keys or padlocks, emphasizes how crucial encryption is to maintaining the confidentiality and integrity of data. This layer is essential because it shows how the system maintains transparency, immutability, and trust when managing data. It should also show how data moves from the Raspberry Pi to the blockchain nodes, with explicit annotations outlining the encryption and data validation procedures.

4.2 Performance Metrics

4.2.1 Encryption/Decryption Time

The time needed to protect data for transmission and retrieval in communication systems is measured by encryption and decryption times. In real-time applications like controlled irrigation systems or climate monitoring networks, where delays can risk system performance, this measure is especially important. Maintaining seamless and effective operations in time-sensitive contexts requires low latency in cryptographic processes.

By drastically cutting encryption and decryption times to just 1.0 seconds, Q-BALAK overcomes this difficulty and outperforms conventional protocols, which normally take 1.5–1.7 seconds. Using lightweight cryptographic libraries designed especially for Internet of Things devices allows for this improvement. Through the simplification of cryptographic procedures, Q-BALAK guarantees continuous and effective data flow, satisfying the demanding requirements of real-time systems while maintaining strong security. This development promotes improved performance and dependability in IoT applications by enabling smooth operations in situations when timing is crucial.

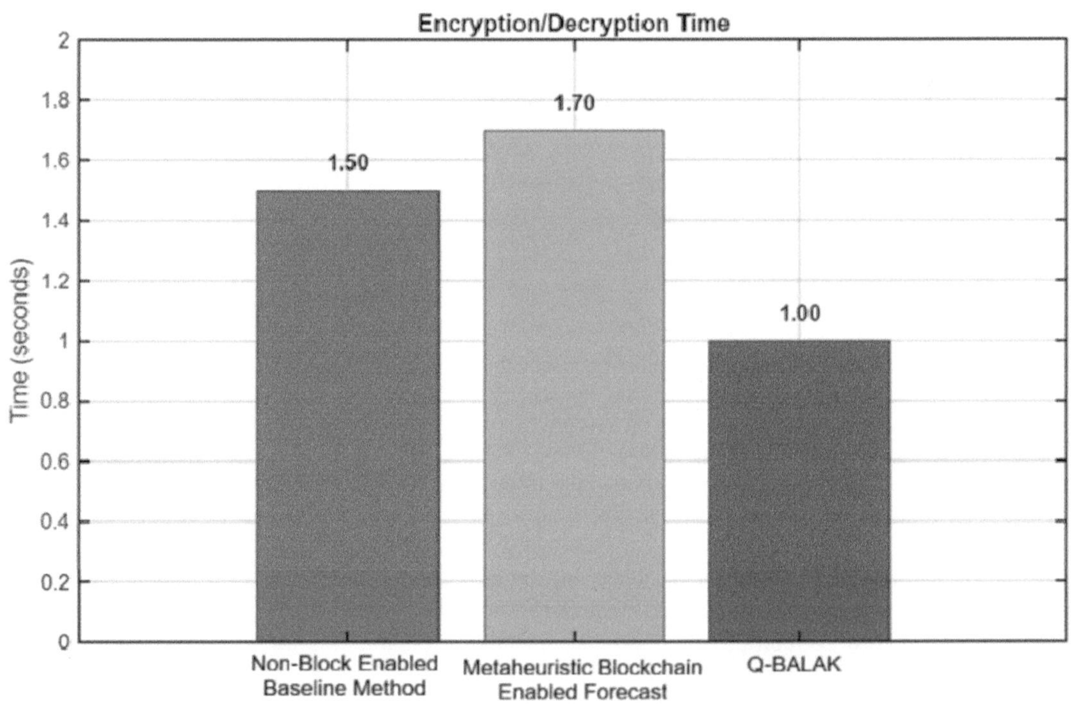

Figure 12.9. Encryption/decryption time-Q-BALAK.

4.2.2 Energy Consumption

The term energy consumption describes how much power is consumed in secure communication systems for data processing, encryption, and decryption. For IoT devices that run on batteries, this statistic is essential, particularly in remote farming settings where routine maintenance or battery replacement can be difficult. Optimizing energy use guarantees extended device functioning, which lowers expenses and improves the system's overall dependability.

When it comes to energy efficiency, Q-BALAK is exceptional. It uses only 80 mAh for cryptographic operations, which is 33% less than the 120–130 mAh needed by current protocols. Effective hardware-software integration, which reduces energy waste while preserving strong security, is responsible for this notable gain. Q-BALAK offers a sustainable solution for Internet of Things applications in resource-constrained situations, such as remote agriculture, by prolonging device lifespan and lowering maintenance frequency, ensuring continuous operation for extended periods of time.

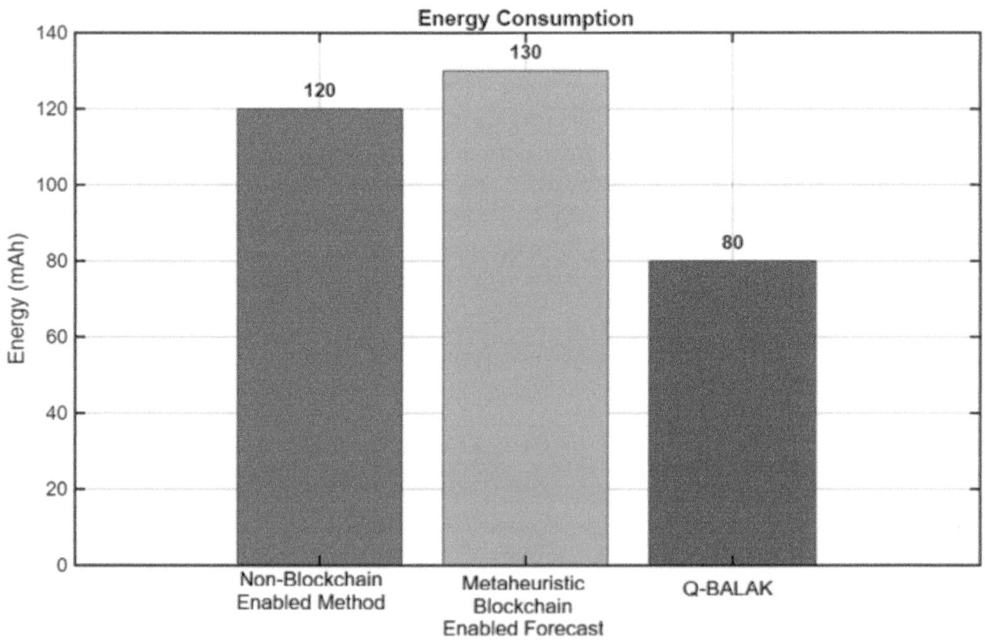

Figure 12.10. Energy consumption comparison.

4.2.3 Data Integrity

Data integrity guarantees that throughout transmission and storage, information is correct, consistent, and impenetrable. This is especially important for IoT systems in agriculture, where private information like weather patterns, crop health, and soil conditions needs to be protected from illegal changes and cyberattacks.

Inaccurate judgments brought on by compromised data integrity could disrupt operations and result in large losses. By utilizing blockchain technology, which offers a decentralized and unchangeable record to completely eradicate instances of manipulation, Q-BALAK ensures uncompromised data integrity. However, during testing, current techniques documented 3–5 instances of tampering, highlighting the weakness of conventional systems. By including strong security features into its design, Q-BALAK guarantees that private agricultural data is safe and reliable, enabling farmers and other stakeholders to confidently make well-informed decisions.

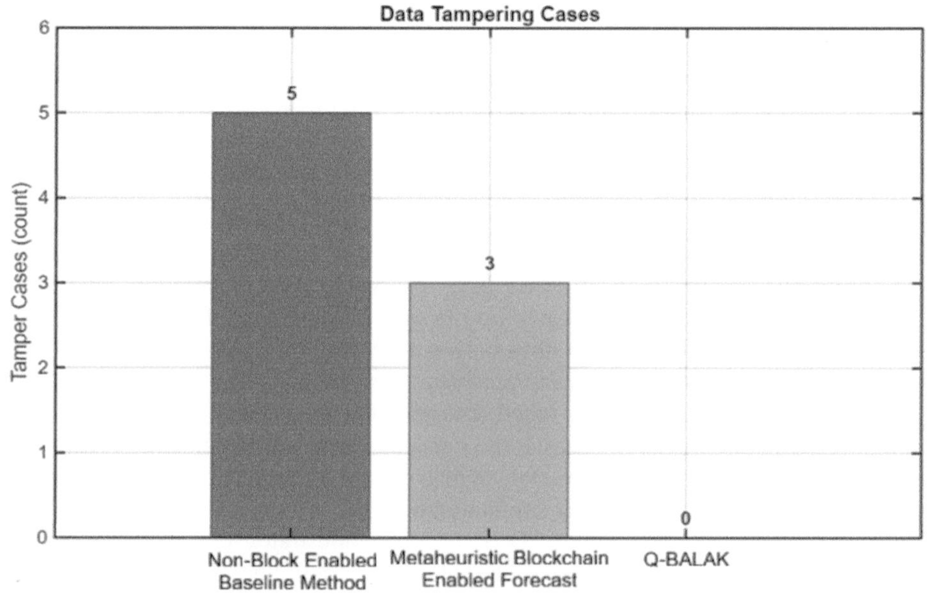

Figure 12.11. Data integrity.

4.2.4 Scalability

The ability of a system to manage increasing data loads and network expansion without sacrificing performance is measured by its scalability. This feature is crucial for Internet of Things applications in agriculture since the solutions must be able to smoothly transition from small-scale farms to large-scale farming operations. As agricultural needs change, a scalable system can accommodate more devices and data flow while maintaining constant efficiency.

With a score of 95%, Q-BALAK exhibits remarkable scalability, far outperforming the 70–75 range of current methods. The distributed design of blockchain technology and its effective

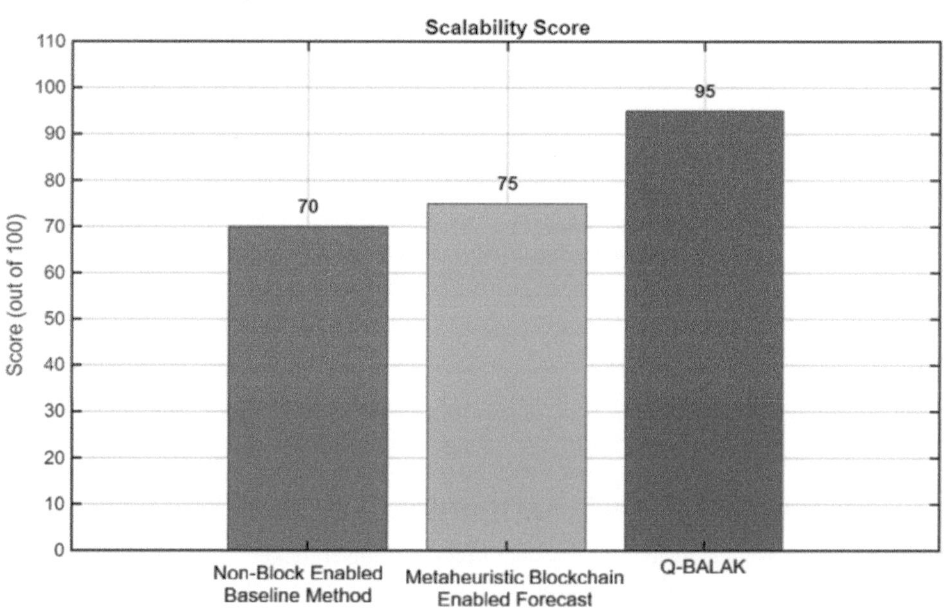

Figure 12.12. Scalability.

transaction processing algorithms, which provide the seamless integration of new nodes and larger data quantities, are responsible for this enhanced performance. Q-BALAK offers a future-ready solution for contemporary farming operations by guaranteeing that the system stays resilient and responsive under rising demands, enabling agricultural scaling with ease and dependability.

4.3 Results and Discussion

The simulation results of Q-BALAK shows that it performs better than other options in guaranteeing effective encryption with little overhead, which is essential for real-time agricultural applications. Q-BALAK surpasses current protocols that have encryption and decryption times of 1.5 to 1.7 seconds by using lightweight cryptographic libraries tailored for Internet of Things devices, resulting in encryption and decryption times of 1.0 seconds. The exact control of irrigation systems, fertilizer schedules, and climate monitoring equipment is made possible by this 33%–41% improvement, which guarantees uninterrupted data transfer. In dynamic farming contexts where decisions made in real time can have a substantial impact on crop output and quality, such efficiency is essential for sustaining continuous operations. The comparison Table 12.1 shows the performance metrics of Q-BALAK better than existing simulation results.

By utilizing blockchain technology, Q-BALAK demonstrates resistance to cyberthreats, including new dangers like quantum attacks. During testing, traditional methods encountered three to five instances of data manipulation, whereas Q-BALAK had a flawless record of zero incidences. The decentralized, unchangeable ledger of blockchain technology guarantees tamper-proof data transfer and storage, protecting private agricultural data from unwanted access. Furthermore, Q-BALAK offers future-proof security thanks to its sophisticated cryptographic algorithms, which strengthen its protection against threats from quantum computing. In addition to safeguarding farmers' data, this resilience promotes confidence among suppliers, distributors, and regulatory agencies, among other players in the agricultural ecosystem.

Q-BALAK was created with affordability in mind; it is highly scalable for big agricultural systems and economical for small-scale farmers. Q-BALAK lowers operating costs for farmers in distant places by extending battery life for IoT devices and using 33%–38% less energy than current protocols. Furthermore, its scalability score of 95, which is far higher than a typical systems' ratings of 70 to 75, demonstrates its capacity to effectively manage growing data loads. Because of its scalability, Q-BALAK can be used in a variety of agricultural contexts and can adjust to both small and large-scale operations. Q-BALAK is well-positioned to propel the agricultural industry's adoption of smart farming technology because to its capacity to handle large transaction volumes and support growing sensor networks.

Table 12.1. Quantitative comparison of Q-BALAK.

Performance Metric	Non Blockchain Enabled Baseline	Metaheuristic Blockchain Enabled Forecast	Q-BALAK	Improvement (%)
Encryption/Decryption Time (s)	1.5	1.7	1.0	~ 33%–41% faster
Energy Consumption (mAh)	120	130	80	~ 33%–38% lower
Integrity	5	3	0	100% tamper-proof
Scalability Score (out of 100)	70	75	95	~ 27%–36% higher

5. Applications of Q-BALAK in Agriculture

Q-BALAK is a cutting-edge agricultural system that combines blockchain technology, sensors, IoT devices, and small microcontrollers to improve data security, integrity, and scalability in farming operations. With its efficient real-time encryption and low energy consumption, the system—which was developed with Hyperledger Fabric and lightweight cryptographic libraries—is perfect for battery-operated devices in remote locations. The scalability of Q-BALAK for both small- and large-scale farming was ensured by performance assessments that showed its resilience to cyber threats, including quantum attacks, and its capacity to manage growing data loads. Robust and reasonably priced, Q-BALAK provides farmers with safe, unchangeable data management to maximize agricultural output.

Precision agriculture: Real-time irrigation and fertilization monitoring and management are made possible by Q-BALAK, which facilitates the safe and effective transfer of sensor data essential to precision agriculture. It protects critical information gathered from IoT devices, like weather, nutrient content, and soil moisture levels, by utilizing blockchain technology and lightweight cryptographic algorithms. Without worrying about data breaches or delays, farmers can improve crop yield, make well-informed decisions, and make the most use of their resources thanks to this safe data flow.

Supply chain tracking: Q-BALAK transforms supply chain monitoring by guaranteeing the accuracy and transparency of produce records through its tamper-proof blockchain platform. From harvest to market, every stage is safely recorded, giving stakeholders and customers verifiable traceability. By reducing fraud, boosting confidence, and simplifying adherence to legal requirements, this method benefits both farmers and consumers by increasing efficiency and responsibility.

Climate monitoring: Accurate weather forecasts and environmental assessments are made possible by Q-BALAK's real-time encrypted data exchange for climate monitoring. The system safely compiles information from dispersed sensors that track humidity, temperature, and other meteorological factors. By preserving the accuracy of this data, Q-BALAK helps farmers predict weather patterns, reduce risks, and modify their plans for sustainable farming methods.

Intelligent agreements: By automating farming transactions with smart contracts, Q-BALAK lessens the need for middlemen and lowers operating expenses. These self-executing contracts provide safe, open agreements for things like renting equipment, selling produce, and buying supplies. Through the simplification of these procedures, Q-BALAK improves mutual trust and efficiency, enabling farmers to concentrate on output while guaranteeing prompt and equitable settlements.

6. Future Scope and Challenges

6.1 Challenges

Adoption hurdles, such as a lack of technical knowledge among farmers, can be a barrier to the successful deployment of sophisticated agricultural technology. This is because users may have difficulty comprehending or making use of these instruments. In example, the initial expenditures of establishing the essential infrastructure, which may include sensors, connectivity, and power systems, can be a hardship on the finances of small-scale farmers who have limited resources. Additionally, scalability continues to be a significant obstacle, as it is necessary to have strong designs, efficient data handling, and seamless integration across a variety of contexts in order to guarantee consistent performance and reliability in large-scale agricultural networks. This can be a challenging endeavor that demands a significant amount of labor and resources. For smart farming solutions to have the greatest possible impact, it is vital that these difficulties be addressed.

6.2 Future Directions

Artificial intelligence (AI) integration for advanced data analytics delivers real-time insights and predictive capabilities, translating raw data into actionable strategies for improved decision-making and efficiency. This is accomplished by changing raw data into actionable strategies. In order to provide smooth multi-platform compatibility and enable varied systems to securely exchange data and interact inside decentralized ecosystems, enhancing the interoperability of blockchain technology is highly recommended. In the meantime, optimizing cryptographic algorithms that are resistant to quantum computing ensures robust security for ultra-low power devices. This protects data integrity and privacy against emerging quantum threats while maintaining energy efficiency, which is an essential factor for widespread adoption in Internet of Things and edge computing environments.

There are still several obstacles that need to be conquered, despite the fact that the possibility of using quantum computing in agriculture is very intriguing. The creation of scalable quantum hardware, the enhancement of quantum algorithms for specialized agricultural applications, and the incorporation of quantum solutions into conventional agricultural processes are some examples of these potential uses. For the purpose of ensuring the successful implementation of quantum technologies in the agricultural sector, it is necessary to strengthen the cooperation that exists between researchers in the field of quantum computing, specialists in the agricultural industry, and software developers.

As a final point of discussion, quantum computing possesses a tremendous potential to revolutionize agriculture by enhancing agricultural output and handling resources. Businesses that specialize in the creation of customer relationship management (CRM) software and billing software play a significant role in the process of utilizing the potential of quantum computing for agricultural applications. These companies may be able to assist farmers in making better decisions, increasing production, and contributing to a more environmentally friendly agricultural future by developing specialized software solutions and making use of quantum capabilities.

7. Conclusion

Through the utilization of blockchain technology and lightweight quantum-resistant cryptography, Q-BALAK provides an innovative solution for the protection of agricultural data. In addition to overcoming the processing limitations of Internet of Things devices, it guarantees the integrity of data, maintains secrecy, and is affordable. Q-BALAK opens the path for secure, efficient, and intelligent agricultural systems that are to the benefit of farmers as well as stakeholders. This is accomplished by planning for quantum risks and exploiting the transparency, which blockchain technology provides.

References

Abdullah Ayub Khan, Zaffar Ahmed Shaikh et al. (2022). A blockchain and metaheuristic-enabled distributed architecture for smart agricultural analysis and ledger preservation solution: A collaborative approach. Journal of Applied Sciences 12(3): 1–19. doi:10.3390/app12031487.

Evstatiev, B. I. and Gabrovska-Evstatieva, K. G. (2020). A review on the methods for big data analysis in agriculture. IOP Conf. Series: Materials Science and Engineering 1032. doi:10.1088/1757-899X/1032/1/012053.

Harshitha, M. S., Shashidhar, R. and Roopa, M. (2021). Blockchain technology for agricultural supply chain management. Global Transitions Proceedings 2(2): 220–226. doi:org/10.1016/j.gltp.2021.08.041.

Krishna Prasad Satamraju and Malarkodi, B. (2020). Proof of concept of scalable integration of internet of things and blockchain in healthcare. MDPI Sensors 20(5): 1388–1404.

Mohammed El-hajj, Hussien Mousawi and Ahmad Fadlallah. (2023). Analysis of lightweight cryptographic algorithms on IoT hardware platform. MDPI Sensors 15(2): 1–29. doi:10.3390/fi15020054.

Mritunjay Shall Peelam, Anjaney Asreet Rout and Vinay Chamola. (2023). Quantum computing applications for internet of things. IET Quantum Computing 103–112.

Sandeep Kumar, M., Maheshwari Venkat, Prabhu Jayagopal and Prasanna Mani. (2021). Applying blockchain in agriculture: A study on blockchain technology, benefits, and challenges. Deep learning and edge computing solutions for high performance computing. Springer Innovations in Communication and Computing 167–181. doi:10.1007/978-3-030-60265-9-11.

Shincheol Lee and Ji Sun Shin. (2023). A new location verification protocol and blockchain-based drone rental mechanism in smart farming. Computers and Electronics in Agriculture, doi:10.1016/j.compag.2023.108267.

Shobha Tyagi and Madhumita Kathuria. (2022). Role of zero-knowledge proof in blockchain security. International Conference on Machine Learning, Big Data, Cloud and Parallel Computing (COM-IT-CON), doi:10.1109/COM-IT-CON54601.2022.9850714.

Silong Li, Yuxiang Chen, Lin Chen and Jing Liao. (2023). Post-quantum security: Opportunities and challenges. Journal of Sensors 23: 139–154. doi:org/10.3390/s23218744.

Wided Moulahi, Imen Jdey, MoatsumAlawida and Tarek Moulahi, (2023). A blockchain-based federated learning mechanism for privacy preservation of healthcare IoT data. Computers and Biology Medicine 167. doi:10.1016/j.compbiomed.2023.107630.

Zhiguo Qu, Zhexi Zhang and Min Zheng, (2022). A quantum blockchain-enabled framework for secure private electronic medical records in internet of medical things. Journal of Information Sciences, Elsevier, 612, 942-958.doi:org/10.1016/j.ins.2022.09.028.

Chapter 13

Investigating the Consequences of Integrating Blockchain and Artificial Intelligence

M. Ramprasath,[1,*] *Elangovan G.,*[1] *Prakash Duraisamy*[2] and *V. Kavitha*[1]

1. Introduction

Blockchain, a decentralized database system with various potential applications, is one of the several emerging technologies now receiving much attention (Baynham-Herd 2017, Maxmen 2018). With its origins in the 1920s, blockchain technology is still a possibly game-changing invention that could change the ways we document transactions, process money, keep tabs on markets, and collaborate (Nakamoto and Bitcoin 2009). By decentralizing the verification of decisions and transactions amongst many people, blockchain technology could be highly beneficial (Taherdoost 2023). All the nodes that participate in mining, cryptographically hash and verify every operation that occurs on the blockchain (Moosavi and Taherdoost 2022). It generates time-stamped documents that cannot be altered, are secure, and are accessible to everyone (Swan 2015). Another obvious system that is experiencing significant growth is artificial intelligence (AI), which enables robots to learn from data and make judgments using this acquired knowledge. According to continuing statistical polling, the artificial intelligence industry might be worth up to USD 13 trillion by 2030 (Pandl et al. 2020).

Information in smart homes is still vulnerable to assaults even though several competing inventions are working towards this goal. Among the areas with the most significant potential for home network security is the development of blockchain technology, that can offer a secure foundation for all networks and thwart command-and-control attacks on secret information appliance to connect to (Lin et al. 2017). All transactions are stored forever in a blockchain because the nodes reach a consensus (Dai et al. 2019). Therefore, it is not feasible to conduct control assaults on data in transit or storage with just one trade. It is necessary to compromise most hubs for the attack to succeed (Salimitari 2017). Figure 13.1 shows the general feature used in real time projects.

The idea of decentralized AI has made strides in recent years. A mix of decentralized artificial intelligence (AI) is built on top of these two technologies (Singh et al. 2020). It enables the

[1] Department of Data Science and Business Systems, SRM Institute of Science and Technology, Tamil Nadu.
[2] Computer Science and Engineering, University of Wisconsin-Green Bay, Wisconsin, USA.
Emails: elangovg2@srmist.edu.in, duraisap@uwgb.edu, kavithav2@srmist.edu.in
* Corresponding author: ramprasm@srmist.edu.in

DOI: 10.1201/9781003620822-13

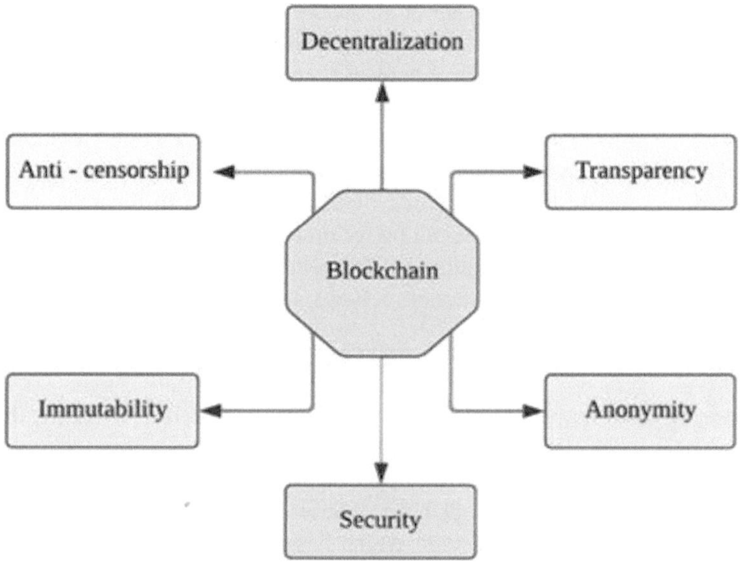

Figure 13.1. Regular feature of blockchain.

operation and preservation of legitimate, meticulously marked, and collaborative information on the blockchain in a distributed and intermediary-freeway (Dinh and Thai 2018). Now, people are looking at cryptocurrency as a reliable platform to store this kind of data, and AI can handle big datasets (Taherdoost 2022). Smart contracts, when integrated into blockchains, enable vetted third parties to keep tabs on user data access and sharing (Wood 2014). All the mining nodes of a blockchain can agree on the accuracy and reliability of their decisions, once they've been exposed to a machine, multiple situations, and an autonomous system (Kumar et al. 2022). Consequently, all parties involved may have faith in and support these choices. Artificial intelligence (AI) approaches, using blockchains, may offer a decentralized solution to the problem of increasing trust and safety in data sharing and decision outcomes among multiple independent supervisors, each of whom can make contributions, plans, and votes on future decisions (Wang et al. 2018). Integrating blockchain with AI has resulted in numerous valuable outcomes (Makarius et al. 2020). By using blockchain technology, medical facilities may store patient details securely. By gaining access to this data, medical professionals may benefit from the insights provided by artificial intelligence algorithms that analyze it. Together, they have made it possible for healthcare providers to combat the COVID-19 epidemic, which is remarkable (Fusco et al. 2020). The innovative BurstIQ is a cryptocurrency-based firm that provides medical data products and a health wallet that manages patient data through artificial intelligence. Doctors and hospitals can use the wallet to view their patient's medical records (Daley 2019). The combination of these two technologies is changing the face of the financial services industry in revolutionary ways, such as increasing the speed of transactions and fostering mutual trust (Soleymani and Paquet 2020). Figure 13.1 shows that blockchains and AI are technically different but can complement one another by addressing each other's shortcomings. Integrating AI and blockchain technology is anticipated to bring about significant changes in various industries, including autonomous vehicles, smart cities, banking, and 6G networks (Moosavi and Taherdoost 2022). Combining AI with a blockchain could have multiple advantages, according to estimates. This allows for the trustworthy dissemination of vast datasets for the purpose of evaluation, education, and taking decisions involving a multitude of stakeholders without depending on a central authority or mediators (Parizi et al. 2018). As a result, several novel and practical large-scale applications might be born (Nakamoto and Bitcoin 2008). The ability of a blockchain to guarantee the authenticity of data makes it a potential storage medium for artificially intelligent system inputs and outputs (Tyagi et al. 2023). Combining blockchain and AI has many benefits as outlined in this article. The integration's real-world applications are the primary emphasis of this study. Also, before the finale, the problems are brought to light.

Boosting the standard of next studies and maintaining with the field's progress are the goals of this examination, which aims to assess the status of the literature on AI and blockchain combos, in a way that new researchers can understand it. Here is the outline of the present research.

In Part 2, we delve deeply into the study's findings approach to locate, sort, and choose the literature. The literature on integrating blockchain technology with artificial intelligence is discussed in Section 3. Here, we offer the most popular articles, discuss how they might be applied, and point out some main issues. The study's final section delves into its findings.

2. Related Work on Blockchain Technology and Artificial Intelligence

The public release of bitcoins by (Nakamoto and Bitcoin 2008) has dramatically increased the visibility of blockchain technology among researchers. Nevertheless, this technological step also heralded a more significant technical shift, expanding blockchain to encompass a wide range of technologies called distributed ledgers (DLTs). These developments have enabled blockchain technology to be used in many other sectors, which has real-world health implications, including computer security and economics (Tyagi et al. 2023). Among the most remarkable achievements is the astronomical market valuation of USD 2.66T (CoinMarketCap 2022) attained by the larger blockchain ecosystems, which comprise more than 2.4 million cryptocurrencies across several tiers. A like pattern is emerging in the corporate world, where around 90% of US, UK, India, and China companies utilize blockchain technology. Authorities worldwide are exploring the use of blockchain technology to improve public services, increase citizen confidence, and streamline procedures for administration (Bhumichai et al. 2024). According to the World Economic Forum, many nations, accounting for nearly all the world's gross domestic product (98%), are considering using CBDCs (Ekbal 2024).

The statistics point to the proliferation of blockchain technology, bringing revolutionary opportunities to the globe to both society and business communities. Alternate methods of agreement have evolved in the field, with Bitcoin's Proof of Work (PoW) evolving into something new and older techniques like Practical Byzantine Fault Tolerance (PBFT) being incorporated (Castro and Liskov 1999). These processes reflect the ongoing flexibility of DLTs and solve various issues, including expansion, effectiveness in energy usage, and quickness of transactions. The continuous attempts to improve DLT features, guaranteeing seamless integration across multiple platforms and businesses, are further underscored by investigating scalability and interoperability concerns. Redefining confidence regarding electronic transactions, influence the systems of society and the methods of government, and going beyond technical gains are all consequences of these technologies (Dodd 2016). The extensive development of blockchain and DLTs highlights their versatile capabilities and signals the beginning of a new age of technological advancement and social upheaval.

Like those used in cryptocurrency transactions, public ledgers can now be securely updated and tracked thanks to blockchain technology (Dinh and Thai 2018). It is worth remembering that "secure" indicates that entries have not been changed, not that they were accurate when recorded. To be more specific, blockchain allows users to link up and share data that has been altered throughout a network by means of a decentralized ledger to which all participants have access. Depending on the circumstances, "players" can include all the various actors involved in the whole bandwidth, including those at the network, regulations, government, and application layers—not just in Bitcoin or any one layer. Concerning the consensus process used by the blockchain solution, the size of the majority and the confidence model used by the intermediates must be determined by these players. A distributed trust system guarantees that transactions are validated and documented by utilizing an agreement procedure involving several individuals. Though decentralization, consensus procedures, and cryptographic algorithms make blockchain systems more secure and transparent, it still depends on beliefs about trust among network participants. To avoid 51% attacks, which might damage the ledger's integrity, Bitcoin's power needs consensus, which relies on the fundamental trust presumption that most nodes are honest.

Consequently, blockchain does not do away with trust altogether; it only changes its location and function inside the system. This is particularly true when the 51% attack, software flaws, or other types of cooperation are possible. Although blockchain technology allows for safe, intermediary-free transactions, the exact level of trust lessness differs across different architectures. While private blockchains require confidence among a few users, public blockchains, such as Ethereum and Bitcoin, are highly decentralized and can promote an atmosphere of distrust as long as protocol rules are respected, and the system is secure. The fact that different networks have made alterations in the past shows that blockchain operators and majority stakeholders can significantly impact how a distributed ledger works. Changes like SegWit (Santos and Kostakis 2018) and Ethereum's DAO fork (Bratspies 2018) show how blockchains change to address concerns such as authoritarianism and security, demonstrating whether the technology is dynamic and vulnerable to influence from influential people. Figure 13.1 and the following text provide a high-level overview of blockchain's essential features. These features are also necessary for any distributed project to run smoothly:

Decentralization: The very nature of blockchain makes it a distributed and decentralized system to enable P2P communication between the nodes that make up the network. This decentralization strengthens the system by removing potential weak spots. Decentralization means that a single entity does not control the system, whereas distribution implies that data and processing are spread across several nodes. Distributed ledger technology (DLT) and decentralization work hand in hand to make sure that no one entity controls the network's operations and that many nodes keep copies of the ledger for more security and redundancy (Narayanan et al. 2016). This principle of decentralization is beneficial in blockchains without authorization, such as Bitcoin, because it strengthens their defense against censorship and manipulation by any one institution. The attack surface, threat model, and protocol/network configuration are the three main determinants of attack resilience.

Resistance to censorship: Filtering resistance is an essential characteristic of permissionless blockchains, which allow anybody to participate. Since the blockchain relies on a distributed system of connections to record communications, it becomes exceedingly impossible for any institution to regulate or censor transactions. While permissioned blockchains—those whereby a service for members determines who may join—may not provide the same degree of opposition to restriction (Solat et al. 2021), they usually offer greater resilience than centralized databases. Part of the reason for this reluctance is the permission aspect of the blockchain, which allows only authorized individuals to make decisions.

Staying the same: Consensus processes and the distributed design of blockchain systems combine to produce immutability, a fundamental property frequently associated with blockchain technology. Two consensus techniques that contribute to the immutability and security of the ledger are Proof of Work (PoW) and Proof of Stake (PoS). Once the information has been registered on the distributed ledger, nothing can be altered unless everyone in the ecosystem agrees. Because of its distributed consensus function, the digital ledger is resistant to manipulation and hacking as a data format.

Openness: All nodes in the network benefit from the blockchain's openness since it makes transaction data publicly available. Blockchain accommodates every node actively participating in the network, unlike centralized databases in which one computer has full access to all user information.

The nodes in this network can see all the details of an exchange. Additionally, all the nodes are kept in sync and updated with the most recent transactions and changes to the blockchain because each node has its copy of the ledger and shares the information of all transactions.

Privacy: Because no outside platforms retain personal data about users, members in blockchain networks can enhance their privacy by creating numerous addresses for access. Although this method provides a high level of anonymity, it cannot ensure complete privacy protection due to technological restrictions.

Safety: Secure record keeping and traceable transactions are two of blockchain technology's many security advantages. To improve this security even further, Merkle Forests ensures that data remains unchanged and allows for the fast and reliable linkage of transaction blocks. These qualities efficiently secure data integrity in susceptible areas such as medical records and banking (Jamaludin and Rohani 2018). Organizations may take security a notch further and safeguard themselves from unlawful entry and manipulation by customizing blockchain systems for specific uses or using artificial intelligence algorithms for improved data analysis.

3. Artificial Intelligence

The development of AI has made it possible for computers to mimic human intellect and reasoning. Independence in artificial intelligence systems is defined as the capacity of AI to function autonomously, and remove the need for human guidance or oversight.

Several techniques have been proposed to define AI, but they can be roughly categorized as follows: (a) understanding rationally, (b) behaving humanely, and (d) performing rationally. Another way to phrase it is that artificial intelligence is "one of the studies of techniques to assign tasks to machines that humans are currently more adept in" (Rich and Intelligence 1991) or "the research of computation that makes it possible to understand, reason and act." While AI approaches use simple algorithms, they set themselves apart from more conventional forms of programming by introducing learning and flexibility. This radical break from predictability contrasts sharply with the more inflexible character of traditional programming methods and enables AI algorithms to deal with more complicated and dynamic scenarios.

Autonomous agents powered by artificial intelligence (AI) have recently demonstrated exceptional proficiency in analyzing and comprehending massive volumes of data, thanks to advancements in Large Language Models (LLMs) such as G5PT and MMLU in Gemini. These state-of-the-art AI models show remarkable proficiency in tasks that require understanding and creating natural language. They were trained extensively on massive datasets.

The complicated topologies of these structures allow for the rapid access and change of significant models, even though they do not store knowledge typically. The application of contemporary methods achieves this. Techniques include timely engineering, training by reinforcement (Ziegler et al. 2019) and improvement. These techniques enhance the systems' adaptability and performance. Artificial intelligence (AI) draws on various disciplines, including computing, theology, language, neurology, psychological research, and electronics engineering. Artificial intelligence algorithms emphasize symbolism more than numerical or textual data. This means that strings convey and alter things, events, and surroundings in the actual world. The next step is to decode the strings into symbols and then arrange them in a hierarchy or lists to show how the symbols relate to one another (Bhumichai et al. 2024). Algorithms developed by artificial intelligence can help machines comprehend and identify things, places, and things in the physical world.

Disruptive methods: Conventional computer programs follow standard algorithms, meaning each process stage requires specific human instructions. On the other hand, AI algorithms solve problems without human intervention. Artificial intelligence (AI) can adapt to changing circumstances without strict human computer programming, simplifying solutions and making them more flexible (Greff et al. 2020). Figure 13.2 shows the uses of the normal features in AI based block-chain model.

Reasoning: Artificial intelligence systems can use deductive and inductive reasoning methods because they can process knowledge instead of just data. Improving the efficacy of machine reasoning relies heavily on this. Several techniques were thus suggested to improve the effectiveness

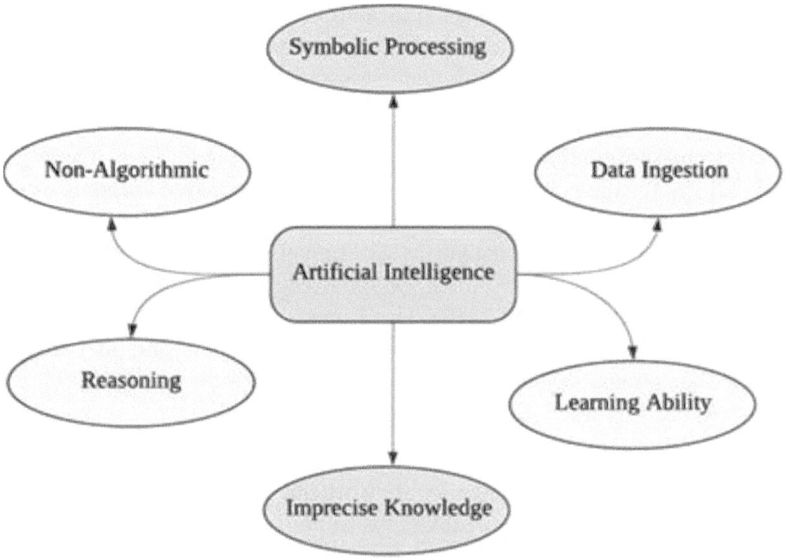

Figure 13.2. Normal feature of Artificial Intelligence.

of computer reasoning, including case-specific reasoning, case-based judgments, and analogous reasoning. These reasoning approaches allow AI computers to mimic human cognition and solve problems more effectively.

Ingestion of data: AI handles bulk datasets from many sources independently using statistical techniques and Machine Learning (ML). Autonomous data input reduces or eliminates human intervention, speeding up operations and ensuring more precise data treatment. All through the data-processing phases, the benefits become apparent in increased efficiency and precision.

Capacity to learn: The primary objective of artificial intelligence is to mimic human cognition through the acquisition of new knowledge, adaptation to changing conditions, and executing activities typically requiring human intellect (Intelligence 2020). Such examples include chess games, predictions of the stock market, and autonomous vehicles. Not only is it essential to mimic human intellect, but collecting data is necessary for AI models to manipulate tasks and improve their learning capabilities.

Unclear knowledge: Artificial intelligence algorithms are excellent at handling unstructured and inaccurate data, unlike conventional applications that rely on precise knowledge. Artificial intelligence is improving at handling incomplete data because of innovations like mathematical morphology, fuzzy set theory, and formal reasoning (Bloch 2006). AI's flexibility makes it a potent tool in real-world settings, where attaining accuracy would be difficult.

3.1 Merged Advantages of Blockchains and AI Integration

Artificial intelligence (AI), robotic process automation (RPA), and blockchains can improve multi-party company procedures by reducing human involvement, increasing throughput, and enabling improved data accuracy. Smart contracts built on the blockchain with machine learning algorithms could do a lot of cool stuff, like choosing the most eco-friendly transporting option, resolving disagreements, reorganizing, paying for or buying stock based on standard cutoff values and measures, and more. Solat et al. (Solat et al. 2021) investigated how a computerized service would fare in the presence of artificial intelligence and blockchain. The investigator chose another data collection technique to acquire relevant information about this research subject. The assets did the same thing, using statistics to simplify their information. The correct effectiveness of AI in controlling robotic activity can also be better understood by reading this study.

Enhancement: Blockchain-based companies have a competitive advantage thanks to AI's lightning-fast and deep data reading, analysis, and correlation capabilities. The expansion of the use of blockchain allows for the possibility of AI, which opens massive volumes of data within the firm. This data can then be used to gain more tortious insights, competently manage the use of data and model sharing, and create a more trustworthy and transparent data market. An architecture that utilizes blockchain technology as a framework for smart contracts and a ledger for robotic control was proposed by (Jamaludin and Rohani 2018). This framework relies on third parties, or oracles, to handle data. Secured event recording, intelligent contract-driven robots, and integration with external artificial intelligence for image processing are all showcased. Production, network administration, and controlling robots are only a few of the many possible applications of the proposed architecture due to its ease of integration, modification, maintenance, and expansion to other domains.

Sincerity: Addressing the challenge of explicable artificial intelligence is made possible by the digital record of blockchain technology, which helps to understand AI's fundamental structure and

Table 13.1. Real-world uses of blockchain-AI integration.

Sector	Applications	Explanations	Current Situation
Healthcare	Maintaining confidentiality when sharing patient information	Artificial intelligence (AI) examines patient data encrypted by blockchain to provide light on diagnosis and therapy.	Computerized health data are protected by MedRec, and cancer therapies are optimized by IBM Watson Health's artificial intelligence.
	The development of new drugs	Drug research can be accelerated by using blockchain technology for safe data exchange and AI to forecast molecules' efficacy.	Under COVID-19, for example, Molecule used blockchain for research collaboration, which aided AI-driven medication discovery.
Finance and Banking	Anti-fraud measures	AI can identify patterns of fraudulent transactions, while blockchain technology prohibits manipulation due to its immutability.	To combat fraud and stay in compliance with regulations, JPMorgan Chase is utilizing AI and blockchain technology.
	Networks of distributed autonomous organizations	Smart contracts run on the blockchain, and AI helps with efficiency and decision-making.	Artificial intelligence (AI) predictions and blockchain-backed transparency enable collective investment decisions on DAOstack.
Supply Chain	Verification of origin	Artificial intelligence (AI) analyzes patterns to predict demand, while blockchain guarantees product origin transparency. With blockchain technology, electricity can be traded between individuals;	While artificial intelligence (AI) forecasts popularity of seasonal commodities, blockchain helps Walmart track the sources of tainted food.
Energy Sector	Intelligent power grids	With AI, grid performance and load balancing may be optimized.	The German platform Power Ledger allows users to trade solar energy, and artificial intelligence checks that supply as well as demand are equal.
Smart Cities	Protecting Internet of Things devices and cities	Thanks to AI data analysis and blockchain security for Internet of Things communications, smarter traffic, waste, and resource management are possible.	To ensure the safety of the Internet of Things (IoT), Smart Dubai employs blockchain technology. Artificial intelligence (AI) is used for better traffic control and construction maintenance.

the data source it is pulling information from. There will be more faith in data and, consequently, AI-generated recommendations. Distributing and storing AI models on a blockchain, especially with AI, can increase data security. An AI data security mechanism based on blockchain was proposed by Rich and Intelligence (1991) for use in 6G networks. Subsequently, they discussed two AI-powered 6G uses: indoor orientation and autopilots. Using a navigation system for indoor use as a case study, they demonstrated how blockchains effectively secure data. An intelligent service assessment and improvement platform that integrates blockchain technology with artificial intelligence is being built. Table 13.1 outlines the many real-world uses of blockchain-AI integration.
Table 13.1 shows the different real world use cases involved in blockchain based AI integration.

4. Use Case for AI and Blockchain

Business process: AI and digital currencies are transforming distribution networks across sectors by digitizing a manual process, providing computerization and information to execute transactions, and assuring the information's reliability and accessibility. Companies might better understand their decarbonization activities by tracking carbon emissions data at the element or consumer level.

4.1 Money and Banking

Several ways artificial intelligence and blockchains are changing the financial services industry include:

- Establishing trust.
- Making transactions involving more than one party easier.
- The quickening of the pace of financial transactions.

Consider the steps involved in getting a loan. All applicants must consent to having their data stored on the blockchain. Integrating data trust with automated application review techniques can lead to faster closings and increased client satisfaction.

4.2 Biological Sciences

Drug manufacturers can improve study success rates and medicine supply chain openness and traceability with the help of artificial intelligence and blockchain technology. When a decentralized structure for clinical trials is combined with solid data analysis, it becomes possible to automate trial participation and data collecting, monitor patients, control consent, and make data open and transparent. Artificial intelligence for healthcare has the potential to enhance nearly every facet of medical care, from therapy discovery and customer assistance to the identification of insights derived from the discovery of patterns and information about patients. Electronic health records (EHR) and other sensitive patient data can be securely shared across medical professionals using blockchain technology.

4.3 Analysis of Social Networks

Only some studies have considered the built-in social media features of the protocol (Ziegler et al. 2019). Several publications used psychological analysis to make predictions about the personalities of characters. The results served as the basis for several theories researchers used to identify personality traits in people. The interconnections between underlying character and mental diseases (Bhumichai et al. 2024), job performance and happiness (Greff et al. 2020), and even relationships can be made sense of with the help of these models. Social media platforms are ideal for studying group dynamics because of their massive user bases and wealth of data. The interactions between blockchain users are crucial to reveal hidden patterns and provide new angles from which to analyze this speculation bubble. Applying principles from social network evaluation to study blockchain links might yield fruitful results. Visualizing a community with nodes representing users identifiable by blockchain addresses and arcs representing transactions between users provides a sense for this assignment. Ziegler et al. (2019) created a method based on social network analysis to discover user behaviour patterns across a bitcoin investment bubble and to collect data about it. Their approach is comprehensive. Therefore, it can be utilized to examine any speculative bubble in cryptocurrencies, whether current or future. They showed that their method can aid in the hunt for speculators and that this skill can encompass existing, potential, and historical bubbles.

4.4 Implications of Merging AI and Blockchains

Businesses and specialized software development firms will substantially embrace these upcoming five to ten years in terms of technology. Leaders in the production who are both innovative and technically adept continue to see the tremendous potential of merging blockchain technology with artificial intelligence. Come with me as I show you how your company may benefit from artificial intelligence and blockchains.

Learning AI's mental processes: The public will only accept AI, if it can supplant human judgment, no matter how sophisticated it gets. One issue that has stalled the majority usage of AI is the difficulty of accounting for the computer's activities. If AI's decision-making processes can be recorded, the public will rapidly trust it. The secret procedures of machines could be exposed if blockchain technology and AI are combined. A ledger with distributed storage could record and make available all the decisions made by AI. Blockchains are perfect for auditing and other applications that require absolute security: since, once recorded, information cannot be changed.

5. Outline of Current AI and Blockchain Uses

Artificial intelligence (AI) and blockchain technology have found widespread and revolutionary uses in many industries. There has been a proliferation of AI applications across various industries. Predicting results for patients, aiding in assessment, and even generating therapies are all areas where AI has found utility in the medical field. Robotic trading, credit scoring, and fraud detection are some of the financial applications of artificial intelligence algorithms (Dwivedi 2019). Improvements in vehicular autonomy are propelled by artificial intelligence (AI) technologies such as deep learning, reinforcement learning, machine learning, and neural networks.

To further aid in the understanding and prediction of climate trends, AI has even found its way into climate modelling.

However, blockchain has been most linked to financial applications since the emergence of currencies. Nevertheless, its usefulness goes beyond that industry. By improving visibility and tracking capabilities, blockchain technology can help supply chain managers verify the legitimacy of products. Blockchain technology allows decentralized energy trading platforms to operate in the energy industry. The technology, decentralization, transparency, and security properties of blockchain have also made their way into voting systems, making them more trustworthy and safer. Although research into the potential uses of AI and blockchain is in its infancy, some exciting possibilities have already been suggested. Using blockchain technology to address one of the main challenges in AI nowadays is creating artificial intelligence models that are more visible and explainable. This is one example of an application. Data utilized in AI models, especially in delicate industries like healthcare, can be protected by the inviolability of cryptocurrency.

5.1 Unified Use of AI and Blockchain Technology

Given its adaptability and potential use in many other industries, blockchain technology has quickly become one of the most widely talked-about breakthroughs of recent times—the astronomical increase in data generated by sensing. Internet of Things (IoT) devices, online communities,

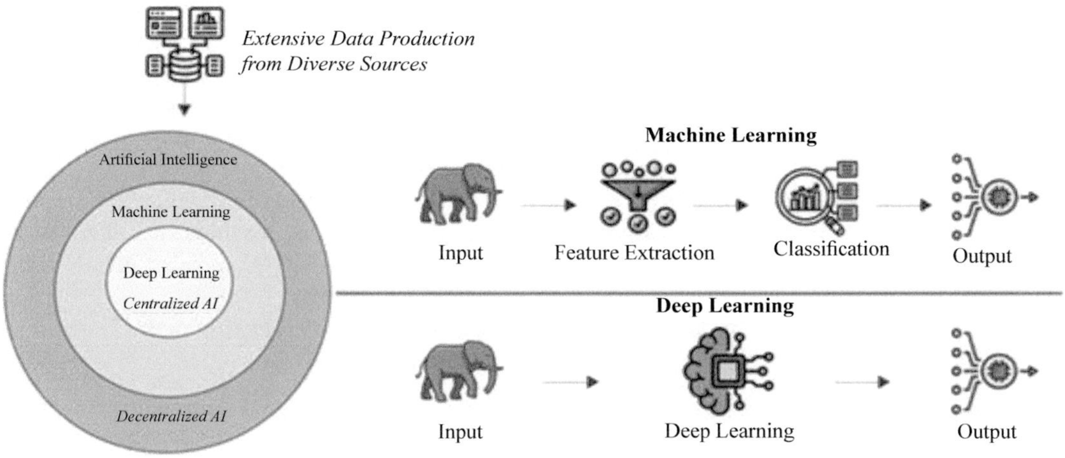

Figure 13.3. Blockchain and AI: centralized and decentralized AI.

web-based applications, and platforms have all been crucial to AI's development. Applying machine learning and deep learning techniques to this data can yield valuable results.

Nonetheless, most AI approaches use a centralized model for training, in which a group of servers runs distinct models using validation and training data sets. Tech behemoths like Amazon, Google, Facebook, and Apple process massive amounts of information to make intelligent judgments. Since data is handled and kept centralized, it is vulnerable to hackers and abuse, a vital issue of centralized AI as shown in Figure 13.3. In addition, there needs to be assurance that the data is accurate or has come from reputable sources. These problems can cause AI to make decisions that are incredibly wrong, harmful, or both. Figure 13.4 depicts the emerging notion of decentralized AI, which merges AI and blockchain technology, in response to these issues.

Transactions and storage of verified, digitally signed, and securely transmitted information on the blockchain enable processing, analytics, and decision-making using decentralized AI. There are no longer any requirements for intermediaries or trustworthy individuals, thanks to this method's

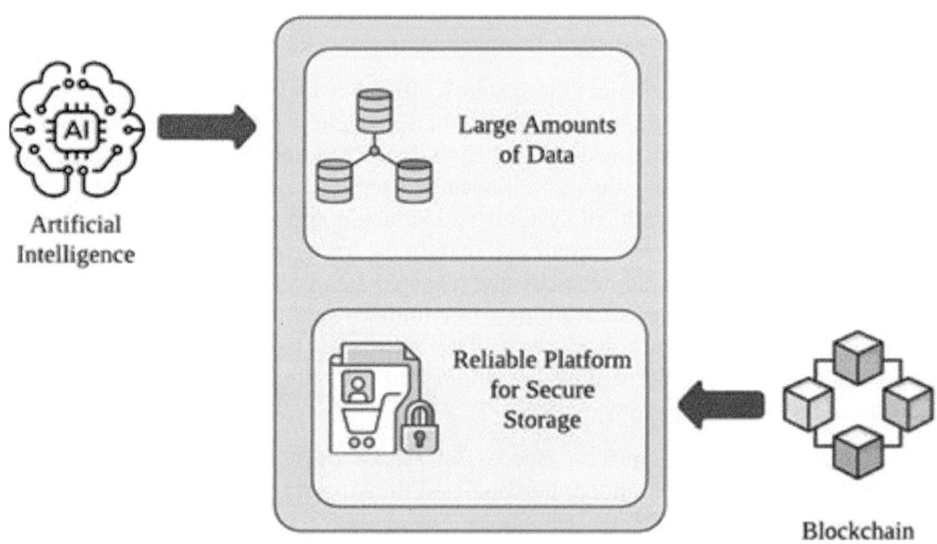

Figure 13.4. Blockchain-based artificial intelligence methods: distributed training.

distributed and decentralized operation. The advent of blockchain as a reliable framework for the secure storage of massive volumes of data has been a boon to AI. Smart contracts are a blockchain component that allow users to program the network to control transactions between parties involved in data generation, access, or making choices. With the help of blockchain, AI methods can provide decentralized learning, which allows for the safe and trustworthy exchange of information and the results of decisions among many separate entities that can work together, coordinate their actions, and cast votes on future choices. The nodes of all the participating miners in the blockchain, may confirm and authenticate the decisions made by autonomous systems and machines based on smart contracts, and these systems and machines can learn and adapt to new situations over time. No one can dispute or deny such choices; everyone involved may check them.

Figure 13.4 dissipates block chain-based AI method in a distributed environment. Here are a few key advantages of decentralized (blockchain) artificial intelligence:

- The system's diskless environment safely saves essential and personal data, enhancing data security.
- Electronic signing data and ensuring private keys are secure allow machine learning algorithms to work on trustworthy data, improving the reliability of their decision-making. Because blockchain's decentralized database records events point-by-point, it enhances confidence in robotic choices by making them more reliable.
- People will have more faith in AI choices made by robots if their reasoning is recorded on the Blockchain.

Communal decisions: Due to decentralized and dispersed decision-making algorithms, numerous robotic applications can now make use of communal decision-making without relying on a central authority. Integrating separate cybersecurity AI agents into a coordinated network overcomes problems with scheduling and delivers complete safety across core networks. This is an example of decentralized knowledge in action.

Maximum effectiveness: By combining AI with blockchain, intelligent Decentralized Autonomous Agents (DAOs) can validate client data and worth, and transfer assets automatically and quickly. Decentralized (blockchain) AI's advantages have far-reaching effects on many AI uses, such as:

Autonomic computing: The design of distributed ledger technology enables for the creation of entirely decentralized autonomous systems by guaranteeing operational decentralization and recording persistent exchanges among users, content, apps, devices, and systems.

Optimization: Decentralization enhances the system's efficiency by processing highly relevant data. This is particularly helpful when there is a requirement to execute several optimization tactics across many apps and systems simultaneously. Blockchain-based decentralized AI planning solutions provide better blueprints with ongoing surveillance and provenance history. Creating crucial and unchangeable plans for mission-critical systems and strategic applications is possible. Regarding thinking, decentralized perception techniques store only the traces of successful perceptions on the blockchain. They do away with the necessity for repeated data collecting in artificial intelligence (AI) applications. When it comes to learning, decentralized learning models are great for both independent and distributed learning systems. This allows for complete coordination of local intelligence in all areas of AI. Regarding developing models, blockchain technology guarantees safe versions and upkeep.

Search: Blockchain can safely preserve successful search traces and traversal paths, offering subsequent operations optimal solutions. Personalized thinking strategies can be developed using blockchain-based distributed reasoning techniques, especially for perception, education, and model deployment.

Decentralized storage: Securing personal and sensitive data securely and solving capacity and scalability problems are both achieved by utilizing a decentralized storage system based on the blockchain.

Information administration: Distributed database management systems deploy information at the node level and consider spatial and temporal properties. For better security and data provenance, metadata can be stored on blockchain.

Educational model development: When a developing learning model is decentralized, it becomes possible to create models for client applications that use resources optimistically.

Model maintenance and deployment: This AI application-centric process keeps track of modifications, ensures that models are versioned consistently, and allows for the safe and reliable sharing of models with other AI applications.

Decentralized perception: To save only the "footprints" generated by effective perceptions on the blockchain, decentralized perception strategies do away with the necessity for repeated data collecting, providing a revolutionary approach for AI applications. By cutting down on repetitive data storage and lowering the amount of data required to be handled and stored, this technique improves data management efficiency. Moreover, the implementation of blockchain technology guarantees improved security and integrity since it offers an impenetrable ledger for the storage of critical perceptual data. By utilizing blockchain's decentralized and immutable nature, this strategy not only maximizes resource efficiency but also improves the dependability and credibility of AI systems.

Distributed reasoning: Personalized reasoning techniques require the development of blockchain-based distributed reasoning strategies, particularly in the phases of perception, learning, and model deployment. These techniques make it possible for data to be securely and transparently shared among numerous nodes by utilizing the decentralized and unchangeable characteristics of blockchain technology. Because of this decentralization, models are more reliable and accurate because the data utilized for perception as well as learning is varied and complete. Furthermore, the trust of participants is increased by blockchain's openness and protective characteristics, which safeguard the data's integrity and privacy. Blockchain-based strategies enable real-time updates and modifications to the argumentation models based on fresh data inputs throughout model deployment, guaranteeing that the customized strategies continue to be applicable and efficient. As a result, this method improves general efficacy and efficiency of the AI system, while also increasing its adaptability to different situations and demands.

6. Decentralized Learning

The seamless incorporation of local information across multiple verticals in AI systems is made possible by decentralized learning models, which are essential for enabling distributed as well as autonomous learning systems. These models enable autonomous learning and decision-making at the local level by spreading the learning process across several nodes. These abilities can be combined to create a coherent and well-coordinated intelligence network. This method guarantees that every node may adjust and improve its learning processes regarding local data and conditions, which improves the scalability and resilience of AI systems. This framework is further strengthened by blockchain technology, which offers an immutable and secure ledger for versions and managing learning models. This keeps the learning process intact and prevents unauthorized alterations by guaranteeing that all improvements and adjustments to the models are openly documented. Blockchain technology combined with decentralized learning models creates a robust, safe, and extremely effective artificial intelligence system that can learn both independently and cooperatively in a variety of contexts.

Decentralized data management: Decentralized data management techniques take temporal and spatial characteristics into account, applying these concerns at the network node level. These schemes make sure that information is precisely timestamped by incorporating temporal properties, which makes it possible to monitor modifications and patterns over time. Conversely, spatial features facilitate the location of data, guaranteeing that the data is pertinent to its place of geographic as well as environmental origin. The granularity and relevancy of the data are improved by this node-level deployment, resulting in more accurate and efficient data handling. This is enhanced by blockchain technology, which stores metadata—detailed information on the history, ownership, and modifications of the data. Increased security is ensured by this safe and unchangeable metadata ledger, which forbids manipulation and unwanted access. It also offers strong data provenance, making it possible to confirm the history and legitimacy of the data. Blockchain technology and decentralized data management systems work together to produce a potent foundation for safe, accurate, and dependable data handling over dispersed networks.

6.1 Positive Aspects of AI and Blockchain Integration

Blockchain and AI together have the potential to create enormous synergies that will capitalize on their respective advantages and lessen some of their disadvantages. Solutions that are not only greater in intelligence but also transparent, safe, and effective may come from this combination.

Data security and privacy: By combining the advantages of both technologies, blockchain and AI greatly improve data security and privacy. Blockchain's decentralized architecture eliminates the possibility of only one single point of a failure, significantly boosting the system's resistance to interruptions and attacks. The blockchain creates a permanent record that is incredibly difficult to change or tamper with by securely encrypting every transaction and data entry and linking it to earlier entries. The integrity and reliability of the data depend heavily on their immutability and transparency. Artificial Intelligence (AI) enhances blockchain security by utilizing sophisticated methods like anomaly detection and pattern recognition. Blockchain activity can be continuously monitored by AI algorithms, which can then quickly detect and neutralize any risks or harmful activity. AI, for example, could identify anomalous patterns that can point to fraud or security breaches, enabling quick action to stop any harm from happening. AI algorithms can anticipate and avert future risks by learning from historical data, thereby increasing the blockchain's security. AI can also improve blockchain network privacy protections. AI may train models on decentralized data without transferring sensitive information by using methods like federated learning, protecting user privacy. Blockchain technology and artificial intelligence work together to build a highly robust, private, and secure system that can handle the demands and complexities of contemporary data management.

6.2 Efficiency and Scalability

Blockchain networks have a hard time growing because of the high processing overhead involved in reaching consensus and confirming transactions. Blockchain network performance can be enhanced by AI algorithms in several ways. First, AI can improve consensus processes by increasing their adaptability and efficiency. Machine learning algorithms have the capability to forecast and optimize the consensus process, hence decreasing the duration and energy requirements for transaction validation. Second, by recognizing and ranking transactions according to their urgency and significance, AI can speed up the validation process and guarantee that crucial transactions are handled more promptly. AI can also distribute compute resources within the network optimally, distributing the load and avoiding bottlenecks. Through ongoing network performance analysis and inefficiency detection, AI may make real-time modifications that improve the blockchain's overall scalability. This integration makes the blockchain more feasible for large-scale along with high-demand applications by enhancing transaction delay and throughput while also ensuring that the blockchain is resilient and responsive to growth.

6.3 Transparency and Trust

The 'black box' issue in deep learning, which is the transparency of AI processes for making decisions, may be resolved with the use of blockchain technology. Blockchain's openness and data immutability can be used to securely record and audit AI choices at every stage of their development. This method improves accountability and trust by giving a transparent, unchangeable record regarding how inputs are handled, which data affects decisions, and the way outputs are produced. Because blockchain is decentralized, it can be widely accessed while retaining security, which makes it a strong option for enhancing the explainability and dependability of AI systems in a variety of industries.

7. Connecting Blockchain with the Unique field of AI: Possibilities and Difficulties

The integration of digital currencies with various branches of artificial intelligence has the potential to alter many areas related to technological applications radically. This section delves into the possibility of combining blockchain technology with multiple branches of artificial intelligence, including deep learning, processing natural languages, neural networks, and machine learning. It also points out some of the challenges that could arise in this area, considering how important both branches are becoming.

7.1 Blockchain and Neural Networks

Possibilities: We can guarantee the legitimacy and provenance of the data needed to train neural networks by combining blockchain technology with neural networks. This is particularly helpful in industries like healthcare and finance where maintaining data integrity is crucial. Additionally, blockchain can support decentralized neural network models, which allow several parties to work together without centralized management and promote transparency and confidence in the model's predictions.

Difficulties: The computational complexity and latency that Blockchain introduces is one of the difficulties. Blockchain confirmation times can cause delays for real-time neural network applications.

7.2 NLP and Blockchain

Possibilities: Blockchain technology may be utilized by NLP systems, particularly chatbots, to verify the legitimacy of information sources and ensure that users receive accurate information. Blockchain technology can also be used to monetize content, allowing writers to get paid directly for their work.

Difficulties: Languages are dynamic, with dialects and semantics that change over time. This can cause synchronization issues in a blockchain setting. Furthermore, the inherent delay in transactions validations of blockchain may cause latency when processing language input in real-time, as in the case of voice assistants.

7.3 Deep-Learning and Blockchain

Possibilities: By confirming the legitimacy of incoming data, deep learning models, particularly in fields like recognition of images, could profit from blockchain's tamper-proof nature. Furthermore, blockchain's transparent and non-repudiable transaction recording feature makes deep learning model monetization easier.

Difficulties: The depth of deep learning models necessitates enormous computer resources. Because of its consensus processes, blockchain might increase costs and make real-time operations difficult. Furthermore, because to storage limitations, storing complex deep learning models on-chain may not be feasible.

7.4 Machine Learning and Blockchain

Possibilities: Blockchain can facilitate the sharing of data and algorithms without worrying about data manipulation, providing a transparent and safe environment for learning models. Additionally, blockchain can support decentralized machine learning, which enables several parties to train and contribute to a model, possibly producing stronger results.

Difficulties: Scalability is an issue. This iterative process must be efficiently supported by the blockchain as machine learning algorithms continue to develop with additional data. Ensuring data privacy is another difficulty, particularly when confidential information is involved.

8. Suggested Methods for Minimizing or Mitigation of Security Issue

Considering the importance of the security risks related to AI and Blockchain, it is imperative to design techniques that effectively mitigate or minimize these risks as shown in Figure 13.5.

The following mitigating techniques for AI can be used.

Creation of a sturdy model: Developing strong AI models entails putting defensive mechanisms in place to identify and fend off adversarial attacks, which try to sway model predictions by slightly changing input data. One important technique in this endeavor is adversarial training, in which AI models are trained using adversarial examples—inputs that are purposefully altered to deceive the model while remaining undetectable to humans—in addition to conventional data. Through this

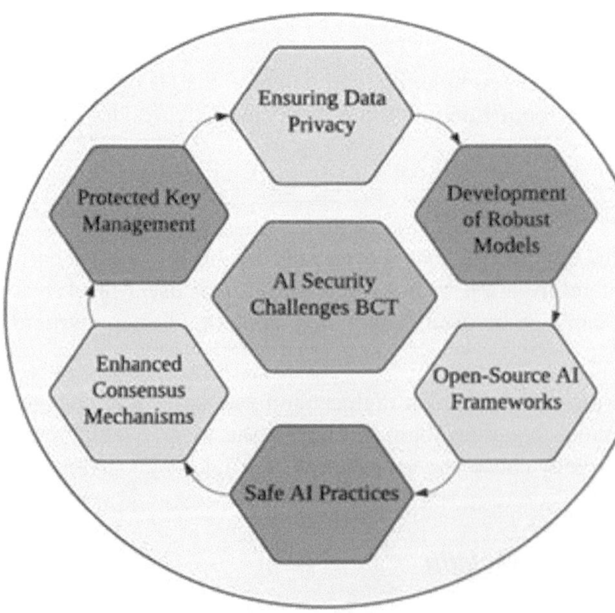

Figure 13.5. Mitigation approaches needed to address the issues.

training, the model gains the capacity to recognize and react to these kinds of adversarial inputs, strengthening its resistance and guaranteeing more dependable performance in real-world situations. In addition to adversarial training, models are further strengthened against potential vulnerabilities by utilizing methods such as input preprocessing, gradient masking, and uncertainty estimation. Organizations can fortify the security and reliability of artificial intelligence (AI) systems across crucial applications by giving priority to strong design principles. This will protect against new dangers in domains like autonomous driving and cybersecurity.

Preserving data privacy: Methods like federated learning and differential privacy are essential for protecting data privacy while utilizing AI. By adding controlled noise to datasets, differential privacy preserves the confidentiality of each data point while maintaining the ability to do correct statistical analysis. Federated learning, on the other hand, allows for direct model training on decentralized devices, eliminating privacy issues by guaranteeing that sensitive data stays on those devices rather than being centrally aggregated. These strategies are necessary to advance AI applications in industries like healthcare and finance where confidentiality of information is crucial, as well as to comply with privacy requirements and foster user confidence. Organizations may effectively utilize AI's promise while protecting individual privacy rights by incorporating these privacy-preserving approaches.

8.1 Open-Source AI Models

Improving the interpretability and explainability of AI models is essential to promoting confidence and openness in their application. Clear understandings of decision-making processes are provided by interpretable AI models, which facilitate stakeholders' comprehension of the methods and rationale behind the conclusions drawn by AI systems. In order to guarantee accountability and match AI outputs with moral and legal requirements, openness is crucial. Achieving explainable AI (XAI) requires the application of strategies like feature relevance evaluation, mechanisms for attention, and model-agnostic explanations, which allow users to recognize significant variables and possible biases in AI systems. Organizations may foster trust among users, authorities, and the general public by encouraging interpretability. This will make it easier for AI to be used responsibly and ethically in a variety of contexts. This method supports equal and impartial outcomes in crucial applications by improving decision-making transparency and reducing the dangers associated with skewed or opaque AI algorithms.

8.2 Secured AI Practices

To prevent artificial intelligence theft and secure sensitive data, secured AI practices are crucial. Robust solutions are provided by methods like homomorphic encryption and secure multi-party computation (MPC), which allow AI models to be trained on encrypted data. Ensuring privacy and utilizing collective insights, secure MPC enables many participants to jointly calculate a function over their own datasets without disclosing the underlying data itself. Analogously, homomorphic encryption permits calculations on encrypted data, enabling AI models to handle private data without having to decrypt it, protecting data privacy all the way through training. These methods reduce the risk of unwanted access or the theft of priceless intellectual property while enabling businesses to take advantage of AI's capabilities without disclosing raw data and complying with strict data privacy laws. The following mitigating techniques for blockchain can be used.

8.3 Smarter Consensus Processes

Proof-of-stake (PoS) and proof-of-authority (PoA) are two alternatives to the conventional proof-of-work (PoW) mechanism that offer strong security features while addressing the main shortcomings of PoW. With proof-of-stake, the energy-intensive mining method is replaced by a consensus mechanism in which validators are selected according to the quantity of cryptocurrency they own and are prepared to "stake" as security. Due to the requirement of a sizable stake to corrupt the system, this method lowers the possibility of 51% assaults, in which a single entity acquires majority authority over the network. However, proof-of-authority relies on verified identities of recognized validators, guaranteeing responsibility and lowering the possibility of hostile attacks. Because PoS and PoA don't require a lot of processing power to mine, they both greatly reduce the environmental effect associated with PoW. These substitutes give blockchain networks sustainable solutions by providing similar security and encouraging efficiency along with ecological responsibility in blockchain technology operations.

8.4 Secure Key Management

Ensuring the security and durability of private key administration in cryptocurrency and blockchain settings requires the use of physical wallets, multi-signature wallets, including key recovery services. Hardware wallets shield private keys from online risks like virus and hacking attempts by storing them offline in safe hardware devices. The possibility of unwanted access is greatly decreased by this physical seclusion. Contrarily, multi-signature wallets require several private keys to authorize transactions, dividing authority among multiple parties and providing an additional degree of protection against malicious activity or single points of failure. Private keys that have been misplaced or compromised can be safely restored with the help of key recovery services, giving back consumers the control of their assets without sacrificing security. When used in tandem, these technologies offer strong security protections for private keys, giving users confidence and peace of mind when it comes to safely managing their cryptocurrency holdings.

8.5 Quantum-resistant Encryption

Using cryptographic methods that are resistant to quantum computing may help blockchain technology be ready for the arrival of quantum computers by reducing the likelihood of security breaches. Because traditional cryptography techniques like RSA and ECC rely on mathematical puzzles that quantum machines can answer quickly, they are vulnerable to quantum assaults. Resilience is provided by quantum-resistant algorithms, such as hash- or lattice-based cryptography, which pose problems that quantum machines are not predicted to be able to solve well. By taking this proactive approach, blockchain networks are protected from the potential threats posed by quantum computing, by maintaining the safety and confidentiality of operations and data.

Smart contracts: By methodically analyzing smart contracts for known weaknesses before they are deployed, automated verification tools are essential to guaranteeing the security and dependability of these contracts. Self-executing contracts, or smart contracts, are agreements that run on the blockchain and have predetermined rules and conditions. They can be vulnerable to coding errors or other security flaws that could result in exploitation or financial losses. Static analysis, computational execution, formal verification, and other approaches are used by automated tools to examine the code of smart contracts and find possible problems such as reentrancy errors, integer overflows, or vulnerabilities that allow unauthorized access. These tools assist developers in identifying and reducing hazards early in the design lifecycle, guaranteeing that smart contracts work as intended and uphold the integrity and trust of blockchain-based payments and applications. They accomplish this by carrying out comprehensive tests prior to deployment.

Cross-chain protocols: Cross-chain communication protocols, such as Polkadot and Cosmos, play a critical role in enabling safe and effective cross-chain transactions by promoting interoperability across various blockchains. By creating frameworks that enable various blockchain networks to easily communicate and trade assets or data, these protocols solve the problem of blockchain silos. Through the use of a relaying chain and para-chains design, Polkadot makes it possible for several blockchains—each with a specialized role—to connect and safely exchange data. Similar to this,

Cosmos has a hub-and-spoke architecture, with a centralized location (Cosmos Hub) that serves as a conduit for communication between separate blockchains known as zones. These protocols facilitate interoperability, which improves the functionality, scalability, and flexibility of blockchain ecosystems. This encourages innovation and broadens the range of industries in which decentralized technologies may be applied.

The strategies mentioned in relation to AI and Blockchain involve a combination of structural (e.g., the development of interoperability standards or the execution of regulations) and technical (e.g., the adoption of quantum-resistant cryptography or the design of robust models). For example, creating more resilient models is one technical way to solve the problem of adversarial assault in AI. Regulations could be put in place on an organizational scale to deter these kinds of attacks. Comparably, with blockchain, the structural concern of compatibility might be resolved by standardized protocols and interfaces, while the technological problem of 52% attacks could be handled by employing alternative consensus processes.

It is crucial to recognize that, even though the tactics described are founded on the most recent research and the best practices in the industry, new problems and solutions may arise due to the speed at which blockchain and artificial intelligence are developing. Staying abreast of current advancements and ensuring that policies and suggestions are based on empirical data and are practically applicable is imperative for scientists, practitioners, and policymakers. Collaboration between regulatory agencies, business, and academics can also promote a thorough understanding of the security issues new technologies raise and provide solid solutions to them.

9. Conclusion

The purpose of this chapter was to examine how AI and BCT may work together, with an emphasis on security issues. We discovered that although these technologies provide new threats and challenges, they also could greatly improve security, efficiency, as well as transparency across a variety of sectors. Concerns about consensus processes, key management, smart contracts risks, and quantum resistance were noted in the context of blockchain, whereas adversarial attacks, data privacy, transparency, as well as bias were noted in the field of AI. The advantages and disadvantages of these technologies are amplified by their integration, resulting in a complicated environment that requires cautious navigation. We offer several mitigating options to respond to these difficulties, such as stronger consensus processes, regulatory actions, and resilient model design. Furthermore, we contend that the safe and moral integration of blockchain technology with artificial intelligence requires the development of a flexible, technologically neutral, international cooperative regulatory framework that actively engages stakeholders.

References

Baynham-Herd, Z. (2017). Enlist blockchain to boost conservation. Nature 548(7669): 523–523.

Bhumichai, D., Smiliotopoulos, C., Benton, R., Kambourakis, G. and Damopoulos, D. (2024). The convergence of artificial intelligence and blockchain: The state of play and the road ahead. Information 15(5): 268.

Bloch, I. (2006). Spatial reasoning under imprecision using fuzzy set theory, formal logics and mathematical morphology. International Journal of Approximate Reasoning 41(2): 77–95.

Bratspies, R. M. (2018). Cryptocurrency and the Myth of the Trustless Transaction. Mich. Tech. L. Rev. 25: 1.

Castro, M. and Liskov, B. (1999, February). Practical byzantine fault tolerance. pp. 173–186. In OsDI (Vol. 99, No. 1999).

CoinMarketCap. (2022). Cryptocurrency prices, charts and market capitalizations.

Dai, Y., Xu, D., Maharjan, S., Chen, Z., He, Q. and Zhang, Y. (2019). Blockchain and deep reinforcement learning empowered intelligent 5G beyond. IEEE Network 33(3): 10–17.

Daley, S. (2019). Tastier coffee, hurricane prediction and fighting the opioid crisis: 31 ways blockchain and AI make a powerful pair. builtin in April, 2020. Available: https://builtin. com/artificialintelligence/blockchain-ai-examples.

Dinh, T. N. and Thai, M. T. (2018). AI and blockchain: A disruptive integration. Computer 51(9): 48–53.

Dodd, N. (2016). The Social Life of Money. Princeton University Press.

Dwivedi, A. D., Srivastava, G., Dhar, S. and Singh, R. (2019). A decentralized privacy-preserving healthcare blockchain for IoT. Sensors 19(2): 326.

Ekbal, A. (2024). Artificial Intelligence and Blockchain Technologies: A Convergence for Future Innovations. Scientific Journal of Artificial Intelligence and Blockchain Technologies 1(1): 1–4.

Fusco, A., Dicuonzo, G., Dell'Atti, V. and Tatullo, M. (2020). Blockchain in healthcare: Insights on COVID-19. International Journal of Environmental Research and Public Health 17(19): 7167.

Greff, K., Van Steenkiste, S. and Schmidhuber, J. (2020). On the binding problem in artificial neural networks. arXiv preprint arXiv:2012.05208.

Intelligence, A. (2020). What it is and why it matters. Access mode: https://www. sas. com/en_us/insights/analytics/what-is-artificial-intelligence. html.

Jamaludin, J. and Rohani, J. M. (2018, November). Cyber-physical system (cps): State of the art. pp. 1–5. In 2018 International Conference on Computing, Electronic and Electrical Engineering (ICE Cube). IEEE.

Kumar, A., Abhishek, K., Nerurkar, P., Ghalib, M. R., Shankar, A. and Cheng, X. (2022). Secure smart contracts for cloud-based manufacturing using Ethereum blockchain. Transactions on Emerging Telecommunications Technologies 33(4): e4129.

Lin, J., Shen, Z. and Miao, C. (2017, July). Using blockchain technology to build trust in sharing LoRaWAN IoT. pp. 38–43. In Proceedings of the 2nd International Conference on Crowd Science and Engineering.

Liu, R., Zhang, M., Chen, G., Ooi, B. C. and Wang, J. (2018). Untangling blockchain: A data processing view of blockchain systems. IEEE Transactions on Knowledge and Data Engineering 30(7): 1366–1385.

Makarius, E. E., Mukherjee, D., Fox, J. D. and Fox, A. K. (2020). Rising with the machines: A sociotechnical framework for bringing artificial intelligence into the organization. Journal of Business Research 120: 262–273.

Maxmen, A. (2018). AI researchers embrace Bitcoin technology to share medical data. Nature 555(7697): 293–295.

Moosavi, N. and Taherdoost, H. (2022, August). Blockchain-enabled network for 6G wireless communication systems. pp. 857–868. In International Conference on Intelligent Cyber Physical Systems and Internet of Things. Cham: Springer International Publishing.

Moosavi, N. and Taherdoost, H. (2022, September). Blockchain and internet of things (IoT): a disruptive integration. pp. 308–315. In International Conference on Emerging Technologies and Intelligent Systems. Cham: Springer International Publishing.

Nakamoto, S. and Bitcoin, A. (2008). A peer-to-peer electronic cash system. Bitcoin.–URL: https://bitcoin. org/bitcoin. pdf, 4(2): 15.

Nakamoto, S. and Bitcoin, A. A Peer to Peer Electronic Cash System. 2009. Available online: https. bitcoin. org/bitcoin. pdf (accessed on 17 January 2022).

Narayanan, A., Bonneau, J., Felten, E., Miller, A. and Goldfeder, S. (2016). Bitcoin and Cryptocurrency Technologies: A Comprehensive Introduction. Princeton University Press.

Pandl, K. D., Thiebes, S., Schmidt-Kraepelin, M. and Sunyaev, A. (2020). On the convergence of artificial intelligence and distributed ledger technology: A scoping review and future research agenda. IEEE Access 8: 57075–57095.

Parizi, R. M., Amritraj and Dehghantanha, A. (2018). Smart contract programming languages on blockchains: An empirical evaluation of usability and security. pp. 75–91. In Blockchain–ICBC 2018: First International Conference, Held as Part of the Services Conference Federation, SCF 2018, Seattle, WA, USA, June 25–30, 2018, Proceedings 1. Springer International Publishing.

Rich, E. and Intelligence, K. A. (1991). McGraw-Hill: New York. NY, USA.

Salimitari, M., Chatterjee, M., Yuksel, M. and Pasiliao, E. (2017, October). Profit maximization for bitcoin pool mining: A prospect theoretic approach. pp. 267–274. In 2017 IEEE 3rd International Conference on Collaboration and Internet Computing (CIC). IEEE.

Santos, F. and Kostakis, V. (2018). The DAO: a million dollar lesson in blockchain governance. School of Business and Governance, Ragnar Nurkse Department of Innovation and Governance.

Singh, S. K., Rathore, S. and Park, J. H. (2020). Blockiotintelligence: A blockchain-enabled intelligent IoT architecture with artificial intelligence. Future Generation Computer Systems 110: 721–743.

Solat, S., Calvez, P. and Naït-Abdesselam, F. (2021). Permissioned vs. Permissionless Blockchain: How and Why There Is Only One Right Choice. J. Softw. 16(3): 95–106.

Soleymani, F. and Paquet, E. (2020). Financial portfolio optimization with online deep reinforcement learning and restricted stacked autoencoder—DeepBreath. Expert Systems with Applications 156: 113456.

Swan, M. (2015). Blockchain: Blueprint for a New Economy. O'Reilly Media, Inc.

Taherdoost, H. (2022). A critical review of blockchain acceptance models—blockchain technology adoption frameworks and applications. Computers 11(2): 24.

Taherdoost, H. (2023). An overview of trends in information systems: emerging technologies that transform the information technology industry. Cloud Computing and Data Science 1–16.

Tyagi, A. K., Dananjayan, S., Agarwal, D. and Thariq Ahmed, H. F. (2023). Blockchain—Internet of Things applications: Opportunities and challenges for industry 4.0 and society 5.0. Sensors 23(2): 947.

Wang, S., Yuan, Y., Wang, X., Li, J., Qin, R. and Wang, F. Y. (2018, June). An overview of smart contract: architecture, applications, and future trends. pp. 108–113. In 2018 IEEE Intelligent Vehicles Symposium (IV). IEEE.

Wood, G. (2014). Ethereum: A secure decentralised generalised transaction ledger. Ethereum Project Yellow Paper 151(2014): 1–32.

Ziegler, D. M., Stiennon, N., Wu, J., Brown, T. B., Radford, A., Amodei, D. et al. (2019). Fine-tuning language models from human preferences. arXiv preprint arXiv:1909.08593.

Chapter 14

Management of Electronic Health Records of Patients Using Blockchain Technology

Ayesha Shaik,[1,*] *Aneesh Rathore,*[2] *Dushyant Singh Bhati*[2] and *Ganesh Chaudhary*[2]

1. Introduction

1.1 Health Record System

The advent of digitalized health care records has undergone an evolution in the health-care industry. Patient-based EHRs combine patient-related information and assist patients in easy access to information about a patient's medical history and provide better decisions with effective care for patients. However, by default, centralized concerns relating to normal management systems of EHR pose risks to the administration of health data in terms of security and privacy and effectiveness.

Centralized architectures hence compromise sensitive health information as they are susceptible to errors such as data breaches, illegal access and single points of failure. Most centralized systems also are not capable of supporting interoperability; this makes the sharing of information between health care providers complex and even impossible. The problems above are aggravated by support through manual processes or compartmentalized electronic solutions, through fractured data, higher error rates, and added delay in patient care. Of its nature, health data is important and must be kept in a format that is secure, integral, and accessible. So, any medical caregiver must meet very strict legal guidelines such as HIPAA and GDPR for protection of the patient's data and for ethical practice. However, traditional systems normally do not achieve this requirement satisfactorily; therefore, there is a significant substitute. Blockchain technology represents the most advanced solution to these problems. The distributed, tamper-proof, and open architecture eliminates all single points of failure and thus enhances the security of data. It is blockchain, ensuring that EHRs are safe, accessible by authorized parties alone, and unchangeable by unauthorized parties because data is distributed among a number of nodes. Besides, other shortcomings of the classic approaches are also attained by scalable management of big health datasets as blockchain is integrated with off-

[1] Centre for Cyber Physical Systems, Vellore Institute of Technology, Chennai.
[2] School of Computer Science and Engineering, Vellore Institute of Technology, Chennai.
Emails: aneesh.rathore2021@vitstudent.ac.in, dhushyant.singh2021@vitstudent.ac.in, ganesh.choudhary2021@vitstudent.ac.in
* Corresponding author: ayesha.sk@vit.ac.in

DOI: 10.1201/9781003620822-14

chain storage technologies like IPFS. In this way, CareBlock was designed precisely to facilitate such goals and alter the environment of managing health data into a new one. It provides a framework through which the EHR can be managed effectively and securely while allowing the patients absolute authority over their respective data. In a nutshell, it is the creative application of blockchain and related technologies in CareBlock aimed at building the first place for a more secure, private healthcare ecosystem that contributes towards paving the road to a more dependable and patient-centric future.

1.2 Problem Statement

Modern health care systems have managed electronic health records. The previous modes of the transfer and filing of medical records are riddled with vulnerabilities and inefficiency. The main drivers of current health infrastructure-based central systems bear some significant challenges, mainly including vulnerability to cyberattacks, unauthorized access to information, and alterations of data. This weaknesses exposes all the patients and professionals in health care to major risks, since the confidentiality and integrity of the confidential patient data are compromised. The above factors that have been advocated through integration of these health care information systems and reliance on manual or compartmentalized digital solutions amplify these problems for the reason that it results in lateness in access to critical pieces of information. Besides motivating medical mistakes, this disintegration becomes impossible for doctors to execute coordinated care. For example, improper interoperability normally hampers free-flowing information that leads to unnecessary tests and poorly-documented patient records. Each use of care must employ a high level of prudence in adherence to the privacy and security regulations included under HIPAA and GDPR because of the sensitivity of health sector data. However, vulnerability to breach and lack of openness to the treatment that the data would receive makes the centralized systems unwilling to uphold these standards at all times. Patients have little or no knowledge or control over how access to their data is carried out and shared, although it is ideal that they own health information. As important as these secure centralized systems are, the price of continuing them presents another challenge. Increasing the volume of health information consumed more resources on these systems and, hence, increased the cost of running the organizations. Of course, aside from that, these central servers are a single point of failure, and catastrophic failures become a special case when they occur through cyberattack or system-wide outages. Considering all of those challenges, much more secure, efficient, and interoperable solutions are very badly needed to remedy the management of health records. A management that would guarantee confidentiality and integrity for sensitive information; in this case, it would make exchange safe and easy for healthcare practitioners while giving patients more control. Blockchain technology is an answer to these major dilemmas: decentralized, transparent, and not penetrable. It may reconfigure the process of managing a medical record by providing blockchain-based solutions and thereby enabling the elimination of the single point of failure, data integrity, and a patient-centric approach in healthcare data.

1.3 Motivation

Due to these reasons, there is an increasing need in the health sector for the development and implementation of safe, effective, and patient-centered management systems for health records. Conventional systems of EHRs have perennially failed to meet even the most basic needs including scalability, privacy, data security, and interoperability despite their huge popularity. All these imperfections compromise the overall effectiveness and reliability of health services besides compromising integrity due to confidential health data. Against such a sensitive backdrop, the CareBlock project erupts using blockchain technology in a way that goes beyond the constraints mentioned above. Three special features of blockchain are decentralization, transparency, and immutability, and it unlocks an unprecedented opportunity for revolutionary healthcare data management. More than that, the blockchain system improves safety and reliability of EHRs since they actually eliminate central points of risk. This further increases the confidence of patients and healthcare providers as, being tamper-resistant makes the medical record unchangeable. Another reason for the project is based on the principle of patient empowerment. Patients are nearly sidelined in the traditional system, in the sense that they have fewer powers over their data. The blockchain helps the patients with fine-grained permissions regarding access to their personal records and whom they might trust such access with, as well as keeping full control in their hands. In addition to patient autonomy, empowerment invites a cooperative health environment, in which patients as well as doctors, start cooperating more effectively. Better data exchange and interoperability also become possible. However, friction in current healthcare exists with clinicians who cannot converse smoothly in many areas thereby creating inefficiencies and lowering the already lowered standard of care delivery due to siloed systems and broken data. Blockchain only addresses a few of these by providing a safe exchange platform for data, real-time team support and no duplicated medical processes. One of the primary drivers of the design of CareBlock is scalability, which, at the same time, is one of the main threats of practically applicable blockchain systems. It should handle really large amounts of healthcare data without overloading the blockchain with options for off-chain storage, such as IPFS. This hybrid approach will probably ensure that the system will remain affordable and useful for practically applicable deployments. As a result, this rising new demand for protecting patients' information with keeping regulatory compliance to HIPAA and GDPR on their radar throws more and more current systems into the limelight due to their deficits. Blockchain-based solutions like CareBlock come with a great promise of doing just that: providing for transparent audit trails, encrypted data storage, and automated management of consent through smart contracts. This project aims to develop a safe, scalable, and holistic health record management system that identifies solutions to problems found today and puts a framework in place for a healthcare system ready for tomorrow. Hence, there will be improvement in patient care due to the safety it ensures regarding security of the data and innovation on health care delivery through blockchain technology.

1.4 Research Challenges

It would require multiple research challenges to be overcome for such a blockchain-based system that is effective and scalable enough to be established in the healthcare sector. Some of the research challenges and technological complexities that were issues of consideration on specific needs of the health sector as well as on blockchain technology.

Scalability data management: Scalability would probably be the major challenge. It would mean the need for enormous computing power and resources for the storage and processing of such humongous amounts of health data that is being produced every day. Constraints in throughput and latency in blockchain networks, especially for those using Proof of Work or similar consensus processes, would pose a problem in achieving the required real-time emergency conditions in the retrieval of health data. This would make the files contained in this case - diagnostics reports and medical images almost impossible to be stored on-chain, because the cost incurred and also the inefficiency. Even if there is off-chain storage systems like the InterPlanetary File System (IPFS) in place, there are other challenges that will be presented in building the intercommunications that shall be seamlessly required between the on-chain and off-chain components.

While many promise blockchain will keep data private in a public or semi-public context, it is very hard to do so; it promises better security through decentralization and immutability. As access to sensitive health information needs to be really so tight to be available only the authorized entities, it's hard to arrive at that while allowing essential data interchange. Moreover, as decentralized networks by definition must come to some form of consensus, this consensus is quite often incompatible with the pressure to keep medical data in confidence.

Interoperability: With many formats, protocols, and standards governing data storage and exchange, healthcare systems are notoriously scattered. Not least of these is that compatibility of an EHR system designed on a blockchain with existing older healthcare IT infrastructures already in place will be the obvious challenge. The promises of blockchain technology may scatter the ecosystem further instead of bringing it together if this cannot be done properly.

Weaknesses of smart contracts: Smart contracts need to be designed so that they enable automatic data exchange and access control. However, it is pretty challenging to engineer such smart contracts that are safe, reliable, and also fault-free. Contract code faults can expose vulnerabilities that might cause the loss of critical information or perhaps unauthorized access. This is yet another reason that a system has to be fully tested and validated, to build confidence in it.

Issues related to regulation and compliance: The health industry is very regulated. Strong regulation exists in protecting the data, privacy of the patient, and proper ethical handling of such information. Considering the decentralized and immutable properties of the blockchain, in terms of compliance with regional laws, HIPAA, GDPR, and the like, it is a significant challenge. It is quite hard to have control for compliance on privacy and accreditation to ensure auditability.

End user adoption and training: Training is required because of less awareness and knowledge of the concept among the different stakeholders in the healthcare facility, be it administrators, patients, or physicians.

Misconceptions regarding blockchain technology and apprehension about being displaced from traditional systems may hinder its adoption. Blockchain needs to be streamlined to be adopted in vast numbers and its benefits proved.

Performance and latency delays can be caused by blockchain reliance on consensus processes mainly in public blockchains due to the fact that the verification of transactions may take some time. Such latency creates an enormous problem in health care because instantaneous access to information may be needed to be just as urgent. The challenge of solving problems in high performance without losing any form of decentralization or security is quite tough and requires innovative solutions.

Implementation cost: This will undeniably save billions of dollars in the future, if minimized breaches and inefficiency can be achieved; however, it costs too much to install and integrate blockchain technology in one go. The initiation cost to build a strong structure, training of the employees, and migration of data from the existing systems discourage health care service providers is enormous.

Accountability and trust: While the transparency of blockchain is quite beneficial, accountability issues still creep into various places. For example, in a decentralized setup, it becomes a bit tricky to attribute data entry mistakes, permissions, or system failure. There needs to be clear measures of accountability while building user trust into this blockchain infrastructure.

1.5 Objectives

The following goals guided the development of CareBlock:

Blockchain for her: Apply blockchain in EHRs and ensure storage in a safe and immobile format so it cannot be accessed or changed illicitly.

Patient empowerment: Personalized granular access controls in enabling patients to manage themselves through their own health care records.

Scalable management of EHRs: The storage of large health files off-chain and alternative off-chain storage solutions serve as substitutes for the inherent scalability limitations found in blockchain technology.

Interoperability and data sharing: It should ensure safe and effective data sharing among healthcare providers so that timely coordinated care is delivered.

Increased privacy and compliance: Privacy is maintained by restricting access, encrypting data, and having open audit trails while also following the legal requirement-HIPAA and GDPR.

2. Background

2.1 Blockchain Basics and Technology Overview

Blockchain is the application of a decentralized ledger with consensus methods, Proof of Work (PoW), and Proof of Stake (PoS), where it validates the transaction while allowing data integrity, where once written, data cannot be erased or changed. As a decentralized system, blockchain does not need a central authority. So, the system is unbreachable and spreads confidence between users on a network. The crypto links assume the form of links among predecessors making up a chain of uninterrupted records of transactions or updates in a block. Much of typical security weaknesses is in single points of failure and tampering with data already available through its design, so it is avoided. Smart contracts are self-executing programs running on the blockchain. They ensure the core part of the blockchain, which ensures efficiency and transparency, mainly through regulatory means and also by automatic intermediaries-free procedures. Smart contracts particularly in CareBlock help manage the consent of patients through choices that enable the patients to decide who will view their records and under what circumstances. In exchange, smart contracts provide a transparently verifiable record of interactions involving that data, managing access or permissions in real-time by recording each authorization event directly on the blockchain at the moment that permission is established. Ironically, however, despite all the strengths that blockchain naturally carries along with it, its innate disadvantages exist most noticeably in respect to the storage of big data because of problems of scalability and affordability. In this context, integration of off-chain storage takes a significant level of importance. CareBlock uses distributed file storage, like the InterPlanetary File System, to deal with health care's issues of enormous data. In this context, each file receives unique hashes from IPFS, and those hashes are stored in blockchain as references. It may be scalable and cost-effective without losing the immutable nature of blockchain and its secured nature. Multiple tools and technologies have been used in the development and deployment of CareBlock. Because Ganache sets up a local blockchain environment, smart contracts and their interactions with the blockchain can be tested during development before such interactions are deployed. A wallet called MetaMask is available, which allows browsers to be connected to the Ethereum network so that transactions can be made safe to sign and authenticate. All these products provide ease of integration and give the user easily used interfaces in the aspect of health record management.

In a scenario where control on EHR is under discussion, blockchain is the perfect framework that will express the aim of decentralization, immutability, and automation. CareBlock envisages revolutionizing health data management, enlisting blockchain and IPFS to ensure the medical record would be safe, tamper proof, and accessible only to a few parties.

3. Literature Review

Frankly, blockchain technology was conceptualized for the first time in the white paper of Bitcoin Nakamoto (2008) developed, making it possible for digital ledgers to become decentralized and unalterable. This invention negating the need for middlemen in transactions opened doors towards using it in various industries in several ways. From this, Hyperledger Fabric that has laid major emphasis on modular architecture has offered and flexibility to various types of applications in permissioned blockchain systems meant for enterprises utilization (Androulaki 2018). The concepts around blockchain have been explained (Pilkington 2016), which have grouped the significance of blockchain outside financial systems into digital revolutions across various domains, be it governance, supply chain, or health.

3.1 Blockchain Applications in Healthcare

In this regard, several researchers pointed out to the exploration of blockchain potential in terms of potential security development, privacy aspects, and further interoperability for the healthcare sector. The blockchain technology as a game-changing solution (Esposito et al. 2018) has been identified for securing the integrity and confidentiality of medical data based on the cloud. In great detail, discuss the principal applications of blockchain in healthcare and state that block chain would be useful in offering safety and privacy for private medical records and, therefore allow safe and transparent access to data (Hölbl et al. 2018).

MedRec, a system based on the blockchain has been proposed, which aims to help patients to control who should or shouldn't access their medical data; manage permissions, and track permissions. Among the very early systems reported more than a decade ago was MedRec (Azaria et al. 2016), which demonstrated how data ownership could be shifted to a patient-centric blockchain. The same fundamental principle has been adopted by Kumar et al. (2019) who even proposed a decentralized architecture for safe document sharing and preserving e-medical information against unwanted access.

3.2 Advanced Architectures and Challenges

A distributed architecture is essential for processing personal health records over the systems. Scalability and cross-platform compatibility are critical in real-world applications of healthcare; therefore, their approach concentrated on such properties is proposed (Roehrs et al. 2017). The security motivation by pointing out the grave privacy issue related to the very concept of smart healthcare and presented blockchain as a viable means to mitigate all unwanted problems like illegal sharing and data breaches (Gupta et al. 2020).

3.3 Focused Applications and Patient Views

A specific application architecture had been designed that made it safe to share medical imaging data, exchanging the same, with decentralized blockchain. Such an exercise depicted the way blockchain technology is used in high-demand use cases that require data integrity and consensus (Patel 2019). What is most important, however, is that adoption into health systems will require an element of high acceptance and trust levels, a fact the authors were able to highlight through the study of patient views on blockchain-based health information exchange platforms (Esmaeilzadeh and Mirzael 2020). The role of block chain in the key supply chain management objectives have been discussed. The block chain usage for protecting the personal data (Zyskind et al. 2015) and adoption of block chain in IoT applications with their challenges (Uddin et al. 2021) has been presented. The white paper with block chain applications in financial and non-financial sector (Crosby et al. 2016) has been discussed with the challenges and business opportunities to revolutionize the digital world.

4. Methodology

A blockchain system for the exchange and storage of medical records works as explained below. All of this contributes to a safe, decentralized, and efficient frame of work handling medical data. The following are mainly involved:

4.1 Implementation of Blockchain

Local blockchain setup: Because Ganache is emulating an environment in which a blockchain can be tested and developed upon, it starts launching a local blockchain. In this regard, within the controlled environment, any action that utilizes the blockchain with a smart contract is tested before being sent to a network.

Account management: The type of browser-based bitcoin wallet, MetaMask, completes user account management. It helps protect access by unauthorized persons and has a measuring tool to note access control in the blockchain network.

4.2 Creation and Implementation of Smart Contracts

Policies: Policies on access to, and management of, medical records are developed in Solidity. One can get rights, access, and integrity of data through the means developed in the contract.

Truffle framework for contract migration: This is used in assembling, testing, and the deployment of the smart contracts. That is to say, it helps in contract management by automating the EVM deployment process.

Decentralization: This is one such decentralized system that can track the contracts and ensure that rules for contracts are followed through. It is often termed as Ethereum Virtual Machine or more commonly known as EVM. That process becomes unbreakable by securing record access and control within itself.

4.3 Data Storage and Access Control

Decentralized storage using IPFS: This is an InterPlanetary File System, or IPFS enables large amounts of data such as images, and medical records to be stored off the blockchain, and only the references to the files are kept on the blockchain. This will scale the cost of data storage because it lowers the costs incurred in storing data in a blockchain.

Access control implementation: Smart contracts control the access rights in such a manner that data can be accessed only by a person having an access-right. Cryptography methods included in the smart contract are used for permission enforcement. In this technique, due to the inclusion of permission enforcement in the access control, accessing data is transparent and safe.

4.4 Frontend Development and User Interface

Frontend stack: HTML, CSS, JavaScript These are HTML, CSS, and JavaScript that make the frontend of the blockchain application intuitive to a user.

Interaction with smart contracts: Web3.js and MetaMask makes it very easy for users to interact with the installed smart contracts by linking their front end to the blockchain. Users get a secure web interface where they view, store, and manage their medical records.

4.5 Proposed Model

The CareBlock system proposed concept combines off-chain storage and blockchain technology to solve the main problems encountered in handling electronic health records. To ensure that the system is both secure, efficient, and also user-centered towards the treatment of health-related data, the design has identified three layers that are necessary; namely, (a) blockchain, (b) smart contracts, and (c) data storage/user interface. The block diagram of the proposed system is shown in Figure 14.1. The figure is an architectural diagram of a decentralized application (dApp) built on the Ethereum blockchain. It consists of multiple layers:

- User Interface Layer – Uses HTML, CSS, and JavaScript for interaction.
- Access Layer – Utilizes MetaMask for managing blockchain accounts.

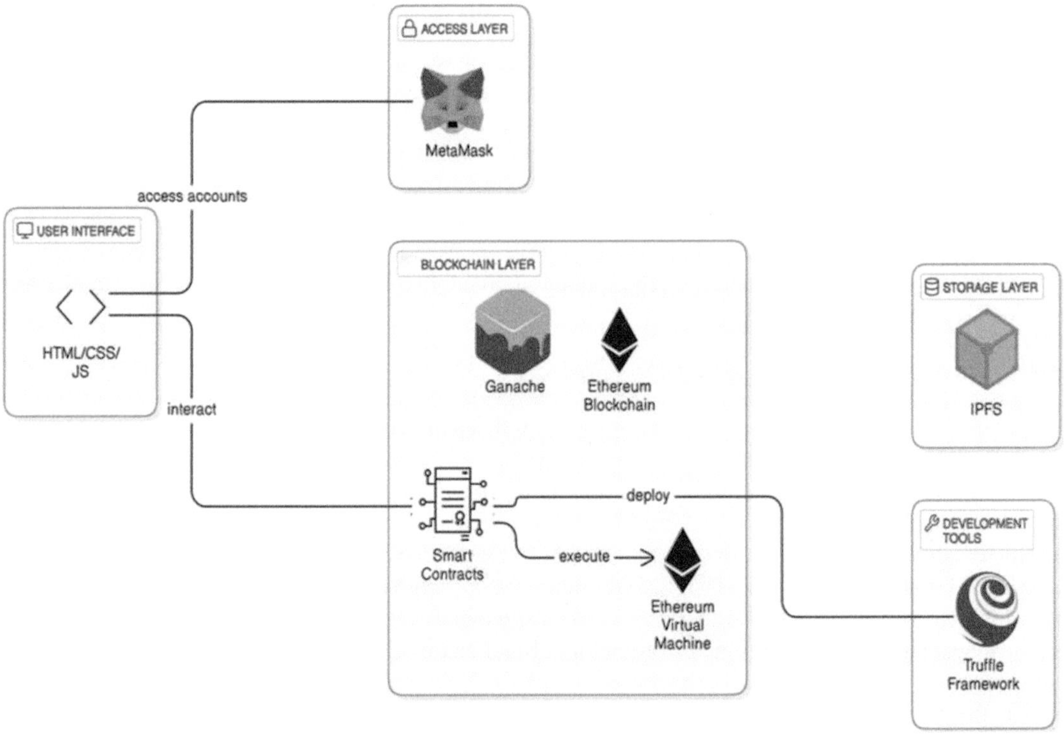

Figure 14.1. Block diagram for the proposed system.

- Blockchain Layer – Uses Ganache for testing locally, Ethereum Blockchain for deployment, Smart Contracts, and Ethereum Virtual Machine (EVM) for execution.
- Storage Layer – Uses IPFS for decentralized storage.
- Development Tools – Employs Truffle Framework for smart contract development and deployment.

The diagram visually represents how these components interact to build a fully functional dApp.

4.6 Blockchain Layer

The layer of the blockchain serves as a base for guaranteed immutability, decentralized control, and safe access, all of which comprise part of the model. Ganache is essentially a local blockchain environment simulating blockchain interactions for testing purposes and for development. It allows one to test all the features before any release into an active Ethereum network. The rights and rules that belong to the system are applied by the running of smart contracts from within the blockchain through the use of the Ethereum Virtual Machine. Through MetaMask, a user enjoys safe authentication of transactions and accounts, making it totally secure and accessible. That layer is an audit trail that is transparent and tamper-proof, meaning every data contact goes to the blockchain.

4.7 Smart Contract Layer

The smart contract layer is established at the base where data access with its administration is highly empowered. Patient consent, data sharing, and permissions for sharing along with it are clearly defined in strait-jacketed guidelines while developing such contracts through Solidity. The Framework of Truffle is used in composing, testing, and deployment of these contracts for live running on the blockchain. Once active, all the rights granted on the patient are applied stringently and only certain people can access certain records. For instance, the patient can temporarily grant access to the health provider; all this is done transparently on the blockchain. The smart contracts layer lowers the probability of mismatching the input and unauthorized access because no intermediaries with probable human interference exist.

4.8 Storage Layer of Data and User Interface

This is the layer which takes care of the usability and scalability of the system. This is the use of a component known as InterPlanetary File System for data storage that manages tremendous datasets containing diagnostic images, medical reports and other EHR files. This allows the blockchain to include only references or hashes of the files, which means a reduction in storage cost and provides it as a scalable option, while the off-chain method keeps medical records safe in place, yet easily accessible for authorized users. HTML, CSS, and JavaScript are used to create the UI for the purpose of making it accessible to the health care providers and patients to convey through a friendly interface with the system. Web3.js combines frontend and blockchain with such an end result that users can upload, review, and manage rights and records. MetaMask is associated with users' safe access as well as the signing of transactions, therefore that they access their records safely and confidently.

4.9 System Workflow

The workflow of the proposed system with the doctor and patient dashboard is shown in Figure 14.2. This represents a patient-doctor access control system for medical records. It illustrates the process of patient and doctor registration, granting and revoking access to medical records, and the doctor's ability to analyze and update records. The flow ensures secure management of patient data while allowing controlled access for diagnosis and updates.

Uploading of records: The patients or doctors upload encrypted health records in IPFS. As a reference, the hash generated by IPFS is uploaded on the blockchain.

Permission management: Patients define permission via user interface. These rights, defining which parties are permitted to access their files for how long, are encoded in the smart contract.

Data access: First of all, unlike a more backendish implementation, the smart contract checks a request's permission access by a healthcare provider against the permissions from the patient. When authorized, the provider requests and pulls down what is needed from IPFS using the hash.

Audit and transparency: Every contact related to changes in permission or any other request is logged in blockchain, so this record of usage of data is imputed and transparent. Thus, that would successfully integrate off-chain storage with the security and the advantage of blockchain technology by the proposed model. CareBlock would provide a strong framework that would probably strengthen the empowerment of patients, improve the security of data, and make health data management more interoperable.

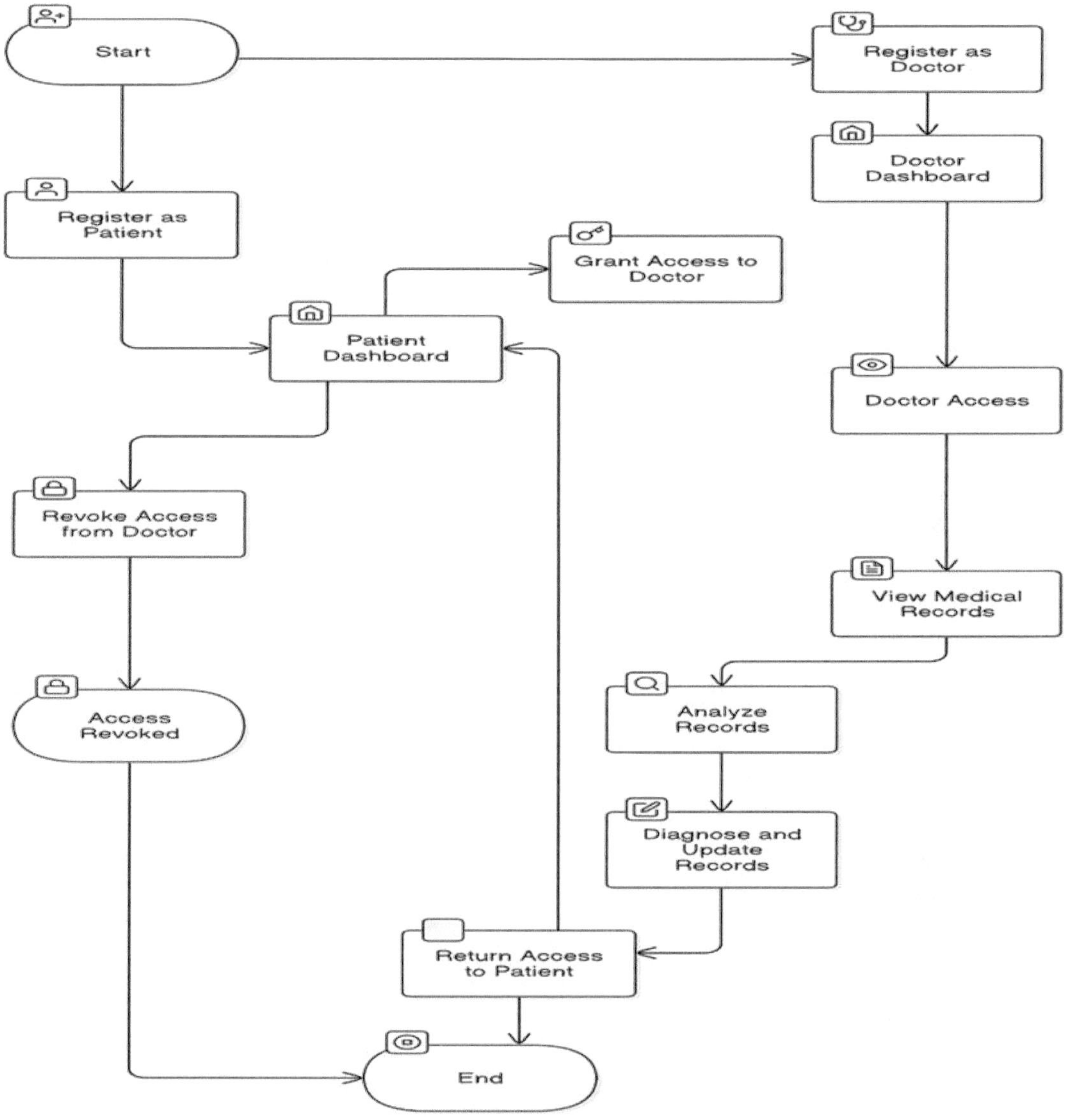

Figure 14.2. Workflow of the proposed system.

5. Implementation

5.1 Implementation in Medicare

The actual implementation of CareBlock within the Medicare framework removes any possible disadvantage tied to traditional EHR systems. CareBlock guarantees the secure, patient-centric, and efficient handling of health data through the CareBlock system due to blockchain and off-chain storage technologies. Many technologies and components are implicit in the application process as a means of having a seamless experience on either side, that is, the consumers or healthcare providers.

Access control: It is based on smart contract. Essentially, the heart and core of CareBlock are the smart contracts that allow for managing access permission control over health-related medical records. There is interaction with an easy interface with the patient to clarify under which circumstance who should have access. For example, in this context, the duration that the physician will allocate to the patient's medical record or for some unusual other purposes or for specific applications, such as the patient's diagnosis or engaging in consultation with the patient. Therefore, smart contracts have been hard-coded in the smart contract and automatically, these rights will be respected. This is based on the fact that each access request will have an immutable and clear audit trail that has been written on the blockchain. In this model, the patients become empowered because it allows the patient full control over his or her information as well as ensuring a secure and reliable means for authorized parties to get access to it.

5.2 Connection to Ganache and IPFS

CareBlock utilizes Ganache to establish a development environment and to conduct testing for local blockchain development. That particular environment allows interactions of simulating blocks with other blocks. And due to that, the feature can be tested well before deploying the system to a real network. Whenever tested, smart contracts including their functionalities get deployed to the Ethereum block chain.

The system should use InterPlanetary File System for the storing of large health files such as reports and images that are applicable in the diagnostics. Each time there is an uploading, encryption, and further storage in IPFS, the system would generate a hash of that file; it is a reference to the said file and should be recorded in the blockchain. The blockchain and IPFS interoperability ensures health-sensitive data are safely stored and always available and retrievable to authorized people without putting pressure on the blockchain.

5.3 MetaMask and User Authentication

MetaMask is a bitcoin wallet available online for providing the connection between the patient and the different health care providers via the blockchain. MetaMask is used to add an option of signing transactions and authenticating to access the system of CareBlock. Therefore, only authenticated ones will be able to enjoy using the system of CareBlock. For instance, a doctor needs to retrieve a medical history regarding a particular patient; hence, he or she logs in through MetaMask, and then it is matched with the one set in the smart contract in the system.

5.4 Front-end to Back-end Interaction

CareBlock's front-end is developed in such a manner that it offers an interface to the users. It will be coded in HTML, CSS, and JavaScript, making it unbelievably hassle-free to upload medical records, set permissions, and even view logs of access. Web3.js will therefore facilitate safe transactions paired with real-time updates, which will be guaranteed by the direct conversation that will occur between the frontend and blockchain. In this way, a health worker can request and retrieve the records based on granted permissions, and the patient may control the permissions on his data. The back end will execute the exchanges and an action will be verified and recorded on the blockchain. The total time consumed for each task such as block chain deployment, smart contracts deployment, migrate contracts and metamask response time is listed in Table 14.1.

Table 14.1. Total time for each process.

Process	Run 1 (ms)	Run 2 (ms)	Run 3 (ms)	Average (ms)
Deploy blockchain	400	450	420	423
Deploy smart contracts	350	370	360	360
Migrate contracts	300	320	310	310
MetaMask response time	250	240	260	250

5.5 Workflow in Medicare Facilities

Patient data upload: Health records get uploaded in the system by a patient or any health care facility. Now, the associated hash is saved into the blockchain while the data gets encrypted and stays in the IPFS.

Access request: The health care professional would do this, directly accessing the patient's record with help of the system. Then, the smart contract would process and verify permission by the patient.

Data retrieval: Once access has been granted, this is retrieved by a health provider in a safe and effective manner using an IPFS hash.

The audit trail: From uploading data to access requests, all contacts record themselves in a blockchain. As such, the audit trail for all data usage is extremely traceable.

Benefits in Medicare:

By implementing CareBlock, Medicare can achieve:

Better data security: The encryption through IPFS and blockchains' immutability gives the idea that confidential medical records can neither be breached, nor can any breach happen for unauthorized access.

Empowering the patient: Patients are endowed with complete control over their data which empowers them in managing their access permissions better and monitoring the data in use.

Interoperability: Blockchain has a standard mode that makes sharing of information across providers easy. In this way, there is better coordination and delivery of care.

Compliance to legislation: Data encryption and audit trail are transparent, in that data will abide by compliance rules such as HIPAA and GDPR.

Through all these components and procedures, CareBlock delivers safe, scalable, and patient centered administration of Medicare EHR.

This would mean that all the transactions or efforts to access have to go through the smart contract, a power play that ensures security of a high grade, where only authorized users can interact with the involved records. It balances out efficiency, scalability, and security by using both on-chain and off-chain storage.

5.6 Architecture

The architecture of the proposed system is shown in Figure 14.3. This figure outlines the development process of a blockchain-based application. It includes deploying a local blockchain with Ganache, writing smart contracts in Solidity, storing records using IPFS, and integrating a user interface with HTML, CSS, and JavaScript. MetaMask is used for account access and interacting with smart contracts.

5.7 Security and Privacy Considerations

As the data related to care is more sensitive, there needs to be a sound security and privacy system in CareBlock. The multiple levels protect the care data of the patient as the access is limited according to a norm ensuring legal requirements like GDPR and HIPAA.

Data encryption: All the EHRs that are going to be in IPFS will be encrypted. This means their data cannot be accessed or misused when intercepted since the interceptor does not have the decryption key correlated to them. The encryption process prevents cyberattacks from illegal disclosures to access key information, like sensitive medical information, diagnostic results, and treatment records.

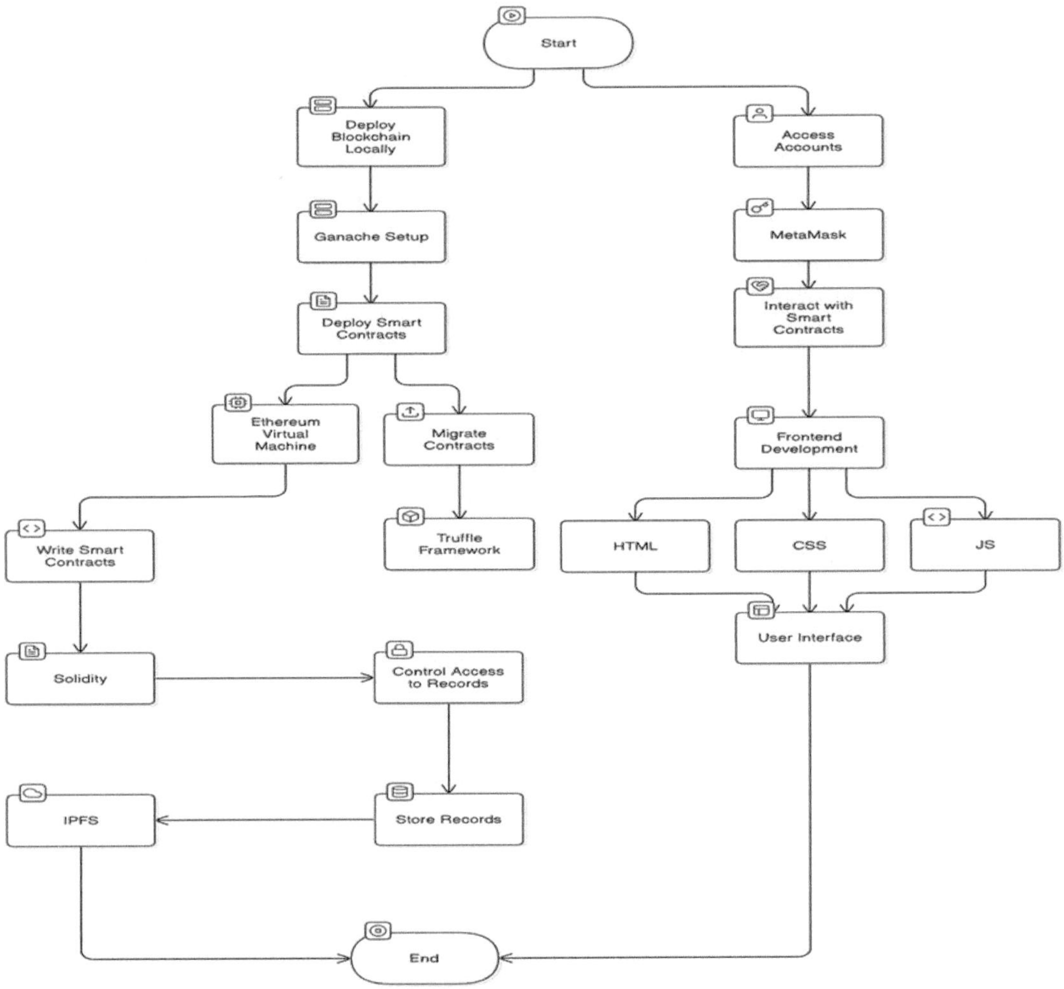

Figure 14.3. Architecture of the system.

Granular-level access control: CareBlock provides the patients granular control technology, so that they know exactly who can read what and when. Smart contracts are automatic gatekeepers, which will encode permissions onto the blockchain so a patient may grant access to some healthcare provider for a time or to a particular sets of data. That is a high degree of personalization because patients will then remain fully in charge of their information while giving access to the entities they need.

Blockchain non-modifiability: The second intrinsic characteristic of blockchain is that it is non-modifiable. This, therefore, means that once any data entry or transaction has been entered into the blockchain, it cannot be deleted or altered. It would then protect it from tampering or unauthorized changes and therefore provide integrity to the medical history. In this case, since the blockchain is non-modifiable, all activities concerning patient data can be freely logged and certified, creating an auditable trail safe.

Authentication and authorization: CareBlock encrypts every transaction as well as any access attempt to authenticate. Any action would have entailed authentication of the identity of the users through MetaMask or similar mechanisms prior to the said action being performed. Because of cryptographic signatures, only authenticated and legitimate users can be used to retrieve or update data in this system. The multi-layered authentications greatly minimize unauthorized access.

Auditability: Yet another critical feature of CareBlock is transparency, and this is enabled fully by the very architecture of the blockchain technology itself. Every permission modulation, access request, uploading of data, and all other actions are logged onto the blockchain. The logs ensure a complete view of usage of the patient's data along with offering accountability between healthcare providers in being tamper-proof while ensuring an easily auditable history of interaction.

Data leak protection: CareBlock reduces the leaks of data in a potential manner as it spreads and disperses across decentralized storage while washing out the single point of failure. In blockchain, there is no central repository of data, and even if the node of the network gets compromised, it has nothing to do with the data. Offchain storages within IPFS ensure huge data sets are properly stored and accessible at the right time.

Compliance with regulatory requirements: The design of CareBlock takes on the deep principle of regulatory compliance. All features in access control, encrypted storage of data, and visible audit trails ensure definite standards are met in the views of various regulations as in the case of GDPR of the EU or HIPAA of the US. For example, it satisfies all requirements of GDPR regarding ownership and control over the data. It falls under the rights of patients to be able to view, edit, or revoke permissions concerning patient data.

Concerns and further improvements: The current version of CareBlock has exemplary protection against security threats and excellent privacy features, but some concerns have been associated with it. Still, a more potent example is an open area of research which shall keep the metadata of transactions private on the blockchain. Data privacy shall be improved in future versions of the system with better integration with state-of-the-art privacy-preserving technology, such as private blockchains and ZKPs. CareBlock creates a trusted environment to manage medical data while accomplishing such an elaborate affair in security and privacy matters. The proposed system is not only curative in overcoming problems present in the current traditional EHR but also evolves a new paradigm of safety and patient-centric management of health information.

5.8 Benefits of Blockchain in Medicare

Blockchain technology offers several advantages for healthcare:

Greater security and privacy: Access control and encryption protect the patients' data, while decentralization eliminates single points of failure.

Patient empowerment and control: Providing full authority over his or her records to the patient gives a patient total control over who can read and amend the information, thus creating openness and confidence.

Data integrity: The immutability offered by blockchain provides tamper-proof records that are accurate and reliable.

Enhanced interoperability: Blockchain will bring about improvements in the life of patients through a universal standard for information between health care providers; henceforth, it avoids duplications that may cause unnecessary testing.

Cost-efficiency: Blockchain reduces long-term costs since data breaches, administrative errors, and redundant recordkeeping are reduced, though the initial setup cost might be very expensive.

5.9 Challenges and Limitations

Despite its potential, implementing blockchain in healthcare presents various challenges:

Scalability: It is very expensive and inefficient to store massive datasets directly on the blockchain. Though it lets use off-chain solutions like IPFS, that in its turn complicates the system.

H.l. compliance with laws: Since blockchains will be an open source, the greatest task will be to ensure compliance under HIPAA and GDPR for the safety and privacy of CareBlock's data.

Legacy systems integration: Many healthcare service providers are still on old systems. This pertains to significant modifications in processes and infrastructure through the integration of blockchain technology into existing systems.

Latency and performance: The transactions in the blockchain are slow essentially due to the consensus processes, thus delaying access to the data in case of an emergency.

Adoption by users: It will assume the shape of educating patients and healthcare providers about the functionalities and merits of blockchain.

The CareBlock login page is shown in Figure 14.4, featuring MetaMask authentication for secure access. The UI includes a navigation bar, a welcome message, and a login button, for a blockchain-based healthcare platform.

This CareBlock registration page is shown in Figure 14.5, where users enter their details to register. The MetaMask transaction request indicates that registration involves a blockchain transaction, for storing user data securely on a decentralized network. Figure 14.6 visualizes the CareBlock patient dashboard, where users can:

- View medical records stored on IPFS.
- Share records by selecting a doctor.
- Manage access by revoking permissions from doctors with existing access.

Figure 14.7 shows the doctor's view of CareBlock. In this view, the EMR's that can be accessible by the doctor is listed. Figure 14.8 shows CareBlock patient dashboard with medical records displayed. It includes:

- Personal details (Name, Age, Public Key).
- Diagnosis information (Doctor's name, Date & Time, Diagnosis result).
- IPFS link for secure storage of medical records.
- Option to share records with selected doctors.

6. Results

Figure 14.4. Home screen of CareBlock.

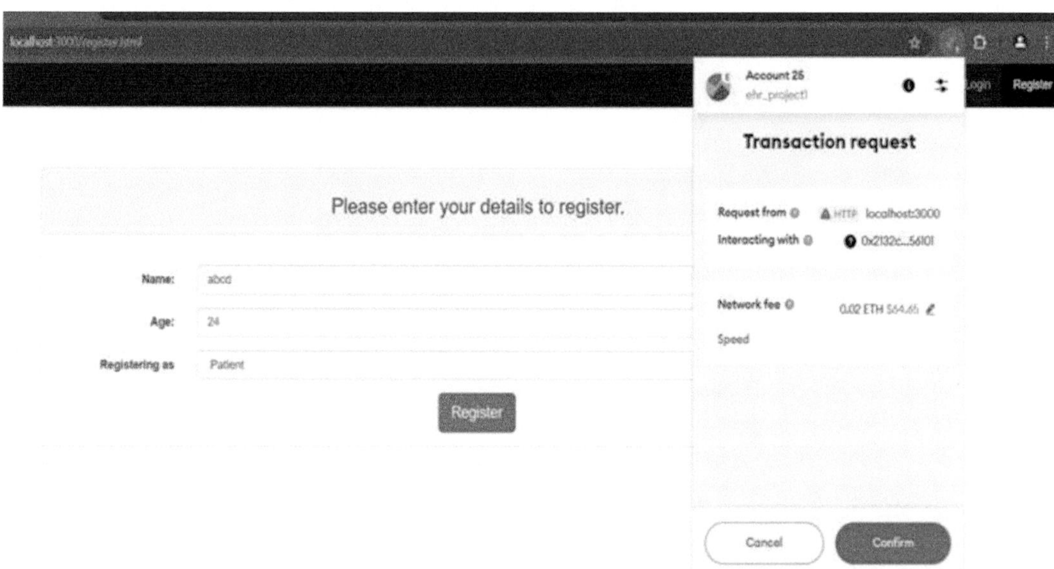

Figure 14.5. Register screen of CareBlock.

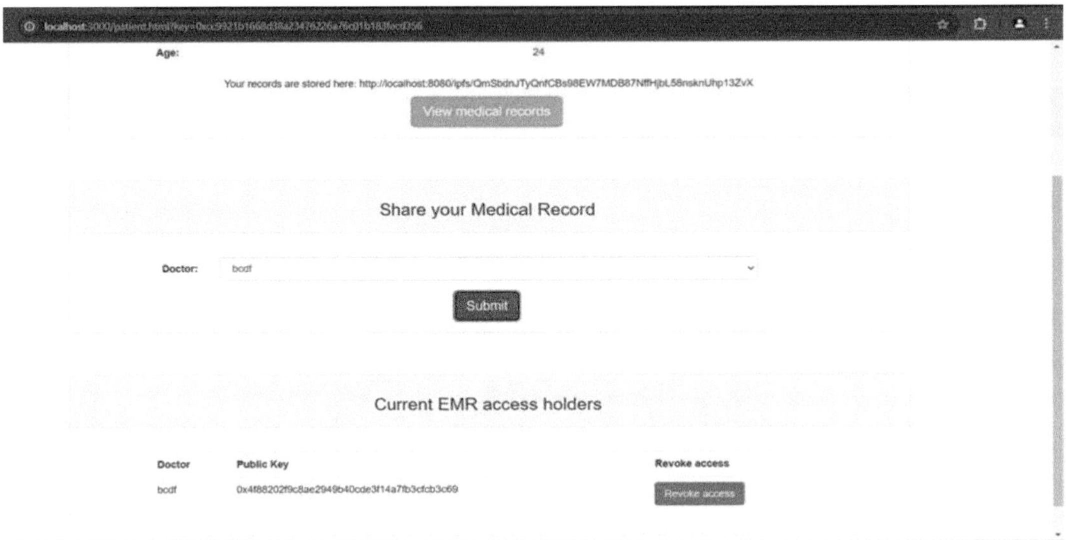

Figure 14.6. Patient's page of CareBlock.

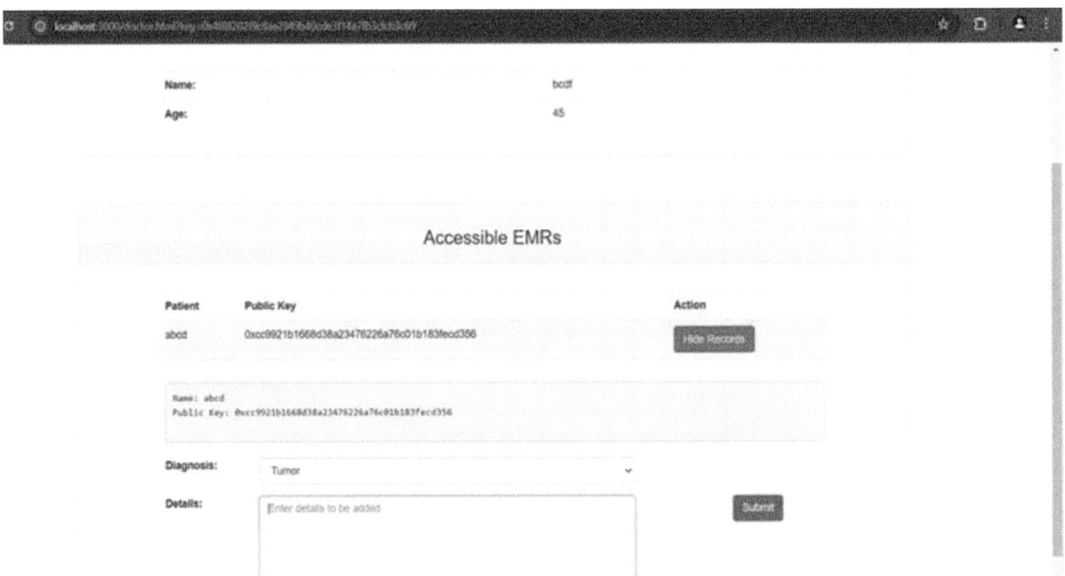

Figure 14.7. Doctor's page of CareBlock.

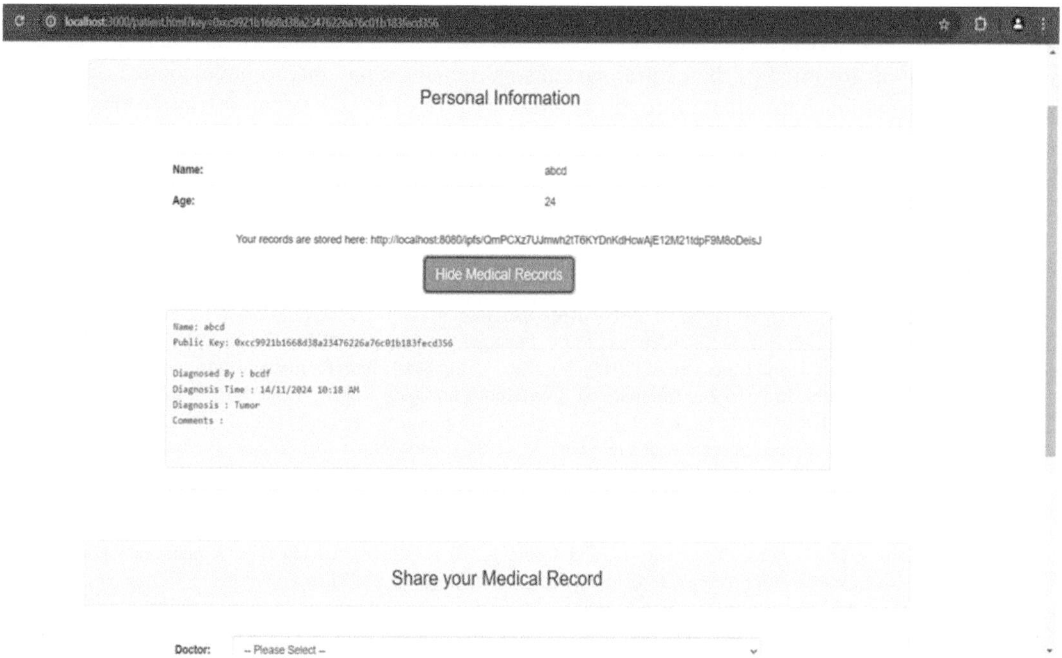

Figure 14.8. Diagnostics results.

7. Conclusion and Future Work

Future Scope

There are far too many good tracks which consult blockchain technology to develop a decentralized medical health record system, through which much further learning and development can take place. This progress in the healthcare sector is made possible through the need for continuously innovative and creative solutions that address emerging challenges or capitalize on new opportunities. Another area that should be improved is the provision for data privacy, more so in the development of more complex algorithms only usable to healthcare data. In this case, the approved healthcare providers will have access to the protected medical data but it will be protected from changing cyber risks.

The other major area of future development is scalability. They are producing more and more large data-like genomic data, high-resolution medical imaging, data from continuous monitoring, etc. Scaling efforts towards improving blockchain topologies and consensus processes are required for this vast growing data load in such a manner that the system allows maintenance of the same speed. This can be realized through novel hybrid storage solutions that encapsulate the efficiency of conventional databases combined with the security enhanced benefits of blockchains.

Conclusion

This is a tremendous innovation from CareBlock in the management of EHRs and indeed represents a very dramatic step in the use of blockchain technology in such administration. It actually manages to address many long-standing issues afflicting traditional healthcare data management systems. CareBlock evolves a sound architecture that duly satisfies the trade-off needs of collision of data privacy, integrity, and scalability in managing healthcare information, with the innovative application of blockchain technology's inherent security features, using state-of-the-art off-chain storage.

CareBlock architecture incorporates smart contracts and fine-grained restriction of access to outline patient-centric data management in a fashion that guarantees sufficient security. This is a great improvement for most of the central systems as it enables patients to have control over their medical records while sharing data through health care providers in a secured way. In this regard, the integration for off-chain storage of IPFS particularly works well as it surmounts the quite often encountered blockchain implementations for scalability issues, which make the system work in the great data needs of the present health care system while preserving efficiency and economy.

References

Androulaki, E. (2018). Hyperledger fabric: a distributed operating system for permissioned blockchains. pp. 1–15. In Proceedings of the 13th EuroSys Conference, Porto, Portugal.

Azaria, A., Ekblaw, A., Vieira, T. and Lippman, A. (2016). MedRec: Using blockchain for medical data access and permission management. pp. 25–30. In 2016 2nd International Conference on Open and Big Data (OBD), Vienna, Austria. doi: 10.1109/OBD.2016.11.

Crosby, M., Pattanayak, P., Verma, S. and Kalyanaraman, V. (2016). Blockchain technology: Beyond bitcoin. Applied Innovation 2(6-10): 71.

Esmaeilzadeh, P. and Mirzaei, T. (2019). The potential of blockchain technology for health information exchange: experimental study from patients' perspectives. Journal of Medical Internet Research 21(6): e14184.

Esposito, C., De Santis, A., Tortora, G., Chang, H. and Choo, K. R. R. (2018). Blockchain: A panacea for healthcare cloud-based data security and privacy?. In IEEE Cloud Computing 5(1): 31–37.

Gupta, M., Abdelsalam, M., Khorsandroo, S. and Mittal, S. (2020). Security and privacy in smart farming: Challenges and opportunities. IEEE Access 8: 34564–34584.

Hölbl, M., Kompara, M., Kamišalić, A. and Zlatolas, L. N. (2018). A systematic review of the use of blockchain in healthcare. Symmetry 10(10): 470.

Kumar, R., Tripathi, R. and Gupta, S. (2019). A decentralized document sharing and e-medical record preservation using blockchain. pp. 1–6. In 2019 10th International Conference on Computing, Communication and Networking Technologies (ICCCNT), Kanpur, India. doi: 10.1109/ICCCNT45670.2019.8944810.

Nakamoto, S. (2008). Bitcoin: A Peer-to-Peer Electronic Cash System. Available: https://bitcoin.org/bitcoin.pdf.

Patel, V. (2019). A framework for secure and decentralized sharing of medical imaging data via blockchain consensus. Health Informatics Journal 25(4): 1398–1411.

Pilkington, M. (2016). Blockchain technology: Principles and applications. pp. 225–253. In Handbook of Research on Digital Transformations.

Roehrs, A., Da Costa, C. A. and Da Rosa Righi, R. (2017). OmniPHR: A distributed architecture model to integrate personal health records. Journal of Biomedical Informatics 71: 70–81. doi: 10.1016/j.jbi.2017.05.012.

Uddin, M. A., Stranieri, A., Gondal, I. and Balasubramanian, V. (2021). A survey on the adoption of blockchain in IoT: Challenges and solutions. Blockchain: Research and Applications 2(2): 100006.

Zyskind, G. and Nathan, O. (2015, May). Decentralizing privacy: Using blockchain to protect personal data. pp. 180–184. In 2015 IEEE Security and Privacy Workshops. IEEE.

Chapter 15

Blockchain-Based Efficient Drone Authentication Scheme for Security Enhancement in Smart Cities

Vincent Omollo Nyangaresi,[1,2,*] *Ahmad A. AlRababah*[3] and *Ali Hamzah Najim*[4]

1. Introduction

The continued developments in the domains of Information and Communication Technologies (ICTs) such as Internet of Things (IoT), Wireless Sensor Networks (WSNs), drone technology and Artificial Intelligence (AI) has resulted in increased remote sensing, automation, as well as adoption of remote operations. As explained by Garg et al. (2024), drones, also referred to as Unmanned Aerial Vehicles (UAVs) or Flying Ad-Hoc Networks (FANET) are extremely popular in IoT applications such as the automation of tasks and remote monitoring. Smart cities encompass the leverage of digital connectivity, AI, ICT, IoT, drones, cloud computing as well as data analytics to streamline services, and boost sustainability, citizen quality of life, as well as resource utilization optimization (Bera et al. 2024, Irshad et al. 2023, Gupta et al. 2023). These cities can perform remote sensing, analysis and integrate critical information from core systems in urban areas. This facilitates appropriate and intelligent responses to be initiated that potentially enhance citizen livelihood, public safety, environmental protection, commercial as well as industrial processes. According to Nyangaresi et al. (2025) and Xie et al. (2021), smart cities use technology and data for informed decision making, as well as efficient management of assets and resources. This requires that data from various sources be collected, processed and analyzed (Jiang et al. 2021). The outcome is a city that is not only sustainable, responsive, more connected, but also offer improved quality of life, services, and opportunities (Yang et al. 2020).

As explained by Yang et al. (2022) and Zhang et al. (2017), the realization of smart cities has been propelled by the emergence and integration of technologies such as 5th generation mobile network (5G), IoT, WSN and FANETs. These technologies facilitate real-time data collection

[1] Department of Computer Science and Software Engineering, Jaramogi Oginga Odinga University of Science and Technology, Bando 40601, Kenya.
[2] Department of Applied Electronics, Saveetha School of Engineering, SIMATS, Chennai, Tamilnadu, 602105, India.
[3] Faculty of Computing and Information Technology in Rabigh, Rabigh, 21911, KSA, King Abdulaziz University.
[4] Department of Computer Technical Engineering, Imam Al-Kadhum College (IKC), Iraq.
Emails: Aaahmad13@kau.edu.sa, alihamza@alkadhum-col.edu.iq
* Corresponding author: vnyangaresi@jooust.ac.ke

DOI: 10.1201/9781003620822-15

(Al Sibahee et al. 2023) that enables the realization of various applications such as environmental monitoring, and traffic management. As such, the intelligent services in smart cities are not only real-time, information-driven, customer-centric, but also aimed at increasing productivity (Alsayaydeh et al. 2024). This helps in the management of various city challenges such as energy consumption, environmental pollution, traffic congestion, and public safety. The domains that benefit from smart cities include medical care, public services, agriculture, emergence services, waste management, energy, and transportation (Rao et al. 2024, Hou et al. 2023). Specifically, the integration of drones into smart cities has led to improved effectiveness and efficiency in domains such as healthcare, traffic management, disaster management, surveillance, and farming (Gour and Yadav 2025, Tanveer et al. 2024, Nyangaresi et al. 2024).

In spite of the numerous benefits of smart cities, there are numerous security and privacy challenges that lurk in this environment. For instance, the drones, sensors and other WSN devices collect sensitive data related to activities such as location, usage patterns, shopping and health. Insufficient protection of this data can render it susceptible to threats such as privacy violations, unauthorized access, data breaches and malicious alteration (Sikder et al. 2021). In addition, reliability, scalability, and Quality of Service (QoS) are issues that need to be addressed in UAV-based smart cities (Gour and Yadav 2025). According to Sharma et al. (2024), node compromise is another serious challenge that must be addressed. In drone-based smart cities, the transmission of sensitive data between these UAVs takes place over insecure communication channels. This renders the communication process prone to numerous security and privacy threats (Ajakwe et al. 2023, Tychola and Rantos 2025, Gul et al. 2025, Kadhim et al. 2025). In addition, smart cities bring together a myriad of sensors and other smart devices, increasing the surface area for launching attacks. Since they sometimes operate in unattended environments, drones are vulnerable to physical capture threats. In smart cities, data exchanges among networks take place using gateway nodes. Therefore, strong security protection must be implemented between the users and the gateway nodes, as well as among the various gateway nodes. As explained by Sucasas et al. (2023), user data and privacy must be protected in smart cities. Mutual authentication is one of the most effective means of ensuring that robust access control is preserved in these networks. In addition, strong encryption must be implemented to thwart cyber threats such as unauthorized access to sensitive data and uphold citizen safety (Hussain et al. 2021). It is also critical to secure IoT systems so as to maintain trustworthiness and resilience in these cities (Sharma et al. 2024). This ensures that only legitimate sensors and nodes can join the network and transmit data to authorized destinations.

However, drones are characterized with limited resources in terms of computation, power, storage and communication capabilities (Derhab et al. 2023, Mahobiya et al. 2024, Nyangaresi et al. 2024). This puts some constraints in the implementation of computationally intensive cryptographic techniques. Therefore, the development of lightweight authentication schemes that can offer perfect security at low computation and communication overheads presents some significant challenges.

1.1 Contribution

The main contributions of our work can be summarized as follows:

- We design a lightweight blockchain-based authentication scheme to enhance security in drone-based smart cities. After the successful login phase, the frequently executed inter-gateway authentication only involves one-way hashing function and exclusive-Or (XOR) operations.
- Conditional revocation phase is incorporated in our scheme to identify and eliminate misbehaving drones from the network. This is accomplished over the blockchain through the gateway node which maintains a revocation list.
- Extensive security analysis is executed to demonstrate that our scheme is robust against typical smart city attack vectors such as ephemeral secret leakage, offline guessing, eavesdropping, impersonation, stolen verifier, man-in-the-middle, forgery, privileged insider, physical capture

threats, replay, session key disclosure, and side-channeling. In addition, it provides anonymity, untraceability, robust authentication, session key setup, and forward key secrecy.

- Elaborate comparative performance to show that our protocol incurs the least computation and communication complexities. Therefore, it is the most efficient among its peers.

The rest of this paper is structured as follows: Section 2 describes the related works while Section 3 details the security analysis of our scheme. On the other hand, Section 4 details the performance analysis of the proposed protocol. Towards the end of this paper, Section 5 provides the conclusion and future research directions.

1.2 Motivation

In order to support secure and efficient data communication in smart cities, robust authentications must be implemented. This is necessitated by the fact that the communication between ground stations and drones takes place over the insecure wireless channels that are prone to attacks. In addition, smart cities are characterized by smart device heterogeneity and this presents an increased surface area for launching attacks. These adversarial security and privacy threats can potentially degrade the performance of smart cities. Over the recent past, various security techniques have been developed to alleviate the numerous security and privacy threats in smart cities. However, most of these techniques are based on highly extensive cryptographic operations that can potentially drain the drone battery. This calls for the development of not only robust but also efficient authentication schemes.

2. Related Works

Numerous security schemes, protocols, systems and techniques have been developed over the recent past to offer protection in smart city environment. For instance, authentication schemes based on Elliptic Curve Cryptography (ECC) have been presented by Hussain et al. (2021), Rangwani and Om (2021), Meher et al. (2022), Wang et al. (2023), Mahmood et al. (2023), Karim et al. (2023), Ayub et al. (2023), Wu et al. (2021a), and Wang et al. (2022). However, perfect key secrecy and untraceability are not supported by the scheme by Rangwani and Om (2021). In addition, this scheme is not resilient against privileged insider attacks (Ali et al. 2024). On the other hand, the protocol by Meher et al. (2022) cannot provide anonymity and is not robust against Ephemeral Secret Leakage (ESL) and replay threats (Bera et al. 2024). Similarly, the cloud-based authentication protocol by Wang et al. (2023) cannot protect against ESL and replay attacks (Bera et al. 2024). On the other hand, the drone authentication technique developed by Mahmood et al. (2023) fails to support forward secrecy (Bera et al. 2024). Although the blockchain-based technique by Karim et al. (2023) offers increased scalability, it cannot support untraceability and anonymity. On its part, the scheme by Ayub et al. (2023) is susceptible to ESL threats, while the protocol by Hussain et al. (2021) has numerous security shortfalls as described by Wu et al. (2021b). Owing to numerous scalar point multiplications, the scheme by Wu et al. (2021a) incurs high computation complexities. On the other hand, the three-factor authentication technique by Wang et al. (2022) is susceptible to forgery, ESL and eavesdropping.

To address some of the above challenges, biometric-based authentication schemes have been developed by Bera et al. (2020), Ghahramani et al. (2020), Darbandi et al. (2025), Banerjee et al. (2019), and Mutlaq et al. (2024). Unfortunately, the scheme by Ghahramani et al. (2020) does not support mutual authentication, while the protocol by Banerjee et al. (2019) fails to provide forward key secrecy and unlinkability. In addition, it is vulnerable to impersonation, and password guessing threats (Xie et al. 2021). To avoid secret keys storage in smart devices, schemes based on Physical Unclonable Function (PUF) are presented by Gope and Sikdar (2020), Xie and Huang (2024), Lounis et al. (2022), Pu et al. (2022), and Nyangaresi and Petrovic (2021). However, PUF-based security techniques have stability challenges. Based on smart cards, authentication method is presented by Moon et al. (2017). Although this scheme is efficient, it cannot offer user anonymity, and is defenseless against impersonation, hidden gateway node and spoofing threats (Gupta et al. 2023). Similarly, the authentication method by Xie et al. (2021) fails to offer mutual authentication and untraceability, as well as protection against impersonation attacks (Gupta et al. 2023). Similarly, the authentication protocol by Xia et al. (2021) does not support mutual authentication and anonymity, as well as protection against impersonation threats. On the other hand, the security framework developed by Alzahrani et al. (2021) is vulnerable to privileged insider and drone capture attacks. In addition, it cannot support confidentiality, message integrity and user anonymity. Similarly, the technique presented by Zhang et al. (2020) cannot withstand Denial of Service (DoS), replay, impersonation, and secret credentials leakage threats (Hussain et al. 2023). In addition, this security approach does not offer mutual authentication. Similarly, fog-cloud authentication scheme by Salami et al. (2023) is not resistant against threat attacks such as forgery, eavesdropping, session key disclosure, and side-channeling. On the other hand, the security technique by Hu et al. (2022) is defenseless against DoS, Man-in-the Middle (MitM) and privileged insider threats (Ma et al. 2023, Umran et al. 2023). In addition, it does not support anonymity. To offer security during the transfer of data between cloud servers and drones, an identity-based scheme has been developed by Khan et al. (2021). This scheme is shown to be efficient and secure, but identity-based security solutions have key escrow problems. Similarly, the three-factor protocol developed by Luo et al. (2020) has numerous security shortcomings (Xie et al. 2021).

3. The Proposed Scheme

The registration authority (*RA*), gateway node (GW_j) and drones (D_i), and operators (U_i) are the main network entities in our scheme, as shown in Figure 15.1.

The *RA* is a trusted entity within the network and is charged with the responsibility of securely registering all the drones prior to their deployment in the field. On the other hand, gateway node GW_j bridges the communication between operators located at the ground stations and the drones. In terms of operations, our proposed scheme comprises of system initialization, registration, login, authentication, and revocation phases. Table 15.1 gives the notations used throughout this paper.

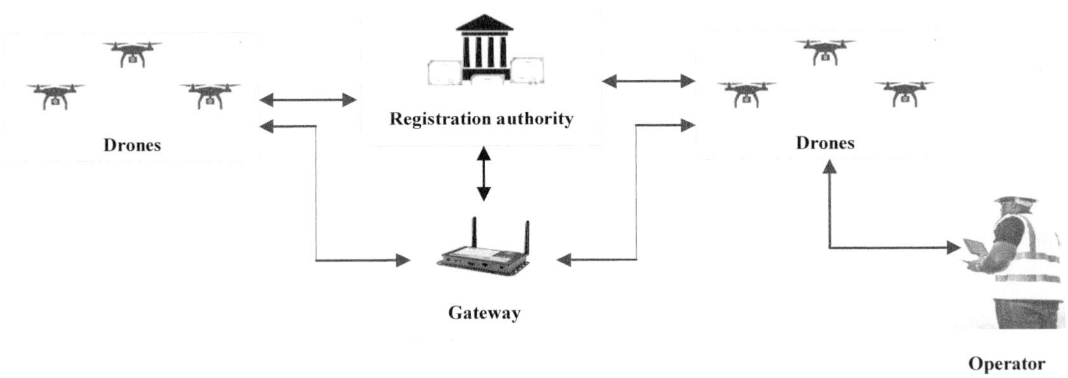

Figure 15.1. Network model.

Table 15.1. Notations.

Symbol	**Description**
RA	Registration authority
GW_j	j^{th} gateway node
D_i	i^{th} drone
IDG_j, ID_{RA}, ID_i	Unique identify for GW_j, RA, and D_i
PID_i	Pseudo-identity belonging to D_i
TD_i	Transient identity of D_i
S_{GW}	Secret keys for GW_j,
P_{GW}	Public key for GW_j,
D_i	i^{th} drone
M_{RA}	The master key for RA
f_v	Fuzzy verifier
$\|\|$	Concatenation operation
k_{GR}	Shared key between GW_j and RA
st_i	Secret token for D_i
T_j	Timestamp
r_i	Random nonce
SK_{DG}	Session key between D_i and GW_j
SK_{Dk}	Session key between D_i and GW_k
$S_G(.)$	Signature of GW_j using ECDSA
h (.)	Collision-resistant one-way hashing function
$\oplus$	XOR operation

3.1 System initialization

The goal of this phase is to have the *RA* generate and disseminate some public security tokens. In addition, the gateway nodes are deployed in this phase. This is accomplished via the steps as summarized in Figure 15.2. For enhanced security, these procedures are carried over secured communication channels.

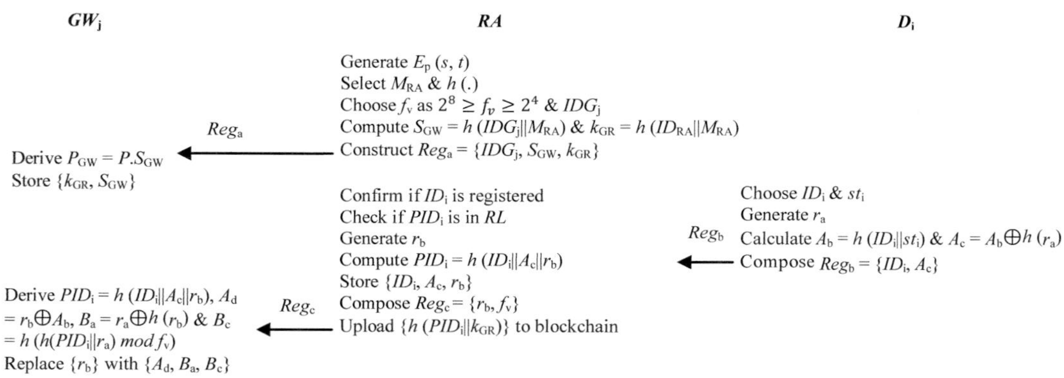

Figure 15.2. System initialization and registration.

Step 1: The *RA* generates some elliptic curve $E_p(s, t)$ whose base point is *P*, and $\{s, t\}$ are constants. Next, the *RA* selects M_{RA} as its private master key, as well as $h(.)$ as a one-way hashing function.

Step 2: Registration authority *RA* chooses fuzzy verifier f_v as $2^8 \geq f_v \geq 2^4$. This is followed by the selection of IDG_j as unique identity for GW_j. Next, *RA* derives the gateway node's secret key as $S_{GW} = h(IDG_j||M_{RA})$.

Step 3: *RA* computes shared key between itself and GW_j as $k_{GR} = h(ID_{RA}||M_{RA})$. Finally, *RA* constructs message $Reg_a = \{IDG_j, S_{GW}, k_{GR}\}$ and forwards it to GW_j.

Step 4: On receiving message Reg_a, the GW_j derives $P_{GW} = P.S_{GW}$. Finally, it stores $\{k_{GR}, S_{GW}\}$ in its repository.

3.2 Registration

In this phase, all drones must be registered at the *RA* to facilitate subsequent authentication and data exchange with the gateway nodes. This is enabled by the execution of the following procedures, across secure channels.

Step 1: The drone D_i selects ID_i as its unique identity. In addition, it chooses st_i as its secret token. This is followed by the generation of random nonce r_a.

Step 2: The D_i derives $A_b = h(ID_i||st_i)$, $A_c = A_b \oplus h(r_a)$. Next, it composes message $Reg_b = \{ID_i, A_c\}$ that is transmitted over to the *RA* as evidenced in Figure 15.2.

Step 3: Upon receiving Reg_b, the *RA* searches its database to establish whether this particular ID_i is already registered. In addition, *RA* determines whether pseudo-identity PID_i related to identity ID_i is in its revocation list (*RL*). On condition that these conditions are positive, D_i is prompted to submit a different identity. Otherwise, *Step 4* is invoked.

Step 4: The *RA* randomly chooses nonce r_b and computes $PID_i = h(ID_i||A_c||r_b)$. Next, it stores $\{ID_i, A_c, r_b\}$ in its database. Meanwhile, *RA* constructs message $Reg_c = \{r_b, f_v\}$ and forwards it to drone D_i. This is followed by the uploading of $\{h(PID_i||k_{GR})\}$ to the blockchain.

Step 5: After getting Reg_c, drone D_i calculates $PID_i = h(ID_i||A_c||r_b)$, $A_d = r_b \oplus A_b$, $B_a = r_a \oplus h(r_b)$ and $B_c = h(h(PID_i||r_a) \bmod f_v)$. Finally, D_i substitutes $\{r_b\}$ with $\{A_d, B_a, B_c\}$ in its memory.

3.3 Login Phase

In this phase, the operators login to the drone D_i. Afterwards, all the verification data is uploaded to the blockchain. This is accomplished through the following procedures, which are summarized in Figure 15.3.

Step 1: The operator at the ground station (*GS*) inputs st_i and ID_i to D_i, after which it derives $A_b = h(ID_i||st_i)$, $r_b = A_d \oplus A_b$, $r_a = B_a \oplus h(r_b)$, $A_c = A_b \oplus h(r_a)$, $PID_i = h(ID_i||A_c||r_b)$. It then checks if $B_c \stackrel{?}{=} h(h(PID_i||r_a) \bmod f_v)$, terminating the session if this condition fails to hold. Otherwise, drone D_i generates nonce r_c and calculates $B_d = (r_b * r_c)*P_{GW}$, $C_a = (r_b * r_c)*P$, $C_b = PID_i \oplus h(IDG_j||B_d||T_1)$ and $C_d = h(PID_i||B_d||T_1)$. Finally, it composes login message $L_a = \{C_a, C_b, C_d, T_1\}$ that is sent over to GW_j.

Step 2: On receiving L_a, the GW_j validates timestamp T_1 and terminates the session if it is invalid. Otherwise, it computes $B_d = C_a * S_{GW}$, and $PID_i = C_b \oplus h(IDG_j||B_d||T_1)$. Afterwards, it confirms whether $\{h(PID_i||k_{GR})\}$ is in the blockchain. In addition, it checks if $C_d \stackrel{?}{=} h(PID_i||B_d||T_1)$. On condition that these conditions hold, then drone D_i is authenticated.

D_i | **GW_j**

Accept st_i & ID_i
Calculate $A_b = h(ID_i||st_i)$, $r_b = A_d \oplus A_b$, $r_a = B_a \oplus h(r_b)$, $A_c = A_b \oplus h(r_a)$
& $PID_i = h(ID_i||A_c||r_b)$
Confirm if $B_c \stackrel{?}{=} h(h(PID_i||r_a) \bmod f_v)$
Generate r_c
Compute $B_d = (r_b * r_c)*P_{GW}$, $C_a = (r_b * r_c)*P$, $C_b = PID_i \oplus h(IDG_j||B_d||T_1)$
& $C_d = h(PID_i||B_d||T_1)$
Construct $L_a = \{C_a, C_b, C_d, T_1\}$ —— L_a ——→

Validate T_1
Derive $B_d = C_a * S_{GW}$, & $PID_i = C_b \oplus h(IDG_j||B_d||T_1)$
Check if $\{h(PID_i||k_{GR})\}$ is in the blockchain &
Confirm whether $C_d \stackrel{?}{=} h(PID_i||B_d||T_1)$
Generate r_d & r_e at T_2
Calculate $D_a = r_d * C_a$, $D_b = r_d * P$, $TD_i = PID_i \oplus h(r_e||IDG_j||k_{GR})$, $D_c = TD_i \oplus h(D_a||PID_i)$,
$SK_{DG} = h(D_a||TD_i||PID_i)$, & $D_e = h(SK_{DG}||T_2)$
Compose $L_b = \{D_b, D_c, D_e, T_2\}$ ←— L_b ——

Verify T_2
Derive $D_a = (r_b * r_c)*D_b$, $TD_i = D_c \oplus h(D_a||PID_i)$, &
$SK_{DG} = h(D_a||TD_i||PID_i)$
Confirm if $D_e \stackrel{?}{=} h(SK_{DG}||T_2)$
Upload $\{TD_i, r_e, IDG_j, S_G(h(TD_i||PID_i||r_e))\}$ to blockchain

Figure 15.3. Login phase.

Step 3: The GW_j generates nonces r_d and r_e at timestamp T_2. Next, it calculates $D_a = r_d * C_a$, $D_b = r_d * P$, $TD_i = PID_i \oplus h(r_e||IDG_j||k_{GR})$, $D_c = TD_i \oplus h(D_a||PID_i)$, session key $SK_{DG} = h(D_a||TD_i||PID_i)$, and $D_e = h(SK_{DG}||T_2)$. At the end, it constructs message $L_b = \{D_b, D_c, D_e, T_2\}$ and forwards it towards D_i.

Step 4: Upon receiving L_b, drone D_i verifies its timestamp using the current timestamp T_3. Provided that it is valid, D_i computes $D_a = (r_b * r_c)*D_b$, $TD_i = D_c \oplus h(D_a||PID_i)$, $SK_{DG} = h(D_a||TD_i||PID_i)$. Next, it establishes whether $D_e \stackrel{?}{=} h(SK_{DG}||T_2)$. Provided that this condition holds, the GW_j uploads $\{TD_i, r_e, IDG_j, S_G(h(TD_i||PID_i||r_e))\}$ to the blockchain.

3.4 Inter-Gateway Authentication

This phase is activated whenever drone D_i moves out of the coverage area of GW_j to which the drone operator has already logged in and verified his/her security tokens. When this happens, D_i must be authenticated by the new gateway GW_k using the following steps.

Step 1: Drone D_i generates nonce r_f and derives $F_a = h(PID_i||TD_i||IDG_k||T_3) \oplus r_f$, and $F_b = h(r_f||PID_i||TD_i)$. Next, it constructs message $Auth_a = \{TD_i, F_a, F_b, T_3\}$ that is sent over to GW_k as evidenced in Fig. 4.

Step 2: On getting $Auth_a$, the GW_k validates timestamp T_3 and retrieves transaction $\{TD_i, r_e, IDG_j, S_G(h(TD_i||PID_i||r_e))\}$ from the blockchain. This is followed by the verification of signature $S_G(.)$ and derivation of $PID_i^* = TD_i \oplus h(r_e||IDG_j||k_{GR})$.

Step 3: The GW_k verifies whether $h(TD_i||PID_i^*||r_e) \stackrel{?}{=} h(TD_i||PID_i||r_e)$. If this condition holds, it calculates $r_f = h(PID_i||TD_i||IDG_k||T_3) \oplus F_a$ and checks if $F_b \stackrel{?}{=} h(r_f||PID_i||TD_i)$. On condition that this condition holds, the GW_k generates nonces r_g and r_h.

Step 4: The gateway node GW_k derives $TD_i^{new} = PID_i \oplus h(r_h||IDG_k||k_{GR})$, $F_c = TD_i^{new} \oplus h(r_f||PID_i)$, $F_d = r_g \oplus h(r_f||TD_i^{new}||PID_i)$, session key $SK_{Dk} = h(r_g||r_f||TD_i^{new}||PID_i)$, and $F_e = h(SK_{Dk}||T_4)$. Finally, it composes message $Auth_b = \{F_c, F_d, F_e, T_4\}$ which is forwarded to D_i.

Step 5: On getting $Auth_b$, drone D_i verifies timestamp T_4 and calculates $TD_i^{new} = F_c \oplus h(r_f||PID_i)$, $r_g = F_d \oplus h(r_f||TD_i^{new}||PID_i)$, session key $SK_{Dk} = h(r_g||r_f||TD_i^{new}||PID_i^{new})$. This is followed by the confirmation of whether $F_e \stackrel{?}{=} h(SK_{Dk}||T_4)$. Provided that this condition holds, drone D_i and gateway node GW_k are mutually authenticated. Finally, sends transaction upload request Req_{block} (*Trans*) towards the gateway node GW_k

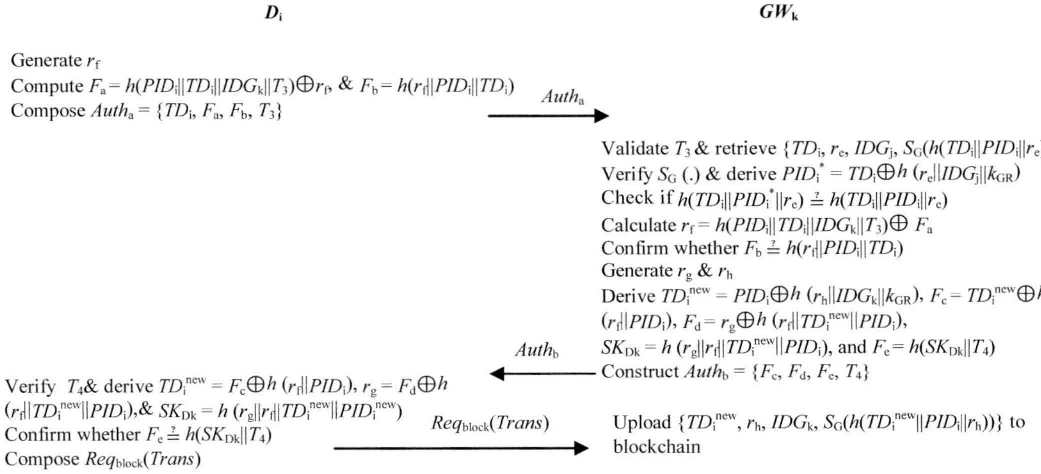

Figure 15.4. Inter-gateway authentication.

Step 6: Upon receiving Req_{block}, GW_k uploads $\{TD_i^{new}, r_h, IDG_k, S_G(h(TD_i^{new}||PID_i||r_h))\}$ to the blockchain.

3.5 *Drone Revocation*

In this phase, any malicious drone D_i can be identified by the gateway node GW_j and its security tokens revoked via the blockchain network. To accomplish this, the GW_j uploads a transaction to retract $\{h\,(PID_i||k_{GR})\}$ is already in the blockchain. Once retracted, its pseudo-identity PID_i is appended to the revocation list. Effectively, drone D_i cannot communicate with other gateway nodes. This is because all gateway nodes share the blockchain transactions and hence can recognize malicious drones. Suppose that this revoked drone D_i attempts to register at the *RA*. This requires that D_i submit $\{ID_i, A_c\}$ to *RA* for effective registration. This will prompt *RA* to retrieve PID_i corresponding to identity ID_i. Since is already in the revocation list, registration request will be denied.

4. Security Analysis

In this section, we demonstrate that the proposed scheme supports numerous salient security functionalities. We also show that our scheme resists a number of typical drone attacks.

Untraceability and anonymity: During the login and authentication phase, messages $L_a = \{C_a, C_b, C_d, T_1\}$, $L_b = \{D_b, D_c, D_e, T_2\}$, $Auth_a = \{TD_i, F_a, F_b, T_3\}$, and $Auth_b = \{F_c, F_d, F_e, T_4\}$ are exchanged. Evidently, none of these messages contain real identities $\{IDG_j, ID_{RA}, ID_i\}$ and hence anonymity is preserved. At the end of the authentication phase, the transient identity of drone D_i is updated as $TD_i^{new} = F_c\oplus h\,(r_f||PID_i)$. As such, attackers are unable to track D_i using TD_i.

MitM, forgery and replay threats: In our scheme, messages $L_a = \{C_a, C_b, C_d, T_1\}$, $L_b = \{D_b, D_c, D_e, T_2\}$, $Auth_a = \{TD_i, F_a, F_b, T_3\}$, and $Auth_b = \{F_c, F_d, F_e, T_4\}$ are exchanged over insecure channels. It is clear that all these messages incorporate timestamp $\{T_1, T_2, T_3, T_4\}$ and hence replay attacks are thwarted. Here, $C_a = (r_b * r_c)*P$, $C_b = PID_i\oplus h\,(IDG_j||B_d||T_1)$, $C_d = h(PID_i||B_d||T_1)$, $D_b = r_d * P$, $D_c = TD_i\oplus h\,(D_a||PID_i)$, $D_e = h\,(SK_{DG}||T_2)$, $TD_i = PID_i\oplus h\,(r_e||IDG_j||k_{GR})$, $F_a = h(PID_i||TD_i||IDG_k||T_3)\oplus r_f$, $F_b = h(r_f||PID_i||TD_i)$, $F_d = r_g\oplus h\,(r_f||TD_i^{new}||PID_i)$, and $F_e = h(SK_{Dk}||T_4)$. It is evident that security tokens are protected either using one-way hashing or elliptic curve primitives, which are computationally infeasible to reverse. As such, forgery and MitM threats are mitigated.

Robust authentication: In our scheme, the operator is authented by D_i by checking if $B_c \stackrel{?}{=} h(h(PID_i||r_a) \bmod f_v)$. On the other hand, the GW_j verifies D_i by confirming whether $\{h(PID_i||k_{GR})\}$ is in the blockchain, and establishing if $C_d \stackrel{?}{=} h(PID_i||B_d||T_1)$. Similarly, D_i authenticates GW_j by verifying whether $D_e \stackrel{?}{=} h(SK_{DG}||T_2)$. During the inter-gateway authentication, the GW_k validates D_i by determining if $h(TD_i||PID_i^*||r_e) \stackrel{?}{=} h(TD_i||PID_i||r_e)$, and $F_b \stackrel{?}{=} h(r_f||PID_i||TD_i)$. Similarly, GW_k is authenticated at the D_i by establishing whether $F_e \stackrel{?}{=} h(SK_{Dk}||T_4)$. In all these authentication instances, the session is halted upon verification failure.

Physical capture threats: Let us assume that adversaries can physically capture drone D_i. Thereafter, security tokens $\{A_d, B_a, B_c\}$ stored in it are retrieved. Here, $A_d = r_b \oplus A_b$, $B_a = r_a \oplus h(r_b)$ and $B_c = h(h(PID_i||r_a) \bmod f_v)$. Clearly, sensitive parameters are either masked or protected by the one-way hashing function. As such, it is cumbersome for attackers to obtain these sensitive security tokens about drone D_i.

Ephemeral secret leakage threats: In these attacks, we assume that nonces $\{r_c, r_d, r_e\}$ of the login phase have been leaked to attackers. Thereafter, attempts are made to retrieve PID_i and calculate session key $SK_{DG} = h(D_a||TD_i||PID_i)$. In addition, we presume that an adversary is capable of accessing nonces $\{r_f, r_g\}$ of the authentication phase. This is followed by attempts to retrieve PID_i and calculate session key $SK_{Dk} = h(r_g||r_f||TD_i^{new}||PID_i)$. Evidently, the attackers still lack $\{TD_i, TD_i^{new}\}$, the transient identity of D_i. As such, these ephemeral secret leakage threats are thwarted.

Side-channeling and offline guessing attacks: Suppose that an adversary has obtained parameters $\{A_d, B_a, B_c\}$ stored in the memory of drone D_i via power analysis. Next, using a dictionary of 10^6 passwords-unique identities pairs, attempts are made to launch offline guessing attacks. This is followed by using bogus st_i^b and ID_i^b to login into drone D_i. However, the drone must compute $A_b = h(ID_i^b||st_i^b)$, $r_b = A_d \oplus A_b$, $r_a = B_a \oplus h(r_b)$, $A_c = A_b \oplus h(r_a)$, $PID_i = h(ID_i^b||A_c||r_b)$. It then checks if $B_c \stackrel{?}{=} h(h(PID_i||r_a) \bmod f_v)$. Due to fuzzy verifier-based masking in B_c, the attacker is unable to discern whether the pair $\{st_i^b, ID_i^b\}$ is legitimate or not. The probability of correctly guessing $\{st_i^b, ID_i^b\}$ is $f_v \times 10^{-12} \cong 2^{-32}$, which is infinitesimal.

Perfect forward key secrecy: Let us assume that GW_j private key S_{GW} and shared key between GW_j and RA, k_{GR} have been leaked to the adversary. Thereafter, attempts are made to derive session keys $SK_{DG} = h(D_a||TD_i||PID_i)$ and $SK_{Dk} = h(r_g||r_f||TD_i^{new}||PID_i)$. Here, $C_a = (r_b * r_c)*P$, $D_a = r_d * C_a$, $PID_i = h(ID_i||A_c||r_b)$, $A_c = A_b \oplus h(r_a)$, $TD_i = PID_i \oplus h(r_e||IDG_j||k_{GR})$, and $TD_i^{new} = PID_i \oplus h(r_h||IDG_k||k_{GR})$. Clearly, nonces $\{r_b, r_c, r_d\}$ are unavailable to the adversary and hence D_a cannot be derived. On the other hand, PID_i cannot be derived devoid of knowledge of nonces $\{r_a, r_b\}$. Consequently, our scheme preserves forward key secrecy.

Eavesdropping and impersonation threats: Suppose that an attacker wants to masquerade as legitimate drone D_i. To achieve this, efforts are made to obtain drone pseudo-identity $PID_i = h(ID_i||A_c||r_b)$ of the login phase and $PID_i^* = TD_i \oplus h(r_e||IDG_j||k_{GR})$ of the authentication phase. However, $\{PID_i, PID_i^*\}$ is never transmitted in publicly exchanged messages $L_a = \{C_a, C_b, C_d, T_1\}$, $L_b = \{D_b, D_c, D_e, T_2\}$, $Auth_a = \{TD_i, F_a, F_b, T_3\}$, and $Auth_b = \{F_c, F_d, F_e, T_4\}$. Therefore, an adversary is unable to eavesdrop $\{PID_i, PID_i^*\}$. Any attempt to derive these values will also flop due to unavailability of $\{ID_i, A_c, r_b\}$ which are protected by the collision-resistant one-way hashing function.

Stolen verifier threats: In this attack, we assume that an adversary has stolen the verification table that is maintained at gateway node GW_j. Afterwards, access to data stored in the blockchain can be obtained. This is followed by attempts to derive $D_a = r_d * C_a$, where $C_a = (r_b * r_c)*P$. However, this requires that the adversary solve the Elliptic Curve Diffie-Hellman (ECDH) problem. Since this is computationally cumbersome, attackers can never correctly derive D_a and hence can never compute session key $SK_{DG} = h\ (D_a||TD_i||PID_i)$. Similarly, the attacker cannot derive session key $SK_{Dk} = h\ (r_g||r_f||TD_i^{new}||PID_i)$ due to inaccessibility of $\{k_{GR}, PID_i\}$.

Privileged insider threat: Let us assume that an attacker can obtain $\{ID_i, A_c, r_b\}$ stored in the registration authority's database during the drone registration phase. Thereafter, an attacker tries to derive session key $SK_{DG} = h\ (D_a||TD_i||PID_i)$. Here, $D_a = r_d * C_a$, $C_a = (r_b * r_c) * P$, $TD_i = PID_i \oplus h\ (r_e||IDG_j||k_{GR})$, and $PID_i = h\ (ID_i||A_c||r_b)$. Due to the difficulty of solving the ECDH problem and reversing the collision-resistant one-way hashing function, an adversary cannot obtain D_a and PID_i required in the calculation of SK_{DG}. Similarly, session key $SK_{Dk} = h\ (r_g||r_f||TD_i^{new}||PID_i)$ can never be derived devoid of knowledge of $\{r_g, r_f\}$.

Session key disclosure attacks: The goal of the adversaries in these attacks is to directly derive session keys $SK_{DG} = h\ (D_a||TD_i||PID_i)$ and $SK_{Dk} = h\ (r_g||r_f||TD_i^{new}||PID_i)$. Here, $D_a = r_d * C_a$, $C_a = (r_b * r_c) * P$, $TD_i = PID_i \oplus h\ (r_e||IDG_j||k_{GR})$, $PID_i = h\ (ID_i||A_c||r_b)$, $A_c = A_b \oplus h\ (r_a)$, $A_b = h\ (ID_i||st_i)$, and $TD_i^{new} = F_c \oplus h\ (r_f||PID_i)$. It is evident that derivation of D_a requires solving of ECDH problem, which is computationally infeasible. In addition, identity IDG_j and shared key k_{GR} are never publicly exchanged in plaintext. Since PID_i uses one-way hashing function to protect $\{ID_i, st_i\}$, these values cannot be accessed for the derivation of session key SK_{Dk}. The unavailability of nonces $\{r_g, r_f\}$ further complicates the derivation of SK_{Dk}.

5. Performance Evaluation

During the comparative performance of a majority of the smart city authentication protocols, computation complexities, supported functionalities and communication complexities are frequently utilized. Therefore, these performance measures are deployed in this section to appraise the developed scheme as described in the sub-sections that follow.

5.1 Computation Complexity

In this sub-section, we use the cryptographic operations carried out during the login and authentication phase to derive the computation overhead of our scheme. Let us denote scalar point multiplication, one-way hashing, symmetric encryption/decryption, point addition as T_m, T_h, T_{en} and T_a respectively. During the login process, drone D_i executes, $8T_h + 2T_m$ operations while the gateway node carries out $5T_h + 2T_m + T_a$ operations. On the other hand, the computation overheads are $5T_h$ and $7T_h$ at D_i and GW_k respectively, during the authentication procedures. As such, the cumulative computation complexity is $25T_h + 4T_m + T_a$. For the implementations, a machine running on a 3.4 GHz Intel Core i7 quadcore processor, Ubuntu 24.04.2 LTS, and 16GB of RAM was utilized. Using the multi-precision Integer and Rational Arithmetic Cryptographic Library (MIRACL), the run times of various cryptographic operations are measures as, $T_m \approx 2.27$ ms, $T_h \approx 0.019$ ms, $T_a \approx 0.825$ ms, and $T_{en} \approx 0.152$ ms. Using these values, the computation complexities of our scheme and other related protocols are derived as shown in Table 15.2.

Table 15.2. Computation complexities.

Scheme	Derivation	Total (ms)
Rangwani and Om (2021)	$17T_h + 6T_m + 10T_a$	22.193
Karim et al. (2023)	$10T_h + 6T_m$	13.810
Wu et al. (2021)	$22T_h + 12T_m$	27.658
Wang et al. (2022)	$25T_h + 6T_m$	14.095
Xie et al. (2024)	$18T_h + 6T_m + T_{en}$	14.114
Salami et al. (2023)	$30T_h + 8T_m$	18.730
Proposed	$25T_h + 4T_m + T_a$	10.380

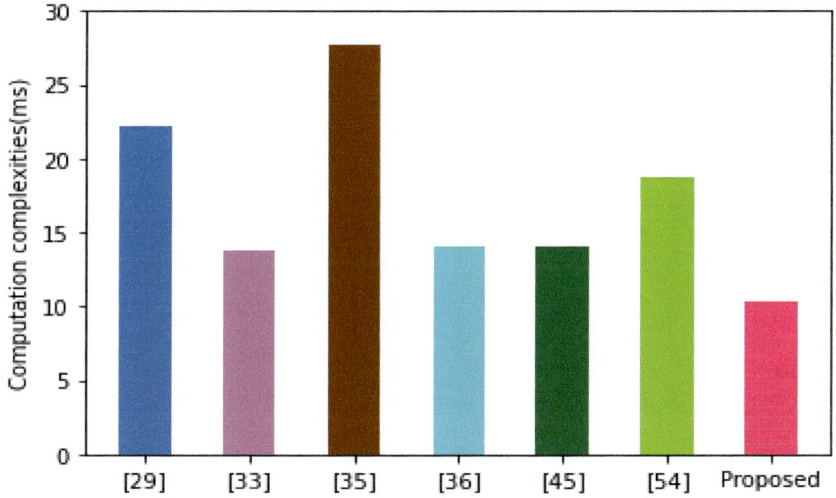

Figure 15.5. Computation complexities.

As shown in Figure 15.5 the protocol developed in Wu et al. (2021) incurs the highest computation complexities of 27.658 ms. This is attributed to the numerous scalar point multiplications executed in this scheme. This is followed by the protocols in Rangwani and Om (2021), Salami et al. (2023), Xie et al. (2024), Wang et al. (2022) and Karim et al. (2023) with computation complexities of 22.193 ms, 18.730 ms, 14.114 ms, 14.095 ms and 13.810 ms respectively.

However, the proposed protocol incurs the computation complexity of only 10.380 ms, which is the lowest. Using the scheme in Karim et al. (2023) as the benchmark, our protocol reduces the computation complexity by 24.84%.

5.2 Communication Complexity

To derive the cumulative communication overheads of our scheme, we deploy the message exchanges during the login and authentication phases. In these phases messages $L_a = \{C_a, C_b, C_d, T_1\}$, $L_b = \{D_b, D_c, D_e, T_2\}$, $Auth_a = \{TD_i, F_a, F_b, T_3\}$, and $Auth_b = \{F_c, F_d, F_e, T_4\}$ are exchanged. Here, the size of hashing, ECC point, timestamp, and identity are 160 bits, 320 bits, 32 bits and 128 bits respectively. As such, L_a ={320+160+160+32=672 bits}, L_b ={320+160+160+32=672 bits}, $Auth_a$ = {160+160+160+32=512 bits}, and $Auth_b$ = {160+160+160+32=512 bits}. This means that the total communication complexity of our scheme is 2368 bits as shown in Table 15.3.

Table 15.3. Communication complexities.

Scheme	No. of messages	Total (bits)
Rangwani et al. (2021)	3	4608
Karim et al. (2023)	3	2400
Wu et al. (2021)	5	3744
Wang et al. (2022)	4	2432
Xie et al. (2024)	4	2976
Salami et al. (2023)	5	3520
Proposed	4	2368

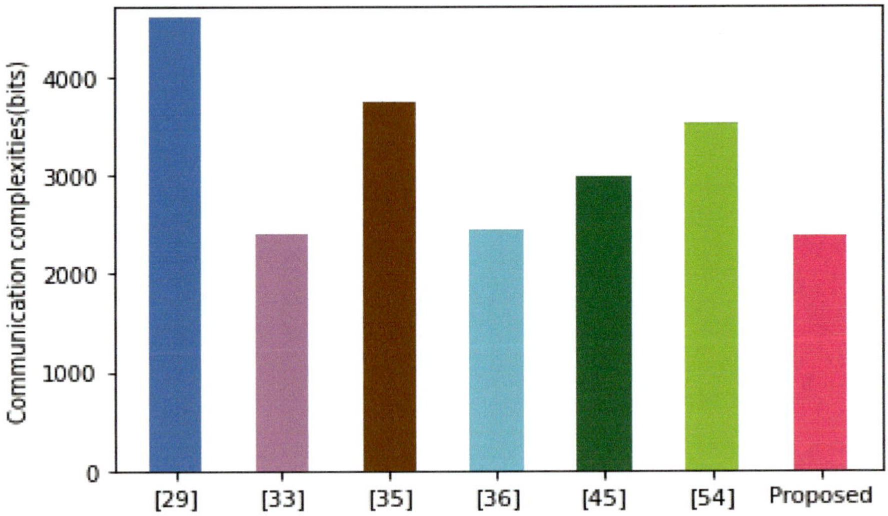

Figure 15.6. Communication complexities.

As shown in Table 15.3, our protocol exchanges 4 messages during the login and authentication phase. Similarly, the schemes in Wang et al. (2022) and Xie et al. (2024) require 4 message exchanges for any successful authentication phase. On the hand, the protocols in Wu et al. (2021) and Salami et al. (2023) require 5 messages, while the techniques in Rangwani et al. (2021) and Karim et al. (2023) exchange only 3 messages. It is evident from Figure 15.6 that the protocol presented in Rangwani et al. (2021) incurs the highest communication complexity of 4608 bits and hence requires the highest amount of bandwidth. This is followed by the schemes in Wu et al. (2021), Salami et al. (2023), Xie et al. (2024), Wang et al. (2022) and Karim et al. (2023) with communication complexities of 3744 bits, 3520 bits, 2976 bits, 2432 bits, and 2400 bits respectively.

On the other hand, our scheme requires only 2368 to fully execute the login and mutual authentication phases, which is the lowest. Using the protocol developed in Karim et al. (2023) as the benchmark, then our protocol reduces the communication complexities by 1.33%.

5.3 Supported Functionalities

The goal of this sub-section is to demonstrate that our scheme offers salient security features such as anonymity, untraceability, robust authentication, session key setup, and forward key secrecy, as evidenced in Table 15.4.

In addition, our scheme mitigates typical smart city security threats such as ephemeral secret leakage, offline guessing, eavesdropping, impersonation, stolen verifier, MitM, forgery, privileged insider, physical capture threats, replay, session key disclosure, and side-channeling. Based on Table 15.4, our scheme supports the highest number of security features. On the other hand, the protocol developed in Karim et al. (2023) offers support for the least security features. In overall, our scheme has been demonstrated to be the most efficient in terms of communication and computation complexities. In addition, we have shown that it supports the highest number of security characteristics. This postures our protocol as the most suitable for deployment in resource-limited smart city scenario.

Table 15.4. Functionalities and security features.

	[36]	[45]	[54]	[33]	[35]	[29]	Proposed
Security features							
Anonymity	√	√	√	×	√	√	√
Untraceability	√	√	×	×	√	√	√
Robust authentication	√	√	√	√	√	√	√
Session key setup	√	√	√	√	√	√	√
Forward key secrecy	√	√	√	√	√	√	√
Resilient against:							
Ephemeral secret leakage	×	×	×	×	√	√	√
Offline guessing	√	√	√	×	×	√	√
Eavesdropping	×	×	×	×	×	×	√
Impersonation	√	√	√	√	×	√	√
Stolen verifier	√	√	√	×	×	√	√
MitM	√	×	√	√	√	√	√
Forgery	×	√	×	×	×	×	√
Privileged insider	√	×	√	×	√	√	√
Physical capture threats	√	√	×	√	×	×	√
Replay	√	√	√	√	√	√	√
Session key disclosure	×	√	×	×	×	√	√
Side-channeling	×	×	×	×	×	×	√
	√ Feature supported; × Feature mising or not considered						

6. Conclusion

Leveraging on communication technologies such as IoT, WSNs, drone technology and artificial intelligence, smart cities have facilitated appropriate and intelligent responses in the provision of urban services. This has seen enhancement in public safety, citizen livelihood, industrial processes, environmental protection, as well as commercial services. However, security and privacy remain thorny issues now that some of the data collected in smart cities is highly sensitive. Although numerous security techniques have been put forward, a majority of them still have security loopholes that can be exploited by adversaries. In addition, some of these security protocols deploy conventional cryptographic primitives that are highly intensive for the resource-limited smart city drones. In this paper, an efficient login and authentication scheme has been developed. When compared with some benchmark schemes, the computation and communication complexities in our protocol are shown to reduced by and respectively. Moreover, our scheme offers protection against typical drone threats such as privileged insider, ephemeral secret leakage, side-channeling, offline guessing, session key disclosure, MitM and stolen verifier. This projects our scheme as the most ideal for deployment in the drone-based smart cities.

References

Ajakwe, S. O., Kim, D. S. and Lee, J. M. (2023). Drone transportation system: Systematic review of security dynamics for smart mobility. IEEE Internet of Things Journal 10(16): 14462–14482.

Al Sibahee, M. A., Nyangaresi, V. O., Abduljabbar, Z. A., Luo, C., Zhang, J. and Ma, J. (2023). Two-factor privacy-preserving protocol for efficient authentication in Internet of Vehicles networks. IEEE Internet of Things Journal 11(8): 14253–14266.

Ali, Z. A., Abduljabbar, Z. A., AL-Asadi, H. A. A., Nyangaresi, V. O., Abduljaleel, I. Q. and Aldarwish, A. J. (2024). A provably secure anonymous authentication protocol for consumer and service provider information transmissions in smart grids. Cryptography 8(2): 20.

Alsayaydeh, J. A. J., Ali, M. F., Al-Andoli, M. N. M. and Herawan, S. G. (2024). Improving the robustness of IoT-powered smart city applications through service-reliant application authentication technique. IEEE Access 12: 19405–19417.

Alzahrani, B. A., Barnawi, A. and Chaudhry, S. A. (2021). A resource-friendly authentication protocol for UAV-based massive crowd management systems. Security and Communication Networks 2021(1): 3437373, 1–12.

Ayub, M. F., Li, X., Mahmood, K., Shamshad, S., Saleem, M. A. and Omar, M. (2023). Secure consumer-centric demand response management in resilient smart grid as industry 5.0 application with blockchain-based authentication. IEEE Transactions on Consumer Electronics 70(1): 1370–1379.

Banerjee, S., Chunka, C., Sen, S. and Goswami, R. S. (2019). An enhanced and secure biometric based user authentication scheme in wireless sensor networks using smart cards. Wireless Personal Commununication 107: 1–28.

Bera, B., Das, A. K., Balzano, W. and Medaglia, C. M. (2020). On the design of biometric-based user authentication protocol in smart city environment. Pattern Recognition Letters 138: 439–446.

Bera, B., Tekchandani, P., Das, A. K., Karuppiah, M. and Sikdar, B. (2024). Continuous authentication for consumer electronics in smart city surveillance. IEEE Consumer Electronics Magazine.

Darbandi, M., Ehsani, A., Bahrami, F., Moghadamnia, E., Abnoosian, K. and Mohammed, B. O. (2025). A new feature-based biometric identification system in IoT-powered smart cities using a hybrid optimization algorithm. Security and Privacy 8(1): e490.

Derhab, A., Cheikhrouhou, O., Allouch, A., Koubaa, A., Qureshi, B., Ferrag, M. A. et al. (2023). Internet of drones security: Taxonomies, open issues, and future directions. Vehicular Communications 39: 100552.

Garg, T., Gupta, S., Obaidat, M. S. and Raj, M. (2024). Drones as a service (DaaS) for 5G networks and blockchain-assisted IoT-based smart city infrastructure. Cluster Computing 27(7): 8725–8788.

Ghahramani, M., Javidan, R. and Shojafar, M. (2020). A secure biometric-based authentication protocol for global mobility networks in smart cities. The Journal of Supercomputing 76(11): 8729–8755.

Gope, P. and Sikdar, B. (2020). An efficient privacy-preserving authenticated key agreement scheme for edge-assisted internet of drones. IEEE Transactions on Vehicular Technology 69(11): 13621–13630.

Gour, R. and Yadav, S. (2025). The transformative role of UAVs in smart city frameworks. pp. 3–24. In Internet of Vehicles and Computer Vision Solutions for Smart City Transformations. Cham: Springer Nature Switzerland.

Gul, O. M., Hamidoglu, A., Khan, M. K. and Yu, S. (2025). Novel game theoretical approach in blockchain aided communication at internet of drones. IEEE Consumer Electronics Magazine.

Gupta, S., Alharbi, F., Alshahrani, R., Kumar Arya, P., Vyas, S., Elkamchouchi, D. H. et al. (2023). Secure and lightweight authentication protocol for privacy preserving communications in smart city applications. Sustainability 15(6): 5346.

Hou, X., Xin, L., Fu, Y., Na, Z., Gao, G., Liu, Y. et al. (2023). A self-powered biomimetic mouse whisker sensor (BMWS) aiming at terrestrial and space objects perception. Nano Energy 118: 109034.

Hu, S., Chen, Y., Zheng, Y., Xing, B., Li, Y., Zhang, L. et al. (2022). Provably secure ECC-based authentication and key agreement scheme for advanced metering infrastructure in the smart grid. IEEE Transactions on Industrial Informatics 19(4): 5985–5994.

Hussain, S., Chaudhry, S. A., Alomari, O. A., Alsharif, M. H., Khan, M. K. and Kumar, N. (2021). Amassing the security: An ECC-based authentication scheme for Internet of Drones. IEEE Systems Journal 15(3): 4431–4438.

Hussain, S., Farooq, M., Alzahrani, B. A., Albeshri, A., Alsubhi, K. and Chaudhry, S. A. (2023). An efficient and reliable user access protocol for Internet of Drones. IEEE Access 11: 59688–59700.

Irshad, R. R., Hussain, S., Hussain, I., Ahmad, I., Yousif, A., Alwayle, I. M. et al. (2023). An intelligent buffalo-based secure edge-enabled computing platform for heterogeneous IoT network in smart cities. IEEE Access 11: 69282–69294.

Jiang, S., Yu, Q.. Chen, X., Chen, L. and Zhang, X. (2021). The application of IoT, big data, and artificial intelligence technologies in smart cities: a review. Journal of Sensors 1–20.

Kadhim, T. A., Abduljabbar, Z. A., AL-Asadi, H. A. A., Nyangaresi, V. O., Ali, Z. A. and Abduljaleel, I. Q. (2025). Lightweight scheme for secure signaling and data exchanges in intelligent precision agriculture. Cryptography 9(1): 7.

Karim, S. M., Habbal, A., Chaudhry, S. A. and Irshad, A. (2023). BSDCE-IoV: Blockchain-based secure data collection and exchange scheme for IoV in 5G environment. IEEE Access 11: 36158–36175.

Khan, M. A., Shah, H., Rehman, S. U., Kumar, N., Ghazali, R., Shehzad, D. et al. (2021). Securing internet of drones with identity-based proxy signcryption. IEEE Access 9: 89133–89142.

Lounis, K., Ding, S. H. and Zulkernine, M. (2022). D2D-MAP: A drone to drone authentication protocol using physical unclonable functions. IEEE Transactions on Vehicular Technology 72(4): 5079–5093.

Luo, H., Wen, G. and Su, J. (2020). Lightweight three factor scheme for real-time data access in wireless sensor networks. Wireless Networks 26(2): 955–970.

Ma, H., Wang, C., Xu, G., Cao, Q., Xu, G. and Duan, L. (2023). Anonymous authentication protocol based on physical unclonable function and elliptic curve cryptography for smart grid. IEEE Systems Journal 17(4): 6425–6436.

Mahmood, K., Tariq, T., Sangaiah, A. K., Ghaffar, Z., Saleem, M. A. and Shamshad, S. (2023). A neural computing-based access control protocol for AI-driven intelligent flying vehicles in industry 5.0-assisted consumer electronics. IEEE Transactions on Consumer Electronics 70(1): 3573–3581.

Mahobiya, C., Iyer, S., Verma, M., Mishra, P. R. and Bohidar, S. K. (2024). Navigating the future: unmanned aerial systems in IoT paradigms. Unmanned Aircraft Systems 355–385.

Meher, B. K., Amin, R., Das, A. K. and Khan, M. K. (2022). KL-RAP: an efficient key-less RFID authentication protocol based on ECDLP for consumer warehouse management system. IEEE Transactions on Network Science and Engineering 9(5): 3411–3420.

Moon, J., Lee, D., Jung, J. and Won, D. (2017). Improvement of efficient and secure smart card based password authentication scheme. International Journal of Network Security 19(6): 1053–1061.

Mutlaq, K. A. A., Nyangaresi, V. O., Omar, M. A., Abduljabbar, Z. A., Abduljaleel, I. Q., Ma, J. et al. (2024). Low complexity smart grid security protocol based on elliptic curve cryptography, biometrics and hamming distance. Plos One 19(1): e0296781.

Nyangaresi, V. O. and Petrovic, N. (2021, July). Efficient PUF based authentication protocol for internet of drones. pp. 1–4. In 2021 International Telecommunications Conference (ITC-Egypt). IEEE.

Nyangaresi, V. O., Abduljabbar, Z. A., Mutlaq, K. A. A., Bulbul, S. S., Ma, J., Aldarwish, A. J. et al. (2024). Smart city energy efficient data privacy preservation protocol based on biometrics and fuzzy commitment scheme. Scientific Reports 14(1): 16223.

Nyangaresi, V. O., Al-Joboury, I. M., Al-sharhanee, K. A., Najim, A. H., Abbas, A. H. and Hariz, H. M. (2024). A biometric and physically unclonable function–Based authentication protocol for payload exchanges in internet of drones. e-Prime-Advances in Electrical Engineering, Electronics and Energy 7: 100471.

Nyangaresi, V. O., AlRababah, A. A., Yenurkar, G. K., Chinthaginjala, R. and Yasir, M. (2025). Anonymous authentication scheme based on physically unclonable function and biometrics for smart cities. Engineering Reports 7(1): e13079.

Pu, C., Wall, A., Choo, K. K. R., Ahmed, I. and Lim, S. (2022). A lightweight and privacy-preserving mutual authentication and key agreement protocol for Internet of Drones environment. IEEE Internet of Things Journal 9(12): 9918–9933.

Rangwani, D. and Om, H. (2021). A secure user authentication protocol based on ECC for cloud computing environment. Arabian Journal for Science and Engineering 46(4): 3865–3888.

Rao, P. M., Pedada, S., Jangirala, S., Das, A. K. and Rodrigues, J. J. (2024). Role of IoT in the ages of digital to smart cities: Security challenges and countermeasures. IEEE Internet of Things Magazine 7(1): 56–64.

Salami, Y., Khajehvand, V. and Zeinali, E. (2023). SAIFC: A secure authentication scheme for IOV based on fog-cloud federation. Security and Communication Networks 2023(1): 9143563.

Sharma, V., Beniwal, R. and Kumar, V. (2024). Towards secure IoT system from a smart city perspective: An optimized algorithm and implementation. Transactions on Emerging Telecommunications Technologies 35(4): e4883.

Sikder, A. K., Petracca, G., Aksu, H., Jaeger, T. and Uluagac, A. S. (2021). A survey on sensor-based threats and attacks to smart devices and applications. IEEE Communications Surveys & Tutorials 23(2): 1125–1159.

Sucasas, V., Aly, A., Mantas, G., Rodriguez, J. and Aaraj, N. (2023). Secure multi-party computation-based privacy-preserving authentication for smart cities. IEEE Transactions on Cloud Computing 11(4): 3555–3572.

Tanveer, M., Aldosary, A., Das, A. K., Aldossari, S. A. and Chaudhry, S. A. (2024). PAF-IoD: PUF-enabled authentication framework for the Internet of Drones. IEEE Transactions on Vehicular Technology.

Tychola, K. A. and Rantos, K. (2025). Cyberthreats and security measures in drone-assisted agriculture. Electronics 14(1): 149.

Umran, S. M., Lu, S., Abduljabbar, Z. A. and Nyangaresi, V. O. (2023). Multi-chain blockchain based secure data-sharing framework for industrial IoTs smart devices in petroleum industry. Internet of Things 24: 100969.

Wang, C., Wang, D., Duan, Y. and Tao, X. (2023). Secure and lightweight user authentication scheme for cloud-assisted Internet of Things. IEEE Transactions on Information Forensics and Security 18: 2961–2976.

Wang, X., Teng, Y., Chi, Y. and Hu, H. (2022). A robust and anonymous three-factor authentication scheme based ecc for smart home environments. Symmetry 14(11): 2394.

Wu, T. Y., Lee, Z., Yang, L. and Chen, C. M. (2021). A provably secure authentication and key exchange protocol in vehicular ad hoc networks. Security and Communication Networks 2021(1): 9944460.

Wu, T., Guo, X., Chen, Y., Kumari, S. and Chen, C. (2021). Amassing the security: An enhanced authentication protocol for drone communications over 5G networks. Drones 6(1): 10.

Xia, X., Ji, S., Vijayakumar, P., Shen, J. and Rodrigues, J. J. (2021). An efficient anonymous authentication and key agreement scheme with privacy-preserving for smart cities. International Journal of Distributed Sensor Networks 17(6): 15501477211026804, 1–13.

Xie, Q. and Huang, J. (2024). Improvement of a conditional privacy-preserving and desynchronization-resistant authentication protocol for IoV. Applied Sciences 14(6): 2451.

Xie, Q., Li, K., Tan, X., Han, L., Tang, W. and Hu, B. (2021). A secure and privacy-preserving authentication protocol for wireless sensor networks in smart city. EURASIP Journal on Wireless Communications and Networking 2021(1): 119.

Yang, J., Kwon, Y. and Kim, D. (2020). Regional smart city development focus: The South Korean national strategic smart city program. IEEE Access 9: 7193–7210.

Yang, P., Cao, X., Quek, T. Q. and Wu, D. O. (2022). Networking of Internet of UAVs: Challenges and intelligent approaches. IEEE Wireless Communications 31(1): 156–163.

Zhang, K., Ni, J., Yang, K., Liang, X., Ren, J. and Shen, X. S. (2017). Security and privacy in smart city applications: Challenges and solutions. IEEE Communications Magazine 55(1): 122–129.

Zhang, Y., He, D., Li, L. and Chen, B. (2020). A lightweight authentication and key agreement scheme for Internet of Drones. Computer Communications 154: 455–464.

Chapter 16

Blockchain as The Future for Industry 5.0

Theories, Algorithms, and Possible Outcomes

Vineet Bhatia,[1,*] *Ranjan Arora*,[1] *Sumati Sidharth*[1] and *Varsha Bhatia*[2]

1. Introduction

The evolution of industry has been defined by technological leaps that have not only changed how we produce goods, but have reshaped the very fabric of society. From the beginning of steam power to the rise of artificial intelligence, each industrial revolution has resulted in a new time of possibilities and challenges. The transition to the fifth industrial revolution marks a fundamental change in manufacturing philosophy, where human creativity and wellness are considered as equally important as technological advancement and innovation.

1.1 The Fifth Industrial Revolution: Integrating Human and Machine Capabilities

The industrial world has been repeatedly reshaped by successive technological evolutions that have revolutionized production and reshaped society. From the advent of steam power to the rise of artificial intelligence, these revolutions have driven unprecedented progress. However, the emerging paradigm, known as Industry 5.0, signifies a critical shift in spotlight. The latest industrial revolution (I5.0) differs from its predecessors by emphasizing the seamless blend of human creativity and technological capabilities (Chakir et al. 2024), moving beyond the earlier focus on pure technical advancement and operational efficiency.

1.2 A Historical Perspective (Groumpos 2021)

Industry 1.0 (1760–1840): Featured by the mechanized processes for production and driven by the invention of the steam engine. This period witnessed the rise of factories and mass production, transforming the manufacturing landscape (Jha et al. 2024).

[1] Department of Technology Management, Defence Institute of Advanced Technology Pune, India.
[2] Sunstone Pune India.
Emails: ranjanarora25@gmail.com, sumatisidharth@gmail.com, varsha.bhatia.in@gmail.com
* Corresponding author: bhatiavineet.in@gmail.com

DOI: 10.1201/9781003620822-16

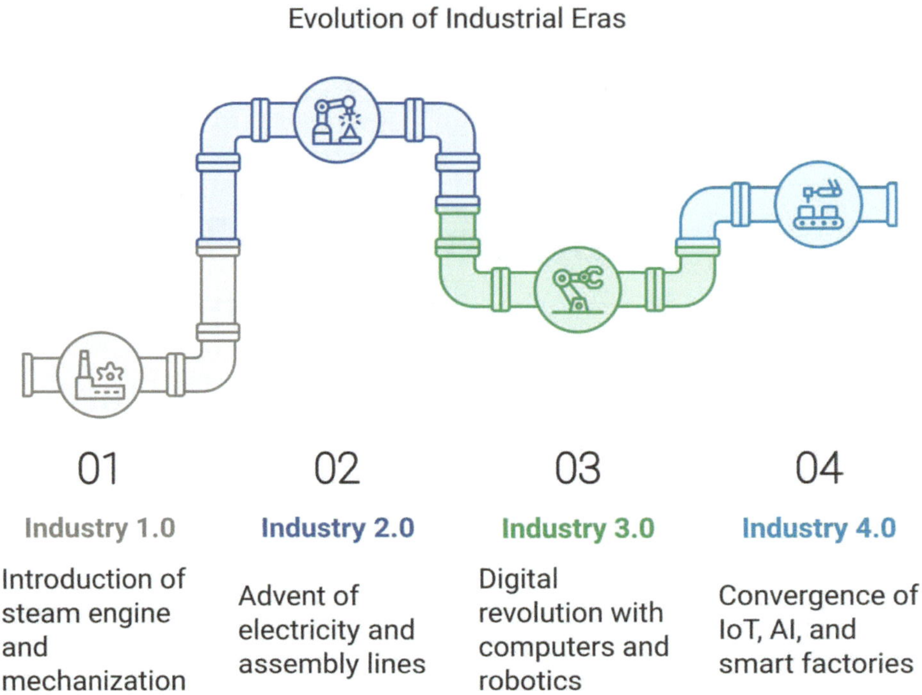

Figure 16.1. Progression of industry eras.

Industry 2.0 (1870–1914): The introduction of electricity powered production on a large scale and enabled new forms of manufacturing, such as assembly lines.

Industry 3.0 (1969–2000): The digital revolution, marked by the emergence of computing devices and automation, gave rise to the maturity of programmable logic controllers (PLCs) and robotics.

Industry 4.0 (2011–Present): The current industrial landscape combines interconnected digital-physical infrastructure, IoT networks, cloud-based platforms, and AI technologies. This phase is marked by digitally-driven manufacturing facilities, self-operating processes, and sophisticated data-driven insights.

The timeline at Figure 16.1 illustrates the evolution of industrial capabilities across five major technological ages - beginning with the basic mechanization of Industry 1.0 and culminating in today's I5.0, which emphasizes the fusion of human expertise with sophisticated technological systems.

1.3 Emergence of Industry 5.0 (Leng et al. 2022)

I5.0 displays a significantly different approach from its predecessors. It recognizes that while technology offers immense potential, human ingenuity, creativity, and emotional intelligence remain irreplaceable (Akundi et al. 2022). This human-centric approach aims to:

Enhance human capabilities: Leverage technology to augment human potential, rather than substitute human workers. This involves developing technologies that empower individuals, improve their skills, and create more fulfilling work experiences (Adel 2022).

Prioritize sustainability: Foster sustainable production processes that minimize environmental impact and conserve resources. This incorporates integrating renewable energy sources, implementing circular economy principles, and reducing waste generation.

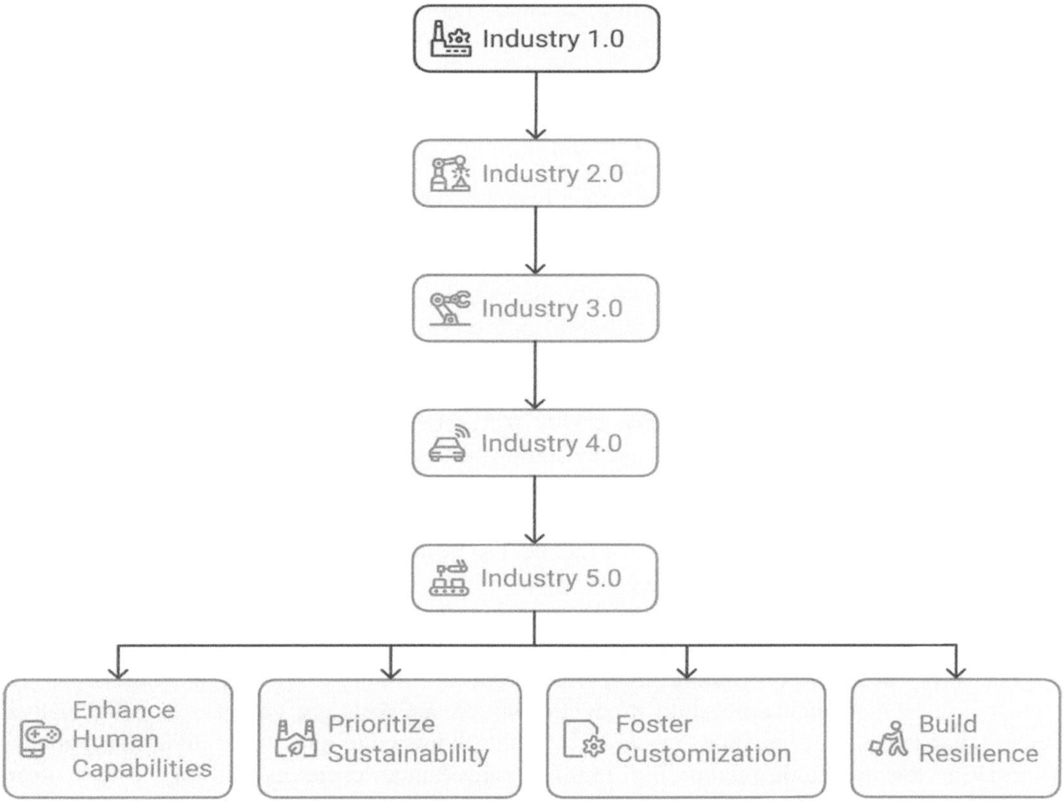

Figure 16.2. Key elements of I5.0.

Foster customization: Manufacturing flexibility in I5.0 allows companies to tailor their output to each customer's unique specifications, moving beyond standardized mass production to deliver individualized products while maintaining operational efficiency. This personalization capability lets manufacturers adapt their processes to meet diverse consumer preferences and specific requirements.

Build resilience: Create resilient systems that can adapt to altering environment, withstand disruptions, and ensure business continuity (Aheleroff et al. 2022).

Figure 16.2 illustrates the key elements that comprise I5.0. This advanced industrial paradigm integrates several interconnected components: human-centric approach, sustainability practices, resilient manufacturing systems, and cutting-edge technologies. The figure demonstrates how these elements work together to create a more advanced and responsible industrial framework that balances technological innovation with human values and environmental considerations.

1.4 The Evolution Beyond Industry 4.0: Prioritizing People in Advanced Manufacturing

The move from Industry 4.0 to 5.0 marks a fundamental reimagining of how we integrate technology into manufacturing and industrial processes. While the fourth industrial revolution focused on connecting machines and systems, this new phase represents a more nuanced understanding of how technology can enhance rather than replace human capabilities. This shift reflects a deeper transformation in industrial philosophy, moving away from purely technical solutions toward a more balanced approach that values both technological advancement and human factors. The change demonstrates an evolving understanding that optimal industrial development comes from combining the precision and efficiency of automated systems with uniquely human qualities like creativity, adaptability, and emotional intelligence.

The new industrial paradigm recognizes that sustainable progress requires more than just technological innovation – it demands careful consideration of human needs, environmental impact, and social responsibility. This holistic approach aims to create manufacturing environments that not only optimize production but also prioritize worker well-being and societal benefits.

Cotransition from I4.0 to I5.0 represents a paradigm shift in the way we approach industrial development and technological integration (Carayannis et al. 2024). While I4.0 focused on automation, data exchange, and the interconnection of smart systems, the center of attention of I5.0 is focused on enhancing the role of humans in collaboration with these technologies.

In the initial phase (2018–2019), research focused on understanding the evolution of I4.0 towards the I5.0 concept. This period emphasized the foundational function of human-machine interaction and underlying technologies, establishing early connections to emerging areas like customer experience. Subsequent research (2020–2021) broadened its scope to encompass societal priorities within digital transformation, incorporating sustainability considerations while building upon the advancements of I4.0. This phase recognized the critical role of the human operator in the transition towards an I5.0 society. Recent studies since 2022 have expanded their focus to explore complex applications, examining how I5.0 principles can be implemented in urban infrastructure and virtual modelling systems. The research has particularly emphasized the development of intelligent urban environments and the deployment of digital replica technologies that mirror physical systems (Digital Twin) (Mathur et al. 2022).

The fifth industrial revolution, I5.0, champions a paradigm shift in manufacturing by reintegrating human intelligence into production processes. This era emphasizes collaborative human-robot interaction, enabling machines to share knowledge and work alongside humans. By fostering this symbiotic relationship, I5.0 leverages human creativity and ingenuity to steer intelligent manufacturing processes. This harmonious collaboration between humans and cobots eliminates the threat of job displacement, paving the way for value-added services and enhanced productivity. This section delves into the core definitions, key features, and current advancements of I5.0

1.5 Definitions of Industry 5.0

As I5.0 is still in its nascent stages, diverse perspectives exist among industry practitioners and researchers regarding its definition (Kovari 2024). Some prominent interpretations are outlined below:

Definition No. 1: Industry 5.0 is a human-cantered evolution steered by the "6R" principles of industrial upcycling ("Recognize, Reconsider, Realize, Reduce, Reuse, Recycle") (Rada 2020), (Mourtzis et al. 2022). This approach emphasizes sustainable practices, innovative product development, and enhanced quality of life.

Definition No. 2: Industry 5.0 reintegrates human workers into the manufacturing process, fostering collaborative human-machine partnerships. By leveraging human intelligence and creativity, this synergy aims to optimize production efficiency and drive innovation.

Definition No. 3 ("European Economic and Social Committee"): Industry 5.0 emphasizes the synergistic integration of "cyber-physical production systems" (CPPS) with human intelligence.

Table 16.1. Concise overview of the diverse perspectives on Industry 5.0.

Definition Source	References	Human-Centricity	Sustainability	Resilience	Digital Technologies	Social Impact	Human-Machine Collaboration	Worker Empowerment	Efficiency & Productivity
European Commission: "Industry 5.0 provides a vision of industry that aims beyond efficiency and productivity as the sole goals, and reinforces the role and contribution of industry to society"	(Julian Müller 2020)	✓	✓	✓	✓	✓	☒	☒	✓
Nahavandi: "Industry 5.0 refers to people working alongside robots and smart machines"	(Nahavandi 2019)	✓	☒	☒	✓	☒	☒	☒	☒
Xu et al.: "Industry 5.0 represents a paradigm shift towards a more sustainable, human-centric and resilient industrial transformation"	(Xu et al. 2021)	✓	✓	✓	☒	☒	☒	☒	☒
Breque et al.: "Industry 5.0 complements the existing Industry 4.0 paradigm by highlighting research and innovation as drivers for a transition to a sustainable, human-centric and resilient European industry"	(Julian Müller 2020)	✓	✓	✓	✓	☒	☒	☒	☒
Amr Adel: "The revolution of industry 5.0 means that humans and machines are working together, improving the efficiency of industrial production"	(Adel 2022)	✓	☒	☒	✓	✓	✓	✓	✓
Leng et al.: "Industry 5.0 can be considered the era of the socially intelligent factory... The core of Industry 5.0 is the harmony of machines, humans, values, tasks..."	(Leng et al. 2022)	✓	☒	☒	✓	✓	✓	✓	☒
Longo et al.: "The Age of Augmentation (Industry 5.0)—focused on cooperation between human intelligence and cognitive computing"	(Longo, Padovano, and Umbrello 2020)	✓	☒	☒	✓	☒	☒	✓	☒

Table 16.1 contd. ...

...Table 16.1 contd.

Definition Source	**References**	**Human-Centricity**	**Sustainability**	**Resilience**	**Digital Technologies**	**Social Impact**	**Human-Machine Collaboration**	**Worker Empowerment**	**Efficiency & Productivity**
Michael Rada: "Change processes directing towards closer cooperation between man and machine, and systematic prevention of waste"	(Rada 2020)	✓	✓	☒	☒	☒	✓	☒	☒
Nahavandi: "Pairing human and machine to further utilize human brainpower and creativity to increase process efficiency"	(Nahavandi 2019)	✓	☒	☒	✓	☒	✓	✓	✓
Kraaijenbrink: "Shift from economic value to societal value, and from welfare to wellbeing"	(Kraaijenbrink 2024)	✓	☒	☒	☒	☒	☒	☒	☒
Momenta: " Industry 5.0 emphasizes a triple-bottom-line of economic, environmental, and societal impact, bringing ESG (Environment, Social and Governance) perspective and balance to what have often been technology-led and economic-driven choices."	(Momenta 2024)	✓	☒	☒	✓	☒	✓	✓	☒
James Jardine: "People working alongside robots and smart machines... adds personal human touch to Industry 4.0 pillars"	(Jardine n.d.)	✓	☒	☒	✓	☒	✓	☒	✓
Marina Ruggieri: "Humans work alongside advanced technologies and AI-powered robots to enhance processes"	(Ruggieri n.d.)	✓	☒	☒	✓	☒	✓	☒	✓
TWI: "Humans working alongside advanced technology with human-centric focus, increased resilience and improved focus on sustainability"	(TWI n.d.)	✓	✓	✓	✓	☒	✓	☒	☒

Note: ✓ indicates the presence of the concept in the definition

This approach aims to attend to the potential dehumanizing effects of I4.0 by prioritizing human-centric design principles (Julian Müller 2020).

Definition No. 4 ("Friedman and Hendry"): Industry 5.0 necessitates a comprehensive approach that involves the human factor alongside technological advancements within industrial systems. This requires collaboration between industry professionals, personnel handling information technologies, and philosophers (Longo et al. 2020).

Definition No. 5: Industry 5.0 ushers in the era of the "Social Smart Factory," where flawless interaction between humans and cobots is facilitated through enterprise social networks.

Definition No. 6: Industry 5.0 represents a step-by-step advancement beyond Industry 4.0, emphasizing regular innovation and a more cautious approach to hyper-connectivity. It prioritizes safety and resilience by incorporating "orthogonal safe exits" into automated systems.

Definition 7: Industry 5.0 prioritizes design centered on humans, fostering collaboration between humans and "ideal human companions" (cobots). This model leverages cobots for repetitive jobs while empowering humans to concentrate on crucial thinking, customization, and complex decision-making.

Table 16.1 provides a conceptual analysis of Industry 5.0. Definitions from literature review: a cross-reference matrix of key themes and components.

1.6 Analysis of Industry 5.0 Definitions

A careful examination of the provided definitions reveals that I5.0 transcends mere evolution; it signifies a transformative shift, a new industrial revolution where both humans and robots collaborate as partners. The merging of human insight with machine intelligence is crucial for navigating this new reality. Crucially, cobots must be trained to adapt dynamically to the evolving cognitive capabilities of human workers.

This human-robot synergy allows for a strategic division of labor: human intelligence excels in critical thinking and developing customized solutions, while cobots efficiently handle labor-intensive tasks. This optimized allocation of responsibilities mitigates workforce shortages by effectively deploying cobots for physically demanding jobs.

Furthermore, advanced manufacturing technologies like Smart Autonomous Manufacturing (SAM) contribute to resource efficiency by enabling layered manufacturing processes, minimizing energy consumption. Proactive Predictive Maintenance (PdM) enhances operational reliability by facilitating timely maintenance and accelerating recovery from unforeseen events.

I5.0 embraces hyper-personalization, delivering customized solutions for every customer through the integration of AI and cognitive systems across the entire manufacturing lifecycle. Cyber-Physical Production Systems (CPPS) leverage cognitive skills throughout the value chain, enabling seamless human-robot collaboration and delivering exceptional customer experiences.

While the acceptance of I5.0 is picking up speed, the development of global standards and policies is still an ongoing process, essential for ensuring a consistent and ethical framework for this transformative industrial revolution. In conclusion, Table 16.2 outlines the primary factors influencing the shift from I4.0 to I5.0.

Table 16.2. Contrasting industrial revolutions: From Industry 4.0 to Industry 5.0.

Aspect	**Industry 4.0**	**Industry 5.0**
Core Focus	Automation, efficiency, smart systems, and data integration	Human-centric partnership, personalization, sustainability, and adaptability
Technology Integration	Cyber-physical systems (CPS), IoT, AI, big data, automation, Cloud computing (CC)	Advanced robotics, AI, IoT, human-robot collaboration, cognitive computing
Human Role	Limited human intervention, machines replace manual tasks	Humans collaborate with machines to enhance decision-making and creativity
Automation	High automation with minimal human involvement	Enhanced collaboration between humans and robots for complex tasks
Personalization	Primarily mass production with some customization options (e.g., configurable products)	Highly personalized manufacturing, tailored to individual consumer needs
Sustainability	Focus on efficiency improvements, but environmental concerns are secondary	Strong focus on sustainable practices, resource optimization, and eco-friendly production
Adaptability	Optimized for fixed, predictable production environments	Highly adaptable systems that respond to changing demands and environments
Production Systems	Smart factories with automated supply chains and processes	Flexible, decentralized systems that allow for rapid reconfiguration and adaptability
Ethical Considerations	Limited center of attention towards effect on workers, ethics of automation	Emphasis on worker well-being, ethics, and societal impact of automation and AI
Customer Experience	Mass production with some level of customization (e.g., user preferences)	High degree of personalization in products and services, driven by consumer feedback and AI insights
Resilience	Efficiency-driven systems, but may struggle with disruptions	Resilient systems that can quickly adapt to disruptions, fostering supply chain robustness
Key Goals	Increased efficiency, reduced costs, and optimized performance	Enhanced human-machine collaboration, sustainability, ethical values, and social responsibility

Figure 16.3 depicts the salient features of Industry 4.0 and Industry 5.0 as follows:

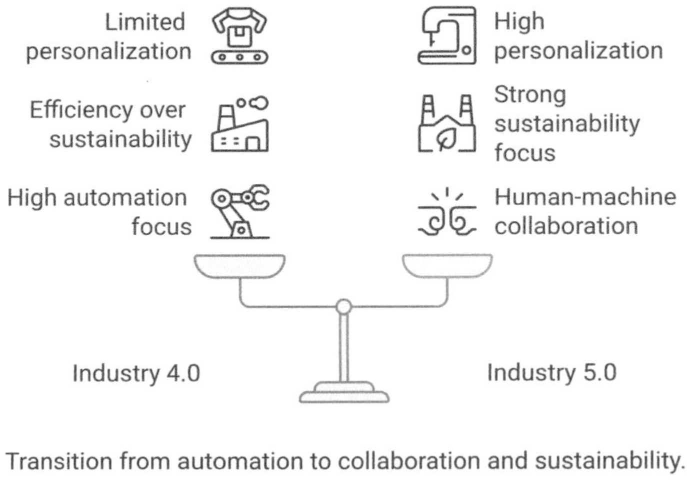

Figure 16.3. Salient features of Industry 4.0 and Industry 5.0.

2. Theories and Concepts Underpinning Blockchain in Industry 5.0

2.1 Blockchain Technology Overview

Blockchain is "an innovative ledger technology that records transactions and protects them by distributing them across multiple computers, requiring network-wide agreement for any retrospective changes." It is characterized by several core principles and comes in various types, each with its own explicit features and use cases (Kaur et al. 2022).

2.2 Core Principles of Blockchain

Three core principles underpin the functionality and security of blockchain technology: decentralization, immutability, and consensus mechanisms (Xu et al. 2019).

Decentralization: Conventional databases generally exist in a centralized form: meaning databases are stored and managed in a particular location by a single entity. This makes them exposed to failure at a single point, censorship, and manipulation. Blockchain, in contrast, functions on a decentralized arrangement. This means the records are dispersed across multiple computers (nodes) in the network. Each node holds a duplication of the blockchain, making the data less vulnerable to attacks and censorship. If one node fails, the other nodes remain functioning, maintaining the integrity of the system. This distributed nature also eradicates the requirement for a centralized agency to control and manage the records. Instead, the network jointly maintains the ledger. This characteristic significantly reduces the risk of failure at single point and control, increasing safekeeping and clearness. Because the blockchain is distributed, no single individual can independently alter the data. Changes require consensus from a significant portion of the network, making it complex for cyber hackers to influence the information.

Immutability: Immutability refers to the lack of ability to change data once the recording of data is carried out on the blockchain. The blockchain has resistance to modification stems by its sophisticated cryptographic architecture. At its core, each data segment receives a distinctive cryptographic signature comparable to a digital thumbprint. These signatures are mathematically interconnected, with every new block incorporating the previous block's identifier. This creates an unbreakable sequence where altering any single element would trigger a chain reaction, disrupting all subsequent connections. Such an intricate security design makes unauthorized changes practically impossible. The immutability of blockchain ensures the reliability and dependability of the recorded data. When absolute data integrity matters like in financial operations, supply network monitoring, and voting mechanisms, this feature becomes an essential technological foundation.

Consensus mechanisms: Blockchain systems require a consensus protocol to validate transactions and maintain ledger accuracy across the distributed network, as there is no single governing entity overseeing the system. This decentralized approach ensures all participants can verify and agree on transaction authenticity without relying on a central authority. This is achieved through agreement mechanisms, which are algorithms that define how the network reaches an agreement on which transactions to add to the blockchain. These mechanisms avoid fraudulent transactions and ensure that all copies of the blockchain remain consistent across the network. Various blockchain networks use separate agreement mechanisms, each having its own advantages and limitations.

The core principles of blockchain are depicted in Figure 16.4.

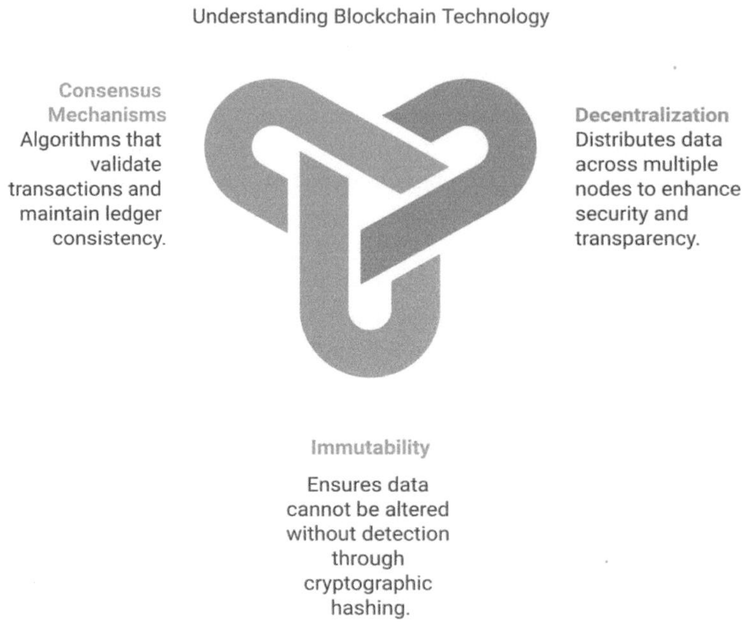

Figure 16.4. Core principles of blockchain.

2.3 Blockchain Types

Blockchains can be largely divided into three main categories: public, private, and consortium (Verma and Dhanda 2023).

Public blockchains: Public blockchains do not require permission: meaning the network can be joined by anyone, contribute to the validation of transactions (based on the agreement mechanism), and view the blockchain data. Bitcoin and Ethereum are examples of public blockchains. These blockchains are characterized by elevated decentralization level and clearness. All transactions are publicly visible and verifiable. This transparency promotes trust and accountability but can also raise privacy concerns as transaction details are often linked to pseudonymous addresses. Public blockchains are generally considered more secure due to their large and distributed network, making them resistant to attacks. However, they can also suffer from scalability issues, meaning they may have restrictions in processing the numbers of transactions per second.

Private blockchains: Private blockchains need permission, meaning only authorized participants have access to the network. These blockchains are typically managed by a solo organization or entity. They offer greater command over data access and privacy compared to public blockchains. Only authorized participants can view and validate transactions. This makes them suitable for applications within organizations where data confidentiality is paramount. Private blockchains are often more efficient and scalable than public blockchains due to the smaller number of participants. However, they are also less decentralized and therefore potentially more vulnerable to manipulation by the controlling entity. Examples of use cases include supply chain management within a company or internal financial transactions.

Consortium blockchains: Consortium blockchains also need permission but are governed by a cluster of companies rather than a single organization. This offers a middle ground between the decentralization of public blockchains and the control of private blockchains. Access to the network is restricted to authorized members, usually consisting of businesses or entities operating within specific industrial sectors or business domains. This allows for collaboration and data sharing while maintaining a degree of control and privacy. Consortium blockchains are suitable for applications involving multiple organizations that need to share information securely and efficiently, such as supply chain tracking across multiple companies or interbank transactions.

2.4 Blockchain Consensus Algorithms

These algorithms are critical to maintain the reliability and safety of blockchain networks. Network members collectively validate transactions and maintain consensus on the distributed ledger's current state through established verification protocols (Zhou et al. 2023). Several consensus algorithms exist, each having its own features and dealings.

2.4.1 Proof of Work (PoW)

PoW is the original consensus algorithm used by Bitcoin. It requires participants (miners) to solve complex computational problems to validate transactions and add new blocks to the blockchain. The first miner to solve the problem broadcasts the new block to the network, and other miners verify its validity. The miner who successfully adds a block is rewarded with cryptocurrency. PoW is known for its high level of security but is also computationally intensive and consumes significant amounts of energy (Hashemi et al. 2023).

2.4.2 Proof of Stake (PoS)

PoS is an alternative consensus algorithm that aims to address the energy consumption issues of PoW. In PoS, validators are chosen to create new blocks based on the amount of cryptocurrency they hold and "stake" as collateral. Validators are incentivized to act honestly because they risk losing their stake if they try to validate fraudulent transactions. PoS is generally considered more energy-efficient than PoW but has its own set of security considerations (Hafid et al. 2023).

2.4.3 Practical Byzantine Fault Tolerance (PBFT)

PBFT is a consensus algorithm designed for permissioned blockchains. It is based on the Byzantine Generals' Problem, a classic computer science problem that deals with achieving consensus in a distributed system where some participants may be faulty or malicious. PBFT requires multiple rounds of communication between nodes to reach consensus. It is known for its high throughput and low latency but is less scalable than PoW or PoS (Kim et al. 2024).

Other consensus algorithms exist, including Delegated Proof of Stake (DPoS), Proof of Authority (PoA), and variations and combinations of these core mechanisms. Different blockchain applications prioritize varying needs ranging from security to scaling capabilities and power consumption, which guides their selection of a consensus mechanism. Each algorithm presents different trade-offs in these areas.

A comparison of consensus algorithms (Auhl et al. 2022) in a matrix form is shown in Table 16.3.

Table 16.3. Comparison of various blockchain consensus algorithms.

Algorithm	Scalability	Energy Efficiency	Security	Decentralization	Use Case	Latency
Proof of Work (PoW)	Low	Very Low	Very High	High	Public blockchains (e.g., Bitcoin)	High (10+ minutes)
Proof of Stake (PoS)	Medium	High	High	Medium	Public and private chains	Medium (minutes)
Practical Byzantine Fault Tolerance (PBFT)	Medium	High	High	Low	Permissioned networks	Low (seconds)
Delegated Proof of Stake (DPoS)	High	High	Medium	Medium	High-throughput chains	Low (seconds)
Proof of Authority (PoA)	High	Very High	Medium	Low	Private/consortium chains	Very Low (milliseconds)

2.5 Blockchain and Human-Centric Manufacturing

Human-centric manufacturing prioritizes the essential role of human workers within sophisticated production environments, integrating them with advanced technologies rather than aiming for complete automation (Mourtzis et al. 2023). Blockchain technology presents several avenues to increase this support between people and machines, leading to improvements in manufacturing processes.

2.6 How Blockchain Enhances Collaboration between Humans and Machines

Blockchain's fundamental traits—distributed control, unchangeable records, and open visibility can significantly reshape how humans and machines operate together in manufacturing (Oliveira et al. 2023).

Improved information exchange and openness: Blockchain functions as a shared, permanent record that can be accessed by authorized individuals and automated systems. This shared data environment promotes transparency throughout the manufacturing journey. Machines can automatically log data about production runs, maintenance activities, and quality checks directly onto the blockchain. Human employees can then use this up-to-the-minute information to track progress, pinpoint potential problems, and make well-informed decisions. This clear flow of information breaks down data silos and encourages better communication between human operators and automated equipment.

For example, data gathered by sensors on machines can be stored on the blockchain, creating a provable history of machine performance and maintenance. Human technicians can then consult this data to diagnose issues more effectively and schedule preventive maintenance actions.

Protected data reliability and tracking: The inability to alter blockchain records ensures that data entered by machines remains secure and trustworthy. This establishes a reliable audit trail for all aspects of manufacturing. This is especially valuable for quality assurance and supply chain oversight. By documenting each stage of production on the blockchain, from the initial sourcing of raw materials to the finishing delivery of goods, manufacturers can trace the place of origin and path of every product (Agrawal et al. 2021). This tracing capability can help determine the underlying cause of defects, raise product quality, and combat counterfeiting.

As an illustration, if a flawed batch of products is discovered, manufacturers can use the blockchain to quickly identify the affected items and trace them back to their source for appropriate action.

Automated operations and processes: Blockchain can automate certain parts of manufacturing using smart contracts. These are automated protocols embedded in code that execute predefined actions when certain blockchain-recorded conditions are met. These digital agreements operate independently, triggering transactions or processes without manual intervention once their programmed criteria are fulfilled.

For example, a smart contract can order new raw materials in an automatic manner when inventory levels drop below a set point, based on data reported by machines on the blockchain. These automated processes reduce the requirement for intervention by workers and increases efficiency.

Distributed decision-making in the workplace: Blockchain supports distributed decision-making in manufacturing through the employment of smart contracts, independent software agents, and digital replicas. Blockchain supports distributed decision-making in manufacturing by employing smart contracts, independent software agents, and digital replicas.

Smart contracts: As previously mentioned, smart contracts can automate various activities within manufacturing. They can be used to handle supply chain logistics, automate payments to suppliers, and enforce quality standards. Regarding human-machine partnership, smart contracts can define the roles and responsibilities of both human workers and machines. For instance, a smart contract could establish the conditions under which a machine should request human assistance, such as when it encounters an error or needs servicing. This automated coordination between people and machines can increase productivity and minimize downtime.

Independent software agents: These are software programs that can operate in an independent manner and make decisions based on defined rules and data. In a blockchain-based manufacturing environment, these agents can interact with smart contracts and other machines to improve production.

For example, an agent could analyze production data stored on the blockchain and adjust machine settings to enhance efficiency or reduce waste. These agents can also communicate with human workers, providing them with real-time information and suggestions.

Digital replicas: A digital replica is a near real model of a physical asset, process, or system. In manufacturing, digital replicas can simulate production, test new designs, and optimize machine performance. The combination of blockchain with digital replicas can lead manufacturers to create a safe and clear record of all modifications and updates made to the digital model (Roumeliotis et al. 2024). This can improve collaboration between human engineers and machines by fostering a shared understanding of the system.

For example, engineers can use the digital replica to explore different scenarios and then establish the outcomes on the blockchain. This information can be then applied to upgrade the actual manufacturing process.

Enabling tailored experiences for consumers through blockchain: Blockchain can also improve the consumer experience by enabling greater personalization, product tracking, and control over personal information.

Personalization: Blockchain can support customized product creation by allowing consumers to specify their preferences and monitor the production of their customized items. By documenting every phase of the customization procedure on the blockchain, consumers gain a clear view of how their product is being made. This can increase consumer confidence and satisfaction. For example, a consumer ordering custom furniture could use the blockchain to track the origin of the materials, the creation of the individual parts, and the final assembly.

Product tracking: Blockchain's inherent tracking capabilities can provide consumers with detailed knowledge about the creator and history of the goods consumers buy. This can be especially valuable for products with complex supply chains, such as food or medications. Consumers can build more trust in brands and make better-informed purchasing decisions thanks to increased transparency. By using blockchain-based apps or scanning product codes, they can get detailed information about how the items are manufactured, what they are made of, and how they are transported. For example, consumers could use the blockchain to confirm the source of their coffee or trace the path of their medication from manufacturer to pharmacy.

Information control: Blockchain can empower consumers to own and manage their personal information related to their purchases. By saving their information on a blockchain, consumers can choose what information they share with manufacturers or other parties. This can address growing issues affecting safety of data and privacy concerns. For example, consumers could use a blockchain-based recognition system to manage their product warranties and service records, choosing to share this information only with authorized service providers.

2.7 Blockchain-Assisted Industry 5.0 Ecosystem Architecture

The generic architecture for a blockchain-enabled I5.0 ecosystem is intended to contribute to a large number of applications, emphasizing the combination of highly developed technologies and diverse functionalities. This architecture for a blockchain-enabled I5.0 ecosystem is designed to contribute to a large number of applications, including supply chain operations, personalized health programs, manufacturing industries, and unmanned aerial vehicles (UAVs). It also facilitates resource management and control through connections with edge, fog, and cloud systems, while integrating with emerging Web 3.0 architectures.

2.7.1 Core Applications

This architecture encompasses several key applications, including:

Supply Chain Management (Sustainable supply chains)
Personalized Health Programs
New Educational Technologies
Manufacturing Industries
Energy
Unmanned Aerial Vehicles (UAVs) for military and disaster management

2.7.2 Resource Management and Control

Using connections with fog, cloud, and edge systems
Collaborating with arising Web 3.0 Architectures

2.7.3 Blockchain Layer

At the core of this architecture lies a robust blockchain layer, which consists of:

Distributed ledger: This is meant to provide a protected and unalterable record of transactions across various applications.

Smart contracts (SCs): These automate processes and payments, enabling decentralized applications (DApps) to operate efficiently and enforce specific conditions.

2.7.4 Data Layer

IoT sensors and devices form the foundational data collection infrastructure, gathering continuous information to enable real-time system monitoring and operational control. This data is securely stored and made accessible through DApps, ensuring that stakeholders can interact seamlessly with the system.

2.7.5 Application Layer

The application layer is critical, as it hosts the DApps responsible to facilitate transactions and interactions among various stakeholders. Key components include:

User interfaces: Designed to enhance user experience, these interfaces allow straightforward interaction with both the DApps and the underlying blockchain.

2.7.6 Integration Layer

Integration within the ecosystem is achieved through:

APIs: These facilitate communication between different components and external services.

Interoperability solutions: They are essential for connecting with emerging Web 3.0 architectures and other decentralized systems, broadening the ecosystem's capabilities.

2.7.7 Network Layer

The network layer maintains reliable services by leveraging fifth-generation (5G) and future network technologies. Key services that ensure efficient data transfer and connectivity include:

Enhanced Mobile Broadband (eMBB)

Ultra-Reliable Low Latency Communications (uRLLC)
Massive Machine-Type Communications (mMTC)

2.7.8 Edge Services

Edge services play a significant role by enabling edge computing, which authorizes data exchanges between users and enterprise applications. This capability reduces latency and improves response times, further enhancing overall system efficiency.

2.7.9 Security Layer

Security is fundamental to the architecture, supported by:

Cryptographic protocols: They ensure data integrity and confidentiality.

Access control mechanisms: They manage permissions for users accessing the blockchain and its applications.

2.7.10 Governance Layer

The governance layer encompasses:

Consensus mechanisms: These validate transactions within the ecosystem, maintaining trust and reliability.

Regulatory compliance measures: They make sure that all components follow legal standards and best approach.

This comprehensive architecture not only emphasizes the integration of advanced technologies but also positions itself as a cornerstone for the future of I5.0, fostering a secure and proficient operational atmosphere. The core elements of this architecture are presented in Figure 16.5.

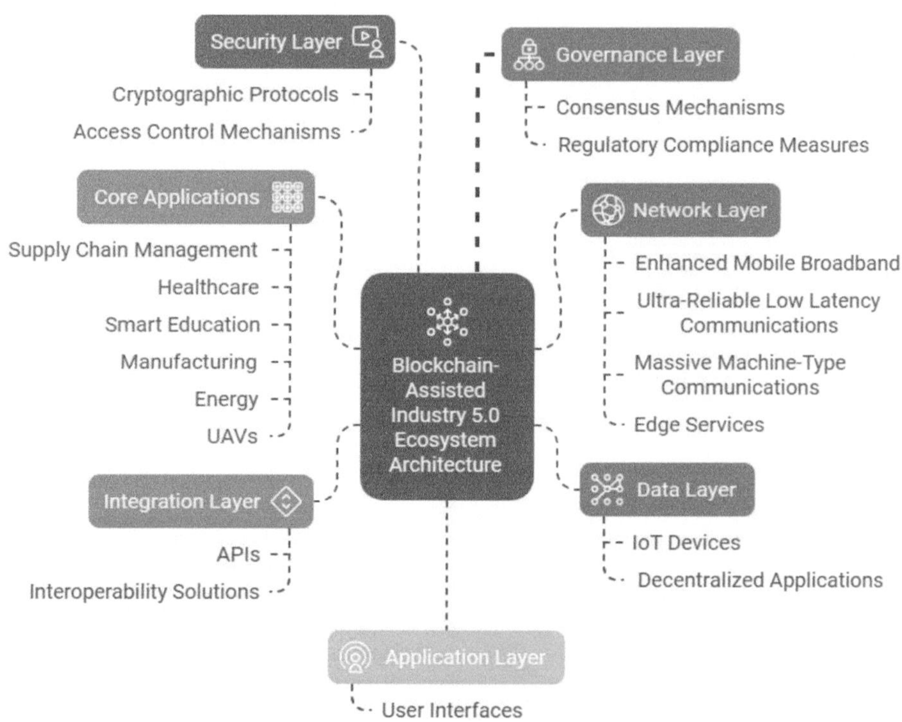

Figure 16.5. Blockchain-assisted Industry 5.0 ecosystem architecture.

2.8 Blockchain Applications in Industry 5.0

Industry 5.0, unlike its predecessor, Industry 4.0, shifts the focus from solely maximizing efficiency and automation to give priority to human focused ideals like sustainability, well-being, and creativity. Blockchain technology, with its decentralized and unalterable nature, can play a pivotal role in achieving these goals. Table 16.4 depicts a detailed summary of key blockchain applications used in I5.0 framework. The table illustrates diverse implementations of blockchain technology across various industrial domains, highlighting how these applications support the main ideologies of human-centric manufacturing, sustainable architecture, and resilience in I5.0.

Table 16.4. Blockchain applications in industrial domain.

Industry Sector	Blockchain Application	Benefits	Specific Areas of Industry 5.0
Smart Cities (Adel and Alani 2024)	Energy management	Secure energy data, prevent tampering	Sustainable energy practices, smart grid management
	IoT security	Enhanced security for IoT devices	Secure data exchange, privacy protection.
	Data privacy	Improved user credibility, data accountability	Data privacy, user confidence.
Manufacturing (Ghobakhloo et al. 2024, Khanfar et al. 2021)	Supply chain management	Enhanced clearness, trackability, and security in the supply chain.	Fair sourcing, sustainable procedures, worker welfare. Sustainability.
	Smart contracts	Automated execution of agreements, reducing manual intervention.	Fair labor practices, compliance with regulations.
	IoT integration	Secure and transparent data exchange between IoT devices.	Worker safety, environmental monitoring.
Healthcare (STL Partners 2024)	Electronic Health Records (EHR)	Secure and immutable patient data, ensuring privacy and integrity.	Patient-centric care, data privacy, personalized medicine.
	Clinical trials	Maintain transparent and unalterable records, ensuring the integrity of research data throughout the trial process.	Ethical research practices, patient consent management.
	Medicines supply chain	Tracking and verification of pharmaceutical products to prevent counterfeits.	Drug safety, ethical pharmaceutical practices.
Energy (Consensys 2024, Bürer et al. 2019)	Peerwise trading in energy sector	Decentralized energy trading options, enabling direct transactions.	Community-based energy systems, renewable energy adoption.
	Renewable Energy Certificates (RECs)	Transparent and secure tracking of renewable energy production.	Sustainable energy practices, carbon footprint reduction.
	Grid management	Improved efficiency and security in energy distribution networks.	Smart grid management, energy conservation (khan et al. 2025)
Finance (Chawla and Mehta 2023)	Cross-border payments	Quicker, less costly, and more safe international money transfers.	Financial inclusion, ethical banking practices (Verma et al. 2023)
	Decentralized Finance (DeFi)	Financial services on blockchain, reducing intermediaries and costs.	Accessible financial services, transparent lending (Conure Technology Services 2024)

Table 16.4 contd. ...

...Table 16.4 contd.

Industry Sector	**Blockchain Application**	**Benefits**	**Specific Areas of Industry 5.0**
	Asset tokenization	Physical assets conversion into digital tokens for easier trading.	Sustainable investments, ethical asset management.
Agriculture (Haloui et al. 2024, Sharma et al. 2024)	Food supply chain	Traceability from origin to eateries, ensuring food security and quality.	Sustainable farming, fair trade practices, worker rights, Food security (Guruswamy et al. 2022)
	Crop Insurance	Smart contracts for automated and transparent insurance claims.	Farmer welfare, risk management.
	Land Registry	Secure and transparent land ownership records.	Land rights, ethical land use.
Supply Chain (Agarwal et al. 2022)/ **Logistics**	Freight management	Transparent and efficient tracking of goods in transit.	Ethical supply chains, reduced carbon emissions, Sustainable Supply chain (Wang et al. 2023), (Laddha and Agrawal 2024)
	Customs clearance	Streamlined and secure customs processes, reducing delays.	Fair trade practices, compliance with regulations.
	Warehouse management	Improved inventory management and tracking.	Sustainable inventory practices, worker safety.
Retail (IBM 2020)	Loyalty programs	Secure and transparent management of customer loyalty points.	Customer-centric marketing, ethical consumer practices.
	Product authenticity	Verification of product authenticity to combat counterfeiting.	Ethical product sourcing, consumer trust.
	Customer data management	Secure and private management of customer data.	Data privacy, ethical data use.
Real Estate	Property transactions	Secure and transparent property buying and selling processes.	Ethical property transactions, sustainable development.
	Smart contracts for leasing	Automated and secure lease agreements.	Fair leasing practices, tenant rights.
	Property management	Efficient management of property records and transactions.	Sustainable property management, community engagement.
Telecommunications, (Infopulse 2024, Chellvamathi and Kulkarni 2023)	Roaming management	Secure and efficient management of roaming agreements.	Fair roaming practices, consumer rights.
	Fraud detection	Improved detection and prevention of telecom fraud.	Ethical business practices, consumer protection.
	Identity Management	Secure and private management of user identities.	Data privacy, ethical identity management.

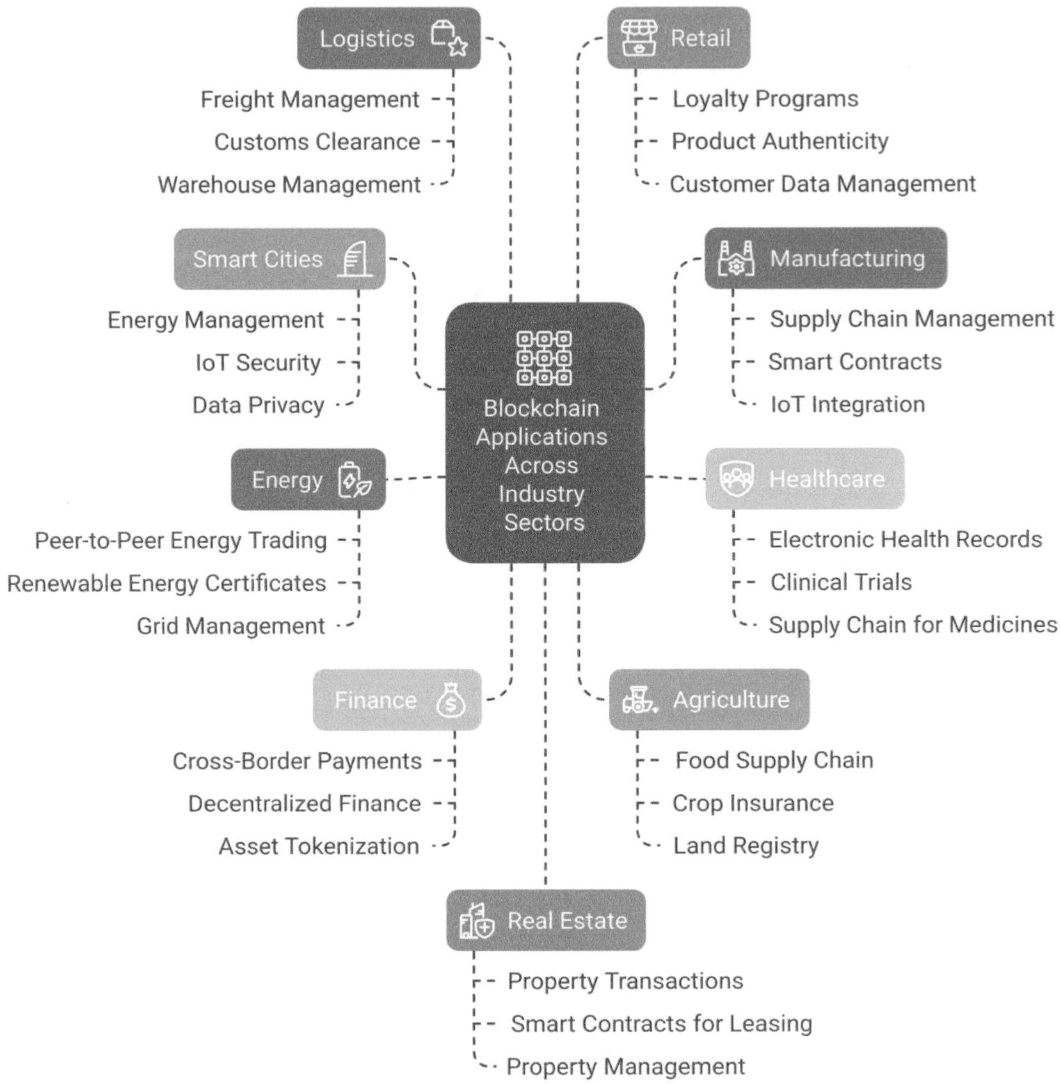

Figure 16.6. Blockchain applications in the industrial domain.

Figure 16.6 illustrates the diverse applications of BT within the industrial domain, highlighting its potential to revolutionize various sectors. The diagram showcases how blockchain can enhance supply chain management, ensure data security, facilitate smart contracts, and integrate with IoT devices. Each application is depicted with its corresponding advantages, like increased transparency, enhanced effectiveness, and cheaper operational costs. This visual representation emphasizes the impact of blockchain on I5.0, laying the way for a more interconnected and secure industrial ecosystem.

2.8.1 Blockchain Technology: The Future of Smart City Management in Industry 5.0

Smart cities aim to create acceptable, greener, and safer urban environments, aligning with the revelation of I5.0. This involves integrating a human touch and personalization through collaboration between humans and robots. Smart gadgets with various sensors collect data to improve traffic, transportation, schools, and libraries. The rise of IoT devices has made the concept of smart cities more prominent, integrating big data, AI, and advanced networking technologies, leading to the Internet-of-Everything (IoE) (Khodadadi et al. 2016). However, deploying smart city projects effectively requires addressing challenges like IoT security, equipment maintenance, user confidence, data center costs, and privacy. Blockchain (BC) technology offers solutions to address these concerns, making it best for the development of a smart city. Management of energy is critical in smart cities, and blockchain can secure energy data from tampering, enhancing big data control and management of energy in smart grids and IoT environments (Rai et al. 2024). Studies have shown that blockchain can improve user credibility, data accountability, and privacy protection. Blockchain, featuring improved storage and processing power, has proven effective at protecting network infrastructure. This solution helps defend against both malicious nodes and unauthorized access points (Drakatos et al. 2022). Challenges in smart city network architecture, such as bandwidth, security, latency, and scalability, have been addressed by developing a hybrid architecture that combines centralized and distributed designs (Sharma and Park 2018). This hybrid approach aims to improve privacy and has been simulated to evaluate its performance and viability in terms of latency.

2.8.1.1 Blockchain and Smart Contracts

Smart contracts agreements that execute automatically through computer code represent one of blockchain's most powerful applications in smart city development. These self-running contracts have their terms directly embedded in the programming code. They facilitate automated processes without the need for intermediaries, streamlining various services such as vehicle registration, public service notifications, and utility billing. For instance, Gupta et al. (2023) explored how automated blockchain protocols could streamline vehicle documentation processes, minimizing bureaucratic overhead while improving processing speed and accuracy (Gupta et al. 2023). By removing the manual interventions need, smart contracts can significantly speed up service delivery and reduce costs.

Moreover, smart contracts ensure data integrity and security, which is very critical for public faith in smart city governance. Transactions executed through smart contracts are recorded on the blockchain, creating an unalterable and clear record of all activities visible at all levels. This transparency is important for building public faith and ensuring accountability in urban governance. Gupta et al. (2023) highlighted how the employment of smart contracts can increase data safety by ensuring a tamper-proof record of all transactions, making it complicated for malignant players to changer or manipulate data (Gupta et al. 2023). This intensity of security is principally essential in areas like public safety, healthcare, and financial services, where data integrity is paramount.

2.8.1.2 Enhancing Urban Infrastructure

Blockchain technology can significantly enhance urban infrastructure by establishing a secure, safe and transparent management system. The decentralized nature of blockchain allows for secure management in city systems, ensuring that data is not under the control of a single entity. This decentralization enhances visibility in areas such as energy management and transportation, as discussed by Srinivas et al. (2023) (Aditya Sai Srinivas et al. 2023). Considering an example of the creation of a decentralized energy grid by blockchain, where energy production and consumption data are recorded on the blockchain. This transparency allows for more efficient management of energy and can smoothen the progress of the combination of energy sources that can be renewed, promoting sustainability.

Additionally, blockchain supports efficient data diffusion across city sectors, improving service delivery and operational efficiency. Khawaja and Javidroozi (2023) highlighted how blockchain can enable real-time data sharing among various organizations in a city, such as the transport sector, waste management department, and public safety. This real-time data sharing is crucial for coordinating services and ensuring that assets are allocated in an efficient manner. For instance, using blockchain to make a real-time traffic management system, where information from cameras and traffic sensors is recorded on the blockchain. Then this data can be used to organize the flow of traffic in the city, reduce cases of congestion on the roads, and provide enhanced transportation system for public in the city.

2.8.1.3 Sustainability in Industry 5.0

Environmental responsibility stands as a fundamental element of I5.0, with blockchain technology serving as a key enabler for implementing sustainable solutions in urban development. The secure, distributed nature of blockchain can help cities track and optimize their environmental initiatives more effectively. Blockchain offers enhanced traceability and real-time data exchange, which are vital for sustainable manufacturing practices. Hameem et al. (2024) discussed how blockchain can be employed to generate a transparent supply chain, where the place of origin and journey of products can be tracked in real-time (Hameem et al. 2024). This traceability is essential to ensure that goods are sourced and manufactured sustainably, fulfilling regulatory needs and customer demands for environment friendly products.

Furthermore, blockchain technology fosters collaboration among various service providers, promoting a holistic approach to urban sustainability. Khawaja and Javidroozi (2023) highlighted how blockchain can facilitate cross-sectoral integration, allowing different service providers to share data and collaborate on sustainability initiatives (Khawaja and Javidroozi 2023). For example, blockchain can be employed to make a shared program towards waste management, where information from waste collection and recycling processes is recorded on the blockchain. This data can then be used to organize efficient waste management practices, reduction in waste, and encourage recycling, contributing to a more sustainable urban environment.

Table 16.5 presents a comprehensive analysis of verified blockchain implementations across various smart city domains within the I5.0 paradigm. The presented examples highlight practical

Table 16.5. Blockchain applications in smart cities in Industry 5.0 context.

Domain	Application	Industry 5.0 Integration	Verified Example	Documented Outcomes	Impact
Smart Governance	Digital Identity	Human-centric digital services	Estonia e-Residency (2014–present)	Over 84,000 e-residents, Digital business registration	Cross-border business access, Digital transformation
Smart Energy	P2P Energy Trading	Sustainable energy with human oversight	Brooklyn Microgrid (2016–present)	Community-based energy market, Solar integration	Community energy autonomy, Reduced carbon footprint
Smart Education	Digital Credentials	Human-centered verification	MIT Digital Certificates (2017–present)	Over 2,000 digital diplomas issued, Instant verification	Credential portability, Enhanced verification
Smart Healthcare	Medical Records	Patient-controlled data sharing	Estonia's KSI Blockchain (2016–present)	95% health data digitized, 99% digital prescriptions	Improved data security, Patient data ownership
Smart Transportation	Vehicle Management	Human-AI collaborative systems	MOBI Vehicle Identity (2019–present)	Standardized vehicle identity, Secure data sharing	Industry standardization, Enhanced security
Smart Supply Chain	Product Tracing	Human-supervised tracking	IBM Food Trust (2017–present)	18,000+ products traced, 20% faster trace time	Food safety, Supply transparency
Smart City Integration	City Services	Citizen-centric services	Dubai Blockchain Platform (2020–present)	24 M documents processed; 8 public services integrated	Enhanced efficiency, Digital inclusion

implementations where blockchain systems effectively support I5.0's key values: enhancing human capabilities, promoting environmental sustainability, and building resilient operations. The selected cases represent operational projects implemented between 2014–present, each with documented outcomes and measurable impacts. The table highlights how blockchain technology enhances smart city operations while maintaining human oversight and control, a key area of I5.0. These implementations demonstrate the realistic function of blockchain in creating well-organized, transparent, and citizen-centric urban services, while ensuring human-machine collaboration remains at the front position towards advancement of technologies.

2.8.2 Blockchain-Enabled Smart Healthcare in the Perspective of Industry 5.0

I4.0 enabled mass customization, but the healthcare sector now demands personalization and a human touch. This necessitates a shift from customized treatments to personalized treatments, allowing healthcare providers to meet the particular needs of patients. As we transition to healthcare 5.0, interactions between doctors, hospitals, and other stakeholders will improve, enhancing the patient experience.

In Healthcare 4.0, analytics primarily relied on centralized patient records stored in healthcare clouds. However, with the rise of decentralization in Healthcare 5.0, analytics will move towards local edge models. This shift requires robust protection for patient information privacy, data security and safety, as data will be out in the open to a variety of threats.

Smart healthcare is evolving towards networks involving body area assisted by sensors, where local nodes observe patients' health and collect information. Doctors, patients, and medical professionals need protected and authorized access to healthcare records. When designing new services in hospitals or recommending doctors to patients, protecting sensitive data during transfer is absolutely essential to the decision-making process.

Blockchain technology (BCT) is bringing revolution in the healthcare sector, aligning seamlessly with the principles of I5.0, which stresses upon the harmonious collaboration of highly developed technologies with human-centric values. In the realm of managing the records of medical documents, BCT can create a safe, decentralized system that not only increases data security and privacy but also assigns power to patients to have control over their records. This alignment with I5.0 ensures that technology serves to augment human well-being and autonomy. Encryption and distribution across multiple nodes make medical records resistant to hacking and cyber threats, while real-time access for healthcare providers improves diagnostic speed and accuracy. This reduction in administrative burden and improvement in patient outcomes exemplifies the human related focus of I5.0.

In telemedicine, BCT enables secure and private sharing of medical records, thereby authorizing patients to control the 'right to use' for their data, which is a keystone of I5.0's emphasis on personal empowerment (Radanović and Likić 2018). Real-time access to records improves the efficiency of consultations, builds patient trust, and enhances interoperability between different platforms, all of which are key goals of I5.0. Additionally, BCT helps streamline regulatory conformity, ensuring that telemedicine follows privacy and security regulations, thereby promoting a more honest and transparent healthcare ecosystem.

For healthcare money transactions, BCT can create a decentralized transaction system that reduces payment costs and enables faster, more transparent processing (Ray et al. 2021). This system improves accountability and enhances security against fraud and cyberattacks, going along with I5.0's focus on efficiency and reliability. Patients gain more control over their healthcare payments, allowing for direct transactions with providers, which is a clear example of how technology can empower individuals in line with I5.0 principles.

In drug supply chain management (SCM), BCT can trace the status of pharmaceuticals from manufacturers to patients, increasing visibility and decreasing the threat of counterfeit or tainted drugs. By enabling every transaction record on a invulnerable ledger, BCT improves traceability and enhances safety, which are critical for ensuring the security and trustworthiness of the supply

Table 16.6. Blockchain enabled healthcare applications: A domain-wise classification.

	Application Areas and Use Cases of Blockchain Technology in Healthcare Industry							
Key Benefits and Value Propositions of Industry 5.0	**Medical Records Management**	**Telemedicine**	**Healthcare Payments**	**Drug SCM**	**Health Insurance**	**Patient Consent Management**	**Personalized Medicine**	**Public Health Surveillance**
Enhanced data security and privacy	✓	☒	☒	☒	☒	☒	☒	☒
Efficient data sharing	✓	☒	☒	☒	☒	☒	☒	☒
Reduced administrative burden	✓	✓	☒	☒	☒	☒	☒	☒
Improved patient outcomes	✓	☒	☒	☒	☒	☒	☒	☒
Streamlined regulatory compliance	✓	✓	☒	☒	☒	☒	☒	☒
Secure and private data sharing	☒	✓	☒	☒	☒	☒	☒	☒
Increased patient trust	☒	✓	☒	☒	☒	☒	☒	☒
Improved Interoperability	☒	✓	☒	☒	☒	☒	☒	☒
Reduced transaction costs	☒	☒	✓	☒	☒	☒	☒	☒
Faster payment processing	☒	☒	✓	☒	☒	☒	☒	☒
Improved transparency and accountability	☒	☒	✓	☒	☒	☒	☒	☒
Enhanced security	☒	☒	✓	✓		☒	☒	☒
Increased transparency	☒	☒	☒	✓	✓	☒	☒	☒
Prevention of fraud	☒	☒	☒	☒	✓	☒	☒	☒
Management of claims	☒	☒	☒	☒	✓	☒	☒	☒
Smart contracts	☒	☒	☒	☒	✓	☒	☒	☒
Improved data security	☒	☒	☒	☒	✓	☒	☒	☒
Administrative costs reduction	☒	☒	☒	☒	✓	☒	☒	☒
Immutable and secure record-keeping	☒	☒	☒	☒	☒	✓	☒	☒
Permissioned access	☒	☒	☒	☒	☒	✓	☒	☒
Transparency and accountability	☒	☒	☒	☒	☒	✓	☒	☒
Streamlined consent process	☒	☒	☒	☒	☒	✓	☒	☒
Interoperability	☒	☒	☒	☒	☒	✓	☒	☒

Table 16.6 contd. ...

...Table 16.6 contd.

Key Benefits and Value Propositions of Industry 5.0	**Medical Records Management**	**Telemedicine**	**Healthcare Payments**	**Drug SCM**	**Health Insurance**	**Patient Consent Management**	**Personalized Medicine**	**Public Health Surveillance**
Secure and standardized data sharing	☒	☒	☒	☒	☒	☒	✓	☒
Immutable and transparent data storage	☒	☒	☒	☒	☒	☒	✓	☒
Efficient clinical trials	☒	☒	☒	☒	☒	☒	✓	☒
Improved patient outcomes	☒	☒	☒	☒	☒	☒	☒	☒
Patient-controlled data sharing	☒	☒	☒	☒	☒	☒	☒	☒
Secure data sharing	☒	☒	☒	☒	☒	☒	☒	✓
Real-Time Monitoring	☒	☒	☒	☒	☒	☒	☒	✓
Standardized data sharing	☒	☒	☒	☒	☒	☒	☒	✓
Privacy protection	☒	☒	☒	☒	☒	☒	☒	✓
Traceability	☒	☒	☒	☒	☒	☒	☒	✓

chain, key tenets of I5.0 (Gomasta et al. 2023). This streamlined process results in better regulatory obedience and improved patient well-being, representing how technology can be efficiently utilized to protect and benefit humans.

For health insurance, BCT can manage claims and payments more efficiently, reducing costs and complexity while enhancing transparency and security (Karmakar et al. 2023). By creating an unalterable record of the transactions, BCT helps prevent scam and streamlines claims processing, improving customer satisfaction. Smart contracts enabled by BCT automate many practices, decreasing the need for mediators and enhancing data security, which all go along with I5.0's focus on efficiency and automation. This increased transparency allows customers to track their claims and understand associated costs, ultimately reducing administrative costs and enhancing the overall healthcare experience.

In patient consent management, BCT provides a safe and unalterable system for saving records of consents, ensuring that patients' wishes are accurately represented and respected. Permissioned access ensures confidentiality and secrecy, while clearness and responsibility in the consent process streamline administrative burdens. This interoperability allows sharing information related to consent securely across diverse companies, ensuring that patients' desires are valued irrespective of where they undergo treatment, which is a clear reflection of I5.0's human-centric approach.

For personalized medicine, BCT enables safe and harmonized sharing of information among healthcare organizations, improving the accurateness and quality of personalized treatments. The unalterable and transparent storing capabilities of patient health records ensure that it is invulnerable and accessible only by persons authorized. BCT also facilitates well-organized clinical trials by making a decentralized network for patient enrolment and data sharing, enabling more widespread acceptance of personalized medicine and improved patient outcomes. Patients gain power to manage their health records, improving trust and engagement in their healthcare, which is a fundamental aspect of I5.0's focus on personal empowerment and well-being.

BCT can create a revolution for how we track and respond to disease outbreaks by creating a secure, decentralized surveillance system. Healthcare providers, public health departments, and other key players can share standardized data in real-time while protecting individual privacy. This improved data exchange not only helps detect outbreaks earlier but also enables faster responses to public health threats. Since the system preserves privacy while ensuring data accuracy, organizations are more willing to share critical information, ultimately leading to more effective disease monitoring and control. The traceability of public health data enabled by BCT improves outbreak response and prevention, going along with I5.0's goal of using technology to enhance public health and safety. Table 16.6 depicts a comprehensive summary of blockchain applications and their implementations in the healthcare sector. This table illustrates various use cases demonstrating how blockchain technology enhances healthcare delivery, data management, and patient care outcomes.

2.8.3 Supply Chain

Blockchain technology is evolving as a key promoter for I5.0, offering enhanced security, transparency, and relationship in supply chains. It supports small and medium-sized enterprises (SMEs) in transitioning to I5.0 by establishing a safe network for supplier collaboration and data sharing (Khan et al. 2023). Blockchain integration with IoT and AI can enhance supply chain efficiency, traceability, and sustainability across various sectors including agriculture, healthcare, and manufacturing (Ahamed et al. 2022, Verma et al. 2022). The technology enables smart contracts, face recognition for authentication, and driverless vehicles in logistics, enhancing trust between producers and consumers. Blockchain also promotes sustainability by incentivizing green consumer behaviour, increasing product lifecycle visibility, improving system efficiency, and facilitating sustainability monitoring across supply chains (Wang et al. 2023). These advancements contribute to the advancement of more sustainable and human focused production systems in I5.0 (Adel et al. 2024).

The integration of blockchain technology into I5.0 offers significant opportunities to enhance supply chain operations and promote industrial sustainability. Key areas of application include:

2.8.3.1 Industrial Sustainability

Supply chain sustainability management: Blockchain can streamline supply chain processes by improving traceability and ensuring global interoperability (Leng et al. 2020). This enables more efficient resource conservation and supports green procurement practices.

Emissions traceability and trading: Innovative blockchain solutions can facilitate the tracking and trading of emissions, providing transparency in environmental impact assessments (Luzzani et al. 2021). These systems also incentivize sustainable practices and help overcome implementation challenges.

Circular economy initiatives: Incorporating blockchain and IoT devices into circular economy frameworks can optimize resource usage, lower transaction costs, and reduce environmental impact (Varriale et al. 2020, Matarneh et al. 2024).

2.8.3.2 Enhancing Industrial Resilience

I5.0 aims to bolster resilience against unexpected challenges, including disruptions in supply chains and cyber-attacks (Ghobakhloo et al. 2023). Blockchain technology plays a critical role in strengthening resilience in both supply chain management and cybersecurity, with the former having significant societal and economic implications (Dacre et al. n.d.).

2.8.3.3 Enhancing User Experience via Interoperability

Research has also highlighted how blockchain technology fosters greater interoperability among industrial and IoT systems. Enhanced interoperability creates a more cohesive and user-friendly environment where integrated data and actions are seamlessly reflected across various systems.

Tang et al. proposed an innovative approach to achieving this through the “IoT Passport,” inspired by traditional international passports. This system enables devices to establish trusted relationships and define specific collaboration rules (Tang et al. 2019). By adhering to SSI verifiable credential standards, the IoT Passport further amplifies system interoperability, offering a more intuitive and user-centered technological experience.

2.8.4 Blockchain Technology: The Future of Industry 5.0 Manufacturing (Chakir et al. 2024)

Blockchain technology is set to play a crucial task in the evolution of manufacturing within the support of I5.0, which emphasizes focus on humans, sustainability, and resilience. By leveraging blockchain’s decentralized and secure environment, manufacturers can increase operational efficiency, ensure data integrity, and foster collaboration among stakeholders. This section explores the key aspects of blockchain’s integration into I5.0 manufacturing, emphasizing its potential to transform manufacturing processes and outcomes.

2.8.4.1 Human-Centric Applications

At the heart of I5.0 lies (Bucci et al. 2025) the idea that technology should enhance human capabilities rather than take over human jobs. Blockchain helps achieve this vision by making production decisions more transparent and inclusive. Through its ability to create permanent, unalterable records, blockchain enables workers to participate more actively in decision-making processes, fostering better communication between humans and technology. Transparent communication empowers workers by giving them a clear understanding of the decision-making rationale and their role in the production process. Fraga-Lamas et al. highlighted how blockchain can increase worker engagement by establishing a platform for transparent communication and collaborative decision-making (Fraga-Lamas et al. 2024).

Moreover, blockchain can be utilized to support skill development and continuous learning among workers. Training programs can be personalized automatically using smart contracts that are self-running agreements having their rules coded directly into them. These programs can be customized to the explicit requirements and skill levels of individual workers, ensuring that they are equipped with the necessary skills for evolving manufacturing roles. Leng et al. discussed how smart contracts can automate the delivery of training content and track worker progress, providing a structured approach to skill development. This not only enhances worker competencies but also stimulates a culture of constant learning and development within the manufacturing environment (Leng et al. 2022).

2.8.4.2 Sustainability and Resilience

Sustainability and resilience are critical elements of I5.0, and blockchain technology can play a pivotal responsibility in achieving these goals. Blockchain has a key advantage and that is its ability to facilitate efficient tracking of resources throughout the manufacturing process. By making a decentralized ledger that records all resource transactions, blockchain can help manufacturers in reduction of waste and promotion of sustainable practices. Visvizi emphasized how blockchain can be used to trace the lifecycle of raw materials, from extraction to final disposal, making sure that resources are used efficiently and sustainably (Visvizi et al. 2024). This level of traceability is essential for meeting regulatory requirements and consumer demands for sustainable products.

In addition to resource management, blockchain can increase safety of data, which is a crucial concern in I5.0. Centralized data management systems are susceptible to cyberattacks and data breaches, which can cause damage to the integrity and security of manufacturing data. Blockchain provides a decentralized and trusted data saving mechanism that enhances information security by distributing data across a network of nodes. Liu highlighted how blockchain can handle the critical issues developed by centralized data management by establishing a secure and tamper-proof data storage solution (Lin et al. 2024).

2.8.4.3 Decentralized Autonomous Manufacturing

The concept of decentralized autonomous manufacturing is gaining traction in I5.0, where manufacturing systems are established to operate autonomously and adaptively. Blockchain technology can facilitate this by enabling decentralized decision-making and continuous improvement. Smart contracts, for example, can be employed for automated decision-making and ensure that manufacturing systems operate efficiently and resiliently. The ManuChain II system, as discussed by Leng, exemplifies how block chained smart contracts can facilitate decentralized decision-making, enhancing resilience against disruptions (Leng et al. 2023). By automating decision-making processes, smart contracts can help manufacturing systems adjust quickly to varying scenarios and maintain operational continuity.

Furthermore, blockchain can be integrated with decentralized deep learning algorithms to enable adaptive learning and process optimization in manufacturing systems. These algorithms are capable of analyzing data collected from a range of sensors and IoT devices, enabling the detection of patterns and the optimization of manufacturing workflows. By incorporating these algorithms within blockchain frameworks, manufacturers can ensure that the data used for learning and optimization is secure, reliable, and tamper-proof.

2.8.5 Blockchain Technology: Revolutionizing Smart Agriculture in Industry 5.0

Blockchain technology is set to revolutionize smart agriculture, particularly within the context of I5.0, which stresses upon sustainability, human-centric approaches, and advanced integration of technologies. By leveraging blockchain, agricultural practices can achieve enhanced transparency, security, and efficiency, addressing critical challenges faced by the sector. This section outlines the main aspects of blockchain's role in smart agriculture, highlighting its potential to transform the industry.

2.8.5.1 Enhanced Data Security and Integrity

Blockchain technology has one of the most significant advantages in smart agriculture as it has the capability of establishing a decentralized ledger that ensures data integrity and safety. In agricultural applications, data is often vulnerable to tampering, which can lead to inaccurate records and compromised decision-making. Blockchain's unalterable environment ensures that data cannot be manipulated after it is recorded, making a trustworthy and secure data management system. Kataev et al. (2024) emphasized the role of blockchain in improving data security within agricultural environments, ensuring the integrity and reliability of essential information (Kataev et al. 2024).

The integration of the Internet of Things (IoT) with blockchain further enhances data security and integrity. IoT devices, such as sensors and drones, can collect actual live data on environmental settings, soil health, and crop status. By recording this data on the blockchain, farmers and stakeholders can make sure that the information is secure and tamper-proof. Tasic explored how integrating IoT with blockchain supports secure data collection, enabling real-time monitoring of agricultural conditions and fostering more informed decision-making (Tasic and Cano 2024). This combination not only strengthens data security but also improves the efficiency and effectiveness of agricultural operations.

2.8.5.2 Improved Supply Chain Transparency

Blockchain technology can significantly increase transparency across the supply chain in the agriculture sector, allowing stakeholders to trace goods right from the place of origin to the end consumer. This traceability is crucial for reducing fraud and increasing consumer trust. By recording each phase of the supply chain on the blockchain, farmers, processors, distributors, and consumers can access a clearly visible and unalterable record of journey of the goods. Maurya emphasized how blockchain can enhance supply chain transparency, ensuring that all key players have authorization to access correct and reliable data about the origin and handling of agricultural products (Maurya et al. 2025).

In countries like India, blockchain has been shown to mitigate issues related to middlemen and excessive commissions, fostering a more transparent market environment. Patti discussed how blockchain can help eliminate intermediaries, reducing costs and making sure that the farmers receive a reasonable price for their products (Patti et al. 2024). Blockchain creates a transparent and secure platform for transactions, fostering trust and fairness within the agricultural supply chain, ultimately benefiting both farmers and consumers.

2.8.5.3 Automation and Efficiency

Blockchain technology, in conjunction with other advanced technologies, can automate various agricultural processes, leading to increased operational efficiency. For example, the use of blockchain with drone application can automate crop monitoring and data collection. Drones equipped with sensors can capture real live data on crop health, soil moisture, and pest infestations, which can then be recorded on the blockchain. This automated data collation allows farmers to make timely and well-informed decisions, improving crop yields and reducing waste. Soliman highlighted how the integration of blockchain and drone technology can increase operational efficiency in agriculture, enabling more sustainable and productive farming practices (Soliman et al. 2024).

Additionally, smart contracts can be utilized to make automated transactions and agreements among farmers and suppliers, streamlining operations and reducing administrative burdens. Smart contracts, as self-running agreements with the terms and conditions written into code, can make processes automated such as payment processing, inventory management, and contract enforcement, reducing the need for manual interventions and intermediaries. Soliman discussed how smart contracts can enhance efficiency in agricultural supply chains, ensuring that transactions are executed smoothly and transparently (Soliman et al. 2024).

3. Conclusion

The collaboration of blockchain technology with I5.0 heralds a transformative era, where decentralized systems, enhanced transparency, and robust security mechanisms are poised to revolutionize industrial processes. Throughout this chapter, we have explored the theoretical foundations, algorithmic innovations, and possible outcomes of integrating blockchain into the I5.0 framework.

Blockchain's inherent qualities, such as immutability and decentralization, offer unparalleled advantages in ensuring data integrity and trust within industrial ecosystems. By leveraging blockchain, industries can achieve greater efficiency, reduce operational costs, and foster a more collaborative environment. Theories discussed highlight the capability for blockchain to address long-standing challenges in supply chain management, data security, and automation.

Algorithmic advancements in blockchain technology are pivotal in driving I5.0 forward. From consensus algorithms that ensure secure and efficient transaction validation to smart contracts that automate multifaceted processes, these innovations are instrumental in creating a seamless and interconnected industrial landscape. The integration of AI and IoT with blockchain further amplifies its capabilities, enabling real-time data analysis, predictive maintenance, and autonomous decision-making.

The possible outcomes of this integration are vast and promising. Industries can expect enhanced traceability and accountability, leading to enhanced product quality and customer satisfaction. Moreover, the ecological benefits of blockchain, such as reduced consumption of energy through optimized processes, go along with the sustainability goals of I5.0.

However, the journey towards full-scale adoption is not without challenges. Regulatory hurdles, interoperability issues, and the need for standardized protocols are some of the obstacles that need to be addressed. Continuous research, collaboration among industry leaders, and policy support will be crucial in overcoming these barriers.

In conclusion, blockchain technology stands as a key pillar of innovation for I5.0, offering a myriad of opportunities to redefine industrial operations. As we continue to explore and refine its applications, the future of industry promises to be more efficient, secure, and sustainable. The synergy between blockchain and I5.0 is not just a vision of the future but a tangible reality that is already taking shape, paving the way for a new industrial revolution.

References

Adel, Amr. (2022). Future of Industry 5.0 in society: human-centric solutions, challenges and prospective research areas. Journal of Cloud Computing 11(1): 40. https://doi.org/10.1186/s13677-022-00314-5.

Adel, Amr, Noor H. S. Alani and Tony Jan. (2024). Factories of the Future in Industry 5.0—softwarization, servitization, and industrialization. Internet of Things 28(December): 101431. https://doi.org/10.1016/j.iot.2024.101431.

Adel, Amr and Noor H. s. Alani. (2024). Human-centric collaboration and Industry 5.0 framework in smart cities and communities: fostering sustainable development goals 3, 4, 9, and 11 in Society 5.0. Smart Cities 7(4): 1723–75. https://doi.org/10.3390/smartcities7040068.

Aditya Sai Srinivas, T., David Donald, A., Dwaraka Srihith, I., Anjali, D. and Chandana, A. (2023). Blockchain: The future of smart city development. International Journal of Advanced Research in Science, Communication and Technology, April, 68–75. https://doi.org/10.48175/IJARSCT-9009.

Agarwal, Udit, Vinay Rishiwal, Sudeep Tanwar, Rashmi Chaudhary, Gulshan Sharma, Pitshou N. Bokoro et al. (2022). Blockchain technology for secure supply chain management: a comprehensive review. IEEE Access 10: 85493–517. https://doi.org/10.1109/ACCESS.2022.3194319.

Agrawal, Tarun Kumar, Vijay Kumar, Rudrajeet Pal, Lichuan Wang and Yan Chen. (2021). Blockchain-based framework for supply chain traceability: a case example of textile and clothing industry. Computers & Industrial Engineering 154(April): 107130. https://doi.org/10.1016/j.cie.2021.107130.

Ahamed, N. Nasurudeen and Vignesh, R. (2022). Smart agriculture and food industry with blockchain and artificial intelligence. Journal of Computer Science 18(1): 1–17. https://doi.org/10.3844/jcssp.2022.1.17.

Aheleroff, Shohin, Huiyue Huang, Xun Xu and Ray Y. Zhong. (2022). Toward sustainability and resilience with Industry 4.0 and Industry 5.0. Frontiers in Manufacturing Technology 2(October). https://doi.org/10.3389/fmtec.2022.951643.

Akundi, Aditya, Daniel Euresti, Sergio Luna, Wilma Ankobiah, Amit Lopes and Immanuel Edinbarough. (2022). State of Industry 5.0—analysis and identification of current research trends. Applied System Innovation 5(1): 27. https://doi.org/10.3390/asi5010027.

Auhl, Zachary, Naveen Chilamkurti, Rabei Alhadad and Will Heyne. (2022). A comparative study of consensus mechanisms in blockchain for IoT networks. Electronics 11(17): 2694. https://doi.org/10.3390/electronics11172694.

Bucci, Ilaria, Virginia Fani and Romeo Bandinelli. (2025). Towards human-centric manufacturing: exploring the role of human digital twins in Industry 5.0. Sustainability 17(1): 129. https://doi.org/10.3390/su17010129.

Bürer, Mary Jean, Matthieu de Lapparent, Vincenzo Pallotta, Massimiliano Capezzali and Mauro Carpita. (2019). Use cases for blockchain in the energy industry opportunities of emerging business models and related risks. Computers & Industrial Engineering 137(November): 106002. https://doi.org/10.1016/j.cie.2019.106002.

Carayannis, Elias G., Rossella Canestrino and Pierpaolo Magliocca. (2024). From the dark side of Industry 4.0 to Society 5.0: looking 'beyond the box' to developing human-centric innovation ecosystems. IEEE Transactions on Engineering Management 71: 6695–6711. https://doi.org/10.1109/TEM.2023.3239552.

Chakir, Aziza, Rohit Bansal and Mohamed Azzouazi (eds.). (2024). Industry 5.0 and Emerging Technologies: Transformation Through Technology and Innovations. Vol. 565. Studies in Systems, Decision and Control. Cham: Springer Nature Switzerland. https://doi.org/10.1007/978-3-031-70996-8.

Chawla, Sonia and Kriti Mehta. (2023). Cryptocurrency adoption in the era of Industry 5.0: opportunities, challenges, and the intellectual landscape – SLR and bibliometric analysis. pp. 240–62. In Advances in Business Strategy and Competitive Advantage, edited by Sumesh Dadwal, Pawan Kumar, Rajesh Verma, and Gursimranjit Singh. IGI Global. https://doi.org/10.4018/978-1-6684-6403-8.ch013.

Consensys. (2024). Blockchain in the Energy Sector | Real World Blockchain Use Cases. Consensys. 2024. https://consensys.io/blockchain-use-cases/energy-and-sustainability.

Conure Technology Services. (2024). Cryptocurrency Revolution in The Age of Industry 5.0. Conure (blog). May 20, 2024. https://www.conurets.com/cryptocurrency-revolution-in-the-age-of-industry-5-0/.

Dacre, Nicholas, Jingyang Yan, Regina Frei, Al-Mhdawi, M. K. S. and Hao Dong. (n.d.) Advancing sustainable manufacturing: a systematic exploration of Industry 5.0 supply chains for sustainability, human-centricity, and resilience. Production Planning & Control 1–30. Accessed December 25, 2024. https://doi.org/10.1080/09537287.2024.2380361.

Drakatos, Panagiotis, Eleni Koutrouli and Aphrodite Tsalgatidou. (2022). Adrestus: Secure, scalable blockchain technology in a decentralized ledger via zones. Blockchain: Research and Applications 3(4): 100093. https://doi.org/10.1016/j.bcra.2022.100093.

Fraga-Lamas, Paula, Tiago M. Fernández-Caramés, António M. Rosado da Cruz and Sergio Ivan Lopes. (2024). An overview of blockchain for Industry 5.0: Towards human-centric, sustainable and resilient applications. IEEE Access 12: 116162–201. https://doi.org/10.1109/ACCESS.2024.3435374.

Ghobakhloo, Morteza, Mohammad Iranmanesh, Behzad Foroughi, Erfan Babaee Tirkolaee, Shahla Asadi and Azlan Amran. (2023). Industry 5.0 implications for inclusive sustainable manufacturing: an evidence-knowledge-based strategic roadmap. Journal of Cleaner Production 417(September): 138023. https://doi.org/10.1016/j.jclepro.2023.138023.

Ghobakhloo, Morteza, Hannan Amoozad Mahdiraji, Mohammad Iranmanesh and Vahid Jafari-Sadeghi. (2024). From Industry 4.0 digital manufacturing to Industry 5.0 digital society: a roadmap toward human-centric, sustainable, and resilient production. Information Systems Frontiers, February. https://doi.org/10.1007/s10796-024-10476-z.

Gomasta, Sarmistha Sarna, Aditi Dhali, Tahlil Tahlil, Md. Musfique Anwar and Shawkat Ali, A. B. M. (2023). PharmaChain: blockchain-based drug supply chain provenance verification system. Heliyon 9(7): e17957. https://doi.org/10.1016/j.heliyon.2023.e17957.

Groumpos, Peter P. (2021). A critical historical and scientific overview of all industrial revolutions. IFAC-PapersOnLine, 20th IFAC Conference on Technology, Culture, and International Stability TECIS 2021 54(13): 464–71. https://doi.org/10.1016/j.ifacol.2021.10.492.

Gupta, Ankur, Rajesh Gupta, Dhairya Jadav, Sudeep Tanwar, Neeraj Kumar and Mohammad Shabaz. (2023). Proxy smart contracts for zero trust architecture implementation in decentralised oracle networks based applications. Computer Communications 206(June): 10–21. https://doi.org/10.1016/j.comcom.2023.04.022.

Guruswamy, Selvakumar, Milica Pojić, Jayashree Subramanian, Jasna Mastilović, Sohail Sarang, Arumugam Subbanagounder et al. (2022). Toward better food security using concepts from Industry 5.0. Sensors (Basel, Switzerland) 22(21): 8377. https://doi.org/10.3390/s22218377.

Hafid, Abdelatif, Abdelhakim Senhaji Hafid and Dimitrios Makrakis. (2023). Sharding-based proof-of-stake blockchain protocols: key components & probabilistic security analysis. Sensors 23(5): 2819. https://doi.org/10.3390/s23052819.

Haloui, Doha, Kenza Oufaska, Mustapha Oudani and Khalid El Yassini. (2024). Bridging Industry 5.0 and Agriculture 5.0: historical perspectives, opportunities, and future perspectives. Sustainability 16(9): 3507. https://doi.org/10.3390/su16093507.

Hameem, A. S., Vinod Kumar, S., Aravinda, K., Navdeep Singh and Abhishek Sharma. (2024). An analysis on the adoption of blockchain technology in Industry 5.0 sustainable supply. pp. 137–53. In Powering Industry 5.0 and Sustainable Development Through Innovation. IGI Global Scientific Publishing. https://doi.org/10.4018/979-8-3693-3550-5.ch010.

Hashemi, Seyed Mohammad, Ruxandra Mihaela Botez and Georges Ghazi. (2023). Blockchain PoS and PoW consensus algorithms for airspace management application to the UAS-S4 Ehécatl. Algorithms 16(10): 472. https://doi.org/10.3390/a16100472.

IBM. (2020). The Retail Industry on Blockchain. December 10, 2020. https://www.ibm.com/blockchain/resources/transparent-supply/retail/.

Infopulse. (2024). Blockchain Use Cases in the Telecom Industry |. 2024. https://www.infopulse.com/blog/blockchain-use-cases-in-the-telecom-industry.

Jardine, James. (n.d.) Top 3 Things You Need to Know About Industry 5.0. MasterControl. Accessed December 28, 2024. https://www.mastercontrol.com/gxp-lifeline/3-things-you-need-to-know-about-industry-5.0.

Jha, Babita, Sarthak Garg and Sarthak Dhingra. (2024). The Industrial Revolution: From Mechanisation (1.0) to Smart Automation (5.0). In Artificial Intelligence and Communication Techniques in Industry 5.0. CRC Press.

Julian Müller. (2020). Enabling Technologies for Industry 5.0: Results of a Workshop with Europe's Technology Leaders. LU: Publications Office. https://data.europa.eu/doi/10.2777/082634.

Karmakar, Amiya, Pritam Ghosh, Partha Sarathi Banerjee and Debashis De. (2023). ChainSure: Agent free insurance system using blockchain for Healthcare 4.0. Intelligent Systems with Applications 17(February): 200177. https://doi.org/10.1016/j.iswa.2023.200177.

Kataev, Mikhail, Larisa Bulysheva and Alexander Krupskiy. (2024). Blockchain-driven IoT solutions in agriculture. Systems Research and Behavioral Science n/a (n/a). https://doi.org/10.1002/sres.3048.

Kaur, Satveer, Neeru Jaswal and Harvinder Singh. (2022). Blockchain technology: principles, applications, and advantages of blockchain technology in the digital era. pp. 204–12. In Applications, Challenges, and Opportunities of Blockchain Technology in Banking and Insurance. IGI Global Scientific Publishing. https://doi.org/10.4018/978-1-6684-4133-6.ch012.

Khan, Prince Waqas, Imene Bareche and Thorsten Wuest. (2023). Towards Industry 5.0: Empowering SMEs with blockchain-based supplier collaboration network. pp. 730–44. In Advances in Production Management Systems. Production Management Systems for Responsible Manufacturing, Service, and Logistics Futures, edited by Erlend Alfnes, Anita Romsdal, Jan Ola Strandhagen, Gregor von Cieminski, and David Romero. Cham: Springer Nature Switzerland. https://doi.org/10.1007/978-3-031-43662-8_52.

khan, Sunawar, Tehseen Mazhar, Tariq Shahzad, Muhammad Amir khan, Ateeq Ur Rehman and Habib Hamam. (2025). Integration of smart grid with Industry 5.0: Applications, challenges and solutions. Measurement: Energy 5(March): 100031. https://doi.org/10.1016/j.meaene.2024.100031.

Khanfar, Ahmad A. A., Mohammad Iranmanesh, Morteza Ghobakhloo, Madugoda Gunaratnege Senali and Masood Fathi. (2021). Applications of blockchain technology in sustainable manufacturing and supply chain management: a systematic review. Sustainability 13(14): 7870. https://doi.org/10.3390/su13147870.

Khawaja, Sarmad and Vahid Javidroozi. (2023). Blockchain technology as an enabler for cross-sectoral systems integration for developing smart sustainable cities. IET Smart Cities 5(3): 151–72. https://doi.org/10.1049/smc2.12059.

Khodadadi, F., Dastjerdi, A. V. and Buyya, R. (2016). Chapter 1 - Internet of Things: An overview. pp. 3–27. In Internet of Things, edited by Rajkumar Buyya and Amir Vahid Dastjerdi. Morgan Kaufmann. https://doi.org/10.1016/B978-0-12-805395-9.00001-0.

Kim, Gwangyong, Jinsung Cho, Min Choi and Bongjae Kim. (2024). Enhanced practical byzantine fault tolerance via dynamic hierarchy management and location-based clustering. Sensors 24(1): 60. https://doi.org/10.3390/s24010060.

Kovari, Attila. (2024). Industry 5.0: Generalized definition, key applications, opportunities and threats. Acta Polytechnica Hungarica 21(3): 267–84. https://doi.org/10.12700/APH.21.3.2024.3.17.

Kraaijenbrink, Jeroen. (2024). What Is Industry 5.0 And How It Will Radically Change Your Business Strategy? 2024. https://www.forbes.com/sites/jeroenkraaijenbrink/2022/05/24/what-is-industry-50-and-how-it-will-radically-change-your-business-strategy/.

Laddha, Seema and Anguja Agrawal. (2024). Unveiling barriers to Industry 5.0 adoption in supply chains: A DEMATEL approach. RAUSP Management Journal 59(2): 123–37. https://doi.org/10.1108/RAUSP-08-2023-0146.

Leng, Jiewu, Guolei Ruan, Pingyu Jiang, Kailin Xu, Qiang Liu, Xueliang Zhou et al. (2020). Blockchain-empowered sustainable manufacturing and product lifecycle management in Industry 4.0: A survey. Renewable and Sustainable Energy Reviews 132(October): 110112. https://doi.org/10.1016/j.rser.2020.110112.

Leng, Jiewu, Weinan Sha, Baicun Wang, Pai Zheng, Cunbo Zhuang, Qiang Liu et al. (2022). Industry 5.0: Prospect and retrospect. Journal of Manufacturing Systems 65(October): 279–95. https://doi.org/10.1016/j.jmsy.2022.09.017.

Leng, Jiewu, Xiaofeng Zhu, Zhiqiang Huang, Kailin Xu, Zhihong Liu, Qiang Liu et al. (2023). ManuChain II: Blockchained smart contract system as the digital twin of decentralized autonomous manufacturing toward resilience in Industry 5.0. IEEE Transactions on Systems, Man, and Cybernetics: Systems 53(8): 4715–28. https://doi.org/10.1109/TSMC.2023.3257172.

Lin, Iuon-Chang, Yi-Hsuan Kuo, Ching-Chun Chang, Jui-Chuan Liu and Chin-Chen Chang. (2024). Symmetry in blockchain-powered secure decentralized data storage: mitigating risks and ensuring confidentiality. Symmetry 16(2): 147. https://doi.org/10.3390/sym16020147.

Longo, Francesco, Antonio Padovano and Steven Umbrello. (2020). Value-oriented and ethical technology engineering in Industry 5.0: A human-centric perspective for the design of the factory of the future. Applied Sciences 10(12): 4182. https://doi.org/10.3390/app10124182.

Luzzani, Gloria, Erica Grandis, Marco Frey and Ettore Capri. (2021). Blockchain technology in wine chain for collecting and addressing sustainable performance: an exploratory study. Sustainability 13(22): 12898. https://doi.org/10.3390/su132212898.

Matarneh, Sandra, Arsalan Zahid Piprani, Riwan Matloob Ellahi, Diep Ngoc Nguyen, Trang Mai Le and Sajid Nazir. (2024). Industry 4.0 technologies and circular economy synergies: enhancing corporate sustainability through sustainable supply chain integration and flexibility. Environmental Technology & Innovation 35(August): 103723. https://doi.org/10.1016/j.eti.2024.103723.

Mathur, Aeshita, Ameesha Dabas and Nikhil Sharma. (2022). Evolution From Industry 1.0 to Industry 5.0. pp. 1390–94. In 2022 4th International Conference on Advances in Computing, Communication Control and Networking (ICAC3N). https://doi.org/10.1109/ICAC3N56670.2022.10074274.

Maurya, Vinay, Vinay Rishiwal, Mano Yadav, Mohammad Shiblee, Preeti Yadav, Udit Agarwal et al. (2025). Blockchain-driven security for IoT networks: state-of-the-art, challenges and future directions. Peer-to-Peer Networking and Applications 18(1): 1–35. https://doi.org/10.1007/s12083-024-01812-w.

Momenta. (2024). Industry 5.0 | Industry 5.0 Strategy | Industry 5.0 Technologies. 2024. https://www.momenta.one/industry5.0.

Mourtzis, Dimitris, John Angelopoulos and Nikos Panopoulos. (2022). A literature review of the challenges and opportunities of the transition from Industry 4.0 to Society 5.0. Energies 15(17): 6276. https://doi.org/10.3390/en15176276.

Mourtzis, Dimitris, John Angelopoulos and Nikos Panopoulos. (2023). Blockchain integration in the era of Industrial Metaverse. Applied Sciences 13(3): 1353. https://doi.org/10.3390/app13031353.

Nahavandi, Saeid. (2019). Industry 5.0—A human-centric solution. Sustainability 11(16): 4371. https://doi.org/10.3390/su11164371.

Oliveira, Miguel, Sumit Chauhan, Filipe Pereira, Carlos Felgueiras and David Carvalho. (2023). Blockchain protocols and edge computing targeting Industry 5.0 needs. Sensors 23 (22): 9174. https://doi.org/10.3390/s23229174.

Patti, Venkata Swathi, Akshith Reddy Musku, Nishanth Sree Vamsi Bahadursha, Deepak Mergu, Krithi Orhi and Deepak Sukheja. (2024). Design and implement decentralized application for agriculture in india using blockchain technology. pp. 1–6. In 2024 International Conference on Advances in Computing Research on Science Engineering and Technology (ACROSET). https://doi.org/10.1109/ACROSET62108.2024.10743329.

Rada, Michael. (2020). Industry 5.0 Definition. Medium. May 26, 2020. https://michael-rada.medium.com/industry-5-0-definition-6a2f9922dc48.

Radanović, Igor and Robert Likić. (2018). Opportunities for use of blockchain technology in medicine. Applied Health Economics and Health Policy 16(5): 583–90. https://doi.org/10.1007/s40258-018-0412-8.

Rai, Hari Mohan, Kaustubh Kumar Shukla, Lilia Tightiz and Sanjeevikumar Padmanaban. (2024). Enhancing data security and privacy in energy applications: integrating IoT and blockchain technologies. Heliyon 10(19): e38917. https://doi.org/10.1016/j.heliyon.2024.e38917.

Rame, Rame, Purwanto Purwanto and Sudarno Sudarno. (2024). Industry 5.0 and sustainability: An overview of emerging trends and challenges for a green future. Innovation and Green Development 3(4): 100173. https://doi.org/10.1016/j.igd.2024.100173.

Ray, Partha Pratim, Neeraj Kumar and Dinesh Dash. (2021). BLWN: Blockchain-based lightweight simplified payment verification in IoT-assisted e-Healthcare. IEEE Systems Journal 15(1): 134–45. https://doi.org/10.1109/JSYST.2020.2968614.

Roumeliotis, Christos, Minas Dasygenis, Vasilis Lazaridis and Michael Dossis. (2024). Blockchain and digital twins in smart Industry 4.0: The use case of supply chain-a review of integration techniques and applications. Designs 8(6): 105. https://doi.org/10.3390/designs8060105.

RS, Chellvamathi and Prasanna Kulkarni. (2023). Blockchain in telecom: challenges and the way forward. pp. 661–67. In 2023 International Conference on Innovative Data Communication Technologies and Application (ICIDCA). https://doi.org/10.1109/ICIDCA56705.2023.10099882.

Ruggieri, Marina. (n.d.) The Innovation behind Industry 5.0. Accessed December 28, 2024. https://www.themanufacturer.com/articles/the-innovation-behind-the-industrial-revolution/.

Sharma, Aishita, Sunil K. Singh, Sudhakar Kumar, Ruchika Thakur, Brij B. Gupta and Varsha Arya. (2024). IoT-enabled smart farming with Industry 5.0. Journal of High Speed Networks 30(3): 477–96. https://doi.org/10.3233/JHS-230258.

Sharma, Pradip Kumar and Jong Hyuk Park. (2018). Blockchain based hybrid network architecture for the smart city. Future Generation Computer Systems 86(September): 650–55. https://doi.org/10.1016/j.future.2018.04.060.

Soliman, Shebl, Ahmed Bendary and Hisham Dahshan. (2024). Enhancing agricultural efficiency with blockchain-orchestrated drone swarms and kubernetes. pp. 567–70. In 2024 6th Novel Intelligent and Leading Emerging Sciences Conference (NILES). https://doi.org/10.1109/NILES63360.2024.10753154.

STL Partners. (2024). 5 Blockchain Healthcare Use Cases. STL Partners (blog). 2024. https://stlpartners.com/articles/digital-health/5-blockchain-healthcare-use-cases/.

Tang, Bo, Hongjuan Kang, Jingwen Fan, Qi Li and Ravi Sandhu. (2019). IoT passport: A blockchain-based trust framework for collaborative Internet-of-Things. pp. 83–92. In Proceedings of the 24th ACM Symposium on Access Control Models and Technologies. Toronto ON Canada: ACM. https://doi.org/10.1145/3322431.3326327.

Tasic, Igor and Maria-Dolores Cano. (2024). An orchestrated IoT-based blockchain system to foster innovation in Agritech. IET Collaborative Intelligent Manufacturing 6(2): e12109. https://doi.org/10.1049/cim2.12109.

TWI. (n.d.) What Is Industry 5.0? (Top 5 Things You Need To Know). Accessed December 28, 2024. https://www.twi-global.com/technical-knowledge/faqs/industry-5-0.aspx.

Varriale, Vincenzo, Antonello Cammarano, Francesca Michelino and Mauro Caputo. (2020). The unknown potential of blockchain for sustainable supply chains. Sustainability 12(22): 9400. https://doi.org/10.3390/su12229400.

Verma, Ashwin, Pronaya Bhattacharya, Nirav Madhani, Chandan Trivedi, Bharat Bhushan, Sudeep Tanwar et al. (2022). Blockchain for Industry 5.0: Vision, opportunities, key enablers, and future directions. IEEE Access 10: 69160–99. https://doi.org/10.1109/ACCESS.2022.3186892.

Verma, Diksha, Pooja Kansra and Pawan Kumar. (2023). Significance of block chain technology and Industry 5.0 in indian banking sector. pp. 263–69. In Opportunities and Challenges of Business 5.0 in Emerging Markets. IGI Global Scientific Publishing. https://doi.org/10.4018/978-1-6684-6403-8.ch014.

Verma, Rajat and Namrata Dhanda. (2023). Chapter 5 - Blockchain types: A characteristic view. pp. 69–85. In Distributed Computing to Blockchain, edited by Rajiv Pandey, Sam Goundar, and Shahnaz Fatima. Academic Press. https://doi.org/10.1016/B978-0-323-96146-2.00013-9.

Visvizi, Anna, Radosław Malik, Gianluca Maria Guazzo and Vilma Çekani. (2024). The Industry 5.0 (I50) paradigm, blockchain-based applications and the smart city. European Journal of Innovation Management ahead-of-print (ahead-of-print). https://doi.org/10.1108/EJIM-09-2023-0826.

Wang, Zhu-Jun, Zhen-Song Chen, Lu Xiao, Qin Su, Kannan Govindan and Miroslaw J. Skibniewski. (2023). Blockchain adoption in sustainable supply chains for Industry 5.0: A multistakeholder perspective. Journal of Innovation & Knowledge 8(4). https://doi.org/10.1016/j.jik.2023.100425.

Xu, Min, Xingtong Chen and Gang Kou. (2019). A systematic review of blockchain. Financial Innovation 5(1): 27. https://doi.org/10.1186/s40854-019-0147-z.

Xu, Xun, Yuqian Lu, Birgit Vogel-Heuser and Lihui Wang. (2021). Industry 4.0 and Industry 5.0—inception, conception and perception. Journal of Manufacturing Systems 61(October): 530–35. https://doi.org/10.1016/j.jmsy.2021.10.006.

Zhou, Sisi, Kuanching Li, Lijun Xiao, Jiahong Cai, Wei Liang and Arcangelo Castiglione. (2023). A systematic review of consensus mechanisms in blockchain. Mathematics 11(10): 2248. https://doi.org/10.3390/math11102248.

Chapter 17

Blockchain for Next Generation Computing

Challenges, Opportunities and Emerging Trends for Modern Generation

Ravinesh Bhadoriya,[1,*] *Narendra Kumar Garg*[2] and *Ajay Kumar Dadoria*[3]

1. Introduction to Next Generation Blockchain and its Role in Society

Background of blockchain: Blockchain technology was introduced in 2008 by an anonymous entity served as the foundational architecture developed by Satoshi Nakamoto, forming the underlying framework behind Bitcoin, the first decentralized cryptocurrency (Tripathi et al. 2023). The core innovation of blockchain lies in its decentralized, distributed ledger, where every transaction is recorded in a block and linked to previous blocks in a chain. This method ensures transparency, security, and immutability of transaction records without requiring a central authority. The decentralized nature of blockchain eliminates the need for intermediaries, making it particularly appealing for peer-to-peer financial transactions. The blockchain is a shared public ledger on which the entire Bitcoin network relies. All confirmed transactions are included in to blockchain.

Eventually, blockchain's capabilities enlarged beyond cryptocurrencies like Bitcoin and Ethereum."These applications have found use in industries ranging from finance and insurance to healthcare and real estate." For instance, supply chain management has benefitted from blockchain by ensuring transparency and traceability at every stage, improving the efficiency of global supply chains and reducing fraud. In digital governance, blockchain is used for secure voting systems and digital IDs, enhancing the integrity of electoral processes and personal data management. However, despite its success, blockchain faces significant challenges that hinder its full potential, particularly as its adoption grows.

[1] Research Scholar (Ph.D.), (ECE Department), Amity University, Gwalior, Madhya Pradesh.
[2] Associate Professor, (ECE Department), Amity University, Gwalior, Madhya Pradesh.
[3] Assistant Professor, (ECE Department), Amity University, Gwalior, Madhya Pradesh.
Emails: nkgarg41@gwa.amity.edu, ajaymanit0@gmail.com
* Corresponding author: singhsinghravi12345@gmail.com

DOI: 10.1201/9781003620822-17

Key Challenges

Scalability issues: As blockchain systems gain popularity, they often struggle with scalability. Traditional blockchain networks like Bitcoin and Ethereum have a limited transaction processing capacity. Bitcoin handles approximately seven transactions per second (TPS), while Ethereum manages about thirty TPS. This limitation causes network congestion, delays in transaction processing, and has high transaction fees during times of increased demand. As blockchain adoption expands to sectors like decentralized finance (DeFi), supply chain management, and beyond, scalability becomes a critical issue.

Transaction speed and throughput: The speed of blockchain transactions is another obstacle. In conventional centralized systems, transactions are validated by trusted intermediaries, resulting in rapid processing times. Blockchain, on the other hand, uses consensus protocols such as Proof-of-Work (PoW) or Proof-of-Stake (PoS), which need nodes to reach an agreement on transaction validity. While these mechanisms ensure decentralization and security, they can be slow and inefficient. The delay in transaction finalization, particularly in PoW systems, limits blockchain's usability for real-time applications like instant payments and large-scale enterprise solutions.

Energy consumption: Another challenge is the high energy consumption required by blockchain networks, particularly in Proof-of-Work systems. Bitcoin mining, for example, demands substantial computational power, consuming energy equivalent to that of small nations. This has raised concerns over blockchain's environmental impact. As the demand for blockchain systems increases, so does the strain on energy resources, making it an unsustainable option for long-term growth unless these energy inefficiencies are addressed.

Security concerns: Although blockchain is considered protected due to its cryptographic features, it is not entirely immune to attacks. Additionally, blockchain systems that rely on smart contracts are vulnerable to bugs or security flaws, as demonstrated in the DAO hack on Ethereum in 2016. As blockchain networks evolve and handle more critical applications, the need for enhanced security measures becomes paramount.

Decentralization and consensus: Blockchain's appeal lies in decentralization, but achieving true decentralization without compromising performance remains a difficult balance. Larger blockchains often struggle to keep their network decentralized while managing the enormous computational resources required for transaction validation. As the blockchain ecosystem grows, maintaining decentralization without creating new inefficiencies remains a challenge.

The Promise of CNTFET and SRAM

The emergence of Carbon Nanotube Field-Effect Transistors (CNTFETs) and Static Random-Access Memory (SRAM) offers a promising solution to these challenges. CNTFETs are based on carbon nanotubes, which have remarkable electrical properties that significantly outperform traditional silicon transistors. These properties enable faster switching speeds, lower energy consumption, and higher transistor densities, which can address blockchain's scalability, transaction speed, and energy efficiency issues. CNTFETs promise to revolutionize blockchain's underlying hardware by enabling faster consensus mechanisms, which are crucial for improving transaction throughput.

SRAM, which is already known for its low-latency and high-speed data access, can further enhance blockchain performance when combined with CNTFET technology. The combination of CNTFET's fast switching and SRAM's rapid data retrieval will dramatically reduce the time required to process transactions and validate blocks. This will improve overall blockchain performance, making it more suitable for high-volume applications like DeFi or supply chain management.

In terms of energy efficiency, CNTFETs consume less power than traditional transistors, which is a crucial factor in reducing the environmental impact of blockchain networks. By integrating

CNTFET-based SRAM into blockchain hardware, nodes can process data more efficiently while consuming significantly less energy, which makes the blockchain systems more sustainable.

Overall, the integration of CNTFETs and SRAM into blockchain infrastructure could enable faster, more scalable, and energy-efficient blockchain systems capable of addressing the growing demands of the next-generation society. This could transform blockchain from a niche technology into a robust infrastructure for industries ranging from finance to governance, helping to establish a more secure, transparent, and sustainable digital future.

2. Understanding the Core Technologies: Blockchain, CNTFET, and SRAM

2.1 Blockchain Technologies

Fundamentals of blockchain: The core principles of blockchain technology revolve around decentralization, immutability, security, and trustless consensus mechanisms, which are essential for ensuring that data stored on the blockchain remains tamper-resistant and verifiable without the need for a central authority.

Decentralization: This is one of the key attributes of blockchain. In traditional centralized systems, data is controlled and arranged by a central entity, such as a bank or government. In blockchain, however, the data is distributed across a network of nodes (individual computers) that all have a copy of the same ledger. This distributed nature makes blockchain resistant to single points of failure, making it more robust and secure compared to traditional centralized systems.

Security: In blockchain, security is achieved through cryptographic techniques. Each transaction on the blockchain is secured with encryption and connected to the preceding one, forming a chain of blocks. This chaining process, combined with cryptographic hash functions, makes it virtually impossible for malicious actors to alter or tamper with data after it has been recorded.

Trustless consensus mechanisms: Lastly, blockchain relies on trustless consensus mechanisms to validate transactions and maintain the integrity of the network. Consensus algorithms protocols are created to ensure that all participants in a network reach a common agreement on the current status of the blockchain. Two widely used methods for achieving this are Proof-of-Work (PoW) and Proof-of-Stake (PoS). These algorithms play a crucial role in maintaining the security and integrity of decentralized systems.

Proof-of-Work (PoW): Proof of Work is the consensus mechanism used by Bitcoin, where miners compete to solve intricate mathematical problems in order to validate transactions and generate new blocks. This process demands substantial computational power, resulting in high energy usage. However, it ensures that transactions are validated through a competitive, decentralized process.

Proof-of-Stake (PoS): This is an alternative mechanism used by networks like Ethereum (after its transition to Ethereum 2.0). In PoS, validators are selected to propose new blocks that are created based on the amount of cryptocurrency participants hold and are willing to "stake" as collateral. PoS is more energy-efficient than PoW since it doesn't involve heavy computational tasks. Despite its many benefits, blockchain technology faces significant limitations that hinder its mass adoption.

Limitations

While blockchain has immense potential, several technical challenges must be addressed for it to reach its full potential:

High transaction costs: The transaction fees associated with blockchain networks, especially those using PoW like Bitcoin and Ethereum, can be prohibitively high during periods of network congestion. This occurs because miners (in PoW) and validators (in PoS) prioritize transactions with higher fees. This makes blockchain less attractive for microtransactions and small-scale applications.

Low throughput: As mentioned, Bitcoin and Ethereum have lower transaction processing capabilities compared to traditional payment systems like Visa, which can manage thousands of transactions per second. Blockchain networks can become congested when the number of transactions exceeds the network's capacity, leading to slower processing times and delays.

Slow processing speeds: Transaction validation, particularly in PoW-based systems, takes time due to the computational complexity of the consensus process. Even in PoS, where transactions are faster, the time required to reach a consensus can still introduce delays in comparison to traditional centralized systems.

Energy consumption: The ecological effect of blockchain technology, particularly in PoW systems, is becoming an increasing issue. Mining cryptocurrencies like Bitcoin require vast amounts of energy, leading to significant carbon emissions. As blockchain networks scale, this energy demand is expected to increase, making it unsustainable in the long term, if not addressed immediately.

Given these limitations, blockchain technology needs significant innovation to achieve high scalability, speed, and efficiency, while reducing costs and environmental impact. This is where emerging technologies like Carbon Nanotube Field-Effect Transistors (CNTFETs) and Static Random-Access Memory (SRAM) come into play, offering potential solutions to these challenges.

2.2 Carbon Nanotube Field-Effect Transistors (CNTFETs)

Introduction to CNTFETs

Carbon Nanotube Field-Effect Transistors (CNTFETs) are an advanced type of field-effect transistor (FET) that use carbon nanotubes (CNTs) as the channel material instead of traditional silicon. CNTs are tube-shaped structures formed by curling sheets of graphene, which possess exceptional electrical, mechanical, and thermal properties (Palai 2018). CNTFETs leverage these properties to achieve faster switching speeds, lower power consumption, and greater miniaturization compared to conventional silicon-based transistors. In a traditional FET, the channel material, usually silicon, is used to control the flow of electrical current. However, the electron mobility of silicon is limited, which results in slower switching speeds and higher energy consumption as transistors become smaller. CNTs, on the other hand, have a much higher electron mobility than silicon, allowing for faster switching times and more efficient operation. Figure 17.1 structural illustration of Carbon Nanotube cells.

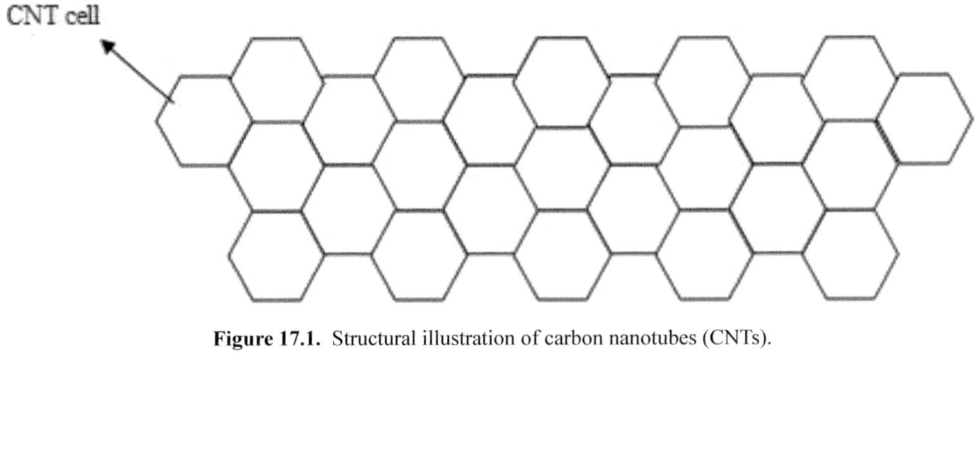

Figure 17.1. Structural illustration of carbon nanotubes (CNTs).

Advantages of CNTFETs

CNTFETs offer several significant advantages over traditional silicon-based transistors:

Faster switching speeds: Due to the high electron mobility of CNTs, CNTFETs can switch on and off faster than silicon transistors. This allows for quicker processing times, which is critical for applications like blockchain, where real-time transaction validation is essential.

Lower power consumption: CNTFETs consume less power compared to silicon transistors, as they have lower capacitance and reduced leakage currents. This is particularly important for blockchain networks, where energy efficiency is a growing concern, especially in mining and validation processes.

Greater miniaturization: The nanoscale nature of CNTs allows for more compact transistor designs. This means that more transistors can fit onto a single chip, increasing processing power without increasing the size of the hardware. This is vital for blockchain systems that require high computational power but need to maintain compact, energy-efficient hardware.

Thermal stability: CNTs are excellent at dissipating heat, making CNTFETs more thermally stable than traditional silicon transistors. This allows CNTFETs to operate efficiently at higher speeds and under more demanding conditions without overheating.

Potential for Blockchain

CNTFETs can significantly enhance blockchain performance in several ways:

Faster transaction processing: The higher processing speeds offered by CNTFETs can enable faster transaction validation, reducing the time it takes to reach consensus. This could dramatically improve transaction throughput in blockchain networks, allowing them to handle a larger number of transactions per second.

Energy efficiency: Blockchain's high energy consumption, particularly in PoW systems, is one of the most significant challenges it faces. By using CNTFETs, blockchain systems can reduce their power consumption, making them more sustainable and scalable.

Improved consensus mechanisms: Faster switching speeds and lower energy consumption will enable real-time consensus mechanisms, making the blockchain more responsive and efficient. This is especially important for decentralized applications that require quick updates and low-latency communication between nodes.

2.3 Static Random-Access Memory (SRAM)

Introduction to SRAM: Introduction to SRAM

Static Random-Access Memory (SRAM) is a type of memory which Stores data as long as power is available, in contrast to Dynamic RAM (DRAM), which needs to be periodically refreshed to retain data, SRAM is faster and more reliable but is also more expensive and less dense. SRAM is used in high-speed applications, such as CPU cache memory, where swift access to frequently used data is critical (www.techtarget.com 2024s).

SRAM consists of flip-flops made up of transistors, which store binary data (0 or 1). Since it does not require refreshing like DRAM, SRAM can provide lower latency and faster access times, making it ideal for high-performance systems that require real-time data processing.

Role of SRAM

In blockchain networks, especially those that involve cryptographic hashing and transaction validation, fast and reliable memory is essential for performance. SRAM is critical for storing frequently accessed data such as the current state of the blockchain or recent transactions. By using

SRAM, blockchain nodes can retrieve and process data quickly, reducing latency and speeding up the transaction validation process.

SRAM's low latency is especially important in consensus mechanisms, where nodes need to access the blockchain state rapidly in order to validate transactions and reach a consensus efficiently. Without fast memory access, blockchain networks can become bottlenecked, slowing down transaction throughput and increasing processing times.

***Integrating SRAM with CNTFET*:** The combination of CNTFET technology with SRAM offers a powerful solution for optimizing both speed and energy efficiency in blockchain systems. By incorporating CNTFETs into the SRAM architecture, it is possible to create a hybrid memory system that takes advantage of the high-speed processing capabilities of CNTFETs while maintaining the low-latency performance of SRAM.

This integration could lead to faster transaction processing, reduced energy consumption, and improved scalability in blockchain networks. The high electrical conductivity of CNTFETs combined with the low-latency access of SRAM can help blockchain systems scale to handle a greater number of transactions per second while keeping power consumption in check.

3. Merging CNTFET-based SRAM with Blockchain: A Paradigm Shift

3.1 Revolutionizing Blockchain Performance

Blockchain technology is inherently decentralized, secure, and immutable, making it a revolutionary technology with applications across industries (Gad et al. 2022). However, despite these advantages, blockchain networks face performance bottlenecks such as high transaction costs, slow processing times, and scalability limitations. These challenges hinder the widespread adoption of blockchain technology, especially in applications requiring high-speed transactions and real-time consensus mechanisms. In this context, CNTFET-based SRAM offers a promising solution to revolutionize blockchain performance by dramatically enhancing transaction speeds, data retrieval times, and overall efficiency.

High-Speed transaction processing: A key aspect of blockchain performance is the ability to process transactions quickly and efficiently. Blockchain networks like Bitcoin and Ethereum have gained global attention for enabling **peer-to-peer transactions** without intermediaries. However, the speed at which these transactions are processed is often slow. In comparison, traditional payment systems like Visa can process **up to 24,000 TPS**. The speed at which blockchain transactions are processed is determined by multiple factors, including the consensus mechanism used. In **Proof-of-Work (PoW)**, miners compete to solve complex cryptographic puzzles, which requires substantial computational power. This leads to slower block generation times and increased transaction confirmation times. **Proof-of-Stake (PoS)** improves upon PoW by selecting validators based on their stake in the network, but it still faces performance issues related to data retrieval and processing. The integration of **CNTFETs** with **SRAM** technology can provide a solution by significantly improving the **switching speeds** and **data access times** of blockchain nodes. **CNTFETs** have much higher **electron mobility** than conventional silicon transistors, meaning they can switch faster, resulting in quicker transaction processing. As a result, blockchain networks could experience **faster block generation** and reduced **confirmation times**.

For instance, in a typical PoW network, miners would benefit from faster processing speeds, enabling quicker validation of transactions and faster block creation. In a PoS network, validators would be able to reach consensus more quickly, reducing delays in transaction confirmation. CNTFETs could also enable real-time consensus mechanisms by improving the speed and efficiency of the processes involved in confirming transactions, whether through PoW, PoS, or other consensus models.

Low-latency data access: Another critical factor affecting blockchain performance is **data access latency**. Blockchain networks rely on distributed ledgers, which means that each node in the network needs to access and validate data stored across multiple locations. This can introduce delays in transaction processing, especially when data retrieval times are slow.

Integrating SRAM with CNTFETs significantly reduces data access latency. SRAM is known for its low-latency and high-speed characteristics, making it ideal for applications that require fast data retrieval, such as caching frequently accessed data. In blockchain, SRAM could be used to store frequently accessed blockchain states or transaction history, allowing for faster access to critical information during consensus.

For example, when validating a transaction, nodes in the network must retrieve the state of the blockchain to check whether the transaction is valid (e.g., verifying whether a wallet has sufficient funds). The combination of SRAM and CNTFETs ensures that this process is executed quickly, reducing bottlenecks and accelerating transaction validation.

Additionally, the low-latency nature of SRAM helps speed up blockchain consensus by allowing nodes to access relevant data almost instantaneously. This reduces the time to consensus, leading to faster transaction processing and improved throughput.

Enhanced user experience: The ultimate goal of blockchain technology is to improve the user experience, particularly in decentralized applications (DApps). In the realm of **Decentralized Finance (DeFi)**, users expect **instant transaction confirmations** and **low transaction fees**. Slow transaction processing can lead to delays and higher costs, which reduces the attractiveness of blockchain-based financial platforms.

The integration of CNTFET-based SRAM in blockchain networks can drastically enhance user experience. With higher throughput and lower confirmation times, blockchain applications can handle a large number of transactions per second (TPS) while ensuring that users experience near-instantaneous transaction confirmation. This would make DeFi platforms, supply chain management, and other DApps far more efficient and user-friendly.

For example, DeFi platforms that rely on blockchain for transactions could use the improved transaction speeds enabled by CNTFETs to offer near-instantaneous settlement of trades and transfers. Similarly, in supply chain tracking applications, the faster transaction processing would allow stakeholders to trace goods in real-time, providing more up-to-date information and improving overall operational efficiency.

In addition, the reduced transaction costs enabled by CNTFET-based hardware would make blockchain applications more accessible to a broader audience, including smaller businesses and users who were previously deterred by high fees. This would drive the adoption of blockchain technologies across diverse industries, enhancing their potential to disrupt traditional business models.

3.2 Improving Blockchain Scalability

The Scalability Problem

One of the most significant challenges faced by blockchain networks is scalability. Blockchain networks are inherently limited by their design, which requires every transaction to be validated by multiple participants (nodes) in the network. This results in a bottleneck when transaction volumes exceed the network's capacity.

Scalability challenges in blockchain are primarily manifested in two ways: The two ways are slow transactions per second (TPS) and high transaction costs. As the number of users on a blockchain network increases, so too does the demand for transaction processing. However, due to the limitations of current consensus mechanisms and the underlying hardware infrastructure, blockchain networks struggle to handle large volumes of transactions without experiencing delays and rising fees.

To address these issues, several solutions have been proposed:

Sharding: This involves breaking up the blockchain into smaller partitions (shards), allowing each shard to process transactions independently, thereby increasing throughput (Faster capital.com).

Off-chain protocols: These aim to move transactions off the blockchain to reduce the load on the network, improving transaction speeds.

Despite these advances, scalability remains a challenge, particularly for applications requiring high transaction volumes, such as Decentralized Finance (DeFi) platforms, gaming applications, and enterprise-level blockchain solutions.

CNTFET-Based SRAM

CNTFET-based SRAM: This technology can significantly enhance the scalability of blockchain systems by improving both transaction throughput and energy efficiency. By leveraging CNTFETs for faster processing and SRAM for low-latency memory access, blockchain nodes can handle a higher number of transactions per second without compromising performance.

For example, DeFi platforms that process hundreds or thousands of transactions daily could benefit from the increased processing power and reduced latencies offered by CNTFETs and SRAM. With these advancements, blockchain networks could scale efficiently to accommodate growing user demand without sacrificing speed or reliability.

In addition to increased transaction throughput, CNTFET-based SRAM could contribute to lower transaction costs. Traditional blockchain networks often struggle with high transaction fees, especially during periods of high demand. By reducing the computational overhead of transaction validation and consensus, CNTFETs can enable blockchain networks to handle more transactions with less power consumption, leading to lower costs for users.

This combination of high throughput and low-cost transactions would make blockchain more viable for real-world use cases, such as e-commerce, IoT applications, and enterprise systems that require scalable and cost-effective solutions.

3.3 Energy Efficiency and Sustainability

Blockchain's energy consumption: One of the most pressing concerns linked to blockchain technology is its energy usage. Proof-of-Work (PoW) consensus mechanisms, especially in cryptocurrencies like Bitcoin, demand miners to carry out resource-heavy tasks to validate transactions and add blocks to the blockchain. This process uses a considerable amount of electricity, raising concerns about the environmental effects of blockchain networks (Alzoubi 2023).

As the popularity of cryptocurrencies and blockchain applications grows, so too does the demand for energy, which in turn increases the carbon footprint of these networks.

Low power consumption with CNTFET: CNTFET technology offers a potential solution to blockchain's energy consumption problem. Because CNTFETs have higher electron mobility and lower power requirements compared to traditional silicon-based transistors, they consume less energy while providing faster processing speeds. When combined with SRAM, which is already known for its low-latency and energy-efficient characteristics, CNTFET-based hardware could dramatically reduce the power consumption of blockchain nodes.

This reduction in power consumption is particularly important for the mining and validation processes. By reducing the energy required for these operations, blockchain networks can lower their carbon footprint, making them more environmentally friendly. In addition, energy-efficient hardware would reduce the overall cost of running blockchain nodes, which could make decentralized networks more accessible to smaller participants.

Sustainable Blockchain Solutions

Incorporating CNTFET-based SRAM into blockchain infrastructure offers a path toward sustainable blockchain solutions. These advancements could enable blockchain networks to achieve the performance, scalability, and low-cost transactions required for mass adoption, while also meeting environmental sustainability goals.

Governments and corporations increasingly face pressure to reduce their carbon footprints and adopt eco-friendly technologies. By transitioning to energy-efficient blockchain systems, powered by CNTFET-based hardware, blockchain technology can align with global sustainability goals. This would not only make blockchain more acceptable to regulatory bodies but also promote its adoption in industries where environmental concerns are paramount.

Moreover, as blockchain technology becomes more energy-efficient, it may attract greater investment and partnerships, especially from organizations looking to leverage blockchain for supply chain tracking, finance, and energy trading, where both performance and sustainability are key drivers.

4. Security and Quantum Resistance

4.1 Security Challenges in Blockchain

Blockchain technology has revolutionized various sectors by providing safe methods for ensuring data integrity, validating transactions, and managing digital assets. However, despite its many advantages, blockchain faces numerous security challenges that can threaten the safety and trustworthiness of decentralized networks. Understanding these security risks is crucial for building more resilient blockchain systems.

The Need for Robust Security

Blockchain's decentralized nature makes it resistant to single-point failures and tampering. However, this very decentralization can also present several vulnerabilities if not properly managed.

Cryptographic vulnerabilities: The integrity and security of blockchain rely heavily on cryptographic algorithms like RSA encryption and Elliptic Curve Digital Signature Algorithm (ECDSA). These encryption methods ensure that transactions are securely validated and that data cannot be tampered with. However, any weaknesses or flaws in these cryptographic systems could leave blockchain open to attacks, such as unauthorized access, data breaches, or malicious modifications.

In addition to these existing threats, blockchain security must also be fortified against future risks, particularly those posed by the advent of quantum computing.

The Quantum Computing Threat

Quantum computing represents a new frontier in computing power that could have profound implications for blockchain security. The most concerning aspect of quantum computing is its potential ability to break existing cryptographic systems used by blockchain networks, such as RSA encryption and ECDSA. These cryptosystems are based on mathematical problems that are currently impossible for classical computers to solve, such as factoring large prime numbers (RSA) or solving elliptic curve discrete logarithms (ECDSA) (Zhu 2019). However, quantum computers have the potential to solve these problems efficiently using Shor's Algorithm, which could break these encryption methods in a fraction of the time it takes classical computers to do so.

If quantum computers are able to crack the cryptographic algorithms that secure blockchain networks, they could compromise the entire system by allowing attackers to forge signatures, impersonate participants, or steal assets. This potential threat has raised alarms in the blockchain community, as many widely used blockchain platforms, such as Bitcoin and Ethereum, rely on cryptographic mechanisms vulnerable to quantum attacks.

Thus, it is essential to develop blockchain systems that are resilient against quantum computing's disruptive capabilities.

4.2 Quantum-Resistant Algorithms and CNTFET-based SRAM

Quantum-Resilient Cryptography

As quantum computers advance, blockchain networks must adopt quantum-resistant or post-quantum cryptography (PQC) to protect data integrity and secure transactions. Quantum-resistant cryptography refers to encryption methods designed to withstand the processing capabilities of quantum computers, ensuring that blockchain organization remain secure even in the presence of quantum threats (Thanalakshmi et al. 2023).

Several cryptographic approaches are considered quantum-resilient and have shown promise for future-proofing blockchain networks:

Lattice based cryptography: Lattice-based cryptographic algorithms are considered one of the most promising solutions to quantum attacks. These algorithms rely on the difficulty of lattice issues, which are considered to be difficult also for quantum computers to address (Kumar 2022). Examples of lattice-based schemes include NTRU and FrodoKEM, both of which offer secure alternatives to traditional RSA and ECDSA.

Hash-based cryptography: Another quantum-resistant approach involves using hash functions to generate digital signatures. Merkle signatures and hash-based signatures such as XMSS (eXtended Merkle Signature Scheme) are designed to resist quantum attacks. These algorithms depend on the assumption that quantum computers cannot break the security of certain hash functions, making them a viable option for blockchain networks looking to maintain integrity in opposition to quantum threats.

Cryptography based on coding theory: Code-based cryptography is based on the complexity of decoding error-correcting codes, an issue that quantum computers have yet to solve efficiently (Sendrier 2017). The McEliece cryptosystem is a widely studied code-based approach that provides both encryption and digital signature capabilities, offering quantum resistance.

These cryptographic algorithms can be integrated into blockchain systems to replace or complement existing cryptosystems, ensuring that blockchain networks remain secure as quantum computing advances. However, integrating these quantum-resistant algorithms into blockchain is not without challenges, including increased computational load, larger key sizes, and compatibility with existing protocols.

Integrating Quantum-Resilient with CNTFET

While quantum-resistant cryptography offers a secure pathway for blockchain networks, it also comes with a significant challenge: the demand for increased computational capacity. Quantum-resistant algorithms, such as or hash-based cryptography, often need more processing resources and memory than traditional cryptographic systems. This increased demand for computational efficiency and low-latency data access makes it difficult for current blockchain infrastructure to handle these algorithms at scale.

This is where CNTFET-based SRAM technology can play a pivotal role. Carbon Nanotube Field-Effect Transistors (CNTFETs) have demonstrated significant improvements in speed, energy efficiency, and miniaturization compared to traditional silicon transistors. By integrating CNTFETs with Static Random-Access Memory (SRAM), blockchain nodes can efficiently manage the increased computational load imposed by quantum-resistant algorithms.

Faster switching capabilities: CNTFETs exhibit significantly higher electron mobility compared to silicon, allowing for faster switching speeds. This means that blockchain nodes could process the more complex quantum-resistant cryptographic functions much quicker, reducing delays in transaction validation and consensus mechanisms.

Low power consumption: One of the key advantages of CNTFET technology is its low power consumption, which is critical when running computationally intensive algorithms. Since quantum-resistant cryptography requires more processing power, the low energy requirements of CNTFET-based systems ensure that blockchain networks can handle these algorithms without excessive energy costs, making blockchain networks more sustainable in the long term.

Efficient data handling: SRAM, when paired with CNTFETs, offers low-latency, high-speed data access, which is crucial for blockchain systems that rely on real-time transaction validation and blockchain state management. As quantum-resistant cryptography requires rapid access to large amounts of data (such as cryptographic keys or signatures), the combination of CNTFET-based SRAM provides a powerful and efficient memory solution for handling the increased data requirements.

With CNTFET-based SRAM, blockchain networks could seamlessly integrate quantum-resistant algorithms, ensuring that they remain secure and performant as quantum computers evolve. The enhanced processing power, energy efficiency, and data access speeds offered by CNTFET technology would allow blockchain systems to scale and adapt to the demands of next-generation cryptographic standards, providing long-term security even in a post-quantum world.

5. A Future Vision for Blockchain with CNTFET and SRAM

5.1 From Current Blockchain to Futuristic Blockchain

"Blockchain technology has made remarkable strides since its inception, with its transformative potential" revolutionizing industries such as finance, supply chain, healthcare, and more. However, as the world becomes more digitally interconnected, blockchain faces increasing challenges in scalability, efficiency, and real-time processing. For blockchain to truly realize its potential, it must evolve beyond its current limitations, particularly as global adoption accelerates and new use cases emerge. In this context, CNTFET-based SRAM presents a pivotal opportunity to drive the next generation of blockchain technology.

The Need for Next-Generation Blockchain

At present, blockchain technology faces several critical limitations that hinder its widespread adoption, particularly in mainstream industries. The most pressing challenges include:

Scalability: Blockchain networks like Bitcoin and Ethereum are currently limited in the number of transactions they can process per second (TPS). As demand for blockchain services grows, these networks are increasingly struggling to handle the volume of transactions required to support applications such as decentralized finance (DeFi), supply chain management, and digital governance. Solutions such as Layer-2 scaling, Sharding, and Off-chain protocols are being explored, but they are only partial solutions to the larger scalability challenge.

Slow transaction speeds: Blockchain's agreement mechanisms, like Proof-of-Work (PoW) and Proof-of-Stake (PoS), often lead to delayed transaction processing and long confirmation times. For many real-world applications, such as online payments or supply chain tracking, the current speed of blockchain transactions is simply too slow.

High power usage: The networks of blockchain, particularly those utilizing PoW (such as Bitcoin), are known for their substantial energy consumption, contributing to environmental concerns (Anglen 2024). As the demand for blockchain grows, so too does the environmental impact, which may undermine the adoption of blockchain in industries focused on sustainability.

Cost and accessibility: The high costs associated with using blockchain networks, particularly in terms of transaction fees, limit their accessibility to a wider population (Sriman 2019). As blockchain seeks to expand into real-world use cases beyond cryptocurrency, these costs must be reduced to make blockchain more viable and accessible to billions of people.

For blockchain technology to scale and adapt to the demands of future industries, it must not only address these limitations but also ensure that it can support a wide range of real-world use cases. From smart cities and decentralized finance to digital identity management and Internet of Things (IoT) applications, the future of blockchain will require instantaneous transactions, robust security, and mass adoption—all of which depend on the ability to rapidly scale blockchain networks in a sustainable manner.

CNTFET and SRAM as a Key to the Future

As blockchain continues to evolve, CNTFET-based SRAM emerges as a transformative technology capable of addressing many of the challenges that current blockchain systems face. The combination of Carbon Nanotube Field-Effect Transistors (CNTFETs) and Static Random-Access Memory (SRAM) offers unique advantages in terms of speed, energy efficiency, scalability, and sustainability, making it the key to unlocking the next generation of blockchain capabilities.

High-speed performance: CNTFETs are a groundbreaking technology that provides significantly faster switching speeds compared to traditional silicon-based transistors. By replacing silicon with carbon nanotubes, CNTFETs offer a drastic increase in processing speeds while consuming less power. This improvement in processing speed would allow for faster block generation and transaction confirmation times in blockchain networks. With CNTFET-based SRAM, low-latency data access is also enhanced, allowing blockchain systems to process transactions in near real-time—an essential feature for applications that require instantaneous or high-frequency transactions, such as IoT networks or smart contracts.

Low-power consumption: One of the primary concerns with current blockchain networks is their energy consumption. Proof-of-Work systems, in particular, are notoriously power-hungry, requiring vast amounts of computational power to secure the network. CNTFETs, with their low-power capabilities, are well-suited to alleviate this issue. The combination of CNTFETs and SRAM reduces overall energy usage while maintaining performance. This low-power design could make blockchain systems significantly more sustainable, ensuring that they do not contribute to the growing environmental concerns associated with high-energy-consuming technologies. In turn, more environmentally-conscious industries and governments may embrace blockchain as a feasible solution to meet sustainability goals.

Scalability: Scalability is a significant bottleneck in current blockchain systems. As more users and transactions flood the network, the ability of blockchain systems to maintain high throughput becomes crucial. CNTFET-based SRAM offers an ideal solution to scalability by enabling blockchain networks to handle more transactions per second (TPS) without compromising efficiency or security. The high-speed performance of CNTFETs, combined with the fast memory access of SRAM, ensures that blockchain can process a much larger volume of transactions, making it possible to scale blockchain systems to handle billions of users—a requirement for global applications such as digital currency, supply chain management, and online voting systems.

Security and reliability: Beyond speed and energy efficiency, blockchain also requires robust security to protect against fraud, hacking, and unauthorized access. CNTFET technology offers enhanced performance, making it possible to run complex cryptographic algorithms at scale with faster execution times and reduced latency. When combined with SRAM, which enhances data integrity and consistency, blockchain systems can secure data more effectively while reducing the risk of vulnerabilities associated with traditional memory storage.

As blockchain systems continue to evolve, the need for speed, scalability, sustainability, and security will only increase. CNTFET-based SRAM is not just a technical improvement—it's a paradigm shift that could propel blockchain into a new era of performance and capability.

5.2 Real-World Applications in Future Societies

Blockchain in Governance and Digital Voting

Blockchain offers transformative potential for e-government applications and digital voting systems. By ensuring decentralization, transparency, and immutability, blockchain can create secure, tamper-proof voting systems, reducing fraud and increasing election trust. Supported by CNTFET-based hardware, blockchain can provide scalable, energy-efficient, and real-time voting processes. Its ability to securely store data without centralized control makes it ideal for driving efficient governance, fostering accountability, and minimizing corruption.

Decentralized finance (DeFi) and supply chain: Blockchain is reshaping decentralized finance (DeFi) and supply chain management. With CNTFET-based hardware, blockchain can support autonomous financial systems and smart contracts, eliminating the need for intermediaries. In supply chains, it can ensure transparent, verifiable tracking of goods from origin to delivery, minimizing fraud and ensuring product authenticity. With higher transaction throughput and lower energy consumption, blockchain technology can power decentralized applications (DApps), driving greater transparency and efficiency across industries, from finance to logistics.

6. Conclusion

The integration of CNTFET-based SRAM with blockchain technology offers transformative potential for addressing the most pressing challenges in blockchain systems, including performance, scalability, energy efficiency, and security. Blockchain technology, which has revolutionized sectors like finance, supply chain management, and digital governance, faces significant challenges such as high transaction costs, low throughput, and environmental concerns due to its energy-intensive nature. The combination of Carbon Nanotube Field-Effect Transistors (CNTFETs) and Static Random-Access Memory (SRAM) promises to address these limitations by significantly enhancing the speed, energy efficiency, and scalability of blockchain systems. Overall, blockchain has unparalleled advantages for optimizing energy efficiency (Ma 2024).

CNTFETs offer faster switching speeds, lower power consumption, and increased miniaturization, making them ideal for high-performance computing systems like blockchain. Their ability to handle large-scale computations more efficiently can lead to quicker transaction processing and faster consensus mechanisms, reducing latency and improving overall network performance. Additionally, CNTFET-based SRAM can provide low-latency memory, crucial for high-speed transaction validation and data storage. This enhanced memory access speeds up blockchain operations, which is vital for applications with high transaction volumes, such as DeFi platforms or supply chain management systems. Moreover, the energy efficiency of CNTFET-based systems addresses the environmental impact of blockchain technology, particularly in Proof-of-Work mining systems, which have been criticized for their high energy consumption. By reducing power requirements and enhancing computational efficiency, CNTFETs contribute to the sustainability of blockchain networks, making them more eco-friendly and capable of scaling without compromising performance.

Lastly, the integration of quantum-resistant cryptography with CNTFET-based SRAM strengthens the security of blockchain systems, protecting them from future threats posed by quantum computing. As quantum computers become more advanced, the ability to implement quantum-resistant cryptographic algorithms efficiently will be critical for maintaining the integrity and security of decentralized networks.

Vision for the Future

The future of blockchain technology is intricately linked to the evolution of hardware, particularly CNTFET-based SRAM. As blockchain networks continue to grow and become integral parts of global economies, the need for high-performance, secure, and energy-efficient solutions will only intensify. The integration of CNTFETs into blockchain infrastructure will likely lead to the development of systems that are faster, more scalable, and less energy-dependent, enabling blockchain to fulfill its true potential. Next-generation blockchain systems, powered by CNTFET-based SRAM, will not only address existing challenges but will also enable a secure, efficient, and sustainable decentralized ecosystem. These innovations will empower smart cities, autonomous financial systems, transparent supply chains, and secure digital governance platforms, making blockchain a central technology in a futuristic, digitally-driven society. The role of blockchain in decentralized applications (DApps), financial technologies, digital voting, and even personal data sovereignty will be significantly strengthened by these hardware innovations.

As we look ahead, the intersection of CNTFET technology and blockchain is poised to define the future of decentralized computing. CNTFET-based SRAM is a key enabler of next-generation blockchain technologies, providing a pathway to a more efficient, scalable, and sustainable blockchain ecosystem. This evolution represents not only a technological leap but also a societal shift toward decentralized, trustless systems that can solve real-world problems across various industries. In the coming years, we can expect blockchain to move beyond its current limitations, embracing innovations in both hardware and software. The power of CNTFETs will continue to shape the blockchain landscape, enabling new levels of performance and energy efficiency that will support the growing demand for decentralized systems. As these advancements unfold, we may witness the emergence of new applications and use cases that were once thought impossible, driving the next wave of technological innovation and social transformation.

In summary, the integration of CNTFET-based SRAM with blockchain technology heralds a future where blockchain systems are faster, more secure, energy-efficient, and scalable. As blockchain continues to evolve, these hardware innovations will be at the forefront of a new era of next-generation decentralized systems, empowering a more secure and sustainable digital society.

References

Ahmad, A., Saad, M., Njilla, L., Kamhoua, C., Bassiouni, M. and Mohaisen, A. (2019). Blocktrail: A scalable multichain solution for blockchain-based audit trails. pp. 1–6. In Proceedings of the IEEE International Conference on Communications (ICC 2019), Shanghai, China, 20–24 May 2019.

Alzoubi, Y. I. and Mishra, A. (2023). Green blockchain – A move towards sustainability. Journal of Cleaner Production 430: 139541, Dec. 2023.

Anglen, J. (2024). Eco-friendly blockchain: green innovations and shaping environmental futures in 2024. Rapid Innovation 2024.

Chauhan, A., Malviya, O. P., Verma, M. and Mor, T. S. (2018). Blockchain and scalability. pp. 122–128. In Proceedings of the 2018 IEEE International Conference on Software Quality, Reliability and Security Companion (QRS-C), Lisbon, Portugal, 16–20 July 2018.

Gad, A. G., Mosa, D. T., Abualigah, L. and Abohany, A. A. (2022). Emerging trends in blockchain technology and applications: A review and outlook. Journal of King Saud University - Computer and Information Sciences 34(9): 6719–6742, Oct. 2022.

Kim, S., Kwon, Y. and Cho, S. (2018). A survey of scalability solutions on blockchain. pp. 1204–1207. In Proceedings of the 2018 International Conference on Information and Communication Technology Convergence (ICTC), Jeju Island, Korea, 17–19 October 2018.

Kumar, M. (2022). Post-quantum cryptography Algorithm's standardization and performance analysis. Array 15: 100242, Sep. 2022.

Li, Y., Qiao, L. and Lv, Z. (2021). An optimized byzantine fault tolerance algorithm for consortium blockchain. Peer-to-Peer Netw. Appl. 14: 2826–2839.

Liu, D., Alahmadi, A., Ni, J., Lin, X. and Shen, X. (2019). Anonymous reputation system for IIoT-enabled retail marketing atop PoS blockchain. IEEE Trans. Ind. Inform. 15: 3527–3537.

Ma, C.-Q., Lei, Y.-T., Ren, Y.-S., Chen, X.-Q., Wang, Y.-R. and Narayan, S. (2024). Systematic analysis of the blockchain in the energy sector: Trends, issues, and future directions. Telecommunications Policy 48(2): Art. no. 102677, Mar. 2024.

Moaiyeri, M., Akbari, H. and Moghaddam, M. (2018). An ultra-low-power and robust ternary static random access memory cell based on carbon nanotube FETs. J. Nanoelectronics Optoelectron 13(4): 617–627.

Palai, A., Vora, M. and Shah, A. (2018). Empowering light nodes in blockchains with block summarization. pp. 1–5. In Proceedings of the 9th IFIP International Conference on New Technologies, Mobility and Security (NTMS 2018), Paris, France, 26–28 February 2018.

Ravindran, R. (2019). Circle of trust: a high volume, energy efficient, stake blind and high attack tolerant blockchain consensus protocol. pp. 1–4. In Proceedings of the IEEE 12th International Conference on Global Security, Safety and Sustainability (ICGS3 2019), London, UK, 16–18 January 2019.

Sendrier, N. (2017). Code-based cryptography: state of the art and perspectives. *IEEE Security and Privacy Magazine* 15(4): 44–50, Jan. 2017, doi: 10.1109/MSP.2017.3151345.

Song-Kyu Kim and Hou Chong Vong. (2025). Secured network architecture based on blockchain technologies: a systematic review. ACM Computing Surveys (CSUR) 57(Issue 7): 1–24.

Sriman, B. (2020). Blockchain technology: consensus protocol proof of work and proof of stake. pp. 469–478. In Intelligent Computing and Applications, Proceedings of ICICA 2019, Advances in Intelligent Systems and Computing, vol. 1012, 2020, doi: 10.1007/978-981-15-5566-4_34.

Thanalakshmi, P., Rishikhesh, A., Marceline, J. M., Joshi, G. P. and Cho, W. (2023). A quantum-resistant blockchain system: a comparative analysis. *Mathematics* 11(18): 3947, doi: 10.3390/math11183947.

Tripathi, G., Ahad, M. A. and Casalino, G. (2023). A comprehensive review of blockchain technology: Underlying principles and historical background with future challenges. Decision Analytics Journal 9: 100344, Dec. 2023

Wang, C. and Zhao, J. (2023). Network approaches in blockchain-based systems: Applications, challenges, and future directions. Computer Communications 212: 141–150, Dec. 2023.

Zhao, S., Li, S. and Yao, Y. (2019). Blockchain enabled industrial internet of things technology. IEEE Transactions on Computational Social Systems PP(99): 1–12, Jul. 2019, doi: 10.1109/TCSS.2019.2924054.

Zheng, Z., Xie, S., Dai, H., Chen, X. and Wang, H. (2017). An overview of blockchain technology: architecture, consensus, and future trends. pp. 557–564. In Proceedings of the 2017 IEEE International Congress on Big Data (BigData Congress), Honolulu, HI, USA, 25–30 June 2017.

Zhu, X. (2019). Research on blockchain consensus mechanism and implementation. In IOP Conference Series: Materials Science and Engineering; IOP Publishing: Bristol, UK 569: 042058.

Chapter 18

Blockchain Technology in Smart Cities

Current Status and Future Directions

R. Ravinder Reddy,[1,*] *M. Venkata Krishna Reddy,*[1] *Ravi Uyyala,*[1]
C. Madan Kumar[2] and *Y. C. A. Padmanabha Reddy*[1]

1. Introduction

A major rise in energy consumption is a consequence of using Proof of Work (PoW) consensus mechanisms in Blockchain systems, which could contradict the goals of sustainability. Compared to other parts of the globe, developing nations in Asia and Africa are experiencing rapid increases in their urbanization levels. A total of 56% of China's population now lives in urban areas, up from 38% in 2001 (Ahmed et al. 2022). In many ways, including health, education, transportation, the economy, and living and working conditions, the urbanization process has considerably raised the living standards of people around the globe (Al Nuaimi et al. 2015). But new challenges and dilemmas are brought about by the rapid increase of the world's urban population. The quality of life for city people is affected by a number of factors, including the scarcity of natural resources, the growth in traffic, air pollution, emissions of greenhouse gases, and ineffective waste management (Alli et al. 2022). Participants in cities, such as people and governments, are compelled to address these difficulties and problems by focusing on wiser methods to sustainable urban development and the enhancement of inhabitants' quality of life.

According to (Kakarontzas et al. 2014), smart cities are being made possible by the continuous improvement of applications that are based on the IoT. Smart housing, intelligent transportation, smart housing, healthcare, Industry 4.0, and banking are just a few of the many benefits that smart cities provide. In order to handle data securely, these applications are essential for raising the quality of life for citizens. Blockchain technology has the potential to make smart cities more secure and private (Alsunaidi 2022). A distributed, verifiable, transparent, and immutable record of international transactions on P2P networks is known as a blockchain. Bitcoin, the original implementation of block chain technology, allowed for the decentralized transmission of digital currency between users (Ioannis 2022). In 2017, the market capitalization of Bitcoin exceeded US $230 billion, indicating its

[1] Department of Computer Science and Engineering, Chaitanya Bharathi Institute of Technology, Hyderabad, India.
[2] Department of Computer Science and Engineering, Vaagdevi College of Engineering, Warangal, Telangana, India.
Emails: krishnareddy_cse@cbit.ac.in, raviuyyala_cse@cbit.ac.in, madan_kumar@gmail.com, padmanabhareddy_cseaiml@cbit.ac.in.
* Corresponding author: rravinderreddy_cse@cbit.ac.in

DOI: 10.1201/9781003620822-18

tremendous success. Apart from the financial sector, blockchain technology has numerous potential applications in areas such as the Internet of Things (IoT), e-Commerce, accounting and auditing, e-Voting, asset and identity management, supply chain, taxes, telecommunication, healthcare (McGh), and countless more (Anagnostopoulos et al. 2017).

There are a lot of different ways to define a "smart city", but one popular idea is that these communities use cutting-edge ICT to make people's lives easier and create more environmentally friendly cities. There are many positive aspects of a smart city, including more transparency in government, more citizen engagement, better traffic and public transportation management, more efficient use of resources, enhanced environmental protection, smarter device control, and better health, energy, and education services. The deployment of smart cities relies heavily on ICT. Here, we zero in on blockchain, a relatively new technology with enormous promise for advancing smart city development and improving smart city services (Batty et al. 2018). Figure 18.1 depicts the characteristics of Smart City.

An ecosystem of smart settings is provided to citizens of a smart city in order to improve their quality of life. In order to improve urban planning, economics, public services, the environment, and public welfare, smart cities use information and communication technologies (Bharat et al. 2020). When it comes to city life, "smart cities" aim to improve everything by integrating digital technologies, both old and new. Among the many vital services that smart cities want to transform are housing, schools, healthcare, transportation, electricity, water, utilities, monitoring, and police enforcement (Christoph et al. 2016). "Smart cities" employ technological integration of the city's social, economic, and physical infrastructure to lessen the negative effects of rapid urbanization and population expansion (Aniello et al. 2023). Smart cities will be more intelligent, connected, liveable, secure, sustainable, and resilient as a result of recent advances in technologies like the IoT, blockchain, Big Data, machine learning, automation, and AI.

Blockchain, the distributed ledger system that underpins Bitcoin and similar cryptocurrencies, emerged from these (Chauhan et al. 2022). Satoshi Nakamoto's (2008) creation of Bitcoin was the initial implementation of blockchain technology. The blockchain, in its most basic form, is an immutable distributed ledger that everyone can view and use. The blockchain records all transactions and anyone involved in the system may see them, send them, and verify their validity. Among the many possible advantages of implementing blockchain technology in smart cities are increased safety, decentralization, automation, transparency, democracy, and distrust. As long as no reliable third party is present, the blockchain system can operate seamlessly as a network of interconnected nodes. The blockchain is totally transparent, since it stores all transaction records that are accessible to the public. Anonymity can be accomplished by obscuring the real-world identities of the nodes

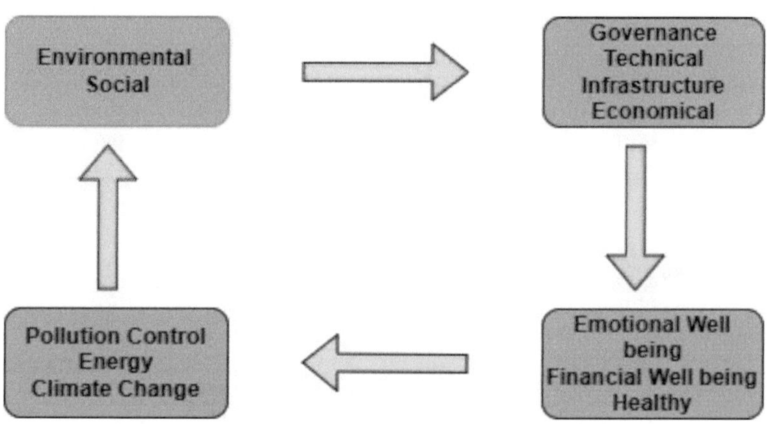

Figure 18.1. Characteristics of a smart city.

and logging transactions using publicly available pseudonymous addresses. Decentralized and peer-to-peer, every node in the blockchain system makes a decision. Data can be automatically stored via smart contracts on the blockchain. Blockchain technology is decentralized, hence in order to ensure consistency, consensus methods must be executed across decentralized nodes. The blockchain system is built around three pillars: trustworthiness, privacy, and permission. Researchers have spent a lot of time studying smart cities and blockchain, but they have mostly dealt with these two important subjects separately until lately.

As far as we are aware, no prior research has examined the intersection of these two crucial domains. We address this deficiency by providing a comprehensive overview of current blockchain technologies in this study and their potential applications in smart cities. Findings from studies on how smart cities might benefit from blockchain technology in terms of performance, intelligence, efficiency, and safety are detailed.

2. Related Work

2.1 Current Status of Blockchain Technology in Smart Cities

Intelligent city research has attracted a lot of attention from academics. The term "smart cities" was defined in a number of ways in (Lei et al. 2018). Batty et al. (2018) have provided a comprehensive review of smart cities, covering everything from research goals and challenges to possible outcomes and areas of attention. Beginning with a definition, Xie et al. (2019) discuss smart cities in depth, touching on every facet such as their designs, areas of application, research challenges, and crucial enabling technologies. There has been a discussion on the functional requirements for smart cities in Davis et al. (2015). Studies on smart city information security have been published in Dehshiri et al. (2024). The research difficulties, methods, and fields of application for trace analysis and mining in smart cities have been surveyed by Nam et al. (2021). Economics models of smart city electric vehicle charging were contrasted and assessed by Nakamoto et al. (2008), who considered both bidirectional and unidirectional energy flows. Additional smart city-specific characteristics that have been studied include software architectures DeLuca (2017), business models Dewan et al. (2020), and people movement analytics. Intelligent transport systems (ITS) are crucial to the implementation of smart cities.

The efficiency of municipal administration and public services could be enhanced with more data collecting and processing. In their study, Musa et al. (2022) looked at the pros, cons, and future applications of big data technology in smart city settings. Innovative methods to enable a fast, efficient, and accurate system in smart cities are discussed in Djahel et al. (2015) thorough review of the technologies (such as machine learning) used in different phases of modern traffic management systems. In their review of deep learning methods, Zhu et al. (2024) looked at smart city video analytics. From a data-centric perspective, they examine technology used for data management, privacy, and security, including deep learning techniques and machine learning. Many additional networking and computer technologies are covered in Feng et al. (2020), including cloud computing, Software-Defined Networking (SDN), and Network Function Virtualization (NFV).

Smart city development can also make use of other promising technologies like fog computing and the Internet of Things (IoT). Fernandez-Anez (2016) examines smart cities that are enabled by the Internet of Things. Notably, Gallo et al. (2018) have investigated the necessary protocols, architecture, and supporting technologies for an urban IoT system. Waste management in smart cities is the subject of Anagnostopoulos et al. (2017), who evaluate six methods that make use of the Internet of Things. The effects of the Cloud of Things (CoT) on smart cities have been examined and a list of the needs, advantages, and disadvantages has been compiled by Ijaz et al. (2016). In Gharaibeh et al. (2017), the authors examine the features and functions of a perfect fog computing architecture, as well as future research problems and directions, and provide a survey of fog computing for sustainable smart cities. Blockchain technology is steadily gaining traction in smart city initiatives worldwide. Its core characteristics like decentralization, immutability, transparency, and security make it ideal for addressing the complex needs of smart cities.

2.2 Key Areas of Blockchain Adoption

Urban governance and transparency: Blockchain ensures transparency in public services by providing immutable records for government processes like land registries, voting systems, and public spending. For example, the goal of the Dubai Blockchain Strategy is to streamline government operations and cut down on red tape so that Dubai may become the world's first blockchain-powered city.

Energy management: Peer-to-peer energy trading is made possible by blockchain technology, allowing individuals and companies to directly purchase and sell renewable energy. For example, Power Ledger projects allow users to trade surplus solar energy using blockchain networks.

Traditional, centralized IT grid architectures struggle to handle the diversity and dispersion of renewable energy sources. Problems with energy management are manageable using blockchain technology. Many energy firms worldwide are investigating blockchain technology because of its great potential in green or renewable energy applications. Goumopoulos (2024) offered a thorough evaluation of thermal energy systems that combine low-temperature heating and high-temperature cooling with optimization techniques to increase the efficiency of the system.

According to studies conducted by Mora et al. (2021), BCT has the potential to tackle energy sector issues such as decentralization, decarbonization, and digitization. Several parts of energy management systems could benefit from blockchain technology, which is discussed in this section. Optimal scheduling and operation can be achieved with the use of a safe energy management strategy in smart cities: the authors of Zhu et al. (2024) highlight the critical need of maintaining a delicate balance in contemporary power systems. To deal with operational risks in managing the associated power generation of wind turbines, they employ a stochastic architecture based on unscented transformations. The use of blockchain technology to ensure the secure flow of data is another aspect of the smart city.

Logistics and supply chain: Ensuring transparency, and the ability to track urban supply chains for food, medical supplies, and goods, blockchain helps reduce fraud and inefficiencies. The example is IBM Food Trust that enhances food safety through end-to-end blockchain traceability.

Smart contracts for automation: Smart contracts automate processes in urban services such as water distribution, waste management, and public transportation. For example, the Ethereum-based solutions are used for real-time billing and resource allocation in utility management.

Digital identity and citizen services: Blockchain enables secure digital identities, allowing citizens to access public services without risking data breaches. To ensure the safety of digital identities and public services, the e-Residency Program in Estonia employs blockchain technology.

Mobility and transportation: Blockchain supports integrated urban mobility by facilitating seamless payments, ride-sharing, and traffic management. We consider the example of Blockchain-based toll systems that allow automated and tamper-proof transactions for smart roads. Two examples of urban transportation-related blockchain applications are peer-to-peer (P2P) vehicle sharing and real-time traffic data. Instead of generating brand-new mobility resources, smart transport maximizes the ones that already exist. One practical alternative to driving alone is ridesharing, in which one person uses another's car to take them to their destination along a predetermined route. Because costs can be spread out over time by making use of vacant seats, profitability is maximized. Because fewer vehicles are using the same route, traffic and pollution are both reduced. Shared travel, wherein a passenger can switch vehicles while rerouting, is supported by the blockchain-based smart transport system (Ibba et al. 2017). Travelers and drivers can coordinate their whereabouts and itineraries via a mobile app. A blockchain records and verifies the transferred data in real time. In exchange for users' location data, the system creates cryptocurrency tokens and rewards them (Kakavand et al. 2017).

A trip planner is the backbone of any intelligent transportation system. A user should be aware of several ways to purchase, pay for, and confirm e-tickets, as well as alternative routes between different transportation providers. At present, mobility services are centralized and serve a crucial purpose in transportation demand-supply management. The diversity of the schemes they control, however, makes it difficult for them to manage connections between different providers in a way that is both trustworthy and scalable. Carriers and travelers can work together in a blockchain-based setting (Kiyomoto et al. 2017). A traveler essentially wants to know the best way to get from point A to point B. Smart contracts are subsequently transmitted to the network together with details regarding potential transportation routes and methods. Based on their route, comfort, and budget preferences, travelers choose and pay for their preferred travel option. The next step is for all vendors to verify and approve the smart contract before it can be recorded on the blockchain. Traffic jams in urban areas are one factor that impacts mobility. A traffic light system that utilizes the consortium blockchain and vehicular ad-hoc networks (VANETs) can enhance city mobility and solve this issue (Liu et al. 2021). Here, a car reports the current traffic conditions on the road it's travelling on. After that, the data is uploaded to the blockchain to be verified and recorded. The activation of smart contracts allows for the dynamic regulation of traffic lights through the use of artificial intelligence and parallelism algorithms.

Waste management: The goal of blockchain-based solutions to this problem is to improve trash management by increasing recycling rates and public understanding of the need for recycling. Solutions that encourage the detection of georeferenced sources of pollution and seek to captivate individuals, in order to increase awareness of the plastic footprint, are available in regard to the first use case (Makani et al. 2022). An open website allows citizens to record locations with plastic trash (OpenLitterMap.com). After receiving the data, it is processed and organized into a structure similar to a blockchain. With the creation of a wallet, users can be incentivized and rewarded with LitterCoin, a virtual money, for their information. Similarly, alternative designs exist to encourage suitable ecological conduct. By offering a crypto currency called Green Coin, this system incentivizes people to participate in an urban garbage collection activity that includes plastics, paper, aluminum, textiles, and batteries. Customers of participating businesses can accept these digital currencies in return for products and services. There are alternative approaches that center on the life cycle tracing of certain waste kinds (Marsal-Llacuna and Maria-Lluïsa 2020). The producer notes down the product's attributes in the initial step. Next, we keep track of every ownership change from the date of creation all the way until it is no longer needed. For a project to be sustainable, a transfer rate must be determined. Each of the previous owners of the electronic gadget receives a percentage of the tax value. The device needs to be sent to a recycling facility when its useful life is over for this to happen. The goal of this plan is to encourage participants to send their products to official recycling facilities. The prevention and reuse of garbage is another crucial aspect of waste management. To prevent unethical acts in disposal, transportation, and recycling, this profitable industry requires openness and auditing. Smart contracts and blockchain technology can solve this management issue by making trash management more traceable (Mu lin et al. 2021). There includes a log of activities with timestamps, as well as information on stakeholders, waste management facilities, waste amount and classification. So, in order to be able to give alerts regarding any observed abnormalities, the audit procedure receives reports of various types. Figure 18.2 shows the example of blockchain smart health care implementation in smart cities. Table 18.1 shows smart healthcare solutions using blockchain.

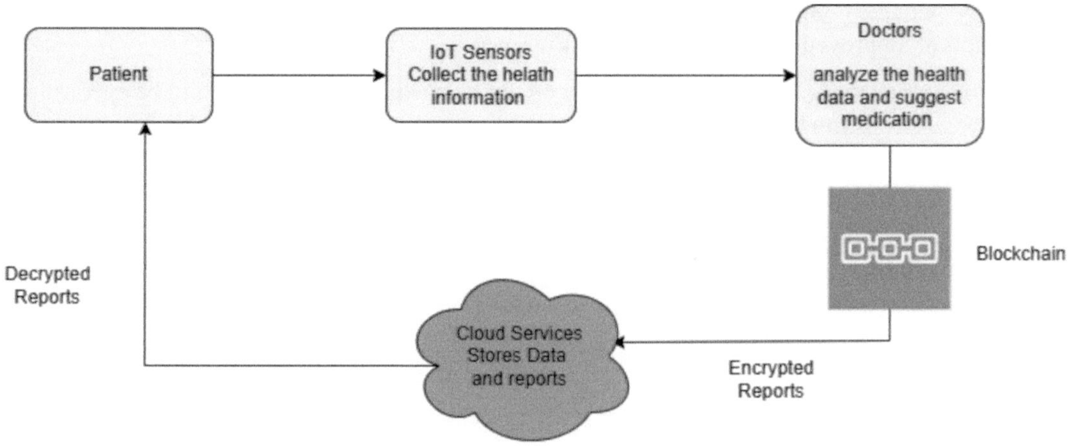

Figure 18.2. Blockchain smart health care.

Table 18.1. Blockchain smart healthcare solutions.

S.No	Blockchain Component	Healthcare solution
1	Ethereum, bitcoin	Public healthcare management
2	Hyperledger fabric	Healthcare data gateway
3	Consortium blockchain	Comprehensive data sharing
4	Bitcoin (PBFT)	Network

2.3 Key Benefits of Blockchain in Smart Cities

Enhanced data security: Blockchain protects sensitive urban data against unauthorized access and cyber threats.

Decentralization: Reduces dependence on centralized authorities, enabling distributed decision-making and resilience.

Transparency and accountability: Immutable records ensure trust in governance, reducing corruption and inefficiencies.

Efficiency: Smart contracts automate urban processes, reducing delays and manual errors.

Citizen empowerment: Facilitates citizen participation in governance and urban services through transparent and decentralized systems.

Blockchain is an appealing technology for smart city problems because of the following properties.

No central authority: Blockchain systems function generally as peer-to-peer networks.

Pseudonym address: Each blockchain node has a public pseudonymous address, which allows users to remain anonymous online. Use situations where user identities need to be kept hidden are well-suited to the inherent pseudonymity.

Transparency: All transaction records are accessible to everyone using blockchain technology, making it transparent.

Decentralized democracy: Before a block is added to the blockchain, all decentralized nodes conduct consensus algorithms to obtain an agreement. Therefore, the blockchain system is decentralized since all nodes make choices in a peer-to-peer fashion.

Why safety: A single point of failure is unlikely with decentralized systems built on the blockchain. The result is an improved level of network security.

Everlasting and unchangeable: All transactions on the blockchain are guaranteed to be everlasting and unchangeable due to the usage of digital signatures. Also, one-way cryptographic hash methods are used to encrypt and link the data blocks. The immutability of the shared ledger is ensured by the fact that every modification generates a unique hash and is immediately identifiable.

3. Obstacles to Implementing Blockchain Technology in Smart Cities

Despite its potential, blockchain faces several challenges in smart city implementations:

3.1 Scalability

Present blockchain networks frequently face difficulties in processing large-scale urban data in real time. The example is high latency and low transaction throughput in popular blockchains like Bitcoin and Ethereum. The exponential rise of blockchain technology during the past ten years was fueled by the considerable enthusiasm for digital currency, particularly among the younger generation. Network congestion, brought on by rapid adoption, is the root cause of latency issues with transaction processing on many systems, including Ethereum 1.0. It is believed that scalability offers the sole viable option to the problem of rising transaction processing costs and delays. Multiple approaches are being researched, designed, and evaluated to reach scalability. Some examples of these include hardware-based methods, parallel mining, sharding, on-chains, cross-chains, and sidechains.

3.2 Energy Consumption

The use of Proof of Work (PoW) consensus processes in blockchain systems results in a substantial elevation of power usage, potentially going against the objectives of sustainability. For example, Bitcoin's energy consumption is a widely debated issue in environmental circles.

3.3 Interoperability

Integrating blockchain with existing city systems and technologies remains a technical challenge due to varying standards and protocols.

3.4 Data Privacy

While blockchain is transparent, certain urban data (e.g., citizen information) requires privacy protection, which blockchain's public nature may compromise. Various smart city applications can benefit from blockchain technology's various characteristics. The decentralized and open access nature of blockchain technology raises privacy issues because anyone can access and read any data recorded on the network. Additional categories for privacy include identity, location, report, and query. A difficult problem is the preservation of users' privacy on blockchains. Many private and legal problems might arise as a result of improperly disclosed personal information. The presence of privacy concerns in smart city applications built on the blockchain will cause citizens to reject them. There have been a lot of recent studies that have suggested different ways to keep blockchain users' privacy intact. On the other hand, numerous cryptocurrencies, like Zcoin and Zcash, have offered alternatives. Table 18.2 delivers the security and privacy concerns of the smart cities.

Table 18.2. Security and privacy challenges of smart cities.

S.No	Security Requirements	Challenges
1	Secure communication	Heterogeneous devices connected together
2	Secure monitoring	Network intruders
3	Secure booting	Various IoT sensors connection
4	Authentication and access control	High computation cost and risk

3.5 Regulatory Uncertainty

While using the blockchain technologies we need to approve some regulatory constraints, which are as follows.

i. Ambiguities in blockchain regulations can delay adoption in government and municipal systems. The field of social science is clearly making its mark on the smart city discussion. It is fundamental to comprehend the opinions of the people regarding their living conditions and emotional state. Evaluations of certain applications of IT systems of currently shown issues can only be addressed by involving citizens in the design stages. One way to include the public is through the idea of e-governance, which is also called civic tech. In conclusion, the new wave of decentralization is essential for trustworthy, low-cost governance. It is important to give serious thought to blockchain concepts when thinking about e-governance. Neo Blockchain is one example of a governance concept that successfully associates digital assets with democracy.

ii. Here we highlight a few of these ideas that have some bearing on the origins of decentralization and e-governance.

iii. Recent developments in the Internet of Things (IoT) and blockchain-based solutions are transforming the Internet of Value (IoV) into a practical instrument for enforcing policies, which has led to increased discussion of decentralized governance models. Here, there have been initiatives to bring the government closer to its residents. As pointed out by Roberts, the SC is driving a strong trend towards decentralizing democracies, which is connected to a new era of information technologies. Multi-Agent Systems (MAS) paradigm concepts have mostly been used to adopt decentralized decision-making, which entails spreading decisions. When it comes to negotiation protocols, the MAS paradigm is associated with devices that reach consensus.

iv. Smart contracts achieve public transparency through the use of information technology, citizens are able to gain cognitive control over the state through democratic involvement. In addition, institutional integration and strategic management are integral parts of the e-governance idea. Efficient management of public places and services is the bedrock of e-governance. At all levels of government, from neighborhoods to cities and beyond, the public can be better represented and engaged in decision-making processes.

3.6 Cost of Implementation

Initial deployment of blockchain infrastructure may be costly, demanding incurring deposits in both technical side and skilled personnel.

3.7 Infrastructure

i. New interactive layers are being added to cities as a result of the smart city innovation wave, which is also streamlining and eliminating obstacles from earlier models. The system's foundational elements include networked sensors, intelligent transportation networks (including metros, trains, and drones), public gardens, efficient and automated lighting, a petrol system, water, telecommunications, and electric vehicle charging infrastructure.

ii. Urban traffic issues can be addressed via means of electronic gadgets and sensors, which can optimize existing transport systems and contribute to the design of future ones. Known as prosumers (producer + consumer), microgrid residences generate power. These decentralized generation systems not only solve problems in some parts of a city, but they also encourage people to interact with one another and explore the possibilities that energy offers. As mentioned earlier, energy can serve as a means to monetize while simultaneously encouraging sustainability, thanks to the rise of cryptocurrencies.

iii. Sustainable development can be facilitated by well-thought-out city planning that incorporates a multi-criteria analysis of residents' needs and wants in relation to city services. Better mobility and more equitable access to opportunities, particularly for urban people, can be enhanced through studies of SC. Air quality, noise, accessibility to city vital points, and security are just a few of the many ways in which logistics impact the quality of life for inhabitants.

iv. Because they touch on people's subjective experiences, these issues are inherently nuanced; what works well for one person may not for another. In this regard, web-based applications might be quite useful in guiding each individual to achieve their objective within a metropolis. Thus, to promote sustainable routing that considers inhabitants' aspirations, both for everyday mobility and for medium-term forecasts regarding the most ideal area to dwell, vehicle routing problems should take these "green" elements into account.

v. Because of the high energy, storage, communication, and compute demands, implementing these solutions in actual smart city systems is difficult. Coming up with effective and dependable unconventional solutions is the key to overcoming this obstacle.

vi. Public services and city management can both benefit from better data gathering and analysis. It is critical that the data be reliable and uncompromised. It is strictly forbidden to alter data without authorization. Smart city applications are becoming more complicated and more and more devices are being used. In smart cities, devices and nodes have the freedom to join and exit the network at will. Decentralized systems are more suited to dynamic settings with fluctuating device numbers and application complexity than centralized ones. The administration of a city affects all citizens. Democracy, openness, and citizen engagement are highly sought after by the general public. City management-related information, including details about environmental impacts, government affairs, and decision-making processes, should be made public by the government. Companies must also be transparent about the usage of client data.

vii. Better municipal administration and decision-making and high-value smart city services could be possible through the sharing of data, including data from the Internet of Things (IoT), company organizational data, and personal citizen data. Governments, organizations, and individuals are unable to share data effectively due to a lack of trust, faith in the market, and incentives.

4. Future Directions for Blockchain in Smart Cities

To maximize blockchain's potential in smart cities, various improvements and tactics are needed:

4.1 Emerging Tech Integration

Integrating blockchain technology with the Internet of Things (IoT), artificial intelligence (AI), and fifth-generation wireless networks (5G) might produce complementary outcomes, paving the way for automated processes, safe data exchange, and real-time data analysis. The example is blockchain-secure IoT devices that can ensure secure communication in smart grids or traffic systems.

4.2 Layer 2 Solutions for Scalability

i. To solve scalability problems, faster and cheaper transactions can be enabled by utilizing Layer 2 technologies, such as Lightning Network.

ii. Smart city blockchain solutions, in particular, need to be scalable and perform well enough to satisfy the needs of the public and private sectors. Regarding this, a number of researchers concentrated on making guidelines to communicate with the devices, this leads to an increase in the number of messages being exchanged, which in turn causes various performance issues like latency and throughput (the number of successful transactions per second). Although PoW consensus improves scalability, it has issues with low throughput and high latency, particularly when solving cryptographically difficult puzzles, which causes resources to be wasted. In addition, PoW uses a lot of CPU power and can be easily tricked into spending twice as much. Because of how long transactions take, it can't be used in real-time. Even when faced with malevolent replicas, the PBFT protocol is able to reach a consensus and is energy efficient, but it cannot scale. There can be no major lag in the smart city's economy if any popular platform can't handle thousands of transactions per second. The performance and scalability of smart city solutions built on blockchain technology are thus important issues that require additional research.

4.3 Green Blockchain Initiatives

Aligning blockchain with environmental goals can be achieved by switching to consensus processes that are energy efficient, such as Proof of Stake (PoS). Ethereum 2.0's transition to PoS as an example, significantly reduces its energy consumption.

4.4 Standardization and Interoperability

Developing universal standards for blockchain integration in smart city systems can streamline interoperability across platforms and regions. We consider the ISO/TC 307 blockchain standard aims to harmonize blockchain frameworks globally. Standardization for blockchain technology is still an area where opinions differ. Standards for blockchain integration, governance, and interoperability are currently being developed by many organizations, including NIST and IEEE. Note that different blockchain systems use different data formats, making it difficult to implement an interoperable system. The fact that independent blockchain systems use different consensus algorithms adds another layer of complication. For instance, for Hyperledger to function smoothly, the PBFT and PoW consensus techniques used by Ethereum must be in sync with one another.

Consequently, for an application development platform to function smoothly, data transmission between blockchains is essential. Accordingly, additional research is required to build interoperable protocols for smart city solutions based on blockchain.

4.5 Citizen-Centric Blockchain Solutions

Designing blockchain applications that prioritize citizen-centric needs, such as participatory governance platforms and community-based energy trading systems. These will support the user level requirements.

4.6 Regulatory Frameworks

i. To promote the use of blockchain technology and guarantee its ethical use and privacy protection, governments should design legislation that is both evident and helpful.

ii. The power of central banks to control economic policy is sometimes diminished by decentralized blockchain platforms. Thus, governments are becoming more wary of the use of cryptocurrency, and blockchain platforms are encountering problems with regularity. Cryptocurrencies are illegal in many countries, including Pakistan, Iran, and Morocco. Particularly, the United States and the European Union (EU), outlined the key regulatory concerns that have a negative effect on blockchains and new distributed technologies. Decentralized blockchain technology eliminates the need for a trusted middleman, necessitating new industry and government laws to prevent conflicts between the parties involved in a transaction. Data generated by smart city devices also comes in a variety of unstructured formats. It is not a good idea to directly store this diverse and unstructured data in blockchain-based systems. Storage standards and data formats must be carefully considered to facilitate seamless data sharing among the many entities of a smart city. Consequently, there is a need for more investigation into the best practices for regulating blockchain-based smart city systems to guarantee data integrity.

4.7 Decentralized Autonomous Organizations (DAOs)

DAOs can democratize decision-making in urban governance, allowing citizens to vote on policies and budget allocations through blockchain platforms.

4.8 Blockchain-Based Digital Twins

Virtual representations of real-life metropolitan settings are called digital twins. Data utilized in digital twins for city management and planning can be safely and securely stored with blockchain technology.

4.9 Education and Awareness

Promoting blockchain literacy among policymakers, urban planners, and citizens can accelerate adoption and foster innovation.

4.10 Corporate-Government Alliances

Government agencies, businesses, and academic institutions working together can drive cost-effective and scalable blockchain solutions for smart cities.

4.11 Storage

The data created by smart city devices is increasing at an exponential rate, making data management and storage a major hurdle. Because of the huge lots of data and processing power available on cloud servers, some studies have concluded that this is the best way to store data. Because data uploads to cloud servers could lead to lengthy delays or even data integrity compromises, cloud storage is untrustworthy and wasteful for smart city applications. In addition, data owners must ensure the authenticity of the outsourced data due to the cloud service providers' potentially harmful actions or lack of trustworthiness. Several plans for centralizing data storage have been put up in an effort to achieve this goal. On the other hand, denial-of-service attacks and single points of failure are possible with these approaches. There have been proposals for decentralized storage methods based on the blockchain to address these concerns. "Koppercoin" is the name of the decentralized token system suggested. Rather than PoW, it employs proof of retrievability (PoR). In this method, individuals are directly rewarded for their file storage contributions by virtue of the digital tokens they receive. In an effort to ensure greater transparency, enhanced security, and faster auditing, a 'BlockStore'—a secure decentralized storage infrastructure based on blockchain technology was proposed. Unused space in the space wallet can be rented out by keeping track of it with BlockStore. Lack of trust, privacy, security, etc. are still problems with decentralized systems, even though they go beyond the problems with centralized storage systems. All the way back to the genesis block, every node in a traditional blockchain system has to be able to process and store full transactions.

Thus, further research is needed to address the difficulty of bringing blockchain technology to smart cities with limited resources.

5. Conclusion

Smart cities could be revolutionized by blockchain technology, which could increase transparency, efficiency, and citizen participation. Its revolutionary potential is on display in cutting-edge uses in digital identification, energy, mobility, and governance, all of which are still in the early phases of adoption. Taking on difficulties like scalability and saving energy, and interoperability will require concerted efforts from stakeholders, including governments, industry leaders, and researchers. As blockchain continues to evolve, its integration with other technologies and alignment with sustainability goals will shape the future of smart cities. By leveraging blockchain, cities can become more resilient, inclusive, and adaptive, meeting the needs of a growing urban population while ensuring environmental stewardship and technological progress.

References

Ahmed, I., Zhang, Y., Jeon, G., Lin, W., Khosravi, M. R. and Qi, L. (2022). A blockchain-and artificial intelligence-enabled smart IoT framework for sustainable city. International Journal of Intelligent Systems 37(9): 6493–6507.

Al Nuaimi, E., Al Neyadi, H., Mohamed, N. and Al-Jaroodi, J. (2015). Applications of big data to smart cities. J. Internet Services Appl. 6(1): 25, Dec. 2015.

Alli, Adam A. and Muhammad Mahbub Alam. (2020). The fog cloud of things: A survey on concepts, architecture, standards, tools, and applications. Internet of Things 9: 100177.

Alsunaidi, Shikah, J. and Fakhri Alam Khan. (2022). Blockchain-based distributed renewable energy management framework. IEEE Access 10: 81888–81898.

Anagnostopoulos, Ioannis. (2018). Implementing a blockchain infrastructure on top of vehicular ad hoc networks. Data Analytics: Paving the Way to Sustainable Urban Mobility: 764.

Anagnostopoulos, T., Zaslavsky, A., Kolomvatsos, K., Medvedev, A., Amirian, P., Morley, J. et al. (2017). Challenges and opportunities of waste management in IoT-enabled smart cities: a survey. IEEE Transactions on Sustainable Computing 2(3): 275–289.

Batty, M., Axhausen, K. W., Giannotti, F., Pozdnoukhov, A., Bazzani, A., Wachowicz, M. et al. (2012). Smart cities of the future. The European Physical Journal Special Topics 214(1): 481–518.

Bhushan, B., Khamparia, A., Sagayam, K. M., Sharma, S. K., Ahad, M. A. and Debnath, N. C. (2020). Blockchain for smart cities: A review of architectures, integration trends and future research directions. Sustainable Cities and Society 61: 102360.

Bösch, C., Erb, B., Kargl, F., Kopp, H. and Pfattheicher, S. (2016). Tales from the dark side: Privacy dark strategies and privacy dark patterns. Proceedings on Privacy Enhancing Technologies.

Castiglione, A., Cimmino, L., Di Nardo, M. and Murino, T. (2023). A framework for achieving a circular economy using the blockchain technology in a sustainable waste management system. Computers & Industrial Engineering 180: 109263.

Chauhan, Bhargavi, K. and Dhirenbhai B. Patel. (2022). A systematic review of blockchain technology to find current scalability issues and solutions. Proceedings of Second Doctoral Symposium on Computational Intelligence: DoSCI 2021. Springer Singapore.

Davis, Kingsley. (2015). The urbanization of the human population. The City Reader. Routledge 43–53.

Dehshiri, Seyyed Jalaladdin Hosseini, Maghsoud Amiri and Seyed Mojtaba Hosseini Bamakan. (2024). Evaluating the blockchain technology strategies for reducing renewable energy development risks using a novel integrated decision framework. Energy 289: 129987.

DeLuca, Lisa. (2017). United Nations: Online data repositories and resources. College & Research Libraries News 78.1: 41–45.

Dewan, Surbhi and Latika Singh. (2020). Use of blockchain in designing smart city. Smart and Sustainable Built Environment 9.4: 695–709.

Djahel, S., Doolan, R., Muntean, G.-M. and Murphy, J. (2015). A communications-oriented perspective on traffic management systems for smart cities: Challenges and innovative approaches. IEEE Commun. Surveys Tuts. 17(1): 125–151, 1st Quart.

Feng, H., Wang, X., Duan, Y., Zhang, J. and Zhang, X. (2020). Applying blockchain technology to improve agri-food traceability: A review of development methods, benefits and challenges. Journal of Cleaner Production 260: 121031.

Fernandez-Anez, V. (2016). Stakeholders approach to smart cities: A survey on smart city definitions. Proc. Smart-CT, Málaga, Spain, pp. 157–167.

Gallo, Pierluigi, Suporn Pongnumkul and Uy Quoc Nguyen. (2018). BlockSee: Blockchain for IoT video surveillance in smart cities. 2018 IEEE International Conference on Environment and Electrical Engineering.

Gharaibeh, A., Salahuddin, M. A., Hussini, S. J., Khreishah, A., Khalil, I., Guizani, M. et al. (2017). Smart cities: A survey on data management, security, and enabling technologies. IEEE Communications Surveys & Tutorials 19(4): 2456–2501.

Goumopoulos, Christos. (2024). Smart city middleware: a survey and a conceptual framework. IEEE Access.

Ibba, S., Pinna, A., Seu, M. and Pani, F. E. (2017, May). CitySense: blockchain-oriented smart cities. pp. 1–5. In Proceedings of the XP2017 Scientific Workshops.

Ijaz, S., Shah, M. A., Khan, A. and Ahmed, M. (2016). Smart cities: A survey on security concerns. Int. J. Adv. Comput. Sci. Appl. 7(2): 612–625.

Jarrett, A. and Choo, K. R. (2021). The impact of automation and artificial intelligence on digital forensics. WIREs Forensic Science 3(6). https://doi.org/10.1002/WFS2.1418.

Kakarontzas, G., Anthopoulos, L., Chatzakou, D. and Vakali, A. (2014, August). A conceptual enterprise architecture framework for smart cities: A survey based approach. pp. 47–54. In 2014 11th International Conference on e-Business (ICE-B). IEEE.

Kakavand, Hossein, Nicolette Kost De Sevres and Bart Chilton. (2017). The blockchain revolution: An analysis of regulation and technology related to distributed ledger technologies. Available at SSRN 2849251.

Kiyomoto, S., Rahman, M. S. and Basu, A. (2017, June). On blockchain-based anonymized dataset distribution platform. pp. 85–92. In 2017 IEEE 15th International Conference on Software Engineering Research, Management and Applications (SERA). IEEE.

Cui, L., Xie, G., Qu, Y., Gao, L. and Yang, Y. (2018). Security and privacy in smart cities: Challenges and opportunities. IEEE Access 6: 46134–46145.

Liu, Z., Chi, Z., Osmani, M. and Demian, P. (2021). Blockchain and building information management (BIM) for sustainable building development within the context of smart cities. Sustainability 13(4): 2090.

Makani, S., Pittala, R., Alsayed, E., Aloqaily, M. and Jararweh, Y. (2022). A survey of blockchain applications in sustainable and smart cities. Cluster Computing 25(6): 3915–3936.

Marsal-Llacuna, Maria-Lluïsa. (2020). The people's smart city dashboard (PSCD): Delivering on community-led governance with blockchain. Technological Forecasting and Social Change 158: 120150.

Mora, H., Mendoza-Tello, J. C., Varela-Guzmán, E. G. and Szymanski, J. (2021). Blockchain technologies to address smart city and society challenges. Computers in Human Behavior 122: 106854.

Mu, L., Liu, J., Zhou, G., Wu, C., Chen, B., Lu, Y. et al. (2021). Obesity prevalence and risks among Chinese adults: findings from the China PEACE million persons project, 2014–2018. Circulation: Cardiovascular Quality and Outcomes 14(6): e007292.

Musa, A. A., Dulawat, S., Saleh, K. T. and Alhassan, I. A. (2022). Applicability of utilizing blockchain technology in smart cities development. Smart City Infrastructure: The Blockchain Perspective 317–339.

Nakamoto, Satoshi. (2008). Bitcoin: A peer-to-peer electronic cash system. Satoshi Nakamoto.

Nam, K., Dutt, C. S., Chathoth, P. and Khan, M. S. (2021). Blockchain technology for smart city and smart tourism: latest trends and challenges. Asia Pacific Journal of Tourism Research 26(4): 454–468.

Xie, J., Tang, H., Huang, T., Yu, F. R., Xie, R., Liu, J. et al. (2019). A survey of blockchain technology applied to smart cities: Research issues and challenges. IEEE Communications Surveys & Tutorials 21(3): 2794–2830.

Zhu, X., Chen, Z., Cheng, T., Yang, C., Wu, D., Wu, Y. et al. (2024). Blockchain for urban governance: Enhancing trust in smart city systems with advanced techniques. Sustainable Cities and Society 108: 105438.

Chapter 19

Intelligent Data Mining

The Role of Machine Learning and Deep Learning in Extracting Insights from Big Data

R. Ravinder Reddy,[1,*] *Allipurapu Meghanath Reddy,*[1] *E. Padma Latha,*[1] *M. V. Krishna Reddy*[1] and *K. Karthik*[2]

1. Introduction

In today's era of the instant availability of information, the explosion of digital data has opened unprecedented opportunities for extracting meaningful information from large databases. The power to analyse and extract insights from large data sets has become a critical component of innovation, from business intelligence to medical diagnosis. However, the complexity, volume, and velocity of data in today's society far exceed the capabilities provided by traditional analytical methods. It is here that the advent of machine learning (ML) and deep learning (DL) has revolutionized the game, offering scalable, automated, and accurate methods for handling large data sets.

Machine learning, a division of artificial intelligence (AI), creates algorithms that enable systems to learn from data and make predictions having no need for explicit programming. Deep learning, a division of ML, uses multiple layered neural networks to detect complex patterns and relationships in high-dimensional data. These approaches together form the paradigm of data analysis in today's era, allowing to detect hidden patterns, predict future patterns, and optimize decision-making processes. This introduction explains the critical role of ML and DL in analyzing large databases, highlighting their approaches, applications, and related challenges. Further, it provides a summary of data-driven insights in a wide range of industries, from finance to healthcare and e-commerce.

2. Literature Survey

Deep learning models—such as convolutional neural networks (CNNs), recurrent neural networks (RNNs), autoencoders, and generative adversarial networks (GANs)—are widely evaluated based on their performance in feature representation, extraction, and learning across domains such as

[1] Department of Computer Science and Engineering, Chaitanya Bharathi Institute of Technology, Hyderabad, India.
[2] CVR College of Engineering, Hyderabad, India.
Emails: meghanathreddyallipurapu@g mail.com, epadmalatha_cse@cbit.ac.in, krishnareddy_cse@cbit.ac.in, karthik.5786@gmail.com
* Corresponding author: rravinderreddy_cse@cbit.ac.in

DOI: 10.1201/9781003620822-19

computer vision, natural language processing (NLP), and time-series analysis. It emphasizes their superiority over conventional machine learning in feature learning of intricate features and enhancing task performance and interpretability. Open issues in deep learning are model interpretability, which prevents decision-making and feature importance understanding and influences trust in high-stakes applications. Models also need a significant amount of labeled data, which poses challenges in data-poor settings. Scalability to heterogeneous, humongous datasets is another major challenge (Hussain 2024). While its significant high computational demands seem to be stopping the use of machine learning and deep learning especially in under-resource environments, it also points out some issues concerning privacy of sensitive data in secured handling with respect to model interpretability and transparency, all of which have effects in various regions such as healthcare, finance due to their trustworthiness challenges (Patil et al. 2024).

A key challenge in applying machine learning models lies in the need for careful data preprocessing to ensure data quality and compatibility, which are essential for achieving accurate and reliable model performance. This article also refers to the complexity of model selection and performance measurement, noting the challenge of finding the right algorithm as well as its efficiency for a specific business task (Kannan et al. 2021). Big data presents several challenges, including data quality, scalability, and interpretability. Machine learning techniques help address these issues by enabling automated data cleaning, efficient processing of large-scale datasets, and the extraction of meaningful patterns, thereby facilitating insights from complex and high-dimensional data. Big data quality brings inconsistencies and inaccuracy, which taint the effectiveness of machine learning models, and thus we require high-quality data for trustworthy insights. Scalability is another issue because machine learning algorithms need to support increasing volumes of data with advanced distributed computing and optimized resource utilization (M. Dr. 2024).

Conventional methodologies of data analysis are constrained by the scale and character of big data's scope, incapable of unleashing its insights. Furthermore, areas of finance, health, and national security are subjected to increasing demand for model interpretableness, as well as explainability, reinforcing methodology shortcomings currently in place (Rane et al. 2024). Conventional methods for data analysis fail to deal with the size and complexity of data, hence reducing the ability to derive useful insights. Additionally, industries like finance, healthcare, and public safety need increased transparency and interpretability of analysis models, hence highlighting the shortfalls of available methods that are incapable of meeting these needs (Sunil et al. 2024). Implementing deep learning in edge computing environments presents several challenges, including limited computational resources, constrained memory, and energy efficiency concerns. Additionally, during model training, issues such as overfitting and underfitting can arise, impacting the generalizability and performance of the deployed models (Rane et al. 2024).

Traditional ML for small data, deep learning for large data. Deep learning models can be categorized as either supervised or unsupervised learning approaches, as noted by (Vaissnave et al. 2024). However, despite their growing application, deep machine learning remains in its early stages, primarily due to the overwhelming resource demands related to big data, high-dimensional parameters, and intensive computational processing (Sarker 2021). Machine learning algorithms allow patterns in big data to be automatically detected in order to make predictions. This study is on decision trees, linear regression, and random forest regression for end data prediction from successive dependent data, using the Scikit-Learn library in Python for execution. Sophisticated algorithms don't necessarily mean increased accuracy rates. Model tuning and configuration are essential to find results (Quevedo 2024). Machine learning is a fast-growing technology that allows computers to learn from data, predict, and enhance performance without direct programming, transformed by one API for cross-architecture programming. No specific limitations are discussed in the abstract. See the full paper for in-depth limitations (Asnicar et al. 2024).

Machine Learning (ML) is a subset of Artificial Intelligence that allows computers to learn from data, discover patterns, and make predictions without being programmed explicitly. It involves methods such as supervised, unsupervised, and reinforcement learning, and a host of algorithms and applications in a wide range of domains. The limitations are privacy, bias, and transparency in ML. Ethical implications are given priority in machine learning applications (Munde 2024). Deep learning, a strong machine learning method, is based largely on sophisticated data mining and pattern recognition by means of use of models such as CNNs and RNNs to automatically learn features from large amounts of data, e.g., images, text, and time series. Data scarcity and annotation difficulty hinder model training. The quality of the annotated data affects the performance of the model and generalization (Lian and Chen 2024). Deep learning is a type of machine learning that uses artificial neural networks with numerous layers to learn complex patterns in complex data. It is especially useful in domains such as computer vision, natural language processing, and speech recognition, with very high success in most domains. Training data biases can lead to discrimination. Complex models are a problem with explainability and transparency (Rai et al. 2024).

The estimation of physical quantities such as radio frequency electric field vector and isotope ratio in plasma, with great advances in speed and accuracy compared to conventional analysis methods. Conventional analysis is computationally costly and time-consuming. There are many free parameters present in spectral data fitting (Ronchi et al. 2024). Complex-valued neural networks improve signal processing and communication modeling performance. We introduce a PyTorch package to perform basic operations such as linear, convolution, and attention modules, facilitating data-driven research and applications of signal processing. Currently, there is a lack of frameworks that support complex-valued neural networks, and the literature reflects limited investigation into manifold-based complex-valued layers. This gap highlights the need for further research and development in this emerging area of deep learning (Lee and Gao 2022).

Future directions of deep learning include enhanced self-supervised learning, hybrid models, multi modal learning, and quantum computing embedding. Explainable AI, federated learning, and edge AI will enable flexibility, privacy, and efficiency across a broad set of applications. Existing models are not capable of ongoing learning and forgetting catastrophically. Complexity makes it hard to explain ML model choices (Rane et al. 2024). The future of deep learning, particularly in artificial electromagnetic materials, is to address challenges like dataset requirements and model interpretability. Incorporating prior knowledge in expert deep neural networks may unlock new physics and problem-solving capabilities in this field. This requires large training datasets to be effective. The nature of the black box hinders the understanding of the earned physics (Wu et al. 2025).

3. Existing Challenges

With rapid advancements in the modern digital era, fast-paced technology has resulted in a phenomenal growth in generation of data. The growth of internet, Internet of Things (IoT), social media, and large-scale digitization has resulted in databases that are typically measured in terabytes and even petabytes. These big datasets consist of various types of data, which include textual, images, video, sensor data, and numerical data. Extracting meaningful information from these large and complex datasets is a challenging task that requires several analytical methods.

Volume: The large size of modern datasets makes manual analysis unfeasible. Previously used data processing methods cannot handle the staggering amounts of data on a daily basis by companies, scientific research, and online services.

Velocity: Data is generated at breakneck speeds. In financial trading, social media monitoring, and industrial IoT, real-time analysis is imperative. Delayed insights can lead to lost opportunities, inefficiencies, or security threats.

Variety: Currently data is in different forms: other than the traditional structured forms, like relational databases. Proper management and analysis of these diverse forms of data require the use of specialized frameworks and tools. It is in various forms, including: (a) Structured (b) Semi-structured (c) Unstructured.

Veracity: Not all data have the characteristics of accuracy, completeness, or reliability. Inconsistencies, biases, missing values, and inaccuracies in data sources may affect decision-making processes. Ensuring data integrity, correcting corrupted records, and handling uncertainty are part of the data management.

4. Role of Machine Learning in Large Databases

Machine learning involves creating algorithms which are capable of processing data, learning new trends, and predicting with little involvement of human judgment. Machine learning enables organizations to automate data processes, enhance decision-making, and discover hidden patterns in big sets of data. The different types of learning are shown in Figure 19.1.

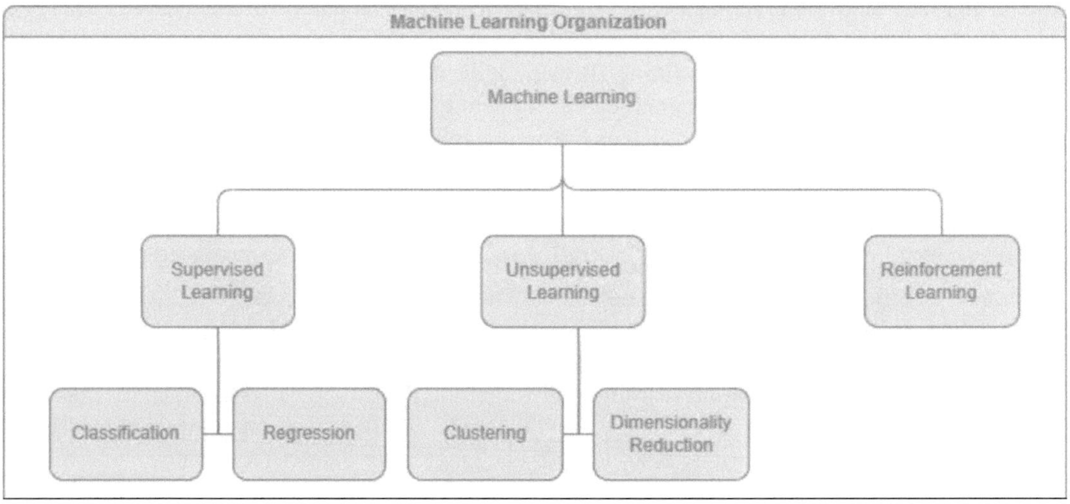

Figure 19.1. Various methods in machine learning.

4.1 Supervised Learning

This type of learning trains a model using labelled data where the input and output is given. These models examine historical data to forecast future outcomes, it completely depends on the given data.

Regression: It works on the numerical data to predict the continuous values, it predicts the new outcome on the historical recorded data. The examples are Linear Regression, Logistic Regression, Support Vector Regression.

Classification: For classifying data into predefined labels. The application is Fraud Detection, it identifies suspicious transaction patterns to identify fraudulent transactions. The examples are Decision Trees, Random Forests, SVMs, Neural Networks.

4.2 Unsupervised Learning

Discovers patterns in unlabeled data without known labels. The algorithm uncovers meaningful patterns in data.

Clustering: This means organizing similar data points into meaningful groups. The application is Customer Segmentation, that is, segmentation of customers based on buying behaviors for targeted marketing.

Anomaly detection: This identifies fraudulent transactions, network intrusions, or defects in manufacturing. The Examples are k-Means, Density-Based Spatial Clustering of Applications with Noise, Hierarchical Clustering.

Dimensionality reduction: This means reducing variables while retaining essential details. Applications are exploratory Data Analysis, it discovers hidden relationships and trends within massive datasets. The examples are Principal Component Analysis, t-Distributed Stochastic Neighbor Embedding, Autoencoders.

4.3 Semi-Supervised Learning

It consists of labeled and unlabeled data. This approach is beneficial in cases where labeling data requires significant time and resources. There are several techniques like self-training approaches, generative adversarial networks, Graph-based techniques, etc. The major applications are listed as follows:

Medical image analysis: Improving diagnosis by leveraging a limited set of annotated scans combined with a vast collection of unannotated images.

Text classification: Constructing NLP models based on a limited labeled dataset in combination with a larger unlabeled data.

Speech recognition: Improving automatic transcription models with a limited amount of labeled speech input.

4.4 Reinforcement Learning

These models learn by interacting with their environment, enhancing decision making through rewards and have feedback mechanisms to update weights. The major techniques are Q-learning, Deep Q Networks, Policy Gradient Techniques. Applications are listed as:

Dynamic pricing: Dynamic price adjustment is based on demand fluctuations, competitive actions, and prevailing market conditions in real time.

Logistics optimization: This is enhancing supply chain efficiency through optimizing delivery routes and warehouse management.

Game AI: This means training AI agents to play video games, board games, and real-world robotics tasks.

Machine learning offers strong techniques for analyzing, processing, and extracting useful insights from large data. The use of these techniques by organizations, researchers can enhance automation, enhance prediction, and facilitate better decision-making. As data keeps increasing in both size and complexity, the strategic use of machine learning remains essential for efficiently handling large amounts of data. The various machine learning approaches are compared in Table 19.1.

Table 19.1. Comparative table that summarizes different types of ML approaches.

Aspect	Supervised Learning	Unsupervised Learning	Semi-Supervised Learning	Reinforcement Learning
Data Requirement	Requires labeled data	Uses only unlabeled data	Mix of labeled and unlabeled	Learns from environment interactions
Key Objective	Predict outcomes based on past data	Discover hidden structures and patterns	Improve learning efficiency with limited labels	Optimize decision-making through trial and error
Dependency on Labels	High	None	Partial	None (relies on rewards)
Techniques Used	Regression, Classification	Clustering, Dimensionality Reduction	Self-training, GANs, Graph-based models	Q-learning, Policy Gradients, Deep Q Networks
Examples of Models	Decision Trees, Neural Networks, SVMs	k-Means, PCA, Autoencoders	GANs, Self-training NLP models	AlphaGo, Robotics, Dynamic Pricing
Use Cases	Customer churn prediction, Fraud detection	Customer segmentation, Anomaly detection	Medical image analysis, Speech recognition	Logistics optimization, Game AI
Handling Large Data	Well-suited, but labeling is costly	Scalable, but interpretation can be hard	Efficient for large datasets with few labels	Computationally expensive, needs simulations
Learning Process	Learns from known input-output pairs	Finds structures in data without labels	Utilizes some labeled data to refine learning	Learns by maximizing rewards over time
Decision-Making	Predictive, based on past knowledge	Pattern-based, exploratory	Hybrid, improving accuracy with less data	Adaptive, continuous improvement through feedback

5. Role of Deep Learning for Complex and High-Dimensional Data

Deep learning, modeled after the brain's structure and functionality, applies multi-layered networks to process as well as analyze complex data. Deep learning is a better approach for processing unstructured data, e.g., images, videos, and text, through the auto-learning of complex patterns and representations (Najafabadi et al. 2016). Deep learning is different from machine learning in that it requires less feature engineering, making it more effective when processing large datasets. The various approaches of deep learning are shown in Figure 19.2.

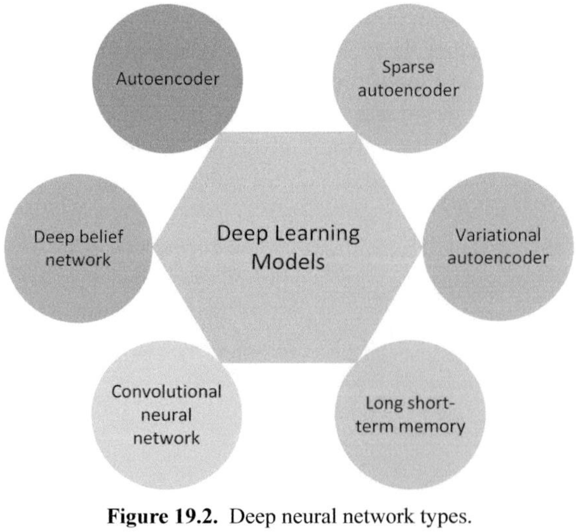

Figure 19.2. Deep neural network types.

5.1 CNNs

CNNs are made specifically to interpret images and videos by learning spatial hierarchies of features. CNNs apply convolutional layers to learn patterns, e.g., edges, textures, and complex objects, making them suitable for tasks involving visual recognition. The key components are:

i. Convolutional layers to capture and analyze spatial patterns in images.
ii. Pooling layers to simplify data without losing vital information.
iii. Fully connected layers to combine the features extracted for classification.

The major applications are:

Image recognition: Used in face recognition, object detection, and security systems.

Medical imaging: Used in tumor detection, disease classification, and radiological examination.

Autonomous vehicles: Enables self-driving cars to read road signs, identify pedestrians, and identify obstacles.

5.2 RNNs

RNNs are built to handle sequential information by learning temporal relationships and context information, making them relevant to activities involving understanding relationships over time. The key variants of RNN are:

Long Short-Term Memory (LSTM): This can overcome the vanishing gradient problem by retaining long-term relations in data.

Gated Recurrent Units (GRU): It simplifies the network while retaining performance levels similar to LSTMs.

Its applications are in various domains are listed as:

Natural Language Processing (NLP): This enables chatbots, sentiment detection, and machine translation.

Speech recognition: Converts speech to text for virtual bots like Siri and Alexa.

Time-Series Forecasting: Predicts stock market trends, weather trends, and sales results.

5.3 Autoencoders

Autoencoders are neural networks designed for tasks such as data compression, anomaly detection, and noise reduction by learning compact and efficient representations of input data. The key components are:

Encoder: Encodes data into reduced dimensions.

Decoder: Extracts baseline data from its reduced form.

The major applications are:

Anomaly detection: Identifies fraudulent transactions, faulty products, or network attacks.

Data denoising: Denoises image, audio, and text data.

Dimensionality reduction: Reduces the feature space without losing relevant information for visualization and clustering operations.

5.4 Transformer Models

Transformer models process sequential and contextual information through self-attention mechanisms, which makes them highly efficient to process long sequences of text and other structured data sets. Its key features are:

Self-attention mechanism: Helps models to focus on different portions of input data based on contextual relevance.

Parallelization: Enables easy processing of large datasets compared to RNNs.

The applications are wide in range with accurate preferences.

Language modeling: Empowers NLP models such as BERT, GPT, and T5 for text comprehension and generation.

Machine translation: Empowers applications like Google Translate and DeepL for text translation across several languages.

Text summarization: Shortens lengthy documents and preserves meaning and context.

5.5 Generative Adversarial Networks (GANs)

GANs concentrate on creating realistic synthetic data through the training of two adversarial networks named the generator and a discriminator. The key components are.

Generator: It generates samples of original data.

Discriminator: Discerns genuineness of the produced data and refines model accuracy step by step. The various applications of GAN are as follows.

Data augmentation: Produces synthetic images to supplement training sets, resulting in enhanced model performance.

Image generation: Generates realistic images, deep fake videos, and art.

Synthetic dataset creation: Fixes the problem of imbalanced datasets by creating rare instances of data relevant for medical studies, fraud detection, and industrial purposes.

Deep learning has transformed the processing and analysis of complex and high-dimensional data. By utilizing particular neural network architectures like CNNs, RNNs, autoencoders, transformers, and GANs, deep learning facilitates revolutionary innovation in AI applications. With improved computing power and data availability, deep learning will be the primary source of innovation in many fields like healthcare, finance, automation, and artificial intelligence.

6. Large Scale Databases

Here are five large databases commonly used in ML, DL, and data analysis to extract meaningful information:

6.1 ImageNet

ImageNet is a large database which is visually based to be used in visual object recognition research. It offers a huge number of images organized by the WordNet hierarchy. The size is more than fourteen million hand-annotated images with labels that describe the objects in the image, distributed over more than 20,000 categories. The major applications are employed extensively to train and test DL models for detecting objects, segmentation tasks, and classification of images. The ImageNet is a Large Scale Visual Recognition Challenge (ILSVRC) that has been a driving force in computer vision research annually.

6.2 Common Crawl

Common Crawl is a data repository available openly for web scraping that can be analyzed and accessed by all users. The repository has raw web page data, text extractions and associated metadata. The size of the repository holds petabytes of data, comprising billions of individual web pages. The major applications are:

a) it is widely utilized for training large-scale language models,
b) it has varied natural language processing tasks, web mining activities, and machine learning field studies that involve large volumes of textual data.

6.3 COCO (Common Objects in Context) Dataset

COCO is a vast image captioning, detection of objects dataset. It contains complex day-to-day scenes with ordinary objects in their everyday settings. Over 330,000 images, of which there are 200,000 images that are labeled and contain over 1.5 million instances of objects that are spread across eighty different object categories and ninety-one categories of other stuff. The main applications are to train and test computer vision models on object recognition, segmentation, and contextual understanding of scenes.

6.4 YouTube-8M

YouTube-8M is an expanded annotated video content-based data consisting of millions of YouTube video IDs and superior quality labels generated by machines. Data size has more than six million videos (around 450,000 hours of video), labeled with a 3,800-word vocabulary of visual entities. These large datasets are essential for training machine learning models, especially in areas like video classification, content analysis, and understanding the relationships between different events in a video over time. This is particularly important for tasks like video recommendations (such as what shows or videos to suggest next) and video retrieval (finding specific videos based on content).

6.5 Wikipedia Corpus

The Wikipedia Corpus is a massive collection of text derived from Wikipedia articles. It is multilingual and covers virtually every topic imaginable, from technology to history to pop culture. It's like a giant, living encyclopedia that is always growing as people contribute new knowledge and edit existing pages. The corpus size contains billions of words and articles, and it continues to expand as more people around the world contribute to Wikipedia. This vast resource is incredibly valuable for language processing tasks like language modeling, classification of textual information, language translations, summarization (condensing information), and entity recognition (identifying important names, places, and terms). It is used to train powerful language models like GPT, BERT which are at the core of many AI applications today.

They provide the diverse and complex data needed to train models that can pick up on patterns, make sense of large amounts of information, and generate insights. Without access to resources like these, developing intelligent systems that can truly understand content, whether video or text would be much more challenging. The various large-scale datasets are listed in the Table 19.2.

Table 19.2. Large scale datasets.

Dataset	No. of Rows (Samples)	No. of Columns (Features)	Size (Rows × Columns)	Description
ImageNet	14M images	1,000 classes	14M × 1,000	A large dataset for image classification, widely used in deep learning.
COCO	330K images	5 annotations per image	330K × 5	Designed for object detection, segmentation, and captioning tasks.
LAION-5B	5.85B image-text pairs	2 (image, text)	5.85B × 2	A massive open-source dataset for training vision-language models.
The Pile	22M text documents	Varies (avg. 40K tokens)	22M × ~ 40K	A diverse dataset used for training large-scale language models.
C4	365M text samples	Varies (avg. 1,800 tokens)	365M × ~ 1,800	A cleaned version of Common Crawl used for NLP tasks.
OpenWebText	8M text samples	Varies (avg. 1,500 tokens)	8M × ~ 1,500	An alternative to OpenAI's WebText dataset for language model training.
Wiki40B	40M Wikipedia articles	Varies (avg. 2,500 tokens)	40M × ~ 2,500	A cleaned Wikipedia dataset widely used for NLP.
MS MARCO	8.8M passages	2 (passage, label)	8.8M × 2	Used for machine reading comprehension and search relevance.
Common Crawl	3.1B web pages	Varies (avg. 5,000 tokens)	3.1B × ~ 5,000	A massive web dataset scraped from the internet.
SQuAD	100K QA pairs	3 (context, question, answer)	100K × 3	A benchmark dataset for question answering models.

7. Techniques to Handle Large Databases with ML and DL

As data size increases, efficient processing and analysis of large databases is a significant issue (Ghadge et al. 2024). The following are some of the most important techniques that optimize performance, improve efficiency, and make large datasets processable at scale.

7.1 Feature Engineering

Feature engineering is a crucial step in enhancing the performance of models. It involves converting raw data into valuable features that make it easier for models to learn patterns, leading to better accuracy and more efficient computations. Here are some key techniques used in feature engineering:

Scaling and normalization: Min-max scaling, standardization, and log transformations help in processing data with heterogeneous ranges and distributions.

Encoding categorical variables: This includes one-hot encoding, label encoding, and target-based encoding to facilitate categorical data making them processed numerically, making it simpler for models to process.

Feature selection: This entails mutual information, recursive feature elimination, and correlation-based selection facilitate selecting most relevant attributes and eliminating redundant or irrelevant features.

Feature creation and polynomial expansion: New features, including polynomial features, interaction terms, and domain-specific crafted attributes can enhance model predictions.

Handling missing data: Mean/median imputation, KNN-based imputation, and predictive modeling replace missing values to preserve data integrity.

7.2 Distributed Computing

Large data processing demands distributed computing environments that decentralize workloads to numerous nodes, which drastically reduces computation time and constrains resources.

Apache spark: A popular distributed computing environment that facilitates parallel big data processing. Spark's MLlib library facilitates scalable machine learning.

Hadoop MapReduce: This is a distributed data processing environment for large dataset processing by dividing them into smaller units and processing them in parallel.

TensorFlow and PyTorch distributed computing: Frameworks like TensorFlow and PyTorch offer distributed training options, including data parallelism and model parallelism, for scalable training of deep networks on large data.

Google Cloud TPUs and AWS GPUs: Cloud-based GPU and TPU (Tensor Processing Unit) offerings speed up deep learning computations, making training much faster.

7.3 Dimensionality Reduction

High-dimensional data slows down computation and can lead to overfitting. Dimensionality reduction techniques help in the choice of most important features with least data complexity.

PCA: This is a standard statistical method that projects high-dimensional data to reduced dimensions with maximum variance retained.

Autoencoders: They are deep learning-based techniques that condense input data to a reduced-dimensional latent representation and reconstruct it, preserving vital information.

t-SNE and UMAP: These are sophisticated non-linear methods for reducing dimensionality used in data visualization and feature extraction.

Feature selection methods: Lasso regression, Decision Tree-based selection, and Genetic Algorithms help in the choice of most relevant attributes from large datasets.

7.4 Real-Time Data Processing

It is essential to process continuously streaming datasets from large databases, allowing for dynamic decision-making and anomaly detection.

Apache Kafka: A vastly used messaging system that enables instantaneous data ingestion, processing, and streaming analytics.

Apache Flink: A robust stream processing platform that enables event-driven architectures and low-latency computations.

Google Cloud dataflow and AWS kinesis: Cloud-based real-time data processing platforms that enable large-scale, event-driven machine learning applications.

Online learning algorithms: Incremental ML algorithms such as Online Gradient Descent and Hoeffding Trees allow models to adapt continuously to new data.

7.5 Transfer Learning

Training deep learning models from scratch on big data is computationally intensive and time-consuming. Transfer learning allows models to leverage pre-trained architectures and minimize the amount of extensive training.

Pre-Trained CNNs for computer vision: ResNet, VGG, and EfficientNet, pre-trained on huge data including ImageNet, can be finely tuned for tasks with little labeled data.

Transformer based models: BERT, GPT, and T5 are pre-trained models that can be fine-tuned for a range of text-based instances with little additional training.

Feature extraction using pre-trained models: Instead of training a model from scratch, pre-trained networks can be used as feature extractors, transferring the learned representation to a new classifier.

Domain adaptation: Transfer learning is particularly effective in adapting models trained on one domain for use in a related but distinct domain, minimizing data labelling tasks.

Efficient processing of big databases calls for a combination of sophisticated ML/DL methods and scalable computational platforms. Feature engineering provides high-quality input data, while distributed computing facilitates parallel processing. Dimensionality reduction minimizes complexity, real-time processing facilitates dynamic decision-making, and transfer learning speeds up deep learning applications. Combining these methods increases model efficiency, minimizes computational effort, and provides scalable solutions to processing huge data.

8. Challenges and Solutions

Effective processing of large data is necessary for ML and DL. Scalability, interpretability, data quality, computational cost, and data privacy are main challenges. These are mentioned below along with their solutions.

8.1 Scalability

The increased size of the data is a challenge to process and this is a major concern these days. Terabyte-sized data processing requires huge computational power. Single-node systems lack scalability in terms of speed, memory, and inefficiency (Rai et al. 2024). The proposed solutions are as follows.

Distributed computing: Apache Spark, Hadoop, and Dask support parallel processing of data.

Cloud services: like AWS, Google Cloud, and Azure offer scalability in storage and computation.

Efficient algorithms: Mini-batch gradient descent reduces memory usage while training efficiently.

Optimized databases: NoSQL databases such as MongoDB and Cassandra optimize data access speed.

8.2 Interpretability

Deep networks are often considered "black boxes" due to their complex decision-making abilities. The need to provide appropriate solutions is a big concern and the existing solutions are listed.

Explainability tools:

LIME: (Local Interpretable Model-agnostic Explanations) gives local approximations.

SHAP: (Shapley Additive Explanations) gives feature importance scores.

Attention mechanisms: In NLP and vision models, attention layers focus on important factors.

Model visualization: TensorBoard and Captum help in determining feature effect.

8.3 Data Quality

Real time data and most of the data quality is a big challenge. Most of the data is Noisy, missing, or biased data that leads to erroneous predictions. Before training the data, we need to apply the following approaches to get the quality data.

Data cleaning: Imputation (mean, median, predictive modeling) removes missing values, and duplicate elimination maintains data integrity.

Data augmentation: Image transformations (rotation, scaling, color changes). NLP processes such as back-translation enhance text datasets. Synthetic data generation (GANs, SMOTE) can be used for handling class imbalance.

Bias mitigation: Statistical fairness measures and re-sampling techniques reduce bias in the training set.

8.4 High Computational Costs

The main challenge is that training of large models is computationally intensive. We need to provide efficient ways of learning mechanisms. Here the solutions are:

Hardware Optimization: GPUs (e.g., NVIDIA A100) speed up matrix operations. TPUs (Tensor Processing Units) optimize deep learning efficiency.

Model Optimization: Quantization compresses models by reducing precision. Pruning reduces unnecessary neural connections. Knowledge distillation enables smaller models to mimic larger models.

Efficient frameworks: PyTorch and TensorFlow support multi-GPU training using tools like Horovod.

8.5 Data Privacy and Security

The major challenge is maintenance of GDPR, CCPA, and HIPAA compliance while preventing data breaches. We can provide the solutions using the following approaches.

Privacy-preserving techniques: Federated Learning learns locally without exposing raw data. Differential privacy introduces noise to protect individual data points. Homomorphic encryption supports computations on encrypted data.

Secure storage and access: Sensitive data is protected through encryption and role-based access control (RBAC), which ensure secure storage, transmission, and access restrictions based on user roles.

Adversarial defense: Adversarial training and input preprocessing block malicious model tampering.

Processing big data in ML and DL requires scalable computation, explainable models, quality data, efficient computation, and robust security. Organizations can obtain accurate and scalable AI models, with compliance and security requirements met through distributed computing, efficient algorithms, and techniques for preservation of privacy.

9. Significance of Machine Learning and Deep Learning in Big Databases

As the digital world expands, organizations and companies generate huge mountains of data every day. Conventional processes for handling and analyzing big data are slow and ineffective. Machine learning and deep learning are acting as cutting-edge technologies that automate data processing, enhance decision-making, and extract valuable information from big databases.

Efficient Data Processing

It is no longer feasible to process big data manually. ML and DL algorithms are capable of processing huge volumes of structured and unstructured data rapidly and accurately. By optimizing data management processes, these technologies will save time and minimize human errors, thus making database operations efficient.

Automating Data Analysis to Make Better Decisions

As an alternative to traditional methods for analysis which depend on human inputs, ML and DL can recognize patterns, predict outcomes, and analyze trends with a minimal amount of human involvement. Companies utilize these features to enhance decision-making in finance, healthcare, and retail sectors, where prompt and accurate insights are the foundation of business success.

Pattern Detection and Identifying Unusual Behaviour

Perhaps the strongest aspect of ML and DL is the capability to recognize patterns and identify unusual behavior in data. This is especially effective in detecting frauds, cybersecurity related issues, and quality control, where early identification of unusual behavior helps in saving costly financial or operational losses. For instance, banks utilize ML models to recognize suspicious transactions and block fraudulent transactions.

Extracting Value from Unstructured Data

Much of the data today—such as emails, social media exchanges, medical scans, and video does not easily fit into traditional database architectures. ML and DL enable businesses to process and analyze unstructured data, and this has enabled speech recognition breakthroughs, sentiment analysis, and medical diagnosis.

Enabling Real-Time Decision Making

Many industries, such as e-commerce, finance, and transport, employs real-time analysis of data to compete. ML and DL power applications such as personal recommendations, stock market predictions, and autonomous car navigation, enable companies to make real-time, data-based decisions.

Improving Scalability and System Performance

As data keeps growing in size, businesses need solutions that scale well. ML and DL help optimize database performance by improving storage, retrieval, and processing rates. This translates to companies being able to handle large volumes of data without impacting system performance.

10. Conclusion and Future Directions

Machine learning and deep learning have revolutionized the manner in which large databases are handled by organizations. The technologies enable the automation of data processing, improve decision making, and retrieve useful insights which were previously difficult to obtain. As AI keeps evolving, the uses of ML and DL in handling big data will continue to expand, driving innovation across industries and shaping the future of data-driven decision-making. Technologies continue to mature and overcome scalability challenges, interpretability challenges, and the cost of computing. It is expected that even more wondrous applications will be seen across industries. With the speedy evolution of Artificial Intelligence, the role of ML and DL in the big data analytics domain is only going to increase.

ML and DL have greatly transformed the art of extracting useful insights from big data. With the automation of complex analytical processes, these technologies allow organizations to realize the true potential of big data. Despite the widespread problems of scalability, interpretability, and data quality, ongoing algorithmic advancements, computing power, and ethics indicate that ML and DL will increasingly be at the center of driving data-driven innovation in industries. This convergence of technology and data is driving better decision-making, greater efficiencies, and innovation breakthroughs in today's digital age.

Future Directions

Federated learning: Training models on distributed data sources while preserving data privacy by not exposing raw information.

Self-supervised learning: Technologies which allow models to learn from a wide range of unlabeled data, reducing dependency on human annotations.

Automated machine learning (AutoML): Automated feature engineering, hyper parameter tuning, and model selection, democratizing ML and streamlining it.

Edge computing: Processes data nearer to devices (e.g., IoT, mobile), reducing latency, bandwidth, and improving real-time decision-making.

Ethical AI: Emphasizes equity, openness, and responsibility to reduce biases and ensure ethical AI implementation.

References

Asnicar, F., Thomas, A. M., Passerini, A., Waldron, L. and Segata, N. (2024). Machine learning for microbiologists. Nature Reviews Microbiology 22(4): 191–205.

Ghadge, N. (2024). Machine learning: enhancing intelligent search and information discovery. Computer Science & Information Technology (CS & IT) 235–243. Retrieved from https://aircconline.com/csit/papers/vol14/csit141021.pdf.

Hussain, N. Y. (2024). Deep learning architectures enabling sophisticated feature extraction and representation for complex data analysis. International Journal of Innovative Science and Research Technology (IJISRT) 9(10): 11. Retrieved from https://www.ijisrt.com/deep-learning-architectures-enabling-sophisticated-feature-extraction-and-representation-for-complex-data-analysis.

Kannan, R., Rosdi, I. S., Ramakrishnan, K., Rasid, H. R. A., Rafy, M. H. I. M., Yusuf, S. et al. (2021). Leveraging business data analytics and machine learning techniques for competitive advantage: Case study evidence from small businesses. International Journal of Management, Finance and Accounting 2(1): 73–87.

Lee, C., Hasegawa, H. and Gao, S. (2022). Complex-valued neural networks: a comprehensive survey. IEEE/CAA Journal of Automatica Sinica 9(8): 1406–1426. doi: 10.1109/JAS.2022.105743. Retrieved from https://www.ieee-jas.net/article/doi/10.1109/JAS.2022.105743.

Lian, J. and Chen, T. (2024). Research on complex data mining analysis and pattern recognition based on deep learning. Journal of Computing and Electronic Information Management 12(3): 37–41. from https://drpress.org/ojs/index.php/jceim/article/view/20559.

M, Dr. (2024). A novel study on machine learning algorithms for big data. International Journal of Scientific Research in Engineering and Management 08: 1–5. doi: 10.55041/IJSREM29690. Retrieved from https://ijsrem.com/download/a-novel-study-on-machine-learning-algorithms-for-big-data/.

Munde, A. (2024). The Machine Learning Pipeline. doi: 10.1201/9781003433309-18. Retrieved from https://www.taylorfrancis.com/chapters/edit/10.1201/9781003433309-18/machine-learning-pipeline-anjali-munde.

Najafabadi, M., Villanustre, F., Khoshgoftaar, T., Seliya, N., Wald, R. and Muharemagic, E. (2016). Deep learning techniques in big data analytics. pp. 1–20. In Deep Learning Techniques in Big Data Analytics. doi: 10.1007/978-3-319-44550-2_5. Retrieved from https://www.researchgate.net/publication/308278086_Deep_Learning_Techniques_in_Big_Data_Analytics.

Patil, D., Rane, N. L., Desai, P. and Rane, J. (2024). Machine learning and deep learning: Methods, techniques, applications, challenges, and future research opportunities. Trustworthy Artificial Intelligence in Industry and Society 28–81.

Quevedo Piratova, D., Villalba, J. and Gomez, A. (2024). Application of machine learning for predictions of consecutive dependent data of type {[(a, b)->c]->d}. Tecnura 28: 66–86. doi: 10.14483/22487638.22094. Retrieved from https://

www.researchgate.net/publication/385102080_Techniques_and_optimization_algorithms_in_deep_learning_A_review.

Rai, A., Rout, B., Bhushan and Mehta, A. (2024). Unveiling the power of data: a journey through machine learning techniques. International Journal of Advanced Research in Science, Communication and Technology 42–47. doi: 10.48175/IJARSCT-17807. Retrieved from https://ijarsct.co.in/Paper17807.pdf.

Rane, N. L., Mallick, S. K., Kaya, O. and Rane, J. (2024). Techniques and optimization algorithms in machine learning: A review. Applied Machine Learning And Deep Learning: Architectures and Techniques 39–58.

Rane, N. L., Mallikarjuna Paramesha, Saurabh P. Choudhary and Jayesh Rane. (2024). Machine learning and deep learning for big data analytics: a review of methods and applications. Partners Universal International Innovation Journal (PUIIJ) 02(03). Retrieved from https://www.researchgate.net/publication/382047488_Machine_Learning_and_Deep_Learning_for_Big_Data_Analytics_A_Review_of_Methods_and_Applications.

Rane, N., Mallick, S., Kaya, Ö. and Rane, J. (2024). Emerging trends and future directions in machine learning and deep learning architectures. doi: 10.70593/978-81-981271-4-3_10. Retrieved from https://www.researchgate.net/publication/385157207_Emerging_trends_and_future_directions_in_machine_learning_and_deep_learning_architectures.

Ronchi, G., Martin, E. H., Lau, C., Klepper, C. C., Goniche, M. and WEST Team. (2024). Complex multicomponent spectrum analysis with Deep Neural Network. Journal of Quantitative Spectroscopy and Radiative Transfer 318: 108925.

Sarker, I. H. (2021). Deep learning: a comprehensive overview on techniques, taxonomy, applications and research directions. SN COMPUT. SCI. 2: 420. doi: 10.1007/s42979-021-00815-1. Retrieved from https://link.springer.com/article/10.1007/s42979-021-00815-1.

Sunil, M., Shekhawat, J., Vanaja, T., Mohammed, V. A., Thaware, S. J. and Singh, V. (2024). Deep learning techniques for advanced large-scale data processing. 2024 15th International Conference on Computing Communication and Networking Technologies (ICCCNT), Kamand, India, 1–5. doi: 10.1109/ICCCNT61001.2024.10723332. Retrieved from https://www.researchgate.net/publication/385548025_Deep_Learning_Techniques_for_Advanced_Large-Scale_Data_Processing.

Vaissnave, V., Nandhini, S., Davamani, K. A., Malathi, P. and Pothumani, S. (2024). Advancements in Deep Learning Algorithms.

Wu, X., Zhao, J., Xie, K. and Zhao, X. (2025). Deep learning design for loss optimization in metamaterials. Nanomaterials 15(3): 178. doi: 10.3390/nano15030178. Retrieved from https://www.mdpi.com/2079-4991/15/3/178.

Chapter 20

Machine Learning and Deep Learning

Necessary Techniques for Extracting Meaningful Information in Large Databases

Soumya Mazumdar[1,*] and *Houssem Hosni*[2]

1. Introduction

Deep learning and machine learning have significantly improved the analysis and interpretation of big datasets, benefiting industries like banking, healthcare, and cybersecurity. These approaches are crucial in processing massive datasets, blockchain applications, and information extraction in digital ecosystems. Artificial intelligence, particularly machine learning and deep learning, has transformed sectors by replicating the human brain learning processes. Machine learning is creating autonomous algorithms and models that can make judgments or predictions without explicit programming (Kayikci and Khoshgoftaar 2024).

Machine learning (ML) is a versatile tool used in various sectors, including banking, healthcare, cybersecurity, and blockchain. Deep learning uses neural networks to process large data volumes, providing insights not possible with conventional methods. Blockchain, the basis for cryptocurrencies like Bitcoin, is a distributed digital ledger for transactional data storage. Integrating ML with blockchain offers new opportunities for smart, decentralized networks, enhancing security, data analysis, and decision-making skills (Kayikci and Khoshgoftaar 2024). This combination benefits both companies and individuals by providing a distributed, safe system for administrative tasks and transaction processing.

AI technologies like machine learning, deep learning, and blockchain are revolutionizing data analysis, risk appraisal, and transaction processing in the financial and banking sectors (Paramesha et al. 2024). These technologies improve efficiency, decision-making, and customer satisfaction, while also enhancing trading strategies, cybersecurity, and fraud detection. Blockchain technology offers transparent, decentralized transactions, boosting confidence and preventing fraud. In the healthcare industry, blockchain technology offers secure storage and exchange of health information,

[1] Dept. of CSBS, Gargi Memorial Institute of Technology [JIS Group], West Bengal, India.
[2] PhD, Senior Digital Engineer, Capgemini Engineering, France.
Email: hosny.houssem@gmail.com
* Corresponding author: reachme@soumyamazumdar.com

DOI: 10.1201/9781003620822-20

addressing privacy and security issues (Ali et al. 2023). Integrating deep learning methods with blockchain could enhance healthcare system scalability and security, overcoming limitations like data security breaches and inefficiencies in diagnosis and treatment.

Blockchain technology is widely used in various sectors due to its features like immutability, decentralization, transparency, and traceability. It consists of blocks with information, a hash, and a timestamp. There are three types of blockchains: public, private, and hybrid. Public blockchains use the Proof of Work algorithm, private blockchains require authorization, and hybrid blockchains offer flexibility (Abu-Elezz et al. 2020). AI technologies like ML and DL have significantly advanced cybersecurity by identifying and reacting to complex threats. AI-powered systems are crucial for threat detection, response, network security, and data protection (Paramesha et al. 2024a). AI also plays a significant role in endpoint security, malware detection, and maintaining digital infrastructure integrity.

Cryptocurrencies like Bitcoin in 2008 disrupted the financial system, leading to the development of blockchain technology (Santos et al. 2022). Blockchain, based on cryptographic principles, enables decentralized transactions, ensuring transparency and security. Ethereum, a Turing-complete blockchain, offers smart contracts for decentralized applications. DeFi's rise signals a shift towards more effective capital allocation. Blockchain has shown promise in data storage, administration, online identity identification, logistics planning, and electoral processes (Abdul 2024). However, it faces issues in scalability, transaction processing capacity, throughput, and latency. Solutions like sharding and off-chain storage may improve efficiency. Legal compliance remains a concern due to blockchain's decentralized structure.

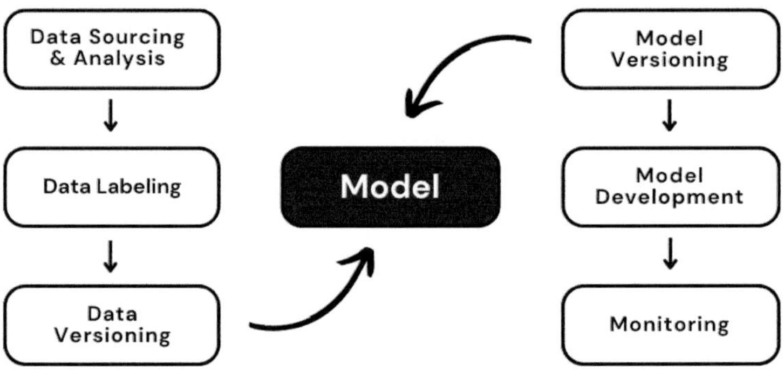

Figure 20.1. End-to-end ML pipeline for analysis.

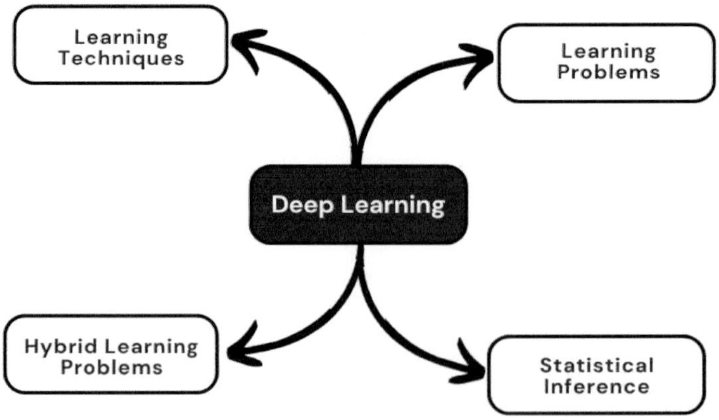

Figure 20.2. Deep learning architecture decision framework.

Table 20.1. Overview of the research questions.

RQ1	In what ways can blockchain technology be combined with ML and DL techniques to improve the security and privacy of data in decentralized applications?
RQ2	What are the potential applications of unsupervised learning algorithms in the analysis of large-scale blockchain transaction data for the purpose of detecting anomalies and fraud?
RQ3	What are the most effective methods for utilizing advanced deep learning models to enhance the scalability and transaction speed of next-generation applications by optimizing consensus mechanisms in blockchain networks?

Machine learning and blockchain integration is gaining popularity in IoT, healthcare, and smart cities due to their ability to analyze large amounts of data for insightful predictions (Azad et al. 2024). The United Nations is concentrating on blockchain's promise for transparent and efficient processes, with funding devoted to its development in many fields. The "ML for Blockchain Data Analysis" topic has witnessed a boom in interest during 2018, with over 1,750 papers in the ACM Digital Library.

The integration of ML and DL with blockchain technology is gaining traction in various sectors, including healthcare, banking, cybersecurity, and logistics. These technologies offer secure, efficient, and scalable data management systems, enabling informed decision-making. This chapter explores the integration of these technologies, analyzing existing models, algorithms, and applications, and presents challenges, innovations, and future directions.

2. Literature Review

This section examines the integration of blockchain technology with ML and DL techniques. It highlights advancements in data processing, security, and efficiency, as well as challenges, limitations, and emerging trends in finance, healthcare, and supply chain management. Key insights are summarized in the research questions.

Table 20.2. Summary of key studies on machine learning and deep learning.

Reference	Key Findings	Relevance to Our Paper
(Kayikci and Khoshgoftaar 2024)	Blockchain and machine learning are being explored in finance, medicine, supply chain, and security industries. Machine learning analyzes data, while blockchain enhances security and privacy, potentially revolutionizing data-driven decision-making through decentralization and efficiency.	The reference emphasizes the advantages of integrating blockchain with machine learning, particularly in extracting valuable information from large datasets, while also highlighting the security and efficiency of blockchain environments.
(Paramesha et al. 2024a)	The research highlights the significant role of AI, particularly ML and DL, in transforming finance and banking industries, with neural networks and support vector machines used for fraud detection and portfolio management.	The source highlights the importance of machine learning, particularly neural networks and deep learning, in managing large financial datasets, while blockchain integration ensures secure data transactions.
(Ali et al. 2023)	The research proposes a framework for scalable and secure healthcare systems using blockchain technology and hybrid deep learning techniques to address challenges in scalability, security, and patient data privacy.	The study utilizes hybrid deep learning models for processing large-scale healthcare data, focusing on extracting meaningful information from big data, and integrates blockchain for secure data storage, aligning with decentralized and secure blockchain technology.
(Abu-Elezz et al. 2020)	The scoping review highlights the eight benefits and threats of blockchain in healthcare, including improved patient data privacy, security, and interoperability.	The paper discusses the scalability and data privacy challenges of blockchain, focusing on healthcare, and highlights the similar technological challenges in securing large datasets using blockchain.

Table 20.2 contd. ...

...Table 20.2 contd.

Reference	Key Findings	Relevance to Our Paper
(AbuHalimeh and Ali 2023)	The implementation of blockchain technology in healthcare faces challenges such as data quality issues, operational issues, and technological challenges, according to a systematic review.	The reference discusses the quality and scalability of data storage in blockchain, addressing challenges related to large databases and operational and technological issues, guiding the integration of machine learning and blockchain.
(Ressi et al. 2017)	Machine learning tools for Ethereum smart contracts can improve detection accuracy by addressing vulnerabilities and dataset flaws, particularly through hybrid models.	This text emphasizes the potential of machine learning (ML) and deep learning (DL) in addressing security gaps and improving the extraction and analysis of information from blockchain databases.
(Christopher et al. 2024)	Classical machine learning methods like KNN, RF, DT, XGBoost, and SVM outperform static tools in smart contract vulnerability detection, while multi-model approaches improve detection precision.	The integration of ML and DL in blockchain is crucial for efficient information extraction and managing large datasets.
(Paramesha et al. 2024a)	ML techniques improve smart contract security by addressing vulnerabilities, combining deep learning and classical ML methods for superior performance.	The study emphasizes the use of ML and DL to improve data security and processing in blockchain databases, emphasizing the importance of extracting valuable information from large datasets.
(Ozkan-Okay et al. 2024)	ML, DL, and RL are vital in cybersecurity for detecting malware and vulnerabilities, but they face challenges from adversarial attacks.	The text highlights the limitations and benefits of machine learning in securely processing large amounts of data, emphasizing the need for improvements in blockchain applications.
(Blessing et al. 2024)	AI-powered systems, such as anomaly detection and predictive modeling, have enhanced cybersecurity threat detection, but ethical concerns and adversarial attacks remain significant challenges.	This study delves into the use of AI and ML models in blockchain data analytics for predictive purposes, specifically in handling cybersecurity threats.
(Santos et al. 2022)	Decentralized ledger of Blockchain, particularly Ethereum smart contracts, has revolutionized financial transactions but also poses technical and economic risks.	This article discusses how machine learning and deep learning can enhance the analysis of decentralized financial transaction data to mitigate risks.
(Abdul 2024)	Blockchain adoption faces scalability and regulatory challenges, but technological advancements like Layer-2 techniques and sharding provide potential solutions.	The discussion revolves around the potential of machine learning and deep learning to enhance blockchain scalability and streamline the handling of large datasets.
(Kasyapa and Vanmathi 2024)	Blockchain is set to revolutionize healthcare by ensuring secure data sharing, but scalability and energy consumption remain challenges, with solutions including consensus processes and sharding.	The study showcases the potential of machine learning and deep learning to tackle scalability issues in large healthcare databases using blockchain technology.
(Belchior et al. 2024)	Blockchain interoperability is crucial for future blockchain development, enabling data exchange across borders, but scalability remains a challenge.	ML and DL are crucial for blockchain's evolution as they enable the extraction of meaningful information from large, decentralized datasets through interoperability and scalability.
(Chi et al. 2023)	The proposed Blockchain Designated Verifier Proof (BDVP) offers a post-quantum solution to enhance blockchain privacy while addressing the vulnerability of ZKP to quantum attacks.	The text explores the potential of ML and DL to enhance privacy and security in large-scale blockchain data systems.

Table 20.2 contd. ...

...Table 20.2 contd.

Reference	Key Findings	Relevance to Our Paper
(Wang and Peeta 2024)	The proposal proposes a secure, privacy-preserving incentive mechanism for collaborative routing in V2X, utilizing secure Multi-party Computation and blockchain technologies for route validation and incentive management.	Chapter focuses on the use of MPC and blockchain for privacy and security, specifically in extracting meaningful information from large databases through secure computations and validations.
(Cholevas et al. 2024)	The survey explores unsupervised learning algorithms for detecting malicious activities on blockchain, categorizing them based on data structures and frameworks for blockchain anomaly detection.	This paper discusses the use of machine learning techniques in anomaly detection using unsupervised learning algorithms, a crucial aspect of large-scale data analysis for blockchain.
(Vlahavas et al. 2024)	The study uses k-means clustering to categorize Bitcoin transaction data, distinguishing between common user transactions, mining activities, and potentially illicit transactions, and validating the results with an online database.	The clustering approach emphasizes the importance of unsupervised learning for large-scale data analysis, demonstrating techniques for extracting meaningful patterns from complex, high-volume data like blockchain transactions.
(Azad et al. 2024)	The study explores the use of machine learning in blockchain data analysis, emphasizing its role in detecting e-crime and predicting trends, and its potential for technological advancement.	The paper explores machine learning for analyzing blockchain data, providing valuable methodologies for extracting meaningful information and detecting fraudulent activities using large datasets.
(Mihus 2022)	The study analyzes economic fraud types and proposes blockchain for fraud prevention in financial reporting, identifying industries most affected by corruption and suggesting anti-fraud tools like External Audit and proactive monitoring.	The paper discusses the use of blockchain and machine learning in fraud detection, highlighting their potential to enhance data integrity and prevent fraudulent activities in large databases.
(Raghavendra and Lingam 2024)	The study explores the potential of blockchain in tackling digital advertising transparency and fraud issues, proposing decentralized and tamper-proof systems to combat these issues.	The chapter highlights the use of decentralized technologies like blockchain for fraud prevention in large-scale data systems, enhancing transparency and reducing fraud in digital systems.
(Höglund et al. 2024)	The text discusses the integration of Federated Learning and blockchain for detecting fraudulent cryptocurrency transactions, highlighting the privacy-preserving nature of FL and the scalability challenges.	The chapter discusses the use of BCFL for fraud detection in cryptocurrency transactions, demonstrating the secure application of machine learning in decentralized data systems.
(Ncube et al. 2024)	The proposal proposes a blockchain-based fraud detection system for healthcare insurance claims, utilizing Ethereum's blockchain and smart contracts to enhance claim processing and minimize fraudulent activities.	The study explores the potential of blockchain in real-world data systems to enhance security, trust, data integrity, and fraud detection.
(Mohanty et al. 2023)	The text discusses the use of blockchain for fraud detection in financial management, highlighting its security, transparency, and tamper-proof capabilities, and outlines future innovations for enhanced detection.	The chapter discusses the integration of machine learning and AI with blockchain for fraud detection, offering valuable insights into advanced data security and fraud prevention techniques.
(Maturi and Meduri 2024).	The study explores the use of Graph-Convolutional Networks (GCNs) to detect fraud in Ethereum transactions, demonstrating their high accuracy in analyzing graph-based blockchain data.	The study explores the use of graph-based models for fraud detection in Ethereum, utilizing advanced machine learning techniques for analyzing large blockchain datasets.

3. Overview of Blockchain Technology

Blockchain technology is a decentralized distributed ledger that revolutionizes data handling, storage, and sharing across networks. It offers transparency, security, and trust, making it crucial in sectors like banking, supply chain, healthcare, and data management. Blockchain consists of blocks linked in a chronological chain, each containing transaction data. Its immutability ensures data integrity and prevents single-point failures (Kayikci and Khoshgoftaar 2024). Decentralization promotes security and trust, as data verification occurs through consensus rather than a centralized authority, ensuring no single entity can control or undermine the system (Kayikci and Khoshgoftaar 2024, AbuHalimeh and Ali 2023).

Peer-to-peer (P2P) networks are the foundation of blockchain systems, enabling members to share computing resources. Each node in a P2P network act as both a provider and consumer, ensuring efficient and fair ledger distribution. Each block in the chain contains transactions, and once validated, it is copied across all nodes in the network (AbuHalimeh and Ali 2023). This method ensures secure transactions by preventing unauthorized changes and ensuring network verification for each transaction (Kayikci and Khoshgoftaar 2024).

Blockchain technology provides security and transparency through cryptographic methods, ensuring data encryption and protection from unwanted access. It connects blocks using a cryptographic hash function, creating a secure chain of blocks. Public and private key cryptography is used to establish digital signatures, validating transactions throughout the network, enhancing system security (AbuHalimeh and Ali 2023).

Merkle trees, also known as mixed trees, are utilized in blockchain to verify the integrity of large databases by enabling the verification of numerous transactions within a block using cryptographic hashes (Kayikci and Khoshgoftaar 2024). Blockchain enhances transparency by allowing all network participants to view the ledger, providing a transparent record of transactions. This is particularly beneficial in healthcare, where data quality and trust are crucial, ensuring a reliable and tamper-proof storage mechanism (AbuHalimeh and Ali 2023, Cholevas et al. 2024).

The decentralized nature of the blockchain technology necessitates consensus among nodes for transaction authenticity. This is achieved through consensus procedures, ensuring all participants agree on the blockchain's current state. Popular consensus techniques include Proof of Work, Proof of Stake, and Practical Byzantine Fault Tolerance (Kasyapa and Vanmathi 2024). These systems perform various functions but achieve the same goal of verifying and certifying transactions without requiring a central authority (Cholevas et al. 2024).

The PoW consensus process involves nodes solving complex mathematical riddles to add blocks to the chain, making it computationally demanding and preventing unauthorized modifications. On the other hand, PoS selects validators based on token ownership, reducing processing power and preserving security while reducing the need for excessive processing power (Kasyapa and Vanmathi 2024). These techniques ensure the safety, transparency, and resistance to alteration of the blockchain (Raghavendra and Lingam 2024).

Blockchain technology, initially designed for Bitcoin, has expanded to various sectors including decentralized finance, supply chain management, healthcare data sharing, and fraud detection. Ethereum, a prominent blockchain system, has introduced smart contracts, self-executing contracts with stipulations encoded directly into the code. These contracts enable automated, secure, and tamper-proof transactions, eliminating middlemen and reducing fraud risk (Raghavendra and Lingam 2024, Ncube et al. 2024).

Blockchain is utilized in healthcare to enhance data security, enhance patient care, and automate electronic medical record maintenance. Its immutability ensures accurate and trustworthy medical information, while smart contracts automate operations like insurance claim validation, reducing fraud and mistakes. This technology enhances patient care and trustworthiness (AbuHalimeh and Ali 2023, Ncube et al. 2024).

Blockchain's decentralized nature makes it ideal for businesses requiring trust and accountability. It can track the origin and movement of items in supply chain management, providing verifiable authenticity and eliminating fraud. This enhances confidence and ensures product quality and safety, making it a valuable tool for businesses (Kayikci and Khoshgoftaar 2024, AbuHalimeh and Ali 2023).

Blockchain designs consist of public, private, and hybrid types. Public blockchains, like Bitcoin and Ethereum, are decentralized, accessible to all members, but have scalability and energy consumption issues. Private blockchains are centralized, used by companies for network control, allowing only chosen participants to verify transactions. Hybrid blockchains combine public and private features, providing flexibility in data storage and sharing for corporate applications (AbuHalimeh and Ali 2023, Kasyapa and Vanmathi 2024).

Blockchain technology offers numerous benefits but also faces challenges in scalability and interoperability. As more blocks are added, the amount of data stored increases, leading to higher storage costs and longer transaction processing times. This is particularly problematic with public blockchains like Bitcoin, as the size of the blockchain rapidly increases due to new blocks (Kayikci and Khoshgoftaar 2024).

Interoperability is a significant challenge in blockchain systems, as they often operate in silos, preventing data or assets from moving between them. This hinders ability the of the organizations to fully utilize blockchain technology across various applications and sectors. Solutions are being explored, including efficient consensus methods and scalable architectures (AbuHalimeh and Ali 2023, Kasyapa and Vanmathi 2024).

Blockchain technology is being utilized in the healthcare sector to enhance data security, patient care, and administrative efficiency. Its decentralized nature eliminates middlemen, ensuring secure data transfer and respecting patient privacy. Blockchain is particularly useful in managing Electronic Medical Records and enhancing fraud detection in healthcare insurance through transparent transactions and smart contracts (Ncube et al. 2024).

Many blockchain systems have emerged in recent years, each offering unique features. Ethereum is renowned for its support of smart contracts and decentralized applications, while Hyperledger, an open-source blockchain created by IBM and the Linux Foundation, is ideal for corporate applications, particularly in healthcare and finance (Kasyapa and Vanmathi 2024, Ncube et al. 2024).

Corda and MultiChain are systems that host private blockchains and enable interoperability across multiple networks. Corda is used in industries like digital assets, insurance, and financial markets for safe data communication. MultiChain allows developers to create their own currency and integrate blockchain technology with existing systems, making it a versatile platform for various applications (AbuHalimeh and Ali 2023, Kasyapa and Vanmathi 2024).

Blockchain technology has revolutionized data handling, storage, and sharing, offering a decentralized, secure, and transparent system suitable for various businesses. Its decentralized design reduces centralized control, enhancing security and trust through consensus methods. The immutability of blockchain ensures data integrity, making it ideal for industries like healthcare and supply chain management (Kayikci and Khoshgoftaar 2024, AbuHalimeh and Ali 2023, Kasyapa and Vanmathi 2024, Cholevas et al. 2024).

Blockchain technology, despite its numerous benefits, still faces challenges in scalability and interoperability. However, as it expands, it is expected to revolutionize different sectors and provide new opportunities for companies and consumers (AbuHalimeh and Ali 2023, Kasyapa and Vanmathi 2024, Ncube et al. 2024). Blockchain's potential for secure, transparent, and tamper-proof systems will remain a crucial part of the digital economy (Kasyapa and Vanmathi 2024, Cholevas et al. 2024).

4. Foundations of Machine Learning and Deep Learning

Machine learning and deep learning are artificial intelligence fields that enable computers to learn without explicit programming. They use algorithms to detect patterns in data and predict or decide. This concept, rooted in work by Mitchell (Nasir et al. 2021), improves system performance through experience. Deep learning has significantly enhanced the analysis and interpretation of big datasets, enabling businesses to extract valuable knowledge from complex data structures. These approaches have been beneficial in fields like banking, healthcare, and cybersecurity. This chapter examines the role of machine learning in processing massive datasets, blockchain applications, and information extraction in contemporary digital ecosystems.

Artificial intelligence (AI), particularly ML and DL, has transformed various sectors by replicating the human brain's learning processes. Machine learning creates autonomous algorithms and models that make judgments or predictions without explicit programming (Kayikci and Khoshgoftaar 2024). Deep learning processes large volumes of data using neural networks to find trends and insights. Blockchain, a distributed digital ledger for transactional data storage, was introduced in 2008 for safe and open data management across various sectors. Integrating ML with blockchain presents new opportunities for smart, decentralized networks, enhancing security, data analysis, and decision-making skills (Kayikci and Khoshgoftaar 2024). This combination will result in a distributed, safe system for handling administrative tasks and transaction processing, benefiting both companies and individuals.

AI technologies like machine learning, deep learning, and blockchain are improving data analysis, risk appraisal, and transaction processing in the financial and banking sectors (Paramesha et al. 2024b). These advancements promote efficiency, decision-making, customer satisfaction, and enhance trading strategies, cybersecurity, and fraud detection. Blockchain technology offers transparent, decentralized transactions, boosting confidence and preventing fraud.

The healthcare industry is undergoing a technological transformation due to the need for improved scalability, security, and data interoperability. Blockchain technology offers a secure solution for securely storing and exchanging health information (Ali et al. 2023), while deep learning methods can enhance system scalability and security, overcoming limitations like data security breaches and inefficiencies in diagnosis and treatment.

Blockchain technology is widely used in various sectors due to its features like immutability, decentralization, transparency, and traceability. It consists of blocks with information, a hash, and a timestamp. There are three types: public (permissionless), private (permissioned), and hybrid (hybrid), each with its own proof of work algorithm and permissions (Abu-Elezz et al. 2020).

AI technologies, including machine learning, deep learning, and natural language processing, have significantly improved cybersecurity by identifying and responding to complex cyber threats. These systems are crucial for threat detection, response, network security, and data protection (Paramesha et al. 2024a). Machine learning algorithms evaluate network data, recognize attack patterns, extract threat information, and provide real-time threat analytics. AI also plays a significant role in endpoint security, malware detection, and digital infrastructure integrity.

The development of cryptocurrencies like Bitcoin in 2008 threatened the established financial environment, resulting in dramatic changes in the financial system (Santos et al. 2022). Blockchain technology, based on cryptographic principles, enables decentralized transactions, ensuring transparency and security. Ethereum, a Turing-complete blockchain, offers smart contracts for decentralized applications, signaling a shift towards more effective capital allocation.

Blockchain technology, primarily created for Bitcoin transactions, has shown promise in data storage, administration, online identity identification, logistics planning, and electoral processes (Abdul 2024). Blockchain, due to its decentralization and immutability, is a robust data audit tool, offering security and transparency. However, it faces challenges in scalability, particularly in transaction processing capacity and latency. Solutions like sharding and off-chain storage could improve efficiency.

The integration of machine learning and blockchain has gained popularity in areas like IoT, healthcare, and smart cities due to its ability to generate large amounts of data that can be analyzed using machine learning algorithms for insightful predictions (Azad et al. 2024). The United Nations is investing in the development of blockchain for transparent and efficient processes. The "ML for Blockchain Data Analysis" topic has witnessed a boom in interest during 2018, with over 1,750 papers in the ACM Digital Library.

The growing need for data processing and analysis in enterprises is a growing issue. ML and DL techniques can extract relevant information from large datasets, enabling informed decisions. Blockchain offers a secure, decentralized framework for data storage and administration, transforming businesses by providing secure, efficient, and scalable systems for data management and decision-making. This combination has great potential in sectors like healthcare, banking, cybersecurity, and logistics, defining the next generation of safe and intelligent systems.

Supervised learning is a popular machine-learning approach, involving trained models with labeled data. Common algorithms include linear regression, decision tree, and support vector machines (SVM). Work by Cortes and Vapnik on SVMs demonstrated high classification accuracy by maximizing the margin between classes (Cortes and Vapnik 1995). This approach has carved out a sweet niche in a lot of areas from image recognition to fraud detection.

Unsupervised learning is a method that mines data for unknown structures or relationships, often using clustering and dimensionality reduction techniques. The k-means algorithm, introduced by MacQueen in 1967, partitions data into similarity clusters. It is particularly useful in marketing to identify customer segments not previously labeled, making it an effective tool in complex datasets.

Reinforcement learning is a crucial method in machine learning that involves agents learning to maximize cumulative rewards over time, particularly in games or robotics. This approach, facilitated by reinforcement learning algorithms like Q-learning, allows agents to learn from their past experiences and improve their performance in future situations (Sutton and Barto 2014). That method has been used to create agents that can play such intricate games as Go and StarCraft.

Deep learning, a part of machine learning, uses deep neural networks with hidden layers to process and analyze data, enabling automatic feature discovery from raw data. CNNs are used for images and RNNs for temporal sequences. LeCun, Bengio, and Haffner's seminal paper discusses how CNNs revolutionized computer vision by allowing exact object recognition in images for the first time (Lecun et al. 1998).

Recent hardware advancements, such as GPUs, have significantly accelerated the training process of deep learning models, resulting in faster training times and higher accuracy in tasks like image classification, according to a study by Krizhevsky et al. (Krizhevsky et al. 2017).

Machine learning and deep learning involve feature engineering, which involves choosing or creating attributes to improve model performance. Traditional methods require extensive domain knowledge to determine important features. Deep learning, however, often uses hidden layers to learn from raw data. Residual architectures make training easier by enabling deep networks to be trained, making feature engineering a crucial aspect of ML and DL (He et al. 2016).

ML and DL face challenges like interpretability and transparency, as many models are "black box" and difficult to understand, which is crucial in fields like medicine and finance. In 2016, Ribeiro, Singh, and Guestrin introduced LIME (Local Interpretable Model-agnostic Explanations), a technique to make models more interpretable by explaining why they made certain predictions at a local level (Ribeiro et al. 2016).

Bias in machine learning models can lead to discrimination or unfairness in decisions if the training datasets are not balanced or representative. Barocas and Selbst emphasize the importance of being mindful of biases when creating and using AI-oriented systems to promote fairness and prevent discrimination in decision-making processes (Barocas and Selbst 2016).

Machine learning and deep learning are increasingly utilized in various industries, including healthcare, finance, and marketing. In the medical field, they are used to diagnose diseases like cancer early, as demonstrated by Esteva et al. who found a neural network can diagnose melanoma accurately from skin images (Esteva et al. 2017).

Machine learning algorithms are increasingly being used in finance to study market trends and detect fraud by recognizing abnormal transactional patterns. These models can enhance early fraud detection by monitoring transactions in real-time. Machine learning is becoming increasingly intertwined with other emerging technologies, such as IoT devices, enabling connected devices to collect and analyze data in real-time, making intelligent decisions without human intervention. However, as these technologies are integrated into daily life, ethical questions arise regarding their use. Researchers and practitioners worldwide must develop models that are responsible, observe individual rights, and minimize negative impacts on society. ML and DL are based on advanced mathematical algorithms and innovative computational approaches that transform our interaction with the digital world. As these technologies continue to grow rapidly, it is crucial to examine the ethical and societal aspects they include.

5. Integration of ML and DL with Blockchain Technology [RQ1]

Blockchain technology is integrating ML and DL, transforming sectors by providing intelligent, secure, and transparent systems for managing large amounts of data while ensuring transaction reliability and integrity. This integration combines advanced ML and DL analysis functionality with unalterable blockchain attributes, enabling autonomous and distributed solutions beyond current limits (Kayikci and Khoshgoftaar 2024).

ML and DL enable quick analysis of vast data sets, revealing hidden patterns (Paramesha et al. 2024). Blockchain ensures open and transparent recording of transactions without centralized entities, making it particularly promising in finance, healthcare, and supply chain management. This combination of intelligence and decentralization is particularly beneficial in these sectors (Ali et al. 2023, Abu-Elezz et al. 2020, AbuHalimeh and Ali 2023).

The integration of ML and DL with blockchain can optimize smart contracts in real-world scenarios. Blockchain stores self-executing protocols, while ML allows for active adjustment of contract conditions based on external data. For example, in a supply chain, a smart contract could automatically function when conditions are met, such as when a product arrives at a delivery point (Ressi et al. 2017, Christopher et al. 2024). Weather conditions, price changes and production lead times are only a couple of variables that algorithms that implement ML can analyze to optimize this process.

The integration of Machine Learning (ML) with blockchain and the integration of deep learning (DL) with blockchain has significant potential in cybersecurity. Blockchain-based systems are resistant to tampering and hacking due to cryptography, unlike decentralized systems. However, they are not completely immune to attacks like Sybil or 51%. ML and DL validate these systems by monitoring network activity for abnormal behavior. Deep learning excels at real-time analysis of complex data, unlike many statistical fields that produce theoretical conclusions (Paramesha et al. 2024a, Ozkan-Okay et al. 2024, Blessing et al. 2024).

The integration of ML and DL in healthcare is beneficial as it allows healthcare professionals to analyze and use patient data more effectively. EMRs, which are fragmented across multiple systems, can be analyzed using ML and DL, while blockchain securely stores these records. This allows for better diagnosis and treatment, as patient data is protected from manipulation. ML can also identify unknown patterns (Ali et al. 2023, Abu-Elezz et al. 2020, AbuHalimeh and Ali 2023).

ML and DL can enhance transaction speed and security in decentralized finance (DeFi) (Santos et al. 2022). Algorithms can predict market moves and detect anomalies, preventing fraud. AI algorithms can ensure optimal transaction conditions with minimal parameters set by a central authority, ensuring no manipulation of triggers. This application area can also benefit from AI algorithms in predicting market moves and detecting anomalies.

Blockchain integration with AI systems faces challenges such as scalability, which is slow and energy-intensive when combined with consensus mechanisms like Proof of Work (PoW). ML and DL algorithms require large compute power for fast data analysis. To overcome this, sharding and second-generation blockchains like Ethereum 2.0 are being studied to make blockchains faster and more efficient while seamlessly integrating complex AI systems (Abdul 2024, Kasyapa and Vanmathi 2024).

Another huge obstacle is interoperability. Currently, many blockchains are incompatible with each other making it hard to integrate ML/DL systems that require one to access data across lots of other blockchains (Belchior et al. 2024). Therefore, for wider adoption of this combined technology, interoperability standards and universal protocols must be developed.

Blockchain technology, known for its transparency, can pose a threat to privacy in sectors like healthcare and finance. Investments are being made to develop solutions that allow data analysis on the blockchain without compromising user privacy, such as secure multiparty computing or Zero-Knowledge Proofs (Chi et al. 2023, Wang and Peeta 2024).

The integration of machine learning and deep learning on blockchain holds significant potential to revolutionize sectors like finance, healthcare, security, and supply chain management. However, challenges like scalability, interoperability, and privacy remain, with further adoption and innovation expected in the future.

6. Analyzing Blockchain Data with Unsupervised Learning [RQ2]

Unsupervised learning is a new field that allows for autonomous and label-free processing of large-scale blockchain transactions. It differs from supervised learning, which uses prelabeled datasets to identify hidden patterns in blockchain data. This approach has multiple applications, especially in the rapidly evolving public, private, and hybrid blockchains, as it leverages massive datasets through artificial intelligence (Cholevas et al. 2024, Vlahavas et al. 2024, Azad et al. 2024).

Blockchain networks record transactions on a permanent, immutable ledger, including cryptocurrencies and smart contracts. This data, including sender and receiver addresses, amount, and time stamps, is often analyzed using unsupervised learning to identify latent clusters of transactions with similar characteristics, detect recurring patterns, or detect anomalies indicating potential fraud (Mihus 2022, Raghavendra and Lingam 2024).

Clustering algorithms, such as K-means or DBSCAN, are widely used to analyze blockchain data using unsupervised learning. These algorithms group similar transactions into clusters, representing different behaviors like regular or abnormal transactions (Vlahavas et al. 2024). DBSCAN is particularly useful for detecting irregular blockchain transactions, as it can identify clusters and "noise" or outliers, allowing for a more accurate analysis of the data.

Autoencoders, an artificial neural network, are used to detect anomalies in blockchain transactions by mapping input data to their compressed reduced representation. This method works well in fixed pattern, normal transaction environments where fraudulent behavior diverges significantly from normal. Unsupervised learning beyond simple clustering and autoencoders is also used to detect different anomalies within blockchain transactions. Isolation Forest is a model used to isolate data points using decision trees, with more isolated points more likely to be anomalies. This method is useful for analyzing blockchain transactions due to the complexity of relationships between different transactional parameters, such as transfer patterns, transaction frequency, and transferred amounts (Cholevas et al. 2024, Vlahavas et al. 2024).

Blockchain monitoring systems, like Bitcoin or Ethereum, use public and pseudonymous transactions with hidden user usernames. Unsupervised learning is used to detect fraud and patterns, similar to detecting anomalies in wallet behavior or flagging suspicious transactions like unusual transfers or high transaction volume over a short time frame. This approach helps detect large-scale fraud and prevent fraudulent activities on the blockchain (Mohanty et al. 2023, Maturi and Meduri 2024).

Smart contracts, self-executing programs that automatically execute predefined clauses, are another application area where unsupervised algorithms can reveal vulnerabilities or detect malicious behavior. Techniques like GANs can simulate fictitious scenarios and identify anomalies from real transactions. These algorithms play a crucial role in automating anomaly detection, especially as the number of transactions monitored increases (Cholevas et al. 2024). Scaling these models to more complex blockchains remains a major challenge. With the growth of blockchain adoption in various applications, more data is produced and generated, making the processing of heterogeneous data flows while maintaining operational efficiency more sophisticated. The processing of this heterogeneous data flow is more sophisticated than that.

Blockchain interoperability is crucial for detecting malicious transactions that cross boundaries. Unsupervised learning algorithms can analyze transactions between multiple networks, providing insights into complex transfers and anomalies. However, unsupervised learning is not a perfect solution, as models must be adjusted for each blockchain network and transaction type. The anonymity of blockchain transactions makes explicit identities or labeled data meaningless to algorithms, as they cannot rely on explicit identifiers. As technology advances, the effectiveness of these algorithms will increase, and they will play a more significant role in detecting fraud and malicious activities within large blockchain networks. Therefore, the use of unsupervised learning algorithms for data analysis in blockchain is not a perfect solution (Airlangga 2024, Heussner et al. 2023).

Unsupervised methods have potential for future digital asset regulation and oversight. They can be used by regulators to monitor cryptocurrency markets for evidence of market manipulation or illicit activity, and private companies to improve internal security measures and fraud prevention in their blockchain systems (Amin et al. 2024, Chatterjee et al. 2023). Unsupervised learning is particularly useful for anomaly detection and fraud prevention in blockchain data analysis (Cholevas et al. 2024). As blockchain scales, these algorithms will automate monitoring, detect suspicious behavior, and enhance security in distributed networks.

7. Optimizing Blockchain Consensus Mechanisms with Deep Learning [RQ3]

Deep learning models can optimize the consensus mechanisms of blockchain networks for scalability, security, and transaction speed. Despite limitations in performance and energy efficiency, Proof of Work and Proof of Stake currently suffer from these issues. By making critical processes smarter and dynamic using deep learning, the aim is to integrate deep learning into these networks (Kayikci and Khoshgoftaar 2024).

Reinforcement learning algorithms optimize the validation processes of blockchain systems by learning from past transactions. For instance, in a PoS mechanism, deep learning can help select better validators based on stake, resulting in faster validation and increased security, as demonstrated in studies (Nguyen et al. 2021, Qiu et al. 2020, Höglund et al. 2024).

Dynamic difficulty adjustment in PoW is crucial for efficient block validation and reducing centralized computational power. Neural networks can adjust difficulty in real-time, based on network conditions, allowing better workload distribution and reduced computational power. Convolutional neural networks can analyze blockchain transactions in real-time, adjusting difficulty to improve network performance (Chin et al. 2022).

Deep learning algorithms are being integrated into blockchains to improve consensus mechanisms, block propagation, and transaction speed. Recurrent neural networks (RNNs) are used to identify and solve bottlenecks in block transmission, minimizing latency. Unsupervised learning algorithms are being developed to identify anomalies within the blockchain network, which are susceptible to attacks like Sybil attacks or double spending. Real-time suspicious transactions and deviant behaviors can be identified using unsupervised learning techniques like clustering, increasing system security and enabling faster reactions to counterfeit or deceptive actions (Cholevas et al. 2024).

Blockchains face challenges in durability and scalability due to increasing transaction volumes. Generative adversarial networks (GANs) can help address this by providing continuous testing and improvement of validation processes through simulation through various consensus configurations. These simulations are essential for large-scale, high-performance requirements in next-generation blockchains, as they offer flexibility and adaptability (Taher et al. 2024).

Deep learning models and blockchain consensus mechanisms offer a new layer of security and fraud prevention by continuously monitoring transactions. This early timed information can help blockchains become more resilient to cyberattacks. Blockchains are scalable, capable of handling large amounts of transactions in minimal time. Deep learning models can improve the transaction capacity while limiting latency by optimizing consensus mechanisms. This is especially important in cross-chain applications where multiple blockchains must exchange information. By analyzing complex interactions, deep neural networks can propose solutions for maximum transaction fluidity, enhancing the overall security and resilience of blockchain networks (Nasir et al. 2021).

The integration of deep learning models into blockchain consensus mechanisms can enhance speed, security, and scalability. Technologies like reinforcement learning, convolutional neural networks, and generative adversarial networks enable real-time optimization of block validation processes and increase transaction capacity. This strategy will enable blockchains to meet the needs of next-generation applications, enhancing their performance and security.

8. Challenges and Limitations

The integration of machine learning (ML), deep learning (DL), and blockchain technology presents significant potential but also presents challenges. One of the main issues is scalability, as blockchain networks are slower due to their decentralized structure. As ML and DL algorithms are integrated, the need for efficient transaction processing becomes more critical. Traditional blockchains like Bitcoin and Ethereum struggle with high transaction times and energy usage. Second-generation blockchains like Ethereum 2.0 aim to address this issue, but the combination with ML/DL requires even more efficient solutions.

The energy consumption issue is also linked to scalability, as deep learning models require large processing capacities to process and evaluate massive datasets. Blockchain technology, which is energy-intensive due to consensus techniques like Proof of Work, places a strain on resources. ML/DL algorithms further compound this problem by requiring more computational resources, making it unsustainable without more efficient models or blockchain architecture.

Interoperability across blockchain systems is another challenge, as blockchain systems often function in isolation, making it difficult for transactions or data to move seamlessly across multiple platforms. Hybrid blockchains aim to integrate public and private networks, but still struggle with smooth data flow. Privacy is a significant concern, as blockchain's openness may disadvantage sensitive data.

9. Future Directions and Innovations

The convergence of machine learning, deep learning, and blockchain technology is a promising area of progress. Sharding, a method that divides a blockchain into smaller "shards," can improve the scalability of decentralized networks like Ethereum 2.0. Off-chain solutions can also be used to analyze massive datasets, reducing computing and energy usage. Next-generation consensus algorithms like Practical Byzantine Fault Tolerance and Delegated Proof of Stake offer energy-efficient solutions for incorporating machine learning algorithms. Reinforcement learning models could further enhance these methods.

Hardware developments like GPUs and Tensor Processing Units have accelerated deep learning model training. Future hardware design breakthroughs, particularly for blockchain-ML/DL integration, could significantly increase system efficiency. Energy-efficient circuits built for both blockchain calculations and AI-based analysis could minimize energy usage and computational expenses.

Quantum computing has the potential to improve both blockchain and machine learning by providing unparalleled processing power. Researchers are exploring the integration of quantum-resistant algorithms with ML/DL models for more secure and scalable blockchain networks. Federated learning offers a decentralized way to teach ML models without compromising privacy, particularly in fields like healthcare.

10. Conclusion

The convergence of machine learning, deep learning, and blockchain technology poses both obstacles and potential. Limitations include scalability, energy consumption, interoperability, privacy, and interpretability. However, developing solutions like sharding, off-chain processing, next-generation consensus algorithms, and hardware advances provide potential avenues ahead. Innovations in quantum computing and federated learning might transform the area, allowing more efficient, secure, and decentralized systems. The combination of ML/DL with blockchain may alter sectors including banking, healthcare, supply chain management, and cybersecurity. Future systems may offer intelligent, safe, and transparent solutions for data management and transaction processing.

Author Contribution

S.M.: Contributed in presenting current research on blockchain technology and its confluence with machine learning and deep learning, as well as identifying important obstacles, restrictions, and future opportunities in this growing area.

H.H.: Contributed on the technical strategies for enhancing blockchain technology applying ML and DL, addressing special research challenges pertaining to data analysis, consensus procedures, and integration.

Both Authors: Collaborated on the overall concept and direction of the research, ensuring the publication presents a cohesive and full review of blockchain's future utilizing advanced ML and DL methodologies.

Data Availability Statement: Data will be made available on request.

Declaration of AI: We therefore certify that the content of this document is completely our own creation. The only AI tool applied in the creation of this work was Quillbot, employed only for the purpose of grammar and spelling checks. No generative AI tools or other AI-assisted technologies were employed in authoring or producing the substantive content of this work. I have meticulously evaluated and, where appropriate, updated every output from Quillbot to ensure that the final text adequately represents my original thoughts and intellectual contribution. I assume full responsibility for the material and conclusions offered herein.

References

Abdul, S. S. M. (2024). Navigating blockchain's twin challenges: scalability and regulatory compliance. Blockchains 2(3): 265–298. https://doi.org/10.3390/blockchains2030013.

Abu-Elezz, I., Hassan, A., Nazeemudeen, A., Househ, M. and Abd-Alrazaq, A. (2020). The benefits and threats of blockchain technology in healthcare: A scoping review. International Journal of Medical Informatics 142: 104246. https://doi.org/10.1016/j.ijmedinf.2020.104246.

AbuHalimeh, A. and Ali, O. (2023). Comprehensive review for healthcare data quality challenges in blockchain technology. Frontiers in Big Data 6. https://doi.org/10.3389/fdata.2023.1173620.

Airlangga, G. (2024). Deep learning for anomaly detection and fraud analysis in blockchain transactions of the open metaverse. Jurnal Informatika Ekonomi Bisnis. https://doi.org/10.37034/infeb.v6i2.865.

Ali, A., Ali, H., Saeed, A., Khan, A. A., Tin, T. T., Assam, M. et al. (2023). Blockchain-powered healthcare systems: enhancing scalability and security with hybrid deep learning. Sensors 23(18): 7740. https://doi.org/10.3390/s23187740.

Amin, M. A., Shah, R., Tummala, H. and Ray, I. (2024). Utilizing Blockchain and Smart Contracts for Enhanced Fraud Prevention and Minimization in Health Insurance through Multi-Signature Claim Processing. arXiv (Cornell University). https://doi.org/10.48550/arxiv.2407.17765.

Azad, P., Akcora, C. G. and Khan, A. (2024). Machine Learning for Blockchain Data Analysis: Progress and opportunities. arXiv (Cornell University). https://doi.org/10.48550/arxiv.2404.18251.

Barocas, S. and Selbst, A. D. (2016). Big data's disparate impact. California Law Review 104(3): 671. https://doi.org/10.15779/z38bg31.

Belchior, R., Süßenguth, J., Feng, Q., Hardjono, T., Vasconcelos, A. and Correia, M. (2024). A brief history of blockchain interoperability. Communications of the ACM 67(10): 62–69. https://doi.org/10.1145/3648607.

Blessing, M., Kolawole, W. and Owen, J. (2024, August). The impact of AI-Powered threat detection systems on modern cybersecurity practices. ResearchGate. Retrieved October 15, 2024, from https://www.researchgate.net/publication/383265005_The_Impact_of_AI-Powered_Threat_Detection_Systems_on_Modern_Cybersecurity_Practices.

Chatterjee, P., Das, D. and Rawat, D. (2023). Securing financial transactions: exploring the role of federated learning and blockchain in credit card fraud detection. TechRxiv. https://doi.org/10.36227/techrxiv.22683403.v1.

Chi, P., Lu, Y. and Guan, A. (2023). A privacy-preserving zero-knowledge proof for blockchain. IEEE Access 11: 85108–85117. https://doi.org/10.1109/access.2023.3302691.

Chin, Z. H., Yap, T. T. V. and Tan, I. K. T. (2022). Genetic-Algorithm-Inspired difficulty adjustment for Proof-of-Work blockchains. Symmetry 14(3): 609. https://doi.org/10.3390/sym14030609.

Cholevas, C., Angeli, E., Sereti, Z., Mavrikos, E. and Tsekouras, G. E. (2024). Anomaly detection in blockchain networks using Unsupervised Learning: a survey. Algorithms 17(5): 201. https://doi.org/10.3390/a17050201.

Christopher, D. B., Suleiman, B., Chitizadeh, A. and Razzak, I. (2024). Vulnerability Detection in Smart Contracts: A Comprehensive Survey. arXiv (Cornell University). https://doi.org/10.48550/arxiv.2407.07922.

Cortes, C. and Vapnik, V. (1995). Support-vector networks. Machine Learning 20(3): 273–297. https://doi.org/10.1007/bf00994018.

Esteva, A., Kuprel, B., Novoa, R. A., Ko, J., Swetter, S. M., Blau, H. M. et al. (2017). Erratum: Corrigendum: Dermatologist-level classification of skin cancer with deep neural networks. Nature 546(7660): 686. https://doi.org/10.1038/nature22985.

He, K., Zhang, X., Ren, S. and Sun, J. (2016). Deep residual learning for image recognition. IEEE. https://doi.org/10.1109/cvpr.2016.90.

Heussner, J., Kramer, M. P. and Hanf, J. (2023). Verified vintages: Leveraging blockchain to stop wine fraud. BIO Web of Conferences 68: 03024. https://doi.org/10.1051/bioconf/20236803024.

Höglund, R., Tiloca, M., Selander, G., Mattsson, J. P., Vučinić, M. and Watteyne, T. (2024). Secure communication for the IoT: EDHOC and (Group) OSCORE protocols. IEEE Access 4. https://doi.org/10.1109/ACCESS.2024.3384095.

Kasyapa, M. S. B. and Vanmathi, C. (2024). Blockchain integration in healthcare: a comprehensive investigation of use cases, performance issues, and mitigation strategies. Frontiers in Digital Health 6. https://doi.org/10.3389/fdgth.2024.1359858.

Kayikci, S. and Khoshgoftaar, T. M. (2024). Blockchain meets machine learning: a survey. Journal of Big Data 11(1). https://doi.org/10.1186/s40537-023-00852-y.

Krizhevsky, A., Sutskever, I. and Hinton, G. E. (2017). ImageNet classification with deep convolutional neural networks. Communications of the ACM 60(6): 84–90. https://doi.org/10.1145/3065386.

Lecun, Y., Bottou, L., Bengio, Y. and Haffner, P. (1998). Gradient-based learning applied to document recognition. Proceedings of the IEEE 86(11): 2278–2324. https://doi.org/10.1109/5.726791.

Maturi, M. H. and Meduri, S. S. (2024). Exploring the use of graph neural networks for blockchain transaction analysis and fraud detection. International Journal of Innovative Science and Research Technology (IJISRT) 564–574. https://doi.org/10.38124/ijisrt/ijisrt24jul532.

Mihus, I. (2022). Possibilities of using blockchain technologies to protect fraud. Scientific Notes of the UniversityKROK 84–94. https://doi.org/10.31732/2663-2209-2022-65-84-94.

Mohanty, D., Voruganti, N. K., Patel, C. and Sahu, N. (2023). Implementing blockchain technology for fraud detection in financial management. BioGecko 12(2): 1914–1921. https://www.researchgate.net/publication/370072691_Implementing_Blockchain_Technology_for_Fraud_Detection_in_Financial_Management.

Nasir, M. H., Arshad, J., Khan, M. M., Fatima, M., Salah, K. and Jayaraman, R. (2021). Scalable blockchains — A systematic review. Future Generation Computer Systems 126: 136–162. https://doi.org/10.1016/j.future.2021.07.035.

Ncube, H. B., Mutunhu, B., Dube, S. and Maguraushe, K. (2024). Blockchain-Based fraud detection System for healthcare insurance claims. European Conference on Innovation and Entrepreneurship 19(1): 540–547. https://doi.org/10.34190/ecie.19.1.2558.

Nguyen, D. C., Ding, M., Pathirana, P. N., Seneviratne, A., Li, J. and Poor, H. V. (2021). Cooperative task offloading and block mining in Blockchain-Based edge computing with Multi-Agent deep reinforcement learning. IEEE Transactions on Mobile Computing 22(4): 2021–2037. https://doi.org/10.1109/tmc.2021.3120050.

Ozkan-Okay, M., Akin, E., Aslan, Ö., Kosunalp, S., Iliev, T., Stoyanov, I. et al. (2024). A comprehensive survey: Evaluating the efficiency of artificial intelligence and machine learning techniques on cyber security solutions. IEEE Access 12: 12229–12256. https://doi.org/10.1109/access.2024.3355547.

Paramesha, M., Rane, N. L. and Rane, J. (2024a). Artificial intelligence, machine learning, and deep learning for cybersecurity solutions: A review of emerging technologies and applications. Partners Universal Multidisciplinary Research Journal (PUMRJ) 1(2): 84–109. https://doi.org/10.5281/zenodo.12827076.

Paramesha, M., Rane, N. L. and Rane, J. (2024b). Artificial intelligence, machine learning, deep learning, and blockchain in financial and banking Services: A Comprehensive review. Partners Universal Multidisciplinary Research Journal (PUMRJ) 1(2): 68–83. https://doi.org/10.5281/zenodo.12826933.

Qiu, C., Ren, X., Cao, Y. and Mai, T. (2020). Deep reinforcement learning empowered adaptivity for future blockchain networks. IEEE Open Journal of the Computer Society 2: 99–105. https://doi.org/10.1109/ojcs.2020.3010987.

Raghavendra, G. and Lingam, M. S. (2024). Blockchain technology in digital Advertising: transparency, fraud Prevention and trust. Educational Administration: Theory and Practice 30(4). https://kuey.net/index.php/kuey/article/view/1477.

Ressi, D., Spanò, A., Benetollo, L., Piazza, C., Bugliesi, M. and Rossi, S. (2017). Vulnerability detection in Ethereum Smart contracts via Machine Learning: A Qualitative analysis. ACM 1(1). https://doi.org/10.48550/arXiv.2407.18639.

Ribeiro, M. T., Singh, S. and Guestrin, C. (2016). Why should I trust you? pp. 1135–1144. KDD '16: Proceedings of the 22nd ACM SIGKDD International Conference on Knowledge Discovery and Data Mining. https://doi.org/10.1145/2939672.2939778.

Santos, S. D., Singh, J., Thulasiram, R. K., Kamali, S., Sirico, L. and Loud, L. (2022). A new era of Blockchain-Powered Decentralized Finance (DEFI) - A review. pp. 1286–1292. 2022 IEEE 46th Annual Computers, Software, and Applications Conference (COMPSAC). https://doi.org/10.1109/compsac54236.2022.00203.

Sutton, R. S. and Barto, A. G. (2014). Reinforcement Learning: An Introduction. The MIT Press. https://web.stanford.edu/class/psych209/Readings/SuttonBartoIPRLBook2ndEd.pdf.

Taher, S. S. H., Ameen, S. Y. and Ahmed, J. A. (2024). Enhancing blockchain scalability with snake optimization algorithm: a novel approach. Frontiers in Blockchain 7. https://doi.org/10.3389/fbloc.2024.1361659.

Vlahavas, G., Karasavvas, K. and Vakali, A. (2024). Unsupervised clustering of bitcoin transactions. Financial Innovation 10(1). https://doi.org/10.1186/s40854-023-00525-y.

Wang, C. and Peeta, S. (2024). Incentive Mechanism for Privacy-Preserving collaborative routing using secure Multi-Party computation and blockchain. Sensors 24(2): 542. https://doi.org/10.3390/s24020542.

Chapter 21

Blockchain Based Smart, Secured Services for Modern Society

Applications, Challenges and Future Directions

V. Karthikeyan,[1,*] *K. Subbulakshmi,*[2] *Y. Palin Visu*[3] and *Ezhil Kalaimannan*[4]

1. Introduction

The merging of AI and ML techniques is a significant turning point in the field of cyber-physical systems (CPS). Networked systems include MCPS, that comprise technical advancement at the forefront toward integration of the physical processes with information transfer and computation. AI and ML are key for these systems as their design and operation are optimized, which makes way for real-time decision-making, increases efficiency, and makes sure they can handle changing situations. Hospitals are using MCPS more and more to offer top-notch, twenty-four care. A networked system of life-critical medical devices that is cognizant of its environment is called MCPS. Achieving high guarantees in system performance, data-driven analysis, sovereignty, safety, confidentiality, and integrity has proven difficult due to the requirement to develop intricate, secure, and effective MCPS. This article provides an illustration of the MCPS system as an example of a newly designed AI healthcare application. It encompasses a range of healthcare systems that use CPS, including tele-healthcare systems that help treat chronic problems like epilepsy and cardiovascular disease, hearing loss, and respiratory disorders, as well as tele-homecare systems that help manage prescription consumption, among others. Often called "smart systems," CPS are cross-functional technologies that have the potential to transform almost every industry. Because of their flexibility and the wealth of information these systems generate, businesses are increasingly using cloud services for CPS archiving (Hamzah et al. 2023, Zaccagnino et al. 2025).

CPSs are being utilized in a numerous application, including autonomous grids, autonomous and intelligent modes of transportation, medical care, emergency response, efficient factories and Industries 4.0, intelligent streets, process management, aviation, and military uses, among others, as

[1] Department of ECE, Mepco Schlenk Engineering College, Sivakasi, Tamilnadu, India.
[2] Department of ECE, Sethu Institute of Technology, Virudhunagar, Tamilnadu, India.
[3] Department of ECE, St. Mother Theresa Engineering College, Thoothukudi, Tamilnadu, India.
[4] Dept. of Cybersecurity, University of West Florida, Pensacola, Fl-32514, USA.
Emails: lakssubbu91@gmail.com, palinvisu@gmail.com, ek@uwf.edu
* Corresponding author: velkarthi85@gmail.com

DOI: 10.1201/9781003620822-21

a result of rapid advancements in fields like cyber security, the data analytics, fog computing, edge computing, and AI.

The advancements in MCPS enhances numerous aspects of our lives, bolstering our essential systems and laying the groundwork for intelligent amenities in the future. The community at large stands to benefit much from CPS.

CPSs evolved from ICSs, which typically use a hierarchical structure and applications based on the ANSI/ISA-95 framework to coordinate information collection and processing elements of NCSs. The concept of CPS originated once Zuse constructed the initial fully functional, programmable mathematical computer for measuring aircraft wings just after developing Z3 in 1941 (Pandey et al. 2024, Karthikeyan et al. 2024a, b, c, Jones et al. 2024). In the nineties, advancements in sensors, devices, social media, and embedded technology prompted a heightened focus on bridging the gap between mechanical and digital systems. Modern ideas like the Industrial Web and the 4th Industrial Revolution revolve around the use of CPSs. These are complex networks that allow cyber-components to remotely operate real-world items or activities (Maria et al. 2024, Tallat et al. 2023).

Figure 21.1 depicts the hierarchical structure of the Medical cyber-physical systems (MCPS), emphasizing its three fundamental components: external memory, cloud memory, and uses. The figure classifies these components according to their main functions: calculation, interface, and management. It highlights the significance of memory systems in enhancing computational efficiency, data processing, and system usability. This structured representation aids in comprehending the interactions among various memory types and their applications in contemporary computing environments.

The purpose of this diagram is to illustrate the flow of information between the interconnected nodes that make up the actual domain and the cyber domain, where the control layer is responsible for ensuring adequate efficiency to meet different dependability, security, and administration standards (Rahman et al. 2025, Karthikeyan et al. 2024d, e, Maurya et al. 2025). Compared to integrated systems, CPS involves more physical equipment. CPS emphasizes the relationship between the physical and cognitive levels more than embedded systems do. A CPS is one in which different hardware components are operated and coordinated by software applications. C3 (Calculation, Communication, with Command) is the acronym used to designate the CPS because of this. With more and more connections being made between the virtual and real worlds, the utilization of cyber-physical structures is expanding. Up until now, automated production groups have been built using the multidisciplinary paradigm, with information technology (IT) and operational technology (OT) having separate control.

Despite the above-mentioned tiers forming an organized design, a CPS typically functions as a network, integrating all material and intellectual components within an industrial setting. Therefore, the relevance of CPSs lies in their complete integration of manufacturing operations and connectivity. Because it enables the network topology to adjust to modifications in hardware settings, developing networks that enable flexibility is essential. Collaboration between communication protocols and complex management systems creates a new ecosystem that proves beneficial for future CPS

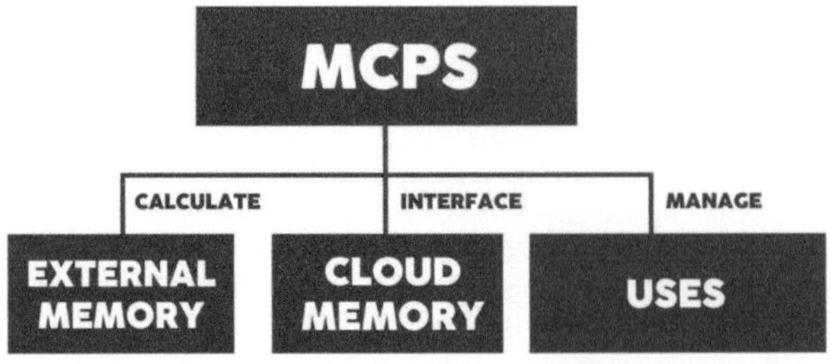

Figure 21.1. Medical cyber-physical systems architecture.

implementations. Consequently, the creation of CPSs may benefit greatly from the Software-Defined Networking (SDN) paradigm (Baiod et al. 2025, Wu et al. 2025, Karthikeyan et al. 2023a, b).

1.1 AI-Enhanced Medical Devices

AI is revolutionizing medical equipment by increasing inventiveness and optimizing processes. AI is incorporated into a lot of modern implants and prostheses, enabling them to adjust to the user's movements for increased usability and convenience. Surgical robots that employ artificial intelligence (AI) enable more accurate and regulated operations with lower client risks and better results overall. Intelligence-powered medical tracking devices can also instantly evaluate patient data, enabling prompt detection of healthcare issues and ongoing monitoring with prompt intervention.

1.2 Remote Patient Monitoring

Access promotion, remote patient monitoring, and healthcare service administration heavily rely on artificial intelligence. Telemedicine is one field where artificial intelligence is already having an impact. Virtual health assistants can now schedule visits, provide health advice, and even remotely check on patients' vital signs. Health monitoring devices can assess critical indicators, spot abnormal changes that point to illness, and speed up therapy by continuously integrating AI. AI also helps with the transition from generic treatment plans to customized ones that are based on the study of patient data. By managing and resolving every issue pertaining to a particular patient's treatment, AI completely changes the way healthcare is delivered (Kaur 2025, Elhadari et al. 2025).

2. Intelligent Medicare

Zhang et al. described a fog and data-analytics model for MCPS. Their framework contains an application service layer, administration, and information gathering. The main objective is to show how cloud computing and big data analysis can optimize healthcare delivery. For Medical CPS, Kaur et al. (2025) presented a model-based paradigm. The patient, device, and MCPS controller are the three main models that are included. These validated models can be used by developers to better craft software test cases. Elhadari et al. (2025) introduced a hierarchical medical assessment framework. Layers for maintaining data, administration, data mining, diagnosis, and knowledge transfer form the backbone of the proposed system. Multiple practical implementations as well as demonstrations of potential applicability were additionally presented. In their design for an integrated medical CPS, Jin and Xiao (2022) primarily considered four layers: physical, interaction, intellectual services, and information assessment (Anitha et al. 2025).

By automating and monitoring therapy, the concept aims to improve the standard of care given to patients. However, the architecture has not yet undergone thorough testing and validation. Jones et al. (2024) presented a new benefit-oriented strategy for the medical sector in order to tackle the challenge of allocating limited healthcare resources. Three forms of happiness are at the heart of the idea: contented resources, contented business owners, and contented patients. The reward-based paradigm is supported by several tests and simulations. Guo et al. (2022) formulated an approach to enhance safeguards for patients via MCPS evaluation and verification. System interactions with regression calculations are employed to produce the critical attributes of a patient. The framework demonstrated accuracy when evaluated against the clinical foundation paradigm, which uses mobile computers to improve communication between healthcare providers, patients, and caretakers. Decision support representatives, manager representatives, controller representatives, caregiver representatives, patient-watching representatives, with confirmation representatives are the six fundamental representatives in the projected multi-agent scheme. When it comes to medical services, Kaur (2025) suggested using big data. They have constructed a comprehensive healthcare intelligence framework, incorporating a rigorous N-ary conceptual evaluation, to enable

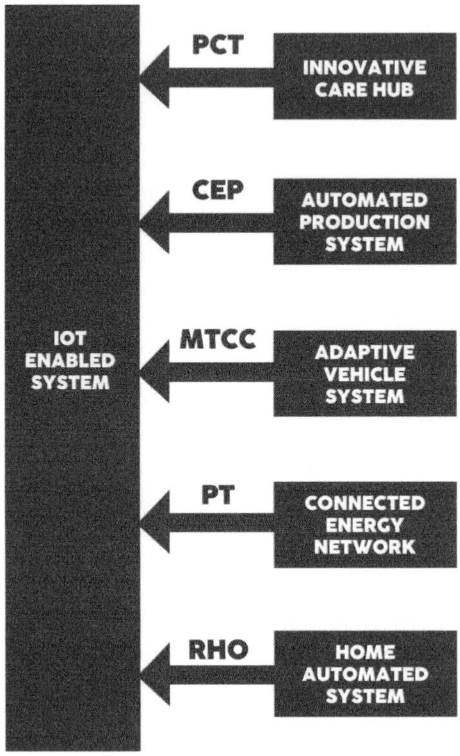

Figure 21.2. Integration of IoT enabled systems with various technological hubs.

comprehensive data collection and interconnectivity (Razaque et al. 2025, Singh and Kumar 2025). Figure 21.2 depicts the integration of an IoT Enabled System with five technological hubs, emphasizing the diverse applications and the interconnected nature of IoT in modern technological advancements.

The model has to be tested in real-world scenarios. Medical CPS systems combine a communication channel, command and control software, and embedded devices. Today's medical technology has advanced so quickly that it has given rise to a new generation of healthcare professionals and alternative therapeutic modalities. MCPS has an extensive array of latent requests, including patient monitoring, implant sensor devices, and analgesic infusion pumps (Velayuthapandian et al. 2024b, c, Tong et al. 2025, Sun et al. 2025). The sensors embedded in such medical equipment gather information from different parts of the individual's body and transmit it to a system for evaluation and treatment planning. The CPS may be highly effective in a range of medicare scenarios: such as in-house and in-clinic client attention because of developments in clinical bio-data's, WSNs, and edge data analysis. These bio-sensors have the potential to gather important patient data, including health information (Karthikeyan et al. 2024e, f, g, Zhao 2025, Kaakandikar et al. 2025).

3. Analysis of Architectural Quality

To determine which of the possible scheme is a virtuous fit for the strategy of MCPS for clinical care, a comparison of important architectural features was conducted. Four categories of architecture were the main focus of this analysis.

Heterogeneity: The range of equipment, especially in a healthcare setting, allows several institutions to circumvent problems caused by design flaws in any one manufacturer.

Reactivity: CPS are frequently employed to maintain a condition or achieve an objective. Monitoring a condition and responding to changes in multiple contexts are common applications of CPS.

Adaptability: The numeral of bases, the quantity of intrusion generated, the locations of the rooms, the responsibilities of the staff, and other physical factors will vary for every CPS installation.

Scalability: The size of CPS deployments will vary significantly. One essential requirement for scaling a CPS is the ability to process data in parallel.

Uncertainty: Different MCPS may need help from a software strategy that can handle some uncertainty because of problems with getting data.

Robustness: Having a CPS that can continue to operate regularly even when there is missing or erroneous data is essential in a healthcare context. Redundancy might be important, but the scheme's flexibility to incorporate sophisticated procedures that can use a variety of information bases to detect incorrect shares is also difficult (Patra and Panigrahi 2025, Alomari et al. 2025).

4. Healthcare CPS Architectures Aided by Big Data and the Cloud 4.0

The modern-day description of MCPS is the integration of a physical object tracking and control with dispensation, announcement, and regulator abilities. "Cyber systems" refer to networks that have embedded computers and feedback loops that are accountable for monitoring and controlling the fundamental corporal processes. The corporal developments will influence the intentions and vice versa (Geetanshi et al. 2025, Igonor et al. 2025, Karthikeyan et al. 2024g, h). The definition of the CPS is widely accepted, but opinions on its key components and how they relate to one another are less widely held. The cyber and physical sections are the maximally recognized mechanisms of the MCPS strategy. While the cyber layer carries out the physical layer's directives founded on its scrutiny, the physical layer collects and evaluates the observed data.

Numerous factors have been suggested to be handled by more than three levels of architecture, as in Dev et al. (2025), including object abstraction, service composition, reliability coordination, and service management.

The requirement for CPS administration and control justifies the structure's intricacy. To account for the many innovations that have been adopted, a new architecture with five deposits—commercial, request, dispensation, broadcast, and discernment—has also been updated (Karthikeyan et al. 2025a, b). Despite varying opinions regarding its structure, CPS can be separated into three phases: awareness, communication, and submission. Strategies with related jobs that model each of these tiers are detailed.

4.1 DCL

This level provides a standardized interface for accessing multi-dimensional diverse information, such as those originating from hospitals, connectivity, and user-generated content. It is collected by information nodes and connecters. To promise the security and availability of information transmission to information administration stage, real-time information can be analyzed in a number of ways and arrangements by means of a converter (Rehman et al. 2025, Prakash et al. 2025, Karthikeyan et al. 2025a, c). Observation, acknowledgement, and sensation make up the first layer. This layer contains an extensive range of smart components, like intelligent devices, controllers, image, geographical location information, digital scanner and, smart tags with bar-code tags. At this level, instruments can process physical input, carry out application layer commands, and gather real-time data for a range of uses (like tracking and monitoring).

Data can be gathered from a wide range of environments, like biology, geography, heat, electrical energy, audio, video, mechanics and more. Collaborative nodes in local and global network domains can help collect and evaluate real-time sensor data at the application stage. Figure 21.3 illustrates the architecture of a decentralized data processing and storage system. It highlights the flow of data

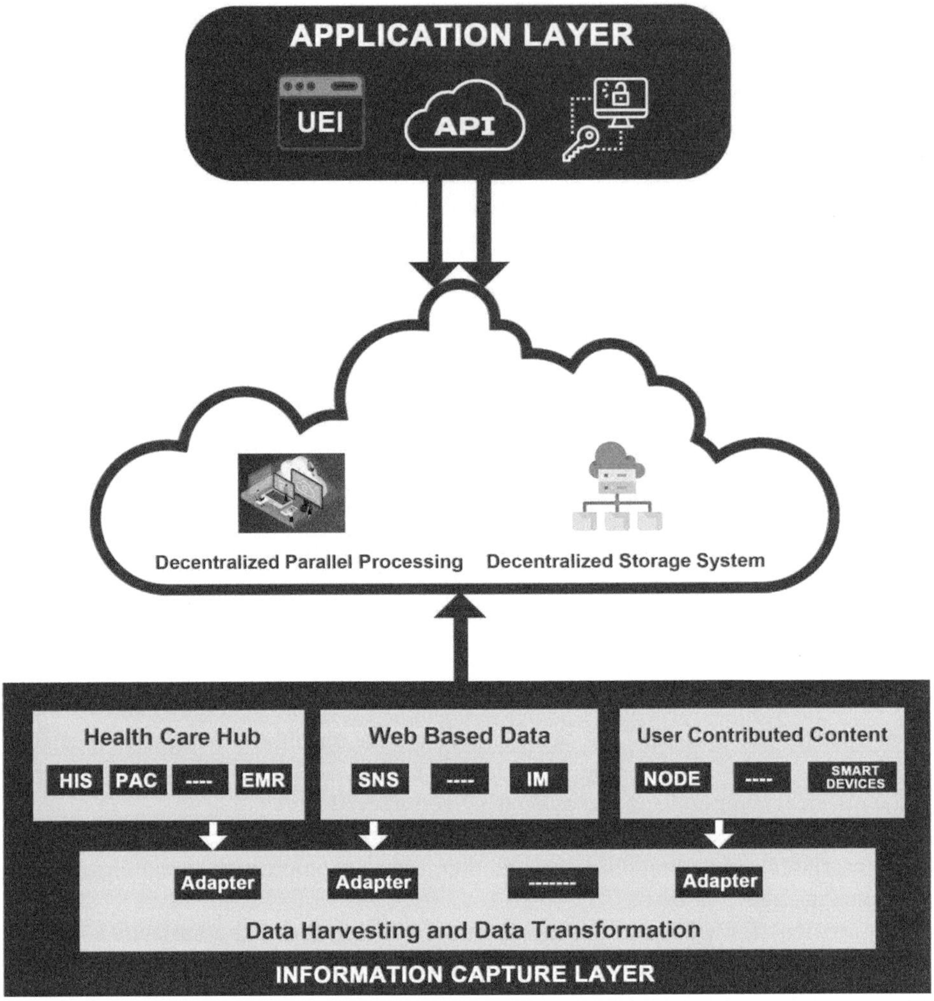

Figure 21.3. Architecture of a decentralized data processing and storage system.

across the information capture layer, decentralized cloud infrastructure (processing and storage), and the application layer, showcasing a streamlined data transformation and utilization process.

Data node: Health-CPS collects standard medical data as well as clinical, research, and medical spending data, as well as details about a person's activities and emotional state. Between the information layer and the scheme is a software layer called an adapter, which performs functions including analysis and encoding collected information (Deepa et al. 2022, Alam 2022, Karthikeyan and Visu 2024, Abdelmaboud et al. 2022). The assessment segment helps with data format transitions, data cleansing, compression, and redundancy reduction. The converter employs a pre-established information stage from the scheme to translate the collected information into the necessary arrangement. The encryption module uses a hierarchical privacy preservation mechanism to encrypt the data in order to guarantee security.

An unauthorized device won't be able to collect the encoded information package, even if it manages to get into the system. Its flexible, useful unit is what gives the adapter its expandable capabilities. The modules of the adapter will be automatically reorganized through the cyberspace if the subsequent conditions are fulfilled.

4.2 DML

By streamlining the exchange of information and processing of medical information of all kinds, DFS will help the medical field run more efficiently by means of the application of technology connected to massive information. According to the significance of the evaluation task and the information's pressure, DPC provides suitable handling and evaluation methodologies. The next level, also known as the communication plane (sometimes called the communication tier or the connection tier), is where information analysis and exchange among the program and its users happens. Incorporating a range of protocols such as zigbee, LoraWAN, IEEE802.15.4, and advanced wireless technologies, this tier enables exchange of information along with distribution across local or regional wireless networks, the WWW, or various previously present infrastructures (Malla et al. 2022, Karuppaiah et al. 2024, Tyagi et al. 2023).

However, most connections use the Internet because of its affordability and accessibility. The use of these networks necessitates real-time activity. Due to the impact of physical environments on communication and the need for dependable message delivery, the communication tier is primarily capable of processing and managing large volumes of data. The primary protocol which is present at this tier, IPv6, is specifically designed to address a more general problem. Additionally, the transport tier is accountable for conveying information across the systems in use, through a range of strategies and smart devices. Cloud computing platforms, IEEE802.11, IEEE802.16, IEEE802.15.4, advanced wireless generations, networking devices, and gateways all work normally at this tier.

DFS module: Supporting efficient data processing and analysis and setting up a productive distributed storage system for massive amounts of data are the two main issues with healthcare data. Among other advantages, Health-CPS leverages DFS techniques to guarantee that vast volumes of medical data are stored securely, accessible, and easily shared. Security Label, Data Type, and Data Description are the three elements that go into creating data standards for the uniform handling of multisource heterogeneous data (Malla et al. 2022).

DPC module: To find insights, the DPC module processes and examines the DFS data. It processes and interrogates physical information, integrates various big-data and AI approaches, and provides real-time analysis for massive volumes of formless data.

4.3 ASL

In this tier, users can review the fundamental findings of the information assessment procedure. It also provides developers a single, standardized interface to use when generating client-focused apps that deliver improved medicare, and personalized healthcare. The ASL tier is the final and busiest tier. The function of this tier is to translate information from the previous layer into commands that ground-based devices, such basic IoT gadgets, can follow. Also, this tier is in charge of rendering the appropriate decisions in light of the information gathered and delivering the required directives for corrective action. The perception layer sends information to this layer, which then assesses it and determines what kind of instinctive replies are required. For programmers and end users alike, the ASL assists like an operating system, and management (Verma et al. 2022, Karthikeyan et al. 2024j, 2023b, Andrew et al. 2023, VarunPrakash et al. 2024).

Management stage: The Health-CPS is in charge of several applications, with conformation administration, distribution administration, parameter administration, surveillance administration, conception administration, and outcome administration.

Progress stage: This domain offers test beds, data access, and a consistent API.

Operating platform: It offers the technology, software, and data needed to execute healthcare applications. Gateway devices communicate with cyberspace via a variety of communication protocols, including cellular, Wi-Fi, and Zigbee.

5. Comprehensive Review of MCPS in the Healthcare Sector

CPSs integrate the physiological analysis of processes with IT and network surveillance. Communication channels that connect to the real operations that affect computing and let you know about the next steps that control the practical operations are very important for every MCPS where networked devices that are built in oversee and control the real processes. Figure 21.4 illustrates the use of an MCPS for controlling, communicating, and computing. The primary objective in every MCPS is to integrate detectors and bots using knowledge acquired through other components of the framework and then processed computationally.

They are, therefore, the foundation of MCPS since they possess the mental and motor abilities required for interpersonal communication. Similar to the CPS, the MCPS is a single framework that combines system management, shareware, and physical operations to create feedback systems. In addition to healthcare, other industries that use these technologies include engineering, manufacturing, robotics, defense, and transportation. Because of this, CPSS needs to be flexible, self-sufficient, skilled, useful, reliable, safe, and easy to use. Great progress should therefore be made because MCPS is evolving quickly and dynamically. Due to the vast number of devices, CPSs have to give scheduling and communication top priority.

Communication: A digital system, interfaces, and physical elements that make up MCPS. The pieces of the actual environment monitor and control the tools that are attached to them (Gopalakrishnan et al. 2024). Contemporary intelligent gadgets are the central controller of MCPS, which process and distribute data rapidly. Linking the CPS to the real world requires the use of the communal connectivity, message link, and additional intermediary parts like interconnected smart nodes, controllers, processors, and other elements. The intelligent input nodes and smart controllers in the system manage both digital and analog signal conversion and transformation.

Scheduling: In networks (SANs), distribution of data and control necessitates careful resource planning (Gad et al. 2022, Indumathi et al. 2024). Patient data is handled by a variety of medical equipment, and these devices are becoming more and more connected and able to communicate

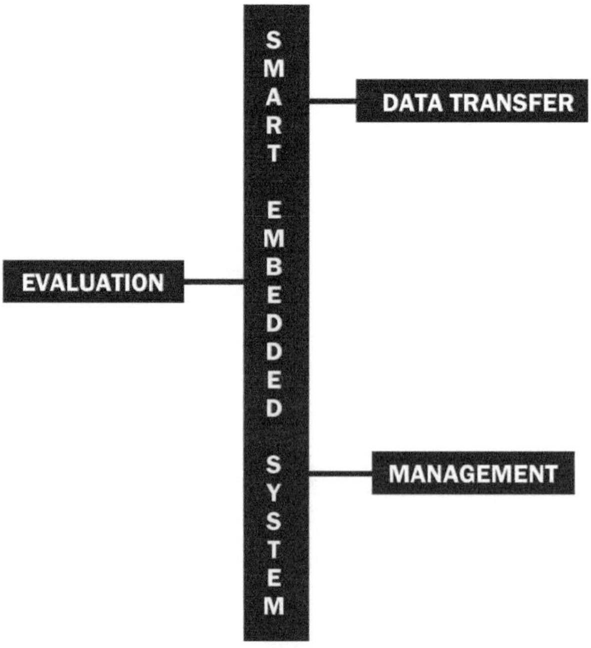

Figure 21.4. Capabilities of cyber-physical system.

with one another online. By guaranteeing effective message arrangement among interrelated healing apparatus, like a network controller, surveillance scheme, and PCA drive, the goal is to condense the economic load of all-purpose health treatment on clients. Through the treatment of a patient-centric health care state that involves medicine distribution, the effects of the therapeutic device's connectivity and interoperability as well as network-enabled patient health control are investigated.

5.1 Clinical Care MCPS Nomenclature

A collection of taxonomies that describe and categorize CPS techniques in healthcare is desirable. We developed such a taxonomy for CPS in healthcare based on an analysis of the body of current literature. Its components are as follows: the first is adaptation; the second is infrastructure; the third is detecting; the fourth is information administration; the fifth is processing; the sixth is connectivity; the seventh is safety; and the eighth is control. Each of these classifications of these elements are covered in detail in this section.

5.2 Categories of MCPS

Research on the possible medical applications of MCPS is still in its early stages. Combining passive client data, like information from biosensors and other smart devices, with active client data, like computerized patient records and the smart feedback system, may help MCPS gather the data needed for efficient decision-making (Malik et al. 2023, Shafik 2024). There needs to be extensive research conducted on the potential benefits of integrating information safety and administration in clinics. Because it necessitates data from a diverse environment, researching its likely claims in clinical facilities and supplementary communal assurance issues is costly. An opportunity to use MCPS in social amenities is given by the interoperability of flexible and self-sufficient devices. The supervision and operation of restorative physical systems may be aided by computation and control, body area networks, programmable devices, small implanted smart devices, and innovative manufacturing processes (Karthikeyan et al. 2024c, f, Rathod et al. 2022, Bellagarda and Abu-Mahfouz 2022).

5.2.1 Architecture

The performance and quality of the system are crucial, hence MCPS design needs special attention. The submission field, customer information needs, and scheme settlement should all be considered when designing the architecture to preserve the structure's functionality. The three foundational ties of MCPS scheme architecture are user-oriented, cloud-oriented, movement-oriented, and instinctive.

5.2.2 Sensing

A network of biological devices collects critical physical information and transmits it for analysis. Service delivery relies heavily on perception, as the structure employs the found features as knowledge variables. One common method for measuring blood sugar concentration is the pricking of the thumb, which sufferers must do frequently. This patient was treated with a harmless approach of blood glucose measurement that relies on radioactive substances and sensors (Krichen et al. 2022).

5.2.3 Data Management

With the use of MCPS's information administration framework, consumer requirements will be satisfied by controlling information from sensors. The preparation of information improves knowledge gathering and dissemination because information gathered will not be employed in its natural state owing to substantial transmission costs and ineffective computation.

5.2.4 Computation

It is employed for monitoring as well as simulation. Physicians/specialists want rapid and simple access to the data they require, as well as the capability to screen and curtain clients at real-time and from any location. Cloud computing can carry out enormous scales of complex computation and communication, making it simpler for experts to get appropriate data from businesses. Utilizing strategies for minimizing data bottlenecks and determining data size is essential to guaranteeing system efficiency (Balamurugan et al. 2024).

5.2.5 Application

MCPS offers a variety of adapted applications, including senior care, assisted living institutions, and medical centers. The complex nature of the system is heavily dependent on the specific application.

5.2.6 Security

Protecting critical health data is a constitutional, judicial, as well as philosophical imperative, making this a crucial component of the MCPS. Therefore, it is crucial to ensure the security of client information when designing infrastructure for healthcare-related apps (Dadhich and Tiwari 2022).

5.2.7 Communication

The two primary methods of communication are collecting patient information with establishing a connection to the central server. The primary channel for information exchange between components is MCPS. Numerous little discussions (data exchanges) take place throughout the system, including the sharing of viewpoints by different stakeholders engaged in healthcare delivery. Recent advances in virtual/augment reality employing advanced wireless connectivity, which allow the delivery of extracted and compressed images in an incredibly effective manner, could increase the MCPS's potential. Perception and monitoring of tolerance and the interpretation of visual information associated to particular biological indicators would benefit from this (Bagga et al. 2022, Ahmad et al. 2022, Karthikeyan et al. 2025a, c, Bhattacharya et al. 2022).

A lot of preparation and adherence to established procedures are necessary for effective communication. However, there are some significant problems with MCPS that must be fixed, most notably the data and communication scheduling. Gateway devices communicate with cyberspace via a variety of communication protocols, including cellular, Wi-Fi, and Zigbee. Several security technologies may also be applied in the physical space to promise the concealment and secrecy of information transfer.

6. Wearable Technology

In CPS, embedded substances are measured via habiliment technology. For occasion, in WSN's, many intelligent gadgets are wired to a transferrable server node. Advances in digital electronics, sensors, and MEMS expertise are expected to lead to a large increase in the quantity of habiliment and attachable gadgets utilized in clinical claims in the near future (Suhail et al. 2022). Additionally, the development of low-cost, low-power, multipurpose intelligent devices with radiocommunication will allow data processing, component communication, and short- and long-distance communication.

For instance, when it comes to ongoing patient monitoring, wireless body area networks (WBAN's) offer circumstantial data for troublesome circumstances. The function of CPS in healthcare could be improved by wearable sensors, which could also help people adapt to new environmental health concerns. Basic activity trackers with pulse monitors have been supplanted in recent years by supplementary classy WSN's for virtual patient-intensive care.

Embedded detectors are an essential part of the WBAN, monitoring neurotransmitter stages regardless of the individual's physical state. The embedded instruments collect data from the patient's bodily systems and transmit it to a conveniently located management device. One example is the use of biological sensors placed underneath the dermis to monitor critical signals like cardiac activity (Farooq et al. 2022). More and more types of embedded clinical implants will become possible in the near future as knowledge of microorganisms continues to progress. On top of that, doctors have observed personally that blind persons can see adequately after getting corneal patch chips implanted in their retinas.

There are embedded computer devices that don't need hands. The manipulator interacts with a touch curtain or keypad, and a miniature presentation shade represents the information. Future portable devices might soon incorporate intelligent devices that will record bio-information like fingerprints, scripts and signatures, faces, videos, and microphones. Among the many potential applications are audio-visual documentation, communication, entertainment, and more.

6.1 Portable Systems

Portable systems in Wireless Body Area Networks (WBANs) are essential for the progression of wearable technology applications, particularly in healthcare and fitness. These systems incorporate lightweight and energy-efficient components to facilitate seamless monitoring and communication functionalities. The fundamental components of portable WBAN systems are sensors, processing units, communication modules, power supplies, and user interfaces. Sensors acquire physiological data, including heart rate, blood pressure, and mobility, whereas the processing units manage data preprocessing, compression, and encryption. Communication modules depend on protocols such as Bluetooth or Zigbee to provide dependable and efficient wireless data transmission. These devices utilize tiny batteries or energy-harvesting methods, engineered for extended operation. User interfaces, such as companion applications or LEDs, offer immediate feedback and interactivity (Karthikeyan et al. 2023a, c, Khalil et al. 2022, Guo et al. 2022).

6.2 Accompanied Devices

It is not possible to wear or implant these devices on another person. Although they can be hidden in other objects, they are not portable.

7. Technological Aspects and MCPS Traits Relevant to Intelligent Clinical Care

The advancement of intelligent systems and medical cyber-physical systems (MCPS) has greatly facilitated the integration of technology into clinical treatment. To improve healthcare delivery, these systems integrate physical, networking, and computational processes. Intelligent clinical care's technological components center on keeping tabs on patients' vitals through the use of cutting-edge sensors, data analytics, and real-time communication. The high standards needed in clinical settings necessitate MCPS characteristics, including security, adaptability and reliability.

Intelligent clinical care relies heavily on sensors and data collection. These days, systems can monitor vital signs, including glucose levels, oxygen saturation, and heart rate, in real time via a network of sensors. The incorporation of these sensors into wearable technology has made continuous monitoring a viable option for patients, as it is both non-invasive and convenient. These tools, when combined with sophisticated algorithms, sift through patient records in search of outliers, health problems, and insights that doctors can put to use.

Interconnection and communication are other crucial components of technology. Strong networking protocols are the backbone of intelligent clinical care systems, allowing for the smooth transfer of data among various devices, healthcare providers, and the cloud. For life-or-death situations like remote surgeries or emergency interventions, technologies like the Internet of Medical Things (IoMT), 5G, and edge computing provide low-latency, high-reliability connectivity. To make sure various devices and platforms can work together smoothly, these systems also prioritize interoperability (Hussain 2024, Peng et al. 2023, Bashir et al. 2023).

Analytics, data, and AI form the backbone of smart clinical care. Artificial intelligence (AI) algorithms and machine learning models sift through mountains of patient data in search of patterns and connections. By predicting how a disease will progress, predictive analytics might pave the way for earlier treatments and more tailored treatment regimens. Improvements in patient outcomes, reductions in diagnostic errors, and the provision of evidence-based recommendations are all ways in which AI-driven decision support systems aid physicians.

Given the importance of integrating physical processes with cyber systems in healthcare, MCPS features are especially pertinent in this context. Critical MCPS qualities include reliability and real-time responsiveness, which guarantee that systems consistently work, particularly in high-stakes settings like surgical theaters or intensive care units. Additionally, these systems must possess the flexibility to adapt to the unique needs of individual patients or modifications in clinical procedures. Systems must also be scalable so that they can handle massive amounts of data and provide support for numerous devices without sacrificing performance (Karthikeyan et al. 2024i, j, Krishnamoorthy et al. 2023, Ramu et al. 2022).

When it comes to smart clinical care, MCPS's security and privacy features are crucial. Compliance with rules like HIPAA or GDPR, as well as strong encryption and secure communication methods, are necessary due to the sensitive nature of health data. To further guarantee security and confidentiality of patient information, intelligent systems use anomaly detection to instantly spot and counteract cyber-attacks.

Intelligent clinical care systems are revolutionizing healthcare by providing real-time monitoring, predictive analytics, and customized therapies. Cutting-edge technologies and MCPS features form the foundation of these systems. Improved patient outcomes, optimized clinical workflows, and a smarter, more efficient healthcare ecosystem could be possible thanks to these systems' capacity to tackle issues like reliability, adaptability, and security.

8. Emerging Challenges in Combining MCPS Combined with AI

Security: When combining AI with cyber-physical systems (CPS), security is essential. Critical industries like manufacturing, transportation, and healthcare are managed by these systems. Strong encryption, safe connection, and continuous monitoring are necessary since AI algorithms are susceptible to cyberattacks.

Interoperability: For CPS to function properly, several hardware, sensors, and software are needed. For seamless data interchange and system integration, standardized standards and protocols must be developed and used.

Reliability and safety: The dependability and security of AI-Integrated Cyber-Physical Systems (CPS) is vital in developed technologies within systems like self-driving vehicles and implanted devices (Bamakan et al. 2022, Karthikeyan et al. 2023a, b, Ullah et al. 2023). AI models require extensive testing and validation to establish their confidence and accuracy, and errors are mitigated by providing mechanisms that ensure safe access and backup systems. Trustworthy AI systems enhance the users' confidence and make identifying and correcting mistakes easier.

Complexity: Combining technology such as sensors, actuators, communication networks, and AI algorithms makes integrating AI into CPS challenging. These systems demand specialized knowledge and abilities to design and manage. Because its components constantly interact with the physical world, the dynamic nature of CPS makes things even more difficult. To manage this complexity, forecast system behavior, and enhance performance, sophisticated system designs, effective data handling techniques, and potent modeling tools are required.

Scalability: The ability of a system to manage more workload or grow quickly is referred to as scalability. Supporting networked devices and processing massive amounts of real-time data present difficulties for AI-integrated CPS. We need to use cloud and edge computing, maximize computing resources, use distributed processing techniques, and build systems that can swiftly incorporate new gadgets and technology.

9. Role of ML/DL in MCPS

Medical Cyber-Physical Systems (MCPS) signify the integration of medical equipment, computing systems, and physical healthcare processes to provide enhanced patient care. These systems frequently incorporate intricate algorithms and real-time monitoring features, and the integration of machine learning (ML) and deep learning (DL) has markedly improved their functionality and dependability. The influence of machine learning and deep learning in medical care and patient services is revolutionary, impacting multiple areas including diagnostics, monitoring, therapy, and system administration (Kumarand Singh 2022, Velayuthapandian et al. 2024a, b, c, Chengoden et al. 2023, Hemamalini et al. 2024).

A principal application of machine learning/deep learning in medical care and public health systems is in diagnostics and predictive analytics. Machine learning models may evaluate extensive patient data, encompassing medical imaging, electronic health records (EHR), and sensor readings, to identify diseases at an early stage. Medical imaging modalities such as X-rays, MRIs, and CT scans extensively employ deep learning methods, particularly convolutional neural networks (CNNs). These models have demonstrated performance comparable to, and at times exceeding, that of human radiologists in detecting anomalies such as tumors, fractures, or infections. Moreover, machine learning-driven prediction models can anticipate illness progression, facilitating prompt interventions and tailored treatment strategies.

A fundamental function of machine learning and deep learning in medical care and patient services is real-time patient monitoring and anomaly identification. Wearable devices and implantable sensors produce continuous data streams about vital signs, glucose levels, heart rates, and additional metrics. Machine learning algorithms can analyze this data in real time, detecting anomalies or trends that suggest potential health concerns. Recurrent neural networks (RNNs) and long short-term memory (LSTM) networks are great at processing time-series data. This makes them good for keeping an eye on patients' conditions and predicting bad things that might happen, like cardiac arrests or seizures, before they do (Alharbi et al. 2022).

In the realm of treatment optimization, machine learning and deep learning facilitate the creation of adaptive systems that customize therapy for particular patients. In robotic surgeries or automated drug delivery systems, machine learning models continuously adapt based on patient responses to dynamically modify dosages or enhance surgical precision. Reinforcement learning, a subset of machine learning, is particularly advantageous for these applications, as it allows computers to enhance performance via trial and feedback processes.

Machine learning and deep learning also enhance the security and resilience of microgrid control and protection systems. Cybersecurity is a crucial concern in MCPS, owing to the sensitive nature of medical data and the necessity for continued operations. MCPS utilizes machine learning models to identify and mitigate cyber dangers such as unauthorized access, data breaches, or system anomalies caused by hostile activity. Machine learning-based intrusion detection systems can detect suspicious activity in real time and respond appropriately by studying network traffic patterns and system logs.

Even with these changes, there are still problems with ML/DL being used in MCPS. For example, they need good data, models that can be understood, and they have to follow rules set by the government. Despite these problems, ongoing research and the use of explainable AI (XAI) methods are trying to solve them, making sure that ML/DL applications in MCPS are reliable and trustworthy (Jin and Xiao 2022).

In summary, machine learning and deep learning are transforming medical care and patient safety by improving diagnostic precision, facilitating real-time surveillance, customizing therapies, and strengthening system security. Advancements in healthcare technology, along with the integration of machine learning/deep learning into medical care processes, promise to improve the efficiency, accuracy, and accessibility of medical services for patients worldwide.

10. Benefits of Natural Language Processing in MCPS

Natural Language Processing (NLP), a branch of artificial intelligence, is essential for improving the functionality and efficiency of Medical Cyber-Physical Systems (MCPS). NLP enhances the understanding and processing of human language, thereby improving the interactions with technology, medical data management, and patient care delivery of healthcare professionals.

A primary advantage of NLP in MCPS is its capacity to derive actionable insights from unstructured data. Medical data, clinical notes, and patient histories frequently reside in unstructured text formats, rendering manual analysis difficult. NLP algorithms can analyze this data, extract pertinent information, and arrange it into structured representations for enhanced understanding, thereby facilitating precise diagnoses and treatment planning (Al-Hawawreh et al. 2023).

NLP enhances patient communication and engagement. Patients can obtain real-time assistance, inquire about information, arrange appointments, or receive prescription reminders with virtual assistants and chatbots utilizing natural language processing. These instruments improve patient experience while alleviating the burden on healthcare professionals. NLP enhances clinical decision-making by synthesizing pertinent material and delivering evidence-based suggestions. It can analyze extensive volumes of medical data, recommendations, and case studies to aid healthcare practitioners in making prompt, informed judgments.

NLP improves interoperability within MCPS by converting medical jargon to standardized terminology, thereby enabling frictionless data flow between systems. This enhances collaboration and efficiency throughout healthcare networks. NLP in MCPS enhances patient outcomes, optimizes healthcare operations, and facilitates the shift toward data-driven, personalized medicine by harnessing the capabilities of written and spoken language.

11. Importance of MCPS-Reliant Rule-Based Expert Systems

When it comes to medical cyber-physical systems (MCPS), rule-based expert systems are crucial because they offer a framework for making decisions based on logic. In order to evaluate data and provide recommendations or alerts, these systems depend on predefined rules, which are typically based on medical knowledge and expert input. Because of their inherent determinism, they shine in situations where clarity and explanation are paramount.

Clinical decision support is a major use case for rule-based systems in MCPS. For the purpose of diagnosis, therapy, or medication adjustment suggestions, these systems analyze patient data using "if-then" logic. For instance, in order to facilitate rapid action, a rule-based system could initiate an alarm in the event that a patient's blood glucose level surpasses a crucial threshold. When there must be strict adherence to clinical guidelines at all times, these systems are priceless (Alharbi et al. 2022).

Rule-based systems can automate routine operations in MCPS, such as monitoring patient vitals or confirming compliance with hospital protocols. In intensive care units, for example, these systems may monitor patient data streams in real time and alert doctors when parameters deviate from the norm, allowing for lightning-fast reactions to any problems that may arise. In addition, MCPS benefit from improved interoperability and integration because of their simple architecture. They allow for communication between the many parts of a system and can be easily integrated into software or medical devices.

Despite the rule-based limitations of expert systems, when it comes to dealing with complicated or ambiguous data, they remain a staple of many MCPS programs due to their simplicity, reliability, and convenience of installation. This, in turn, guarantees the consistent and safe delivery of healthcare. These systems utilize established "if-then" rules based on medical expertise, guidelines, and organizational regulations to analyze data and provide actionable insights. Their roles in diagnosis, treatment, and management render them essential in contemporary healthcare systems.

11.1 Therapeutic and Diagnostic Use Cases

Rule-based systems are essential in improving clinical decision-making in diagnosis and therapy. By examining patient data, including symptoms, laboratory results, and imaging findings, these systems can detect probable illnesses and offer diagnostic recommendations. For example, a guideline might state, "If a patient's fasting blood glucose exceeds 126 mg/dL on two separate occasions, consider them as potentially diabetic." These qualities facilitate the early identification of diseases, especially in intricate situations where human observation may overlook small trends. Moreover, rule-based systems facilitate treatment personalization by suggesting therapies according to particular patient characteristics such as age, weight, allergies, or comorbidities. A system that keeps an eye on anticoagulant drugs might change dosages based on changes in a patient's INR (International Normalized Ratio) readings. This would lower the risks of over- or under-dosing. These systems contribute to the enforcement of clinical protocols, reduction of errors, and improvement of patient outcomes.

11.2 Management Use Cases

Rule-based solutions automate routine and resource-intensive operations in administrative functions, markedly enhancing operational efficiency in healthcare environments. They optimize appointment scheduling by aligning patient requirements with available healthcare providers or facilities. A rule-based system may evaluate parameters such as urgency, location, and specialist availability to optimally assign appointments. Likewise, these systems manage insurance verification and claims processing by corroborating patient information and treatment specifics with established reimbursement criteria. This diminishes manual errors, expedites claim approvals, and alleviates administrative bottlenecks (Ullah et al. 2023).

11.3 Resource Management Use Cases

Resource management is an area in which rule-based systems demonstrate exceptional proficiency. Hospitals frequently have difficulties with bed assignment, personnel scheduling, and inventory management. Rule-based systems can allocate resources dynamically depending on real-time data. A regulation might allocate intensive care unit (ICU) beds to severely ill patients according to their health metrics while suggesting transfers for stabilized individuals. Furthermore, these systems facilitate adherence to legal and regulatory mandates by guaranteeing consistent compliance with documentation and procedural standards.

11.4 Other Use Cases in MCPS

In addition to direct applications, rule-based solutions enhance interoperability inside MCPS by standardizing data and facilitating seamless communication among diverse components. A rule-based engine can correlate clinical terminologies across different systems, facilitating seamless data flow and minimizing ambiguity. Notwithstanding their benefits, rule-based systems possess inherent limits. They necessitate comprehensive knowledge engineering to establish and uphold rules, and they may encounter difficulties with ambiguous or unanticipated situations. Nonetheless, their simplicity, transparency, and predictable characteristics render them exceptionally dependable for well-defined tasks (Alharbi et al. 2022).

Rule-based systems are essential to MCPS, facilitating precise diagnoses, customized treatments, and efficient administrative processes. These systems improve clinical care and organizational efficiency by automating processes and ensuring compliance with standards; hence, they enhance patient outcomes and decrease healthcare expenditures.

12. Conclusion and Future Scope

MCPS represents the next big leap in healthcare delivery through integration with cutting-edge technologies like Artificial Intelligence and the Internet of Things. By harnessing networked systems, robust data analysis, and AI-driven decision-making, MCPS can form a seamless platform to better improve patient outcomes, allow continuous care, and increase operational efficiency. The vast collection of medical data through IoT sensors and subsequent processing with intelligent algorithms enable healthcare professionals to take well-informed decisions, ensure timely interventions, and deliver personalized treatments. With healthcare systems becoming increasingly digitalized, rigorous cybersecurity measures are required to protect sensitive patient data, medical devices, and critical infrastructure.

There is huge potential for further development of MCPS, especially in the health-care sector. Advances in AI will enable more accurate, predictive, and diagnostic work by MCPS, especially in detecting diseases much earlier and possibly monitoring patients more effectively than ever. It will further solve the issue of bias as well as data integrity concerns in developing an ethical framework for AI. Adoption of active safety mechanisms, such as AI for cyber threat prevention, will strengthen the robustness of these systems against potential attacks. Integration of emerging technologies such as 5G, edge computing, and blockchain will enhance the speed, security, and scalability of MCPS.

References

Abdelmaboud, A., Ahmed, A. I. A., Abaker, M., Eisa, T. A. E., Albasheer, H., Ghorashi, S. A. et al. (2022). Blockchain for IoT applications: taxonomy, platforms, recent advances, challenges and future research directions. Electronics 11(4): 630.

Alam, T. (2022). Blockchain-based internet of things: review, current trends, applications, and future challenges. Computers 12(1): 6.

Alharbi, S., Attiah, A. and Alghazzawi, D. (2022). Integrating blockchain with artificial intelligence to secure IoT networks: Future trends. Sustainability 14(23): 16002.

Al-Hawawreh, M., Aljuhani, A. and Jararweh, Y. (2023). Chatgpt for cybersecurity: practical applications, challenges, and future directions. Cluster Computing 26(6): 3421–3436.

Alomari, M. A., Al-Andoli, M. N., Ghaleb, M., Thabit, R., Alkawsi, G., Alsayaydeh, J. A. J. et al. (2025). Security of smart grid: cybersecurity issues, potential cyberattacks, major incidents, and future directions. Energies 18(1): 141.

Ahmad, R. W., Salah, K., Jayaraman, R., Yaqoob, I. and Omar, M. (2022). Blockchain in oil and gas industry: Applications, challenges, and future trends. Technology in Society 68: 101941.

Andrew, J., Isravel, D. P., Sagayam, K. M., Bhushan, B., Sei, Y. and Eunice, J. (2023). Blockchain for healthcare systems: Architecture, security challenges, trends and future directions. Journal of Network and Computer Applications 215: 103633.

Anitha, A., Anitha, K., Amala, J. and Kamaraj, B. (2025). Blockchain technology in healthcare: advantages, challenges, and impact on Health 4.0. pp. 177–214. In Blockchain Technology Applications in Knowledge Management. IGI Global Scientific Publishing.

Aounzou, Y., Boulaalam, A. and Kalloubi, F. (2025). Convergence of blockchain, IoT, and machine learning: exploring opportunities and challenges—a systematic review. International Journal on Smart Sensing and Intelligent Systems 18(1).

Baiod, W., Light, J. and Hussain, M. M. (2025). Blockchain application in the accounting information system: advantages, challenges, and future research directions. pp. 193–219. In Technological Horizons. Emerald Publishing Limited.

Bagga, P., Das, A. K., Chamola, V. and Guizani, M. (2022). Blockchain-envisioned access control for internet of things applications: A comprehensive survey and future directions. Telecommunication Systems 81(1): 125–173.

Balamurugan, K., Alexander, T. J. and Karthikeyan, V. (2024). Application of solar energy in sustainable mobility. pp. 250–264. In Energy Efficient Vehicles. CRC Press.

Bamakan, S. M. H., Malekinejad, P. and Ziaeian, M. (2022). Towards blockchain-based hospital waste management systems; applications and future trends. Journal of Cleaner Production 349: 131440.

Bashir, A. K., Victor, N., Bhattacharya, S., Huynh-The, T., Chengoden, R., Yenduri, G. et al. (2023). Federated learning for the healthcare metaverse: Concepts, applications, challenges, and future directions. IEEE Internet of Things Journal 10(24): 21873–21891.

Bellagarda, J. S. and Abu-Mahfouz, A. M. (2022). An updated survey on the convergence of distributed ledger technology and artificial intelligence: Current state, major challenges and future direction. IEEE Access 10: 50774–50793.

Bhattacharya, S., Victor, N., Chengoden, R., Ramalingam, M., Selvi, G. C., Maddikunta, P. K. R. et al. (2022). Blockchain for internet of underwater things: State-of-the-art, applications, challenges, and future directions. Sustainability 14(23): 15659.

Chengoden, R., Victor, N., Huynh-The, T., Yenduri, G., Jhaveri, R. H., Alazab, M. et al. (2023). Metaverse for healthcare: a survey on potential applications, challenges and future directions. IEEE Access 11: 12765–12795.

Dadhich, M. and Tiwari, H. (2022). Quantum blockchain for smart society: applications, challenges, and opportunities. Advancements in Quantum Blockchain With Real-Time Applications 178–198.

Deepa, N., Pham, Q. V., Nguyen, D. C., Bhattacharya, S., Prabadevi, B., Gadekallu, T. R. et al. (2022). A survey on blockchain for big data: Approaches, opportunities, and future directions. Future Generation Computer Systems 131: 209–226.

Dev, A., Kumar, V., Khare, G., Giri, J., Amir, M., Ahmad, F. et al. (2025). Advancements and challenges in microgrid technology: a comprehensive review of control strategies, emerging technologies, and future directions. Energy Science & Engineering.

Elhadari, Z., Zougagh, H., Idboufker, N. and Ech-chebaby, M. (2025). Survey on the adoption of blockchain technology in internet of things environments: techniques, challenges and future research directions. IAENG International Journal of Computer Science 52(1).

Farooq, M. S., Ahmed, M. and Emran, M. (2022). A survey on blockchain acquainted software requirements engineering: model, opportunities, challenges, and future directions. IEEE Access 10: 48193–48228.

Gad, A. G., Mosa, D. T., Abualigah, L. and Abohany, A. A. (2022). Emerging trends in blockchain technology and applications: A review and outlook. Journal of King Saud University-Computer and Information Sciences 34(9): 6719–6742.

Geetanshi, Manocha, H., Babbar, H. and Mangla, C. (2025). Securing the Internet of Things: Cybersecurity challenges, strategies, and future directions in the era of 5G and edge computing. Current and Future Cellular Systems: Technologies, Applications, and Challenges 89–106.

Gopalakrishnan, K., Karthikeyan, V. and Vanathi, P. T. (2024). A texture classification system based on an adaptive histogram equalized shearlet transform. Integrating Metaheuristics in Computer Vision for Real-World Optimization Problems 299–322.

Guo, X., Zhang, G. and Zhang, Y. (2022). A comprehensive review of blockchain technology-enabled smart manufacturing: a framework, challenges and future research directions. Sensors 23(1): 155.

Hamzah, M., Islam, M. M., Hassan, S., Akhtar, M. N., Ferdous, M. J., Jasser, M. B. et al. (2023). Distributed control of cyber physical system on various domains: a critical review. Systems 11(4): 208.

Hemamalini, V., Mishra, A. K., Tyagi, A. K. and Kakulapati, V. (2024). Artificial intelligence–blockchain-enabled–internet of things-based cloud applications for next-generation society. Automated Secure Computing for Next-Generation Systems 65–82.

Hussain, I. (2024). Secure, sustainable smart cities and the internet of things: Perspectives, challenges, and future directions. Sustainability 16(4): 1390.

Igonor, O. S., Amin, M. B. and Garg, S. (2025). The application of blockchain technology in the field of digital forensics: a literature review. Blockchains 3(1): 5.

Indumathi, G., Karthikeyan, V. and Raj, V. A. (2025). A preview of cybersecurity measures in healthcare applications using 5G. pp. 114–134. In Cybersecurity in Healthcare Applications. Chapman and Hall/CRC.

Jin, H. and Xiao, J. (2022). Towards trustworthy blockchain systems in the era of "Internet of value": development, challenges, and future trends. Science China Information Sciences 65: 1–11.

Jones, G. M., Winster, S. G., Maheswari, M. and Sundar, R. (2024). Enhanced cyber-physical system in health care using LSTM. Intelligent Cyber-Physical Systems for Healthcare Solutions: From Theory to Practice 401.

Kaakandikar, R. M., Darware, S. A. and Senthilkumar, K. R. (2025). Future trends and innovations: emerging trends in blockchain technology relevant to libraries. Leveraging Blockchain for Future-Ready Libraries 377–410.

Karthikeyan, V., Kishore, M. N. and Sajin, S. (2024a). End-to-end light-weighted deep-learning model for abnormality classification in kidney CT images. International Journal of Imaging Systems and Technology 34(1): e23022.

Karthikeyan, V., Kirubakaran, G. and Prakash, R. V. (2024b). The rise of electric vehicles–2023 status and future directions. Energy Efficient Vehicles 186–211.

Karthikeyan, V. and Visu, Y. P. (2024c). Attention-based lightweight deep hybrid CNN framework for image restoration. The Imaging Science Journal 1–26.

Karthikeyan, V., Dani Reagan Vivek, J., Gopalakrishnan, K. and Kalaimannan, E. (2024d). Industrialized Control and Automation System (ICAS) a software-defined analysis framework for Industry 4.0. Knowledge Management and Industry Revolution 4.0: 299–332.

Karthikeyan, V., Kirubakaran, G., Prakash, R. V. and Visu, Y. P. (2024e). Trends and research potential with the use of ChatGPT and other AI tools: application and challenges of generative AI tools. pp. 30–55. In Applications, Challenges, and the Future of ChatGPT. IGI Global.

Kaur, G. (2025). Internet of things in healthcare: services, applications, challenges and emerging trends. Beyond Boundaries 13.

Khalil, U., Malik, O. A. and Hussain, S. (2022). A blockchain footprint for authentication of IoT-enabled smart devices in smart cities: State-of-the-art advancements, challenges and future research directions. IEEE Access 10: 76805–76823.

Krichen, M., Ammi, M., Mihoub, A. and Almutiq, M. (2022). Blockchain for modern applications: A survey. Sensors 22(14): 5274.

Krishnamoorthy, S., Dua, A. and Gupta, S. (2023). Role of emerging technologies in future IoT-driven Healthcare 4.0 technologies: A survey, current challenges and future directions. Journal of Ambient Intelligence and Humanized Computing 14(1): 361–407.

Kumar, M. and Singh, A. (2022). Probabilistic data structures in smart city: Survey, applications, challenges, and research directions. Journal of Ambient Intelligence and Smart Environments 14(4): 229–284.

Maria Jones, G., Godfrey Winster, S., Maheswari, M., Sundar, R., Kalaivani, A., Menaka, D. et al. (2024). Enhanced cyber-physical system in health care using LSTM and Bi-LSTM. pp. 401–418. In Intelligent Cyber-Physical Systems for Healthcare Solutions: From Theory to Practice. Singapore: Springer Nature Singapore.

Malla, T. B., Bhattarai, A., Parajuli, A., Shrestha, A., Chhetri, B. B. and Chapagain, K. (2022). Status, challenges and future directions of blockchain technology in power system: A state of art review. Energies 15(22): 8571.

Malik, H., Anees, T., Faheem, M., Chaudhry, M. U., Ali, A. and Asghar, M. N. (2023). Blockchain and Internet of Things in smart cities and drug supply management: Open issues, opportunities, and future directions. Internet of Things 23: 100860.

Maurya, V., Rishiwal, V., Yadav, M., Shiblee, M., Yadav, P., Agarwal, U. et al. (2025). Blockchain-driven security for IoT networks: State-of-the-art, challenges and future directions. Peer-to-Peer Networking and Applications 18(1): 1–35.

Pandey, A. K., Das, A. K., Kumar, R. and Rodrigues, J. J. (2024). Secure cyber engineering for IoT-enabled smart healthcare system. IEEE Internet of Things Magazine 7(2): 70–77.

Patra, I. K. and Panigrahi, S. (2025). Securing the future of IoT-based smart healthcare: challenges, innovations, and best practice. pp. 129–141. In Explainable IoT Applications: A Demystification. Cham: Springer Nature Switzerland.

Peng, X., Zhao, Z., Wang, X., Li, H., Xu, J. and Zhang, X. (2023). A review on blockchain smart contracts in the agri-food industry: Current state, application challenges and future trends. Computers and Electronics in Agriculture 208: 107776.

Prakash, P. S., Karthic, S. and Saravanan, M. (2025). Public safety management in smart Society 5.0: A blockchain-based approach. Networked Sensing Systems 273–295.

Rahman, A., Islam, J., Kundu, D., Karim, R., Rahman, Z., Band, S. S. et al. (2025). Impacts of blockchain in software-defined Internet of Things ecosystem with Network Function Virtualization for smart applications: Present perspectives and future directions. International Journal of Communication Systems 38(1): e5429.

Ramu, S. P., Boopalan, P., Pham, Q. V., Maddikunta, P. K. R., Huynh-The, T., Alazab, M. et al. (2022). Federated learning enabled digital twins for smart cities: Concepts, recent advances, and future directions. Sustainable Cities and Society 79: 103663.

Rathod, T., Jadav, N. K., Alshehri, M. D., Tanwar, S., Sharma, R., Felseghi, R. A. et al. (2022). Blockchain for future wireless networks: A decade survey. Sensors 22(11): 4182.

Razaque, A., Bektemyssova, G., Yoo, J., Hariri, S., Khan, M. J., Nalgozhina, N. et al. (2025). Review of malicious code detection in data mining applications: challenges, algorithms, and future direction. Cluster Computing 28(3): 1–37.

Rehman, Z., Gregory, M. A., Gondal, I., DOng, H. and Ge, M. (2025). Eclipse attacks in blockchain networks: detection, prevention, and future directions. IEEE Access.

Shafik, W. (2024). Blockchain-based internet of things (B-IoT): Challenges, solutions, opportunities, open research questions, and future trends. Blockchain-based Internet of Things 35–58.

Singh, R. and Kumar, S. (2025). A Comprehensive Insights into Drones: History, Classification, Architecture, Navigation, Applications, Challenges, and Future Trends. arXiv preprint arXiv:2501.10066.

Suhail, S., Hussain, R., Jurdak, R., Oracevic, A., Salah, K., Hong, C. S. et al. (2022). Blockchain-based digital twins: Research trends, issues, and future challenges. ACM Computing Surveys (CSUR) 54(11s): 1–34.

Sun, P., Wan, Y., Wu, Z., Fang, Z. and Li, Q. (2025). A survey on privacy and security issues in IoT-based environments: Technologies, protection measures and future directions. Computers & Security 148: 104097.

Tallat, R., Hawbani, A., Wang, X., Al-Dubai, A., Zhao, L., Liu, Z. et al. (2023). Navigating industry 5.0: A survey of key enabling technologies, trends, challenges, and opportunities. IEEE Communications Surveys & Tutorials 26(2): 1080–1126.

Tong, X., Hamzei, M. and Jafari, N. (2025). Towards secure and efficient data aggregation in blockchain-driven IoT environments: a comprehensive and systematic study. Transactions on Emerging Telecommunications Technologies 36(2): e70061.

Tyagi, A. K., Dananjayan, S., Agarwal, D. and Thariq Ahmed, H. F. (2023). Blockchain—Internet of Things applications: Opportunities and challenges for industry 4.0 and society 5.0. Sensors 23(2): 947.

Ullah, H., Manickam, S., Obaidat, M., Laghari, S. U. A. and Uddin, M. (2023). Exploring the potential of metaverse technology in healthcare: Applications, challenges, and future directions. IEEE Access 11: 69686–69707.

VarunPrakash, R., Karthikeyan, V., Vishali, S. and Karthika, M. (2024). Multi-level LSTM framework with hybrid sonic features for human–animal conflict evasion. The Visual Computer 1–17.

Velayuthapandian, K., Veyilraj, M. and Jayakumaraj, M. A. (2024a). An intelligent parking allocation framework for digital society 5.0. Intelligent Decision Technologies 18(3): 2145–2159.

Velayuthapandian, K., Murugan, N. and Paramasivan, S. (2024b). End-to-end CNN conceptual model for a biometric authentication mechanism for ATM machines. Discover Electronics 1(1): 26.

Velayuthapandian, K., Karuppiah, G., Vadivel, S. R. S. and Joseph, D. R. V. (2024c). Mammogram data analysis: Trends, challenges, and future directions. pp. 1–38. In Computational Intelligence and Modelling Techniques for Disease Detection in Mammogram Images. Academic Press.

Verma, A., Bhattacharya, P., Madhani, N., Trivedi, C., Bhushan, B., Tanwar, S. et al. (2022). Blockchain for industry 5.0: Vision, opportunities, key enablers, and future directions. IEEE Access 10: 69160–69199.

Wei, Q., Pan, S., Liu, X., Hong, M., Nong, C. and Zhang, W. (2025). The integration of AI in nursing: Addressing current applications, challenges, and future directions. Frontiers in Medicine 12: 1545420.

Wu, D., Zheng, A., Yu, W., Cao, H., Ling, Q., Liu, J. et al. (2025). Digital twin technology in transportation infrastructure: a comprehensive survey of current applications, challenges, and future directions. Applied Sciences 15(4): 1911.

Zaccagnino, R., Malandrino, D., Zaccagnino, G., Benevento, G. and Napolitano, M. M. (2025). Healthcare and cyberspace: from cyber-physical systems to medical digital twins. pp. 47–72. In Artificial Intelligence Techniques for Analysing Sensitive Data in Medical Cyber-Physical Systems: System Protection and Data Analysis. Cham: Springer Nature Switzerland.

Zhao, B. (2025). Towards a secure blockchain ecosystem: current vulnerabilities and future directions in smart contract security. In ITM Web of Conferences (Vol. 73, p. 03014). EDP Sciences.

Chapter 22

Machine Learning

A View from a Learning Perspective

R. Ravinder Reddy,[1,*] *E. Padma Latha,*[1] *M. V. Krishna Reddy,*[1] *K. Karthik*[2] and *Ravi Uyyala*[1]

1. Introduction

In the early ages, computers were used to solve complex problems, today, with the advent of artificial intelligence, computers are being used to learn and gain knowledge from data and acquire new insights from data to promote decision making. As a result of the growing amount of data, the requirement of automation, and the need for intelligent decision-making, machine learning plays a very significant role in the present world. When decisions are uncertain, we can easily categorize data, but when decisions are complex then machine learning comes. Machine learning techniques can arrive at more quality decisions from the huge amount of data that is available now. It can analyze the data and produce quality decisions (Haw et al. 2024).

Though AI helps machines to act and think like humans by simulating cognitive functions, such as decision-making, problem-solving, learning, perception, and reasoning, it still lacks true consciousness, emotions, and creativity. However, AI augments human abilities by improving efficiency, automation, and problem-solving in various fields. Machine learning will train a computer by using numerous examples and data, enabling it to learn from patterns and make intelligent decisions (Bishop and Nasrabadi 2006). Computers learn from experience and that is precisely what machine learning is. The past experiences of computers have a name: it is called Data. Machine learning is not just a revolutionary development in the history of computer science, but also one of the most prominent achievements of human intelligence. Its applications are wide-ranging across many fields including robotics, data analysis, autonomous navigation, bioinformatics, speech recognition, text processing, and web data management.

During the initial stage of computers, they took input and instructions from the user to generate output. Machine learning has changed the concepts of this early age: it takes the past experiences in a structured format, from these experiences it learns the model to predict the new samples in the future as shown in Figure 22.1. It mimics the human brain in learning perspective.

[1] Department of Computer Science and Engineering, Chaitanya Bharathi Institute of Technology, Hyderabad, India
[2] CVR College of Engineering, Hyderabad, India
Emails: epadmalatha_cse@cbit.ac.in, krishnareddy_cse@cbit.ac.in, karthik.5786@gmail.com, raviuyyala_cse@cbit.ac.in.
* Corresponding author: rravinderreddy_cse@cbit.ac.in

DOI: 10.1201/9781003620822-22

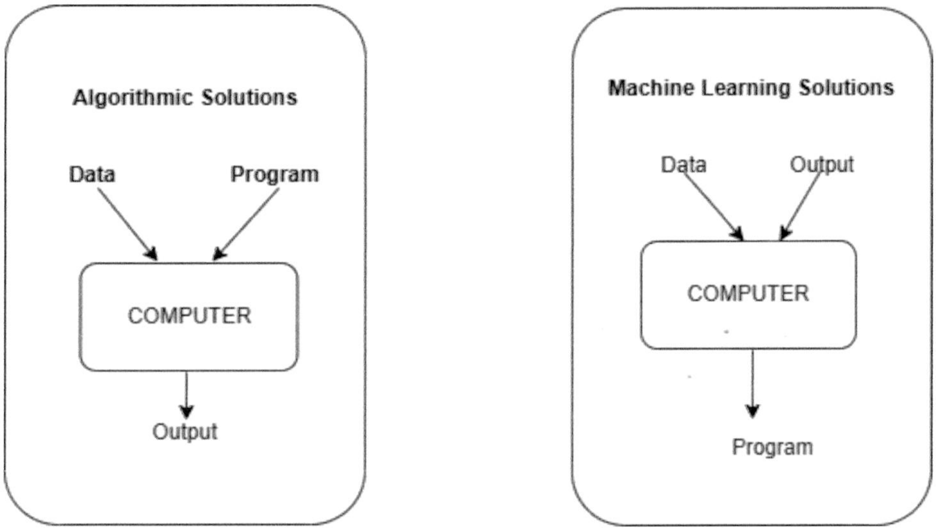

Figure 22.1. Traditional models vs. ML models.

1.1 Advantages of Machine Learning

Machine learning makes things simpler and more accurate when compared to what the human brain does. Due to this advantage it is applied in many domains. The major advantages of machine learning in the different domains are listed here.

Automation of tasks: Machine learning reduces humans in repetitive processes (e.g., chatbots, email filtering).

Improved accuracy and efficiency: Machines learn from data to make better predictions over time (e.g., medical diagnosis, fraud detection).

Faster decision making: Machines analyze large datasets quickly for real-time decisions (e.g., stock market predictions, traffic control).

Handling large and complex data: Machines easily processes big data that is too large for traditional methods.

Continuous improvement: Machine models learn and adapt from new data, improving over time.

Personalization: Machines provide tailored recommendations (e.g., Netflix, Spotify, e-commerce suggestions).

Better customer experience: Machines enhance interactions via chatbots, voice assistants, and predictive analytics.

Fraud detection and security: They identify suspicious transactions in banking and cyber-security.

Predictive maintenance: Help industries prevent equipment failure, reducing downtime.

Versatile applications: Machines are used in healthcare, finance, education, retail, self-driving cars.

ML algorithms attempt to comprehend human brain and mimic the structure to solve and understand real world problems. Computational biology serves as a foundational domain for applying machine learning techniques, offering opportunities to evolve and address complex biological problems through data-driven approaches (Mitchell and Mitchell 1997). Machine learning is highly interdisciplinary: it collaborates and draws knowledge and techniques from multiple subjects. It is a blend of mathematics, statistics, computer science, biology, data mining and so on. It uses domain specific expertise to build intelligent models. Machine learning has a large impact on many applications. Learning algorithms have a deep influence on numerous segments of science and industry, driving advancements, automation, and innovation. ML algorithms enhance efficiency, decision-making, and innovation across nearly every field. As AI keeps advancing, its influence will also increase, resulting in safer, efficient and smarter industries.

The current state-of-the-art machine learning algorithms provide robust solutions for a wide range of problems, enabling the development of intelligent systems tailored to specific application needs. ML has come a long way in the past decade with breakthroughs in deep learning, reinforcement learning, and generative models. It is also linked to the fields of artificial intelligence, computational statistics, and pattern recognition (Paluszek and Thomas 2024). ML is a huge field and there are diverse extensions and specialized subfields that explore the different applications. ML is constantly being enhanced with modern and new techniques and a lot of extensions are emerging to tackle and meet real-world challenges (Ali et al. 2024). Depending on the need, different ML techniques and architectures are applied in different situations. A general assumption in the fields of statistical and computational learning theory is that learning the correct probabilities does not necessarily imply the ability to compute these probabilities exactly, nor does it guarantee convergence to them (Kim and Kang 2024).

1.2 Types of Learning

ML encompasses unsupervised learning, reinforcement and supervised learning: each with unique algorithms and applications. ML is a subdomain of artificial intelligence; it is the backbone of artificial intelligence. It plays a pivotal role in learning. The learning is done as per the different methods and the types of data. The fundamentals of machine learning involve understanding the various types of learning paradigms. A machine can learn through different approaches, primarily categorized as supervised learning, unsupervised learning, and reinforcement learning. The major types of learning are classified as follows.

1.2.1 Supervised Learning

The model gains knowledge by analyzing data that has predefined labels, classified data (mapping input-output pairs). It learns from the given input and corresponding output, and based on this data, it is classified into different groups. These models learn from the data, and use 66% of data for training and 33% of data to test the model. Supervised learning is further classified into two types based on the data that is trained. If the trained data is numerical then regression is used, if the trained data is categorical the classification method is used.

Regression: It is used to predict the numerical outcome from the training data. Examples are predicting house prices, stock prices.

Classification: It identifies the different classes from the training data, it learns from the training to classify the samples into different classes. Examples are email spam detection, disease diagnosis.

Some of the most commonly used algorithms in supervised learning include Decision Trees, Support Vector Machines (SVM), Neural Networks, Linear Regression, and Logistic Regression.

1.2.2 Unsupervised Learning

The model finds hidden patterns in unlabeled data, it groups the data based on the similarity score. When the given data is unlabeled, the learning model may identify novel patterns or hidden structures within the dataset. Unsupervised learning is further divided into:

Clustering: It groups the data based on the similarity score, which data points are more similar or nearer are grouped into the one cluster. Examples are customer segmentation in marketing, land classification.

Anomaly detection: In this the data points which are more dissimilar are compared to the majority of the data points. The examples are fraud detection in banking, outliers in the data.

Commonly used algorithms in unsupervised learning include Hierarchical Clustering, K-Means, DBSCAN, and Principal Component Analysis (PCA).

1.2.3 Semi-Supervised Learning

This learning method is used in situations where the data is partially labelled and the rest is unannotated. These are used when the availability of the training data is very little. It helps improve model performance while reducing the cost of manual labelling. These models are commonly used in image recognition, speech recognition and medical diagnosis (Wang et al. 2024).

1.2.4 Reinforcement Learning (RL)

Knowledge is acquired through the interactions between the model and the environment for optimizing reward. Generally, these models are worked, based on the reward and punishment and trial and error-based approaches. The major examples are self-driving cars and Game AI. Some of the most common algorithms employed are Q-Learning, Deep Q-Networks (DQN), and Proximal Policy Optimization (PPO). Different types and their subfields of learning are shown in Figure 22.2.

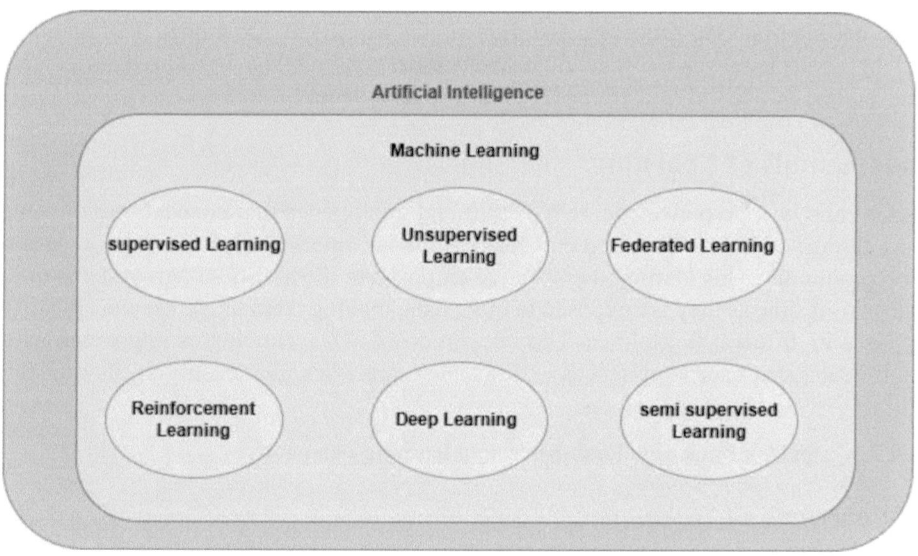

Figure 22.2. The sub-fields of Machine Learning.

1.2.5 Deep Learning (DL): An Extension of ML

Deep learning models are used when huge quantities of data are available to achieve higher accuracy. This approach uses artificial neural networks (ANNs) for processing and learning intricate patterns received from the data. These models solve many problems, which are not solved by the machine learning models. Current advancements in deep learning have achieved progress in many areas, particularly in the domains of Computer Vision and Natural Language Processing (NLP). A few of the top algorithms responsible for this progress are Convolutional Neural Networks (CNNs), Recurrent Neural Networks (RNNs), and Transformer models. The utilization of these technologies has significantly improved features such as image recognition, language translation, and text generation: thus transforming the artificial intelligence frontier.

1.2.6 Transfer Learning

Using a pre-trained model on a new task improves performance and reduces time. The new model does not train from scratch. It has three types of transfer learning, which are described as follows.

Positive transfers: Learnings from the source task benefits the target task's performance.

Negative transfers: Negative transfer occurs when the application of knowledge from a previous task adversely affects the performance on a new task.

Neutral transfers: Learning from the source task imposes negligible or no impact on the target task.

Transfer learning is used in various applications like speech recognition, object detection, image classification and other applications. The example is: Applying a medical AI model trained on X-rays to CT scans.

1.2.7 Federated Learning

It learns from the different nodes in a distributed manner. IT is a privacy-preserving ML approach where models are trained without sharing raw data to keep the data locally and only share the updates to the model. The examples are Google's Gboard and Healthcare AI.

1.2.8 Explainable AI (XAI)

Ensuring that machine learning models are transparent and interpretable is essential for understanding the internal decision-making processes and the features learned by the model. The examples are it explains why a bank loan was rejected and helps in understanding why an AI predicts a disease. The popular techniques are SHAP, LIME, and attention mechanisms.

2. Fundamentals of Learning

Machine Learning is a pivotal subdomain of artificial intelligence that concentrates on developing algorithms through which computers can learn from information given and from data without explicit programming. This learning perspective emphasizes the ability of ML systems to improve performance over time as they are exposed to more data, making them universal and able to perform complex tasks in different disciplines. The integration of ML technologies into everyday life has significantly enhanced user experiences and business strategies, demonstrating its transformative potential.

Below are key aspects of machine learning from a learning perspective.

2.1 Key Concepts and Techniques

The key concepts of the machine learning process are like dataset, training and testing the model. Here the key concepts of machine learning are listed.

Dataset: The learning that is done through examples in an organized manner is called a dataset. It is used for all types of learning. The performance of the machine learning models depends on the dataset which is used to build the model through training. The model will learn from the trained dataset and will be tested. The dataset is divided as given below.

Training data: Training data is the data used to train the model and typically consists of about two-thirds of the total dataset.

Testing data: Testing data is the data used to test the performance of the model and is normally one-third of the total dataset.

Features and Labels

To improve its performance, the model needs to extract the appropriate features, which are more significant in the model performance evaluation. The labels are required in supervised learning to incorporate the data to improve accuracy.

Feature (X): Input variables (e.g., age, salary, temperature).

Label (Y): Output variable (e.g., house price, spam/not spam).

Model Training and Testing

Machine learning algorithms build the model from the dataset. It is done in two phases, training and testing.

Training: The model learns important information and patterns from data.

Testing: The model is assessed using data that it has not encountered during training.

Overfitting and Underfitting: The performance of the model depends on the size of the data and its training process. It is impacted majorly by the number of samples by which it is trained.

- **Overfitting:** The model is picking up the noise in the training data, so it does well on the training set but fails on new, unseen data.
- **Underfitting:** The model is oversimplified and does not pick up the patterns in the data, so it does poorly on both training and new data.

Performance metrics: The performance need measures the learning model. These measures will tell the model working, it majorly impacts the model working. Accuracy, Precision, Recall, F1-Score are used for classification. Mean Squared Error (MSE), R^2 Score measures are used for regression.

Model training and evaluating the performance: The process involves training models on datasets, evaluating their performance, and deploying them for real-world applications. This iterative process is crucial for refining model accuracy and effectiveness.

Applications and impact: user experience and business optimization: ML technologies are used to predict user behaviors, analyze reviews, and optimize investments, such as mining for bitcoins. These applications improve user experience and allow businesses to improve their marketing strategies.

Diverse domains: ML is applied in natural language processing, image recognition, recommendation systems, speech recognition and predictive analytics, showcasing its versatility and impact across sectors.

Emerging trends: *Deep learning and transfer learning:* These advanced techniques are gaining prominence, offering solutions to complex problems by leveraging large datasets and pre-trained models (Mandalapu et al. 2023).

Adversarial and federated learning: These are recent topics in ML, focusing on improving model robustness and privacy through distributed learning approaches (Mendonça et al. 2024).

While machine learning facilitates great progress, it also comes with challenges, including ensuring the accuracy of models and addressing ethical issues of data privacy and bias. Theoretical perspectives, like those challenging traditional statistical assumptions, highlight the complexity and evolving nature of ML, suggesting that learning true probabilities may not always equate to correct calculations (Hastie et al. 2005).

Humans learn from their past memories and experiences, but machines mostly adhere to the commands provided by humans to execute predefined tasks. But machines can be trained to learn from past data through a process called machine learning. Learning at its core encompasses the processes of understanding and reasoning. ML is transforming various industries by enhancing decision making based on the data and automating these processes. As a part of AI, it enables computers to gain information from data and make decisions or predictions without explicit programming. Systems process data, create a predictive model, and when new data is presented, it can make precise predictions. The more data the model is trained on, the better it learns, resulting in more precise predictions.

2.2 Importance of Machine Learning

Machine Learning (ML) is about building prediction models that learn patterns from data and enable automated decision-making at scale. Machine learning is about algorithms producing algorithms and optimization.

Handling Massive Data

ML models are trained from the data, if the size of the data is big the learned model performance will increase. Traditional methods struggle with big data, ML efficiently processes and extracts insights from vast amounts of information. Deep learning methods will produce more accurate results for the huge datasets.

Automation of Tasks

ML reduces human effort by automating repetitive tasks like fraud detection, customer support (chatbots), and recommendation systems. In the automation industry and IoT huge quantities of data are generated, where these machines are gaining important information, learning and performing well to do different jobs.

Improved decision-making: Machine Learning algorithms analyze patterns and trends, helping businesses and organizations make data-driven decisions: it builds more accurate and efficient models.

Enhanced accuracy and efficiency: Unlike rule-based systems, ML improves over time, making predictions and classifications more accurate and efficient. These models are more accurate than human based learning.

Real-time applications: Due to its greater accuracy, the machine learning model is deployed in many domains. It is used in autonomous vehicles, healthcare diagnostics, stock market predictions, and language translation, making real-time decisions possible. The memory of computers has grown exponentially, allowing them to handle vast amounts of data without any delay. With more computational power, ML has been used in different fields. For example, in the healthcare sector, for predicting diagnoses for the physicians to examine. Tech firms are using sentiment analysis to identify social media trends. Another major application is fraud detection, used in the financial sector.

Healthcare: It learns from previous users medical reports and based on that it predicts the diseases and medical imaging analysis. Using the recommendations it recommends drugs (Gopalakrishnan and Kumaran 2022).

Finance: It learns financial data from banks and markets to detect fraud transaction analysis, stock market prediction, etc. (Noriega et al. 2023).

Retail: It learns from user preferences and builds recommendation systems: examples are Amazon, Netflix.

Autonomous systems: Self-driving cars, robotics.

NLP: These days the intervention is replaced everywhere by artificial intelligence, especially in voice and hand written documents, for example: Chatbots, speech recognition.

Personalization and user experience: Due to their various advantages, ML models can be used for personalized features. For example, recommendation engines (Netflix, Amazon, Spotify), targeted ads, and customer insights for a better user experience.

Cyber security and fraud detection: Cyber security and fraud detection applications have huge quantities of data to analyze fraud and threats. ML helps in detecting anomalies and preventing cyber threats, ensuring safer digital environments. These days, machine learning applications are inbuilt in many of the cyber security software's (Khalid et al. 2024).

3. Understanding the Learning Process

ML can be seen as a computational approach to mimic the process of learning exhibited by humans or other intelligent systems. Just as humans learn from experience, machine learning models learn from input data to improve their performance on specific tasks. Representation of data to learn the model is very crucial in the learning process. However, ML simplifies this process by formalizing it into three key stages:

Data representation: Input data must be appropriately represented for the machine to process it, often in the form of features or structured formats.

Model learning: Algorithms identify patterns, relationships, or functions from the data.

Generalization: The learned model applies its knowledge to unseen data, ensuring adaptability beyond the training set.

From a learning perspective, ML is categorized into distinct paradigms depending on the category of data and supervision involved. Learning perspectives involves the different learning approaches that were discussed in the previous section.

3.1 Key Elements of Learning in Machine Learning

Data as the Foundation of Learning

Data is the raw material from which models learn. The quality, diversity, and quantity of data significantly affect learning outcomes. Feature engineering, data pre-processing, and augmentation are critical for effective learning.

Learning Algorithms

Algorithms define how a system learns from data. These can range from linear models such as logistic regression to complex non-linear systems (neural networks) (Al-Tameemi et al. 2025). Gradient-based optimization techniques, like stochastic gradient descent (SGD), are essential in adjusting model parameters during the learning process.

Loss Functions and Feedback

Loss functions measure the gap between original and predicted outcomes, directing the learning of the model. For example, Mean Squared Error (MSE) is commonly used in regression tasks, while Cross-Entropy Loss is used for classification problems.

Generalization and Overfitting

The model's capacity to generalize unseen data is crucial for successful learning.
Regularization methods, such as L1, L2, and dropout, are employed to prevent overfitting. These techniques ensure that the model focuses on learning meaningful patterns rather than memorizing the training data

Model Evaluation

Evaluation is an essential part of learning. Metrics such as accuracy, precision, recall, and F1-score are used to measure a model's performance. Cross-validation helps verify that models perform well across various data subsets, ensuring better generalization. The major key elements of machine learning are summarized in Table 22.1, they influence the learning process in machine learning.

While learning the model we need to consider the important aspects of ML in perspective of learning to improve the learning quality. Some of the important aspects listed here.

Data-driven learning: The core principle is that the system learns by analyzing vast datasets, extracting meaningful insights from the input data to make classifications or any predictions.

Iterative improvement: As the system encounters more data and feedback, it continuously refines its internal model, adjusting parameters to optimize its performance over time.

Generalization: It is the ability to apply known patterns learned from training sets, unseen instances, thus making the model's ability to make accurate predictions in real-life scenarios.

Feedback loop: Essential for learning, where the system receives feedback on its predictions, allowing it to identify errors and adjust accordingly to improve future results.

Table 22.1. Key elements of learning in machine learning.

S.No	Key Element	Description
1	**Data**	The foundational input consists of structured or unstructured data used for training models.
2	**Features**	The measurable properties or characteristics of the data that influence model learning.
3	**Labels**	The expected output or ground truth used in supervised learning for training models.
4	**Model**	The mathematical representation or algorithm that learns from data to make predictions.
5	**Training**	The process where the model learns patterns from data using optimization techniques.
6	**Loss Function**	A mathematical function that quantifies the error between predicted and actual values.
7	**Optimization**	Techniques like Gradient Descent used to minimize the loss function and improve learning.
8	**Hyper parameters**	Configurable parameters (e.g., learning rate, batch size) that control the learning process.
9	**Evaluation Metrics**	Measures like accuracy, precision, recall, and F1-score to assess model performance.
10	**Generalization**	The ability of a model to perform well on unseen data, avoiding overfitting or underfitting.

3.2 Challenges in Machine Learning from a Learning Perspective

The major challenges of machine learning in the learning perspective are crucial in consideration for enhancement of the accuracy of the model. Before training the model one has to consider these aspects to improve the quality of the learning model (Barbierato and Gatti 2024).

Bias and variance trade-off: Striking a balance between bias (under fitting) and variance (overfitting) is central to achieving optimal learning.

Scalability of learning: Large datasets and high-dimensional data require efficient algorithms and computational resources.

Interpretability and explainability: Understanding the learning process of complex models (e.g., deep neural networks) remains a challenge.

Data quality and labeling: Incomplete, noisy, or biased data can hinder learning outcomes. Manual labeling is often resource-intensive.

Ethical and fair learning: Ensuring fairness and avoiding biases in machine learning models is crucial for equitable learning outcomes.

3.3 Future Directions in Learning-Centric Machine Learning

Machine learning has been developed in many directions to improve performance to further levels. To enhance the accuracy in various aspects the learning is done in various ways (Baidwan et al. 2024).

Transfer learning: Utilizing knowledge learned in one domain to improve learning in another. Example: Pre-trained models like GPT and ResNet.

Few-shot and zero-shot learning: Enabling models to learn effectively with few labelled examples or generalize to new classes with zero examples.

Meta-learning (learning to learn): Focusing on building models that learn in order to adapt quickly to new tasks with little training. The examples are algorithms that aim to improve the performance of other learning algorithms.

Federated learning: Distributed learning paradigm where models learn across multiple devices or locations without sharing raw data.

Uncertainty-aware learning: Integrating probabilistic models to quantify uncertainty in predictions, enhancing robustness (Murphy 2012).

4. Learning Performance Measures

One should guarantee the performance of ML in choosing the best for a specific task. The following are the most critical performance metrics divided depending on the type of the problem. While measuring the performance of a machine learning model some key aspects need to be considered.

Bias and fairness: The model may inadvertently generate biased or discriminatory output if biases exist in the training data. It thus emphasizes the need for the presence of high-quality, unbiased data and ethics considerations during model development.

Model complexity: Choosing a model with the correct order of complexity is crucial to avoid overfitting (learning training data too well) and underfitting (not capturing essential patterns).

Interpretability: Understanding how a model makes decisions is important to build trust and address potential issues, especially in high-stakes applications.

Performance measures of the ML models vary from one model to another, here the different measures for different types of learning are listed.

4.1 Classification Performance Metrics

The supervised learning models are used to predict discrete labels like true or false (spam vs. not spam, disease vs. no disease). The following are various classification evaluation metrics, many of which rely on the confusion matrix—a tool that summarizes the actual versus predicted outcomes of a model.

Machine learning models learn from approximations and probabilistic patterns derived from past data, enabling them to predict outcomes for new samples based on the learned parameters. Here the learning is different in the sample classes. The predicted data contains the values like True Positive, False positive, True Negative and True Positive.

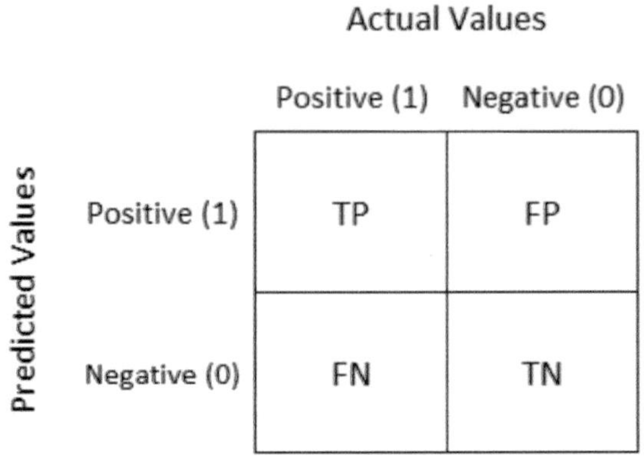

Figure 22.3. Confusion matrix, classifying the data.

Accuracy

The classification performance of the model is measured mostly in the form of accuracy. It is measured between the original and predicted values. Accuracy is measured in between the correctly predicted and the total samples.

$$Accuracy = \frac{TP + TN}{TP + FP + TN + TP}$$

$$Accuracy = \frac{Correct\ Predictions}{Total\ Predictions}$$

It works well with balanced datasets and is not suitable for imbalanced datasets. The performance varies based on the dataset and size used.

Precision (Specificity)

Precision indicates the positive rate model. It measures how many positive predictions were correct. Useful in cases where false positives are costly.

$$Precision = \frac{TP}{TP + FP}$$

Recall (Sensitivity)

It measures the proportion of correctly predicted positive samples out of all actual positive samples. Measures how many actual positives were correctly identified. It is useful when false negatives are costly.

$$Recall = \frac{TP}{TP + FN}$$

F1-Score

This is the harmonic mean of recall and precision, it is used for imbalanced datasets. When accuracy is not giving good results, this measure can be used to estimate the model.

$$F1 - score = 2 * \frac{Precision * Recall}{Precision + Recall}$$

ROC Curve & AUC

ROC (Receiver Operating Characteristic) Curve is used to plot in between True Positive Rate (Recall) vs. False Positive Rate. The AUC (Area under Curve): Higher AUC = Better Model.

4.2 Regression Performance Metrics

The outcome of the regression analysis is continuous, so the same measures can't be used for regression analysis. It is used to predict continuous values.

Mean Absolute Error (MAE)

It is the average of absolute difference between the original and predicted values. It is less sensitive to outliers.

$$MAE = \frac{1}{n}\sum(yi - yi')$$

Mean Squared Error (MSE)

The MSE measures the squared difference between the original and predicted values. It gives a penalty to big errors more heavily than the MAE and is more sensitive to outliers due to the squaring of the errors.

$$MSE = \frac{1}{n}\sum(yi - yi')^2$$

Root Mean Squared Error (RMSE)

It is similar to Mean Squared Error (MSE), but it is expressed in the same units as the target variable, providing error values within a more interpretable and limited range.

$$RMSE = \sqrt{MSE}$$

4.3 Clustering Performance Metrics

The unsupervised learning measures vary from the supervised learning models; here the outcome is not measurable. This is used by models to cluster similar data points together.

Silhouette Score

It measures how well-separated clusters are. Clustering evaluation primarily involves measuring intra-cluster similarity and inter-cluster dissimilarity. The range: –1 to 1 (higher = better).

Davies-Bouldin Index (DBI)

t evaluates both the compactness and separation of clusters. A lower DBI indicates better clustering performance.

Adjusted Rand Index (ARI)

It compares clustering outcomes with the actual ground truth labels.

Model complexity and generalization metrics help prevent overfitting and ensure the model generalizes effectively. The key factors to consider include the bias-variance trade off and cross-validation. A summary of these measures is provided in Table 22.2.

Table 22.2. Overall view of ML model performance measures.

S.No	Metric	Best for	When to Use?
1	Accuracy	Classification	Balanced datasets
2	Precision	Classification	Avoiding False Positives
3	Recall	Classification	Avoiding False Negatives
4	F1-Score	Classification	Imbalanced datasets
5	ROC-AUC	Classification	Evaluating overall performance
6	MAE	Regression	Less sensitive to outliers
7	MSE	Regression	Penalizes large errors
8	RMSE	Regression	Same units as target variable
9	R^2	Regression	Model explanatory power
10	Silhouette Score	Clustering	Cluster separation
11	Cross-Validation	Generalization	Prevents overfitting

4.4 Applications

There is a widespread usage of ML. It is involved in all domains and has many applications. Here we list the major applications of ML (Tufail et al. 2023).

Healthcare: Disease diagnosis, prediction and estimate and recommend medicine. Personalized treatment and drug discovery.

Finance and banking: Fraud detection and risk assessment in the finance market and stock market prediction.

E-commerce and retail: Product recommendations and customer sentiment analysis.

Self-driving cars: Object detection and autonomous navigation; traffic pattern prediction.

Natural Language Processing (NLP): Chatbots and virtual assistants; language translation and processing.

Cybersecurity: Anomaly detection in network traffic and Spam and phishing email detection.

Manufacturing and industry: Predictive maintenance and detecting equipment failures. Quality control and defect detection.

Social-media and marketing: Personalized ad recommendations and fake news and hate speech detection.

Education and e-learning: AI-based tutors and automated grading and personalized learning paths (Jalil et al. 2019).

Agriculture: Crop disease prediction and soil monitoring and smart irrigation using AI sensors.

These are the major areas where machine learning is used, but in almost every field it is being used to achieve better results from past experiences.

5. Conclusion

Machine Learning is transforming industries by making systems smarter, faster, and more efficient. As data continues to grow, ML's role will become even more vital in shaping the future. From a learning perspective, machine learning represents a computational embodiment of how systems acquire knowledge, adapt, and generalize to new scenarios. It bridges theoretical foundations with practical applications, driving advancements across industries and disciplines. Understanding the intricacies of learning, spanning data representation, model optimization, and evaluation enables the development of more scalable, interpretable and robust systems. As this field evolves, innovative paradigms like self-supervised learning, meta-learning, and federated learning continue to redefine the boundaries of what machines can achieve, offering unprecedented opportunities to shape the future of intelligent systems.

References

Ali, Y., Hussain, F. and Haque, M. M. (2024). Advances, challenges, and future research needs in machine learning-based crash prediction models: A systematic review. Accident Analysis & Prevention 194: 107378.

Al-Tameemi, M. N. (2025). Machine Learning: Algorithms, Real World Applications and Research Directions Pembelajaran Mesin: Algoritme, Aplikasi Dunia Nyata, dan Arah Penelitian.

Baidwan, R. S., Singh, T., Kumar, S. and Kumar, R. (2024). Machine learning: a multifaceted exploration of trends, regulations, and global impact. i-Manager's Journal on Future Engineering & Technology 19(4).

Barbierato, E. and Gatti, A. (2024). The challenges of machine learning: A critical review. Electronics 13(2): 416.

Bishop, C. M. and Nasrabadi, N. M. (2006). Pattern Recognition and Machine Learning (Vol. 4, No. 4, p. 738). New York: Springer.

Gopalakrishnan, S. and Kumaran, M. S. (2022). IIoT Framework based ML model to improve automobile industry product. Intelligent Automation & Soft Computing 31(3).

Hastie, T., Tibshirani, R., Friedman, J. and Franklin, J. (2005). The elements of statistical learning: data mining, inference and prediction. The Mathematical Intelligencer 27(2): 83–85.

Haw, S. C. (ed.). (2024). Perspectives on machine learn. Journal of Telecommunications and the Digital Economy 12(3): 1–6.

Jalil, N. A., Hwang, H. J. and Dawi, N. M. (2019, July). Machines learning trends, perspectives and prospects in education sector. pp. 201–205. In Proceedings of the 3rd International Conference on Education and Multimedia Technology.

Khalid, A. R., Owoh, N., Uthmani, O., Ashawa, M., Osamor, J. and Adejoh, J. (2024). Enhancing credit card fraud detection: an ensemble machine learning approach. Big Data and Cognitive Computing 8(1): 6.

Kim, J. and Kang, J. (2024). A Theory of Machine Learning. arXiv preprint arXiv:2407.05520.

Mandalapu, V., Elluri, L., Vyas, P. and Roy, N. (2023). Crime prediction using machine learning and deep learning: a systematic review and future directions. IEEE Access 11: 60153–60170. https://doi.org/10.1109/ACCESS.2023.3286344.

Mendonça, M. O., Netto, S. L., Diniz, P. S. and Theodoridis, S. (2024). Machine learning: Review and trends. Signal Processing and Machine Learning Theory 869–959.

Mitchell, T. M. and Mitchell, T. M. (1997). Machine Learning (Vol. 1, No. 9). New York: McGraw-Hill.

Murphy, K. P. (2012). Machine Learning: A Probabilistic Perspective. MIT Press.

Noriega, J. P., Rivera, L. A. and Herrera, J. A. (2023). Machine learning for credit risk prediction: A systematic literature review. Data 8(11): 169.

Paluszek, M. and Thomas, S. (2024). An overview of machine learning. MATLAB Machine Learning Recipes: A Problem-Solution Approach 1–19.

Tufail, S., Riggs, H., Tariq, M. and Sarwat, A. I. (2023). Advancements and challenges in machine learning: A comprehensive review of models, libraries, applications, and algorithms. Electronics 12(8): 1789.

Wang, J., Liu, S. and Zhang, W. (2024). Visual analytics for machine learning: A data perspective survey. IEEE Transactions on Visualization and Computer Graphics 30(12): 7637–7656.

Chapter 23

Top Algorithms in Machine Learning for Predicting Insights in Smart Applications

E. Padma Latha,[1] *K. Karthik,*[2] *R. Ravinder Reddy,*[1,*] *Bandi Rambabu*[2] and *Ravi Uyyala*[1]

1. Introduction

Computers have come quite a long way from their early years when they could only follow pre-coded instructions to perform specific functions. In those times, machines required careful programming for every new problem they encountered. With the introduction of machine learning (ML), however, computers are now able to learn from information, improve their decision-making, and adapt to new situations bringing in an element of flexibility that was originally the domain of human beings.

Machine learning is the core of artificial intelligence (AI), enabling systems to move beyond pre-programmed instructions and make autonomous, intelligent choices. This has unlocked the possibility of intelligent applications that can handle real-time data, adapt to changing circumstances, and continuously improve their performance. Intelligent thermostats that learn to optimize user habits, and autonomous vehicles that can safely navigate roads, are just a few examples of machine learning-based applications greatly improving convenience and efficiency. The impact of smart applications driven by ML can be seen in a majority of sectors. In the medical field, wearable fitness monitors track vital signs and offer personalized health information. Education has seen AI-driven learning systems that acquire personal learning patterns, improving learning. Energy systems use predictive analysis to optimize power supply, and virtual personal assistants simplify routines by predicting user needs. These are some of the ways machine learning improves technology and life.

This research examines the underlying machine learning algorithms that drive intelligent applications, how they learn, and how they are applied practically in real-world situations. Learning how machine learning makes technology adapt and predict outcomes will provide priceless insight into the growing power of artificial intelligence in building a wiser and effective society.

[1] Department of Computer Science and Engineering, Chaitanya Bharathi Institute of Technology, Hyderabad, India.
[2] CVR College of Engineering, Hyderabad, India.
Emails: epadmalatha_cse@cbit.ac.in, karthik.5786@gmail.com, rambabubandi@gmail.com, raviuyyala_cse@cbit.ac.in
* Corresponding author: rravinderreddy_cse@cbit.ac.in

DOI: 10.1201/9781003620822-23

2. Related Work

Machine learning (ML) and deep learning (DL) have revolutionized sectors like healthcare, finance, smart cities, and big data analytics. These new computational models enhance data representation, feature extraction, and predictive performance over conventional statistical and rule-based approaches. Despite this, challenges such as model interpretability, scalability, high computational complexity, and data quality remain, necessitating continued research and innovation. This survey outlines state-of-the-art ML and DL methods, their applications, and the challenges that limit their extensive use.

ML and DL have mainly contributed to the development of smart cities, revolutionizing urban life with intelligent systems. These technologies lead to innovations in healthcare, pollution control, energy management, and transportation. With the processing of large volumes of sensor data, ML models enable proactive decision-making, predictive maintenance, and resource optimization. For example, ML-based traffic management systems optimize traffic lights and minimize congestion, and energy management systems forecast consumption patterns, allowing dynamic pricing and demand-response. In spite of these developments, data security, privacy issues, and the absence of standardized frameworks are major challenges (Saranya and Amutha 2024).

Machine learning has revolutionized search engine operations and information retrieval systems from the conventional keyword-matching model to context-sensitive recommendation processes. Supervised, unsupervised, and reinforcement learning methods improve search quality by natural language processing (NLP) methods and tracking user behavior patterns. Personalized recommendation systems improve the precision of search results, hence improving user engagement. Data sparsity, algorithmic bias, and privacy concerns, however, need to be addressed to develop responsible and transparent AI-powered search engines (Ghadge 2024).

Predictive analytics is one of the most significant ML applications that allow industries to forecast trends and make sound decisions. Decision trees, random forests, support vector machines (SVMs), k-nearest neighbors (KNN), and deep learning models are algorithms that are applied in various applications including healthcare, finance, and business forecasting. In healthcare, ML allows for early disease diagnosis, whereas in finance, it identifies fraud and optimizes investments. Predictive analytics is challenged by over-fitting, model generalization problems, and the requirement of high-quality labeled data. Ongoing development is required to create flexible models that make good predictions with minimal data (Alladi 2016).

The combination of ML with IoT has transformed automation, real-time processing, and intelligent decision-making. IoT produces vast streams of data, which are processed by ML to ensure optimal use of resources and operational efficiency in industries. Techniques like deep learning, reinforcement learning, and clustering improve IoT applications in healthcare, smart homes, automated industries, and agriculture. Predictive maintenance in manufacturing, for instance, saves downtime and maintenance expenses by detecting potential failures before they happen. Data heterogeneity, real-time processing limitations, and security attacks are challenges researchers are overcoming with edge computing and federated learning (Mehata and Devnani 2021).

The big data phenomenon presents challenges and opportunities for machine learning implementations. The ability to process massive data sets is critical to achieve useful insights; however, data quality problems, scalability, and computational complexity represent performance impediments. Machine learning algorithms must cope with noisy and high-dimensional data in terms of accuracy and interpretability. Distributed computing environments such as Apache Spark and Hadoop provide large-scale data processing capabilities; however, data quality fluctuation and dataset bias remain an issue. Future research should focus on developing adaptive learning models that can cope with sophisticated data sets (Zhang 2024).

Improvements in ML have resulted in advanced models like decision trees, SVMs, XGBoost, CNNs, and RNNs. These models perform well in applications like image recognition, time-series prediction, and NLP. Issues like hyper-parameter tuning, over-fitting, and computational expense remain. AutoML and optimization methods like Bayesian optimization and genetic algorithms mechanize the development process, making it more scalable and efficient (Rane et al. 2024).

Deep learning models are superior in feature representation and extraction compared to conventional ML methods. CNNs, RNNs, autoencoders, and GANs are some of the models that have revolutionized computer vision and NLP, and time-series analysis. Hierarchical feature representation learning is automatically performed by deep learning, eliminating the need for hand-crafted feature engineering. Nonetheless, its dependence on immense datasets and high computational power is still a limitation. Transfer learning, self-supervised learning, and NAS are some of the methods under investigation to make deep learning models more data-independent and efficient (Shah et al. 2024).

With the growing importance of ML in decision-making, there is increased concern regarding trustworthiness, interpretability, and ethics. Deep learning's black-box nature, biases in training data, and privacy risks are barriers to the application of ML across sensitive domains like finance and healthcare. Explainable AI methods and federated learning enable high transparency and data security, allowing cautious ML application. The above issues need to be addressed in developing moral and responsible AI systems (Aslam 2023).

The future of ML and DL is in technologies such as self-supervised learning, hybrid models, and multi-modal learning. Blending ML with quantum computing, edge AI, and federated learning will make it more efficient, private, and interpretable. Research is also being done to minimize catastrophic forgetting in neural networks and maximize model explainability. As ML continues to develop, interdisciplinary collaboration and ethics will dictate its effect on society (Anitha et al. 2024). Deep reinforcement learning (DRL) is central to the optimization of energy supply in smart grids. Techniques like Q-learning, Deep Q-Networks (DQN), and Proximal Policy Optimization (PPO) improve demand forecasting and load balancing, reducing wastage of energy and optimizing grid stability. DRL-based approaches are far more efficient and environmentally friendly than traditional optimization methods (Khan and Li 2023).

Ensemble learning methods like Random Forest, XGBoost, and LightGBM improve predictive ability in smart farming. Hybrid stacked models offer better performance compared to single-algorithm models for yield forecasting in crops and pest infestations. These algorithms work very effectively in poor-resource environments, where precise forecasting results in higher agricultural productivity. Ensemble IoT sensor data with ensemble models further improves decision-making in precision agriculture (Patel and Joshi 2024).

Graph Neural Networks (GNNs) have revolutionized urban traffic congestion prediction dynamics by embedding complex spatial-temporal relationships. GNNs are more precise in capturing location-to-location relationships compared to CNNs and LSTMs. The models facilitate real-time traffic management and allow optimization of road utilization, reduce congestion, and improve transport efficiency (Chen and Wang 2023). Explainable AI (XAI) techniques, i.e., SHapley Additive exPlanations (SHAP), improve predictive maintenance model transparency. The utilization of gradient-boosted trees with SHAP allows for identifying key drivers resulting in equipment failure and thus allows predictive maintenance action. The approach, besides better fault detection precision, also makes a contribution towards actionable explanations and thus allows AI-based prediction to be linked to human decisions (Kisten et al. 2024).

Federated learning offers a privacy-protecting environment for AI model training from decentralized patient data gathered by wearable devices. This allows for the improvement of the accuracy of disease prediction while preserving privacy according to regulations like GDPR. Adaptive aggregation algorithms enhance model performance, which enhances the security and dependability of AI-based healthcare solutions. Federated learning plays a crucial role in the construction of smart healthcare and data privacy challenges (Almeida and Ribeiro 2024). Transfer learning allows for artificial intelligence model adaptation to diverse smart home environments with minimal retraining. Pre-trained models, for example, ResNet-50 for image-related tasks and BERT for natural language, boost the performance of activity recognition systems in heterogeneous living spaces. This eliminates the need for copious amounts of annotated training data while maintaining high accuracy in anomaly detection, thereby increasing the scalability of smart home automation (Nguyen and Kim 2023).

Quantum computing speeds up smart material discovery through an enhancement of the conventional ML models. Quantum-fortified support vector machines (SVMs) and neural networks allow quicker simulation of material behavior, which enhances the effectiveness of material design. Quantum algorithms allow the discovery of new photovoltaic materials and the creation of energy-efficient compounds, allowing next-generation sustainable material production (Zhang and Li 2024). Automated machine learning (AutoML) streamlines predictive analytics in intelligent retail through the auto-picking of top algorithms for processes such as demand forecasting and stock optimization. AutoML tools such as AutoGluon and H2O.ai eliminate model and hyper parameter selection through automation, leading to higher accuracy at lower human effort. Employing top-tier models such as CatBoost and Prophet, AutoML optimizes inventory management and customer demand forecasting (Singh and Roy 2023).

Generative Adversarial Networks (GANs) produce synthetic urban mobility data, which solves the problem of data unavailability in urban planning in smart cities. Trained on small real datasets, GANs produce data that appears real, which enhances the robustness of traffic forecasting models. It is especially useful where access to real data is expensive or restricted by privacy issues, making urban planning more effective (Wang and Zhang 2024).

The integration of metaheuristic algorithms, such as genetic algorithms (GA) with deep learning models such as long short-term memory (LSTM) networks, significantly improves water quality forecasting in smart utility systems. By leveraging evolutionary computation for parameter tuning, GA-LSTM hybrids improve the accuracy of contaminant detection while also lowering computational expenses. The method maximizes the utilization of resources in water treatment processes and hence makes smart water management more sustainable and trustworthy (Fernández and García 2023).

Advantages of ML in Smart Applications

Machine learning has various advantages in the different fields, which are listed as follows.

Automation: Minimizes human intervention in operations such as traffic control or customer service.

Accuracy: More accurate with time, i.e., in the fields of disease identification or fraud identification.

Real-time decision-making: Analyses information in real-time for uses such as self-navigating.

Scalability: Supports massive datasets, IoT and smart city needs.

Adaptability: Learns anew from fresh data, hence boosting personalization on services such as Netflix or Spotify.

These benefits are the result of ML's capacity to learn, a process that is based on computational theories, statistics, and cognitive science, which is shown in Figure 23.1.

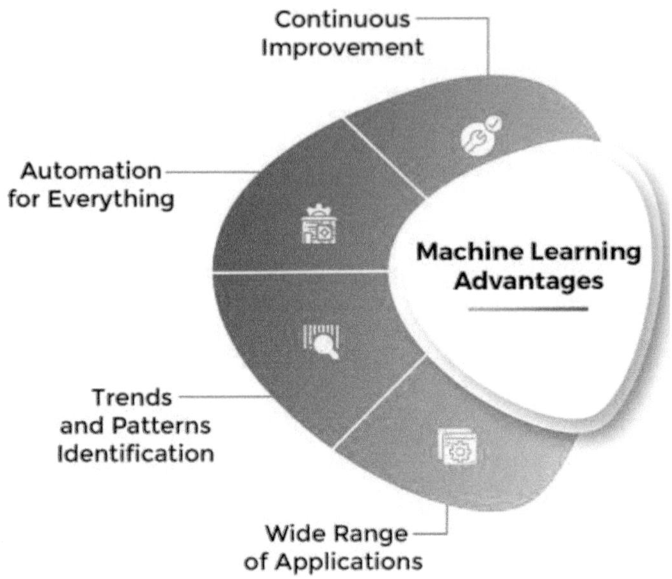

Figure 23.1. Advantages of ML.

3. Types of Learning and Top Algorithms

ML combines various paradigms of learning—supervised, unsupervised, and more—each serving a particular purpose in smart apps. In this section, we describe these paradigms and list the most prominent algorithms in them, with emphasis on their learning mechanisms. Basic learning is classified as shown in Figure 23.2.

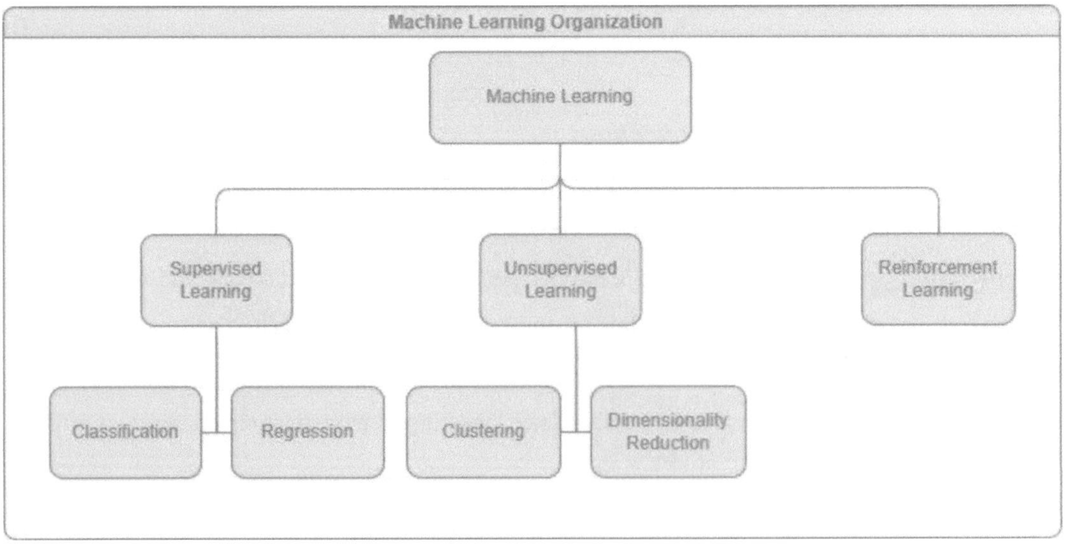

Figure 23.2. Types of ML algorithms.

3.1 Supervised Learning

Supervised learning involves training models upon labeled data-input-output pairs with the purpose of prediction. Supervised learning can be compared with a teacher taking a student along through examples. Supervised learning is heavily adopted in the context of smart applications for the completion of tasks such as classification (picking one between categories) and regression (estimating continuous values).

3.1.1 Linear Regression

Linear regression data fits into a line, reducing the disparity between predicted and actual data points (with metrics such as Mean Squared Error). It has the assumption of a linear, direct relationship between the input features and the output target. This model is normally used for continuous outcome forecasting. For instance, in the home automation case, linear regression would forecast energy consumption as a linear function of temperature and time. Linear regression is simple, intuitive, and computationally efficient, and thus highly appropriate for resource-constrained devices. It has difficulty with non-linear information and intricate relationships. For instance, it could not predict dynamic, mood-based music preferences.

3.1.2 Logistic Regression

Logistic regression is a type of linear regression, but with a logistic transformation, in order to produce a probability between 0 and 1, and thus also appropriate for binary classification tasks. It is also trained to reduce a loss function like Cross-Entropy. This method is usually applied for event classification. For instance, a smart locking system can apply logistic regression to distinguish between valid and invalid access. It is quick and efficient, especially with small datasets. It assumes that the features are independent of one another, which might be restrictive in more complex situations. A health wearable may apply logistic regression to accurately predict whether a heart rate is "normal" or "abnormal".

3.1.3 Decision Trees

Decision Trees partition the data by feature thresholds, as exemplified by the question "Is it raining?" for classification. The tree is structured to reduce uncertainty at each partition, usually by employing measures like the Gini Index. Decision trees can be used in systems like smart irrigation, where they decide if plants need to be irrigated based on weather and soil conditions. Decision trees are explicit and easy to understand, enabling users to perceive the decision-making process. They can overfit the data very easily and need pruning to prevent it. A decision tree for a smart irrigation system can be as follows: "If humidity < 50% and temperature > 30°C, water the plants."

3.1.4 Random Forests

Random Forests are created by aggregating many decision trees, with every tree trained on a random set of the data. By averaging the predictions of all the trees, they suppress over-fitting and enhance accuracy. Random Forest algorithms are widely used in smart building environments, where they predict occupancy levels from sensor data. They are robust, resistant to noise and outliers. But they are more computationally expensive and more difficult to interpret than a single decision tree. An intelligent HVAC system could employ Random Forests to adjust the settings by requesting several trees to search for trends in occupancy.

3.1.5 Support Vector Machines (SVMs)

SVMs function by determining the best hyper-plane with maximum margin in between two classes. They can even deal with non-linear data using kernels (e.g., RBF). In intelligent cameras, SVMs are utilized to assist in object classification in video—i.e., in separating a "person" from an "object". SVMs are highly accurate, particularly for small- and medium-sized data. However, they can have a slow training process and might suffer from difficulties to fit extremely large data sets. An SVM would be able to detect anomalies in a smart grid's power flow, detecting irregularities in real-time.

3.1.6 Gradient Boosting (XGBoost)

Gradient Boosting constructs the models sequentially, as each new model will rectify the mistakes of the past one. XGBoost is a tuned version that makes the process more efficient and scalable. This algorithm is utilized to forecast occurrences such as machine breakdowns in intelligent manufacturing units, utilizing the history of breakdowns to forecast potential future breakdowns. It is highly accurate and highly effective with structured data. It is difficult to tune and less intuitive than more basic models. XGBoost can potentially be used to precisely predict the lifespan of a smart drone's battery.

3.2 Unsupervised Learning

Unsupervised learning is used in situations where the data given is not labeled results. The algorithm runs to discover the patterns or the structure of the data set, similar to a child at play in the environment without guidance. In smart applications, unsupervised learning is typically used for processes like clustering and anomaly detection.

3.2.1 K-Means Clustering

K-Means separates data into "k" clusters by minimizing the distance of data points and cluster centroids. It keeps on running until the clusters stabilize. K-Means is employed for segmenting users within applications to suggest personalized content, such as ads. It is fast, scalable, and easy to execute. It has to decide the number of clusters beforehand ("k") and may have difficulty with outliers. K-Means can cluster patterns of energy consumption in a smart home, such as in the morning and evening hours we can identify the peak consumption patterns.

3.2.2 Alternative Unsupervised Approaches

Aside from K-Means, there are other techniques like DBSCAN (Density-Based Spatial Clustering of Applications with Noise) and Principal Component Analysis (PCA), but K-Means remains in common usage in intelligent applications due to it being simple to apply.

3.3 Computational Methods and Support Systems

3.3.1 K-Nearest Neighbours (KNN)

KNN is an easy algorithm that keeps all the points in memory and makes predictions based on the "k" nearest points to a new point. KNN can be applied in intelligent cities to forecast traffic flow by comparing the existing situation with past data. It is simple to comprehend and can learn new data without being retrained. KNN is slow for large data because it does a real-time search for the closest neighbors. In an intelligent parking system, KNN can potentially identify empty lots by learning from the nearest available historical information on parking space vacancy.

3.3.2 Naive Bayes

Naive Bayes applies Bayes' Theorem to learn probabilities assuming that the features are independent of one another. This algorithm is especially well-suited for applications such as spam filtering in intelligent email clients. Naive Bayes performs well in and is suitable for small datasets. The feature independence assumption is usually unrealistic and therefore not very useful in more complicated instances. A smart helper might use Naive Bayes to classify a user query as “weather-related” or “music-related” based on keywords.

3.3.3 Deep Learning (Neural Networks)

Neural networks consist of node layers, each of which alters its weights in order to decrease error. Complex data types like images and sound can be dealt with by Deep learning models including Convolutional Neural Networks (CNNs) and Recurrent Neural Networks (RNNs). Deep learning provides capabilities like speech recognition in intelligent speakers, where the model is trained to identify spoken commands. Neural networks work best with unstructured information such as pictures and audio. They require enormous amounts of data and intensive computations. A CNN used on a smart doorbell could potentially recognize faces with great accuracy, enabling secure entry.

The classification and key functionalities of the different learning algorithms are listed in Table 23.1.

Table 23.1. Different algorithms used in smart applications.

Algorithm	**Key Idea**	**Example in Smart Systems**	**Advantage**	**Limitation**
Linear Regression	Fits a line for predictions	Predicting energy usage	Simple and efficient	Struggles with non-linearity
Logistic Regression	Classifies using probabilities	Smart access control	Fast for small data	Assumes feature independence
Decision Trees	Splits data by conditions	Automated irrigation	Easy to interpret	Can overfit easily
Random Forests	Uses multiple trees for accuracy	Smart HVAC adjustment	Reduces overfitting	High computational cost
SVM	Finds best classification boundary	Object detection in cameras	High accuracy	Slow for large data
K-Means Clustering	Groups data into clusters	User segmentation in apps	Fast and scalable	Needs pre-defined clusters
KNN	Classifies based on neighbors	Predicting parking availability	Learns dynamically	Slow with large data
Neural Networks	Learns patterns from data	Face recognition in security	Great for images/ audio	Needs large datasets

4. Performance Measures

Machine learning model evaluation is crucial to guarantee accuracy and reliability. Various models have particular metrics used to gauge their effectiveness. These performance measures are classified and shown in Table 23.2.

4.1 Classification Metrics

Classification models output discrete labels, i.e., fraud vs. not fraud, disease vs. not disease. The primary performance measures are:

Accuracy: Evaluates overall accuracy but is unreliable for unbalanced datasets.

Precision: Deals with the proportion of correctly predicted positives, which is useful in fraud detection because false positives are costly.

F1-Score: It weighs precision and recall equally, which is suitable for situations where both false negatives and false positives are of concern, like in medical diagnosis.

Table 23.2. Metrics to evaluate performance.

Metric Type	Metric	Purpose	Example Application
Classification	Accuracy	Measures overall correctness	Spam detection
	Precision	Focuses on correctly predicted positives	Fraud detection
	F1-Score	Balances precision and recall	Medical diagnosis
Regression	Mean Squared Error (MSE)	Penalizes large prediction errors	Stock price forecasting
	Root Mean Squared Error (RMSE)	Interpretable error measure	Energy consumption prediction
Clustering	Silhouette Score	Evaluates cluster separation	Customer segmentation

4.2 Regression Metrics

Regression models forecast continuous values, i.e., prices of stocks or consumption of energy. Important measures are:

Mean Squared Error (MSE): Computes the average of squared prediction errors and true values, punishing large errors.

Root Mean Squared Error (RMSE): Offers a more understandable form of MSE by squaring the root, aiding in the estimation of prediction accuracy in energy forecasting.

Clustering metrics: Clustering models classify similar data points without prior labels, such as customer segmentation. One commonly used evaluation measure is:

Silhouette Score: Indicates how distinct clusters are with respect to intra-cluster homogeneity and inter-cluster separation, commonly used in market segmentation.

5. Key Concepts and Techniques

Machine learning (ML) is the underlying paradigm for most contemporary intelligent applications, and from an educational point of view, some of the most important concepts and methodologies are crucial to the success of these systems. This paper will elaborate further on the primary factors that constitute machine learning in the context of smart technology:

5.1 Data as a basis

Data is the raw material on which machine learning models learn. Data can arrive in the intelligent application in various forms sensor readings in real-time from sensors, user inputs, environmental signals, or interactions. Treat data as the building block on which all intelligence is constructed.

The integrity of this data is crucial to facilitate effective learning outcomes. If the data is characterized by noise, incompleteness, or errors, the model's capacity to predict accurately will be adversely affected. Pre-processing is necessary to enhance data quality through techniques such as normalization (scaling data to a standard range), cleaning (elimination of duplicates or errors), and feature engineering (converting raw data into useful attributes). For instance, in a smart home environment, it is crucial that temperature sensor data, motion sensor data, and user input be standardized and cleaned so that the system can efficiently read and respond to this data.

5.2 Model Training

Model training forms the core procedure through which machine learning algorithms learn patterns from data. During the training period, the model tunes its internal parameters to minimize the gap between its predictions and the actual outcomes. For instance, in Linear Regression, the model adjusts the weights with the goal of minimizing the Mean Squared Error (MSE), a measure that reflects the difference between the predictions and actual values. In the case of Neural Networks, this process of adjustment is achieved through gradient descent, an algorithm that iteratively adjusts weights by calculating the gradient of the loss function.

Training operations in intelligent applications must be very efficient because real-time deployment at all times is required. For example, a self-driving car needs to be able to handle training data and quickly adapt to changing road situations so that the model can make split-second decisions without hesitation. But the process is more than one training sample. Real-world models often need to be retrained from time to time as new data sets become available. This brings the models up to date and ready to respond to changing patterns of behavior shifts.

5.3 Generalization

Generalization is a core machine learning concept. Generalization can be described as the ability of a model to apply the learning from training data to make correct predictions on novel, unseen data. In other words, a well-generalizing model does not merely memorize the training data but rather learns patterns that generalize to a larger context. Generalization in the context of intelligent systems is important as it allows applications to excel in a wide variety of real-world scenarios. For instance, an intelligent thermostat should be able to predict heating needs by learning from past temperature patterns and also adapt to shifting environmental parameters as they change over time.

To promote generalizability capacity, different methods like regularization are utilized. As an example, L2 regularization in Support Vector Machines (SVMs) prevents the process of making extremely complex models that can overfit the training set. Similarly, dropout in Neural Networks turns off some of the nodes randomly while training, forcing the network to learn more generalized and robust features, thereby preventing overfitting to specific features in the training set.

5.4 Feedback Mechanisms

One of the most significant advantages of intelligent systems is their ability for long-term improvement and learning, similar to how human beings sharpen their understanding from feedback and experience-based learning. The cycles of feedback are central to the adaptive nature of intelligent application.

Feedback from intelligent systems can take the form of new data or physical results from the outside environment that the system uses to inform better decision-making in the future. A good example is a smart thermostat, which does not take an informed guess at heating needs and then ignores the data; instead, it continues to alter its forecasts as it learns more about the users' habits, seasonality, and outside meteorological conditions. Through learning from this new data, the accuracy of its forecasts is greatly enhanced.

This learning cycle is similar to how people learn from experience adjusting approaches through feedback. Feedback loops allow models to adjust and optimize themselves as intelligent applications run in changing environments, with ongoing performance improvement.

6. Applications and Challenges

Machine learning is revolutionizing sectors through automation, optimizing efficiency, and facilitating intelligent decision-making. It ranges from medical care to smart cities, simplifying operations, enhancing user experience, and saving costs. The following are the ways in which ML is fueling innovation in sectors. The various applications of machine learning algorithms are shown in Figure 23.3.

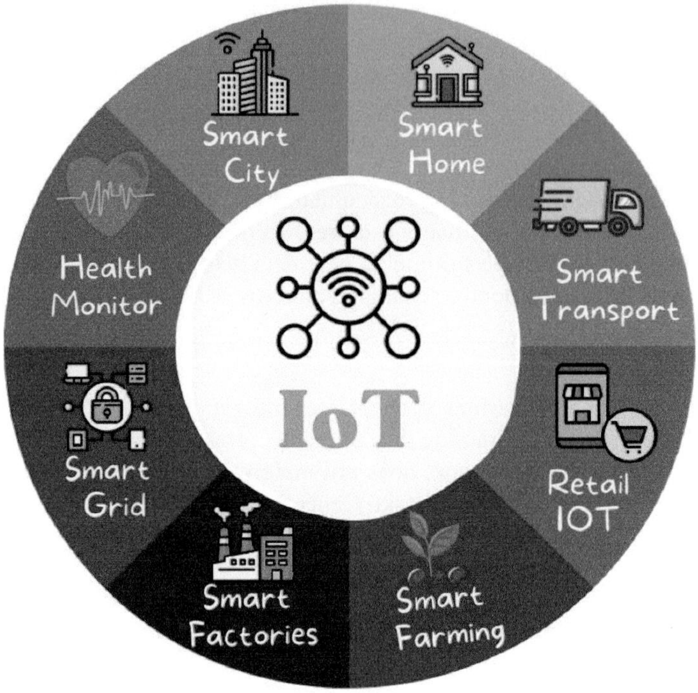

Figure 23.3. Smart applications.

6.1 Applications

Machine learning is adopted in many domains. These days machine learning is applied in almost all fields. Here we identify the major applications of machine learning. The various applications of machine learning are listed as follows.

Healthcare: Machine learning is transforming healthcare by offering improved diagnosis, predictive analysis, and individualized treatment. Deep learning models like Convolutional Neural Networks (CNNs) aid in analyzing medical images, identifying diseases like tumors or fractures with higher accuracy. Logistic regression aids in disease prediction, identifying abnormalities in patient records for early intervention. Support Vector Machines (SVMs) aid in genetic analysis, indicating potential risk factors for diseases. These developments result in an earlier diagnosis, improved treatments, and improved patient outcomes.

Smart homes: Machine learning increases automation and energy efficiency in smart homes to make daily living easier. Random Forest models maximize energy use through consumption pattern analysis so that smart thermostats can set adjustments for enhanced efficiency. K-Means clustering identifies usage patterns, maximizing resource management. Reinforcement learning also increases personalization of automation so that smart lights, security systems, and home appliances learn people's behavior. The technology not only maximizes energy conservation but also enhances comfort and security.

Autonomous vehicles: Autonomous vehicles are dependent on machine learning for real-time decision-making and navigation. SVMs assist in the detection of obstacles, identification of objects such as pedestrians and signs to improve road safety. Gradient Boosting enhances navigation through road surface prediction and route planning. Neural networks interpret sensor information from LiDAR and cameras to aid in lane-keeping, automatic braking, and adaptive cruise control. These technologies lower accidents, enhance traffic flow, and maximize transport efficiency.

Smart cities: Machine learning is essential to urban development, enhancing transit, energy delivery, and public service quality. The K-Nearest Neighbors algorithm forecasts traffic congestion based on real-time automobile traffic analysis, thus facilitating enhanced signal management and traffic flow. Naive Bayes models evaluate citizen feedback to enable local governments to respond more effectively to issues affecting communities. Reinforcement learning enhances energy supply in smart grids by forecasting patterns of varying demand. Collectively, such AI-driven processes promote sustainability, enhance infrastructure, and enhance the quality of life.

Retail and e-commerce: Retailers leverage machine learning to customize shopping, manage inventory, and identify fraud. Decision Trees scan purchase history to make recommendations based on personal tastes, enhancing customer interaction. Neural Networks improve recommendation engines by identifying latent shopping behaviors. Automated ML (AutoML) technology enhances demand forecasting, maintaining ideal stock levels, and preventing wastage. Also, fraud detection software identifies dubious transactions, avoiding financial losses. These improvements make customers more satisfied and business processes better.

6.2 Impact

Machine learning advantages permeate industries. A smart factory with predictive maintenance from ML saves 30% of downtime, cutting costs and increasing productivity. AI-based diagnostics improve diagnosis with accuracy, lowering the rate of misdiagnosis and thus enhancing treatments. Usage in smart cities lowers traffic and energy, making the city livable. Retail employs AI-driven analytics to improve customer engagement and optimize inventory, resulting in higher revenue.

The application of machine learning in these applications enables industries to attain higher efficiency, reduced operating expenses, and improved user experience. With continued development in artificial intelligence, its contribution to creating even more intelligent and automated systems will increase.

6.3 Challenges from a Learning Perspective

Pmobias-variance tradeoff: Eit-page and under generalization might track themselves over one another in between Random Forests and linear regression.

Scalable: Both KNN and SVM lag with big data, while Neural Networks require heavy resources.

Interpretability: While interpretable, Neural Networks and Random Forests remain far more opaque than Decision Trees.

Quality of data: Noisy sensor data in smart systems distorts the learning process.

Constraints for real-time: Edge devices will require lighter-weight models such as Naive Bayes over Neural Networks.

7. Conclusion and Future Scope

The most popular ML algorithms like Linear Regression, Logistic Regression, Decision Trees, Random Forests, SVMs, KNN, Naive Bayes, Neural Networks, Gradient Boosting, and K-Means, power predictive insights in smart applications by learning like humans do from data. They simplify tasks, make them more accurate, and learn to address new challenges, revolutionizing industries from healthcare to transportation. From a learning perspective, their achievement is in the balance between complexity, scalability, and generalization. As smart tech keeps advancing, these algorithms will enable more intuitive, more powerful systems, revolutionizing the way we engage with the world.

7.1 Future Scope

Machine learning continues to evolve, with innovations constantly upgrading to enable AI to become quicker, smarter, and safer. Key developments that are shaping the future include efficiency improvement, privacy protection improvement, and improved transparency.

Tiny ML: Artificial intelligence for small devices, Tiny ML brings AI to low-power devices like microcontrollers and sensors and allows real-time processing without cloud computing. This reduces latency, enhances privacy, and enables smart applications in healthcare, security, and industrial automation with minimal power and memory.

Federated learning: Privacy-focused AI, Federated learning enables AI models to train on many devices without exposing raw data. The decentralized model enhances privacy in sensitive use cases such as healthcare and finance and is compliant with data protection laws. It enables more intelligent personalization without exposing user data.

Explainable AI: Transparent decision-making, Most AI systems are "black boxes," and it is hard to explain their output. Explainable AI (XAI) encourages transparency so that users can comprehend how models such as Decision Trees and Logistic Regression generate predictions. This gives confidence to AI systems in healthcare, finance, and the justice system.

Hybrid models: Merging strengths to boost accuracy, hybrid approaches combine a variety of ML methods like Random Forests and Gradient Boosting to attain highest accuracy and reliability. They find extensive applications in fraud detection, smart agriculture, and demand forecasting, where accuracy is of top priority.

References

Alladi, D. (2016). Machine Learning Algorithms for Predictive Analytics: A Review and Evaluation. Available at SSRN 5064585. Retrieved from https://papers.ssrn.com/sol3/papers.cfm?abstract_id=5064585.

Almeida, F. and Ribeiro, B. (2024). Federated learning for privacy-preserving smart healthcare. Nature Machine Intelligence 6(1): 45–58. Retrieved from https://pubmed.ncbi.nlm.nih.gov/35696470/.

Anitha, M., Kumari, V. S., Pillai, N. M., Jayarin, P. J. and David, D. B. (2024, July). Exploring cutting-edge machine learning and data mining techniques for enhancing big data management with advanced algorithmic strategies for optimal data processing and analysis. pp. 1–5. In 2024 Second International Conference on Advances in Information Technology (ICAIT) (Vol. 1). IEEE.

Aslam, F. (2023). Advancing Intelligence: Unveiling the power of advanced machine learning algorithms for real-world applications. Journal of Engineering Research and Reports 25(7): 159–165. Retrieved from https://journaljerr.com/index.php/JERR/article/view/949.

Chen, Y. and Wang, L. (2023). Graph neural networks for urban traffic flow prediction. ACM Transactions on Intelligent Systems and Technology 14(2): 1–24. Retrieved from https://ieeexplore.ieee.org/stamp/stamp.jsp?arnumber=10077454.

Fernández, A. and García, S. (2023). Metaheuristic-optimized neural networks for smart water management. Environmental Modelling & Software 169: 105813. Retrieved from https://www.adb.org/sites/default/files/publication/614891/artificial-intelligence-smart-water-management-systems.pdf.

Ghadge, N. (2024). Leveraging Machine Learning to Enhance Information Exploration. Available at SSRN 4872025. Retrieved from https://papers.ssrn.com/sol3/papers.cfm?abstract_id=4872025.

Ghadge, N. (2024). Machine Learning: Enhancing Intelligent Search and Information Discovery. Retrieved from https://aircconline.com/csit/papers/vol14/csit141021.pdf.

Kisten, Melvin, Absalom El-Shamir Ezugwu and Micheal O. Olusanya. (2024). Explainable artificial intelligence model for predictive maintenance in smart agricultural facilities. IEEE Access 12: 24348–24367.

Khan, A. R. and Li, S. (2023). Deep reinforcement learning for autonomous smart grid management. IEEE Transactions on Sustainable Energy 15(3): 1124–1135. Retrieved from https://www.researchgate.net/publication/373700164_Deep_Reinforcement_Learning_for_Smart_Grid_Operations_Algorithms_Applications_and_Prospects.

Mehata, R. and Devnani, J. (2021). Integrating IoT with machine learning: A path towards ubiquitous smart applications. Available at SSRN 4991102. Retrieved from https://papers.ssrn.com/sol3/papers.cfm?abstract_id=4991102.

Nguyen, T. and Kim, J. (2023). Transfer learning for cross-domain smart home automation. Neural Processing Letters 55(4): 3219–3238. Retrieved from https://www.mdpi.com/2076-3417/13/13/7882.

Patel, R. and Joshi, M. (2024). Ensemble learning techniques for IoT-driven smart agriculture. pp. 89–112. In Smart Farming and Artificial Intelligence. Springer. Retrieved from https://www.researchgate.net/publication/350536874_IoT_Based_Smart_Agriculture_Automation_in_Artificial_Intelligence.

Rane, N. L., Mallick, S. K., Kaya, O. and Rane, J. (2024). Techniques and optimization algorithms in machine learning: A review. Applied Machine Learning and Deep Learning: Architectures and Techniques 39–58.

Saranya, M. and Amutha, B. (2024). A survey of innovative machine learning approaches in smart city applications. pp. 231–241. In Innovative Machine Learning Applications for Cryptography. IGI Global. Retrieved from https://www.igi-global.com/chapter/a-survey-of-innovative-machine-learning-approaches-in-smart-city-applications/340982.

Shah, S. M. A., Soomro, A., Hussain, K., Rahu, M. A. and Karim, S. (2024). Innovative machine learning solutions: Investigating algorithms and applications. Journal of Applied Engineering & Technology (JAET) 8(1): 32–51.

Singh, P. and Roy, A. (2023). AutoML-driven predictive analytics for smart retail. Expert Systems with Applications 223: 119876. Retrieved from https://www.researchgate.net/profile/Marc-Schmitt-7/publication/367365130_Automated_machine_learning_AI-driven_decision_making_in_business_analytics/links/6633abb408aa54017ad53b63/Automated-machine-learning-AI-driven-decision-making-in-business-analytics.pdf.

Wang, X. and Zhang, Y. (2024). Generative adversarial networks for synthetic smart city data generation. IEEE Access 12: 23456–23470. Retrieved from https://www.researchgate.net/publication/356809414_Generative_Adversarial_Networks_for_Synthetic_Data_Generation_A_Comparative_Study.

Zhang, H. and Li, Q. (2024). Quantum machine learning for smart material design. Advanced Materials 36(18): 2300123. Retrieved from https://inspirehep.net/literature/2736793.

Zhang, S. (2024). Leveraging machine learning algorithms for predictive analytics in big data: challenges and opportunities. Insights in Computer, Signals and Systems 1(1): 10–70088. Retrieved from https://www.researchgate.net/publication/388763105_Leveraging_Data_Mining_and_Machine_Learning_for_Predictive_Analytics_in_Big_Data.

Index

For Product Safety Concerns and Information please contact our EU representative GPSR@taylorandfrancis.com Taylor & Francis Verlag GmbH, Kaufingerstraße 24, 80331 München, Germany